Livestock Animals: A Veterinary Science Perspective

Edited by Johann Casini

SYRAWOOD
PUBLISHING HOUSE

New York

Published by Syrawood Publishing House,
750 Third Avenue, 9th Floor,
New York, NY 10017, USA
www.syrawoodpublishinghouse.com

Livestock Animals: A Veterinary Science Perspective
Edited by Johann Casini

International Standard Book Number: 978-1-64740-078-1 (Hardback)

Cataloging-in-Publication Data

Livestock animals : a veterinary science perspective / edited by Johann Casini.
    p. cm.
Includes bibliographical references and index.
ISBN 978-1-64740-078-1
1. Livestock. 2. Livestock systems. 3. Livestock--Diseases. 4. Veterinary medicine. I. Casini, Johann.
SF140.L65 L58 2022
636--dc23

TABLE OF CONTENTS

PREFACE

It is often said that books are a boon to mankind. They document every progress and pass on the knowledge from one generation to the other. They play a crucial role in our lives. Thus I was both excited and nervous while editing this book. I was pleased by the thought of being able to make a mark but I was also nervous to do it right because the future of students depends upon it. Hence, I took a few months to research further into the discipline, revise my knowledge and also explore some more aspects. Post this process, I begun with the editing of this book.

The domesticated animals that are raised in specific agricultural settings are known as livestock. They are used for various purposes such as labor and production of commodities including meat, milk, leather, fur and wool. Livestock is a broad term that can refer to any species of animals that are kept by humans for domestic or commercial purposes. It includes domestic animals like camel and cattle as well as semi-domestic animals such as fallow deer and reindeer. Livestock production plays a vital role both economically and culturally in various rural communities. It is necessary to provide them with good husbandry, appropriate feeding and hygiene to get maximized production of commodities. They must be checked regularly by veterinarians. This book provides comprehensive insights into livestock animals. It provides significant information of this discipline to help develop a good understanding of it. This book will serve as a reference to a broad spectrum of readers.

I thank my publisher with all my heart for considering me worthy of this unparalleled opportunity and for showing unwavering faith in my skills. I would also like to thank the editorial team who worked closely with me at every step and contributed immensely towards the successful completion of this book. Last but not the least, I wish to thank my friends and colleagues for their support.

Editor

A longitudinal study on the performance of in vivo methods to determine the osteochondrotic status of young pigs

Christian P. Bertholle[1], Ellen Meijer[1], Willem Back[2,3], Arjan Stegeman[1], P. René van Weeren[2] and Arie van Nes[1*]

Abstract

Background: In today's porcine industry, lameness has a major welfare and economic impact, and is often caused by osteochondrosis (OC). The etiological factors of the disease have been studied in depth, however, to this day, little is known about the natural course of the disorder and how it can be detected at an early stage in pigs. The aim of this pilot study was to assess the potential of three non-invasive techniques for the detection and monitoring of early OC processes in piglets. A group of weaned piglets ($n = 19$) were examined longitudinally using radiographs, a visual lameness scoring scheme and a quantitative pressure-mat based locomotion analysis system to detect OC in the humeroradial, femoropatellar and tarsocrural joints. At several time points, a selection of animals was euthanized for post-mortem examinations, including histology, which was the gold standard.

Results: In this study, clear signs of subclinical signs of OC were observed, however, we were unsuccessful in producing clinical OC. Lesions were observed to be commonly bilaterally symmetric in the joints examined in 80 % of cases. The radiographic examinations showed a clear correlation with the gold standard, particularly when subclinical lesions were of a high histological score. Moreover, radiography was also able to detect the early repair processes, which appeared to take place at least until 14 weeks of age. Both visual scoring and pressure mat analyses showed good intra-assay reproducibility, with the pressure mat showing intra-class correlation values between 0.44 and 0.6 and the inter-observer agreement of visual scoring method was between 88 and 96 %, however their correlation to OC lesions detected by histology was very weak, with only 2 out of 12 traits for the visual scoring method showing significant and biologically logical relations to a specific joint having histological OC lesions. For the pressure mat, only a maximum of 5 associations for specific joints with histological OC lesions were found out of a possible 8.

Conclusion: All tested in-vivo methods showed good reproducibility. Radiography was the most reliable technique to detect and monitor longitudinally the earliest signs of OC in these piglets. It also demonstrated that the "Point of No Return" (PNR) of the disease, when repair processes end, might be later than anticipated, after 13 weeks of age. All in all, our study shows that the timing of the use of these in-vivo methods is critical to detect and monitor OC, especially in the early phases of the disease. It also shows the difficulty in producing OC regardless of the optimization of the experimental settings in relation to the etiological factors known to induce OC.

Keywords: Pig health, Osteochondrosis, Lameness, Gait analysis, Radiography, Histology, Early detection

* Correspondence: A.vanNes@uu.nl
[1]Department of Farm Animal Health, Faculty of Veterinary Medicine, Utrecht University, Yalelaan 7, NL-3584 CL Utrecht, The Netherlands
Full list of author information is available at the end of the article

Background

Osteochondrosis has often been reported to be the main cause of lameness in pigs [1–3] and is still currently causing substantial economic losses and alarming welfare concerns for both sows and slaughter pigs [4]. It is estimated that 80 % of the pigs in today's porcine industry show superficial to mild signs of OC [1]. OC has been defined as a multi-focal disturbance in the endochondral ossification process that occurs during skeletal growth, leading to lesions in both the articular and physeal cartilage [1, 5]. Many advances have been achieved in the understanding of the disease process, for example with respect to the identification of etiological factors. However, little is known still on the natural course of the disease and especially on the early stages [5]. Nevertheless, the fact that the molecular and cellular structure of epiphyseal cartilage and that the mechanism of disruption of the endochondral ossification process are similar in different species, lead us to believe that these early stages could also be comparable in the pig [6–8]. This means that recent findings in equine medicine on OC, where it is suggested now that there is a dynamism of the interplay process between lesion initiation and early repair, could be extrapolated to porcine OC [9, 10]. Additionally, the existence of a Point of No Return (PNR) has also been discovered in the horse. In the pig, this PNR is anticipated to be before 13 weeks of age, after which lesions can no longer be repaired, as vascularization of the cartilage tissue has disappeared [11, 12]. The understanding of OC has for a long time principally relied on data that was acquired by either radiography or post-mortem analysis [13], however more recent techniques have also shown promising results, such as micro-CT [14, 15]. In the pig, good correlations have been reported between radiographic results and post-mortem macroscopic and microscopic data, although this was in animals showing more the later clinical stages of the disease [13], than the early stages or subclinical signs of OC. In spite of the valuable performance of radiography, there is a need in the field to have other effective in-vivo detection methods able to detect the early signs of OC in a higher through-put way. In other species, such as the horse, the dog and cattle, the development of locomotion in relation to leg and claw disorders has been studied successfully using modern gait analysis techniques [16–18]. In this pilot study, we investigated the frequency of OC in our pigs and per joint. We studied also the bilateral symmetry of OC, and the efficacy and suitability of three non- invasive techniques to detect early signs of OC, by comparing them to the gold standard (histological examination). These in-vivo techniques included the visual scoring of the conformation and gait of the animals, the use of a pressure mat to quantify gait and radiographic screening.

Results

Health-prevalence of OC-bilateral symmetrical aspect

Average daily weight gain of the 19 pigs was 253 g/day (+/− 246 g) and average body weight was 6.5 kg (stdev. +/− 0.64) and 34.2 kg (stdev. +/− 0.79), at 4 and 14 weeks of age respectively. At necropsy, no macroscopic OC lesions were detected in any of the joints in any of the pigs. At histology, most of the lesions were observed in the femoropatellar joint, which featured the most severe lesions (Fig. 1). The OC status appeared to be highly symmetrical in left and right joints for all three joints with an identical histological score (H score) or an H score differing only by one unit in over 80 % of cases (Table 1).

In-vivo versus post-mortem gold standards (Radiographic versus Histological examination)

The comparison between histology and radiography was done using the data from both techniques from the 10 radiographed pigs at the last time point. Fig. 2 shows a receiver operating characteristic (ROC) curve describing the sensitivity/specificity of radiographic results for all joints with different cut-offs of histology, considered as the gold standard, Table 2 shows the statistical difference in sensitivity and specificity between the radiographs and histology. Results show that the sensitivity increases and the specificity decreases, as the threshold setting for histology increases.

When examining the different areas under the curve (AUC), which measure the general accuracy of the test, for each threshold setting, the range is chronological from 72.3 % (for H and R scores >1) to 89.1 % (for scores above 3). This demonstrates a good overall specificity and sensitivity of radiographs compared to the gold standard, which appears to increase as the histological lesions become more severe. Concerning the correct identification of negative joints for OC, radiography has

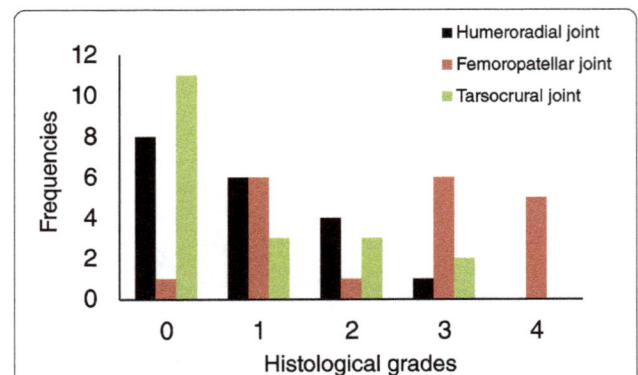

Fig. 1 Histology scores for n = 19 pigs (using both left and right sides of the joint were combined and the overall worst score from all investigated locations within the joints)

Table 1 Left/right comparison of OC histological status (n = 19)

	Humeroradial (n = 32[a])	Femoropatellar (n = 39[a])	Tarsocrural (n = 25[a])
Bilateral lesions (both scores >1)	9.4 %	33.4 %	8.0 %
Bilateral no lesions (both scores 0)	56.3 %	23.1 %	68.0 %
1 side score 0, the other score >1	31.3 %	25.6 %	12.0 %
1 side score 0, the other score >2	3.1 %	18.0 %	12.0 %

[a]Only the outcomes where left and right histology scores were available for each location

a good specificity from R score 0 up to R score 2, but is accompanied by a low sensitivity.

Above R score 2 (R scores 3–4) the opposite occurs: the sensitivity dramatically increases up to 81.2 % compared to 67.8 % (for R score 2), and specificity decreases, demonstrating that radiography identifies positive histological joints better from R score 3 onwards (Table 2). Overall, from these results we can establish that there is a good relation between histology and radiography and additionally both sensitivity and specificity vary with the cut-off. Figure 3 shows typical examples of histological lesions with the corresponding radiographs.

Longitudinal radiographic monitoring
Our results showed that lesions remained stable (80–95 % of lesions) except for the femoropatellar joint, where progression, regression and even resolution of lesions were apparent (Table 3). Resolution occurred in all joints after the hypothetical timing of the PNR (7 weeks of age < PNR < 13 weeks of age).

Visual scoring reproducibility
Inter-observer agreement was good to excellent (88–96 %) for all scored parameters but one: (sickled-buckled fore, Table 4).

Pressure mat reproducibility
Table 5 gives intra-class correlation (ICC) data on the pressure mat kinetic factors showing overall excellent reproducibility (ICC values 0.44 < x < 0.6).

Comparison of visual scoring and histology
When considering visual scoring and histology, 12 pigs were selected on the basis that they were the only ones showing the highest number of results present for both techniques, in a total of 9 joint locations (2 in the humeroradial joint, 4 in the femoropatellar joint and 3 in the tarsocrural joint). This comparison showed that the hind pasterns was the only element that showed a significant relationship at joint level, once the Benjamini

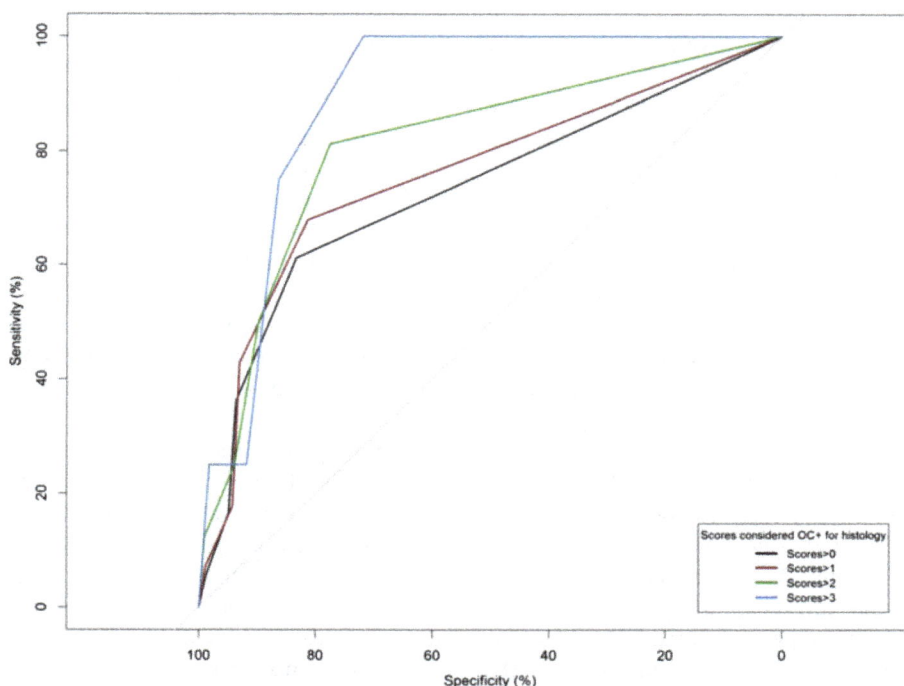

Fig. 2 ROC curves showing the difference in sensitivity and specificity of radiography versus the gold standard, histology using various thresholds

Table 2 Description of the sensitivity and specificity of radiography for the detection of OC as defined by the gold standard, histology ($n = 10$)

	Specificity (%)	Sensitivity (%)	AUC	Confidence interval
Roc 1 (Both H&R scores >0)	83.3	61.1	72.3	63.61–82.22 %
Roc 2 (Both H&R scores >1)	81.4	67.8	75.3	65.66–85.79 %
Roc 3 (Both H&R scores >2)	77.5	81.2	80.9	69.49–92.24 %
Roc 4 (Both H&R scores >3)	71.8	100.0	89.1	79.9 %–98.2 %

R scores radiographic scores, *H scores* histological scores, *AUC* area under the curve, *ROC*

and Hochberg false discovery rate correction was applied to all the Spearman correlation test results (Table 6).

Comparison of pressure mat and histology

Using the same set of pigs as for the visual scoring/histology comparison, four comparisons between the ASIs (Asymmetry Indexes) and histology were tested. The first two were based on the ASIs and the transformed H scores. Only Fore/Hind ASIs showed relations with Animal and Femoropatellar levels (Table 7).

Thirdly, an "ASI animal level score" (sum of all ASI absolute values) was compared to the H score on animal level, but no associations were found (results not shown). Finally, a direct comparison between ASIs of severe histologically OC Positive joints (H scores >3) and those of OC negative joints was tested. Only ASI Pmax fore (Right/Left setting) showed a significant link with the presence and absence of severe histological OC lesions (results not shown).

Discussion

Frequency, severity of and bilateral symmetrical aspect of OC lesions

In the present study, 64.9 % of the animals showed microscopic signs of OC, with 24.6 % having histologically severe lesions (H scores > 3; Fig. 1). The femoropatellar joint had the highest incidence of lesions and the highest number of high grade lesions. Nevertheless, in clinical terms, only 7 % of animals lesions of a high grade radiographically (R scores > 3) and at necropsy, no lesions were macroscopically visible in any of the pigs. We were therefore unexpectedly unsuccessful in creating clinically OC, despite using environmental conditions (slippery flooring) and nutritional regimens (high protein diets fed ad-libitum) designed to promote the development of the disorder [5]. The advantage of this situation was that we were able to assess our non-invasive methods under challenging conditions. Few studies have looked at the the frequency of OC in pigs on a longitudinal scale and in the early subclinical phases of OC. Some studies reported frequencies from 41.4 % using macroscopic evaluation to 78 % using microscopy [18]. There are also large differences with respect to the frequency per joint, with some studies reporting the

tarsocrural joint as the joint to be most commonly affected [19], while others mention the medial condyles of the femur and humerus [13]. The techniques used to detect OC were similar in nature, thus these differences are most likely linked to the multifactorial aspect of this diseases, which can affect one population of identical genetic pigs differently if they were exposed to different environmental conditions for example. Regarding the bilateral symmetrical effect, the humeroradial joint was the most consistent in showing this effect in our study (Table 1). OC is generally seen as a multifocal disease with a relatively strong bilateral occurrence of lesions [5, 20, 21]. This finding has even prompted some studies to analyse solely one side of the animal [13]. Our results confirm these previous findings.

In-vivo radiography as a diagnostic tool for OC detection compared to histology

Unsurprisingly, the different AUC results measuring the accuracy of radiographs versus histology showed that radiography performed best for detecting the most severe histological lesions (Table 2). Additionally, in terms of specificity, radiography showed that all histological lesions from H score 2, 1 and 0 were identified as OC negative. This was expected, as these lesions are mostly mild subclinical ones and would be therefore challenging to visualize with a clinical tool, such as radiography. All things considered, from these results, we can conclude that the most reliable threshold for radiography, to identify correctly the OC status of these joints, is situated around H score 3, when lesions are at high score histologically. This observation is important, as it demonstrates the ability of radiography to detect only the more severe subclinical OC lesions in pigs reliably. Although radiography is considered currently as the standard imaging method to diagnose clinical OC [22], and most studies have investigated OC in other species [7, 21–23], not many have focused on pigs [24, 25]. Other studies have also shown significant correlations between both methods, which were joint dependent [13, 26], however these studies were performed on older pigs and foals. Our results show a similar relevant correlation between both methods. In terms of monitoring, the results showed that lesions occur already at 7 weeks of age, and

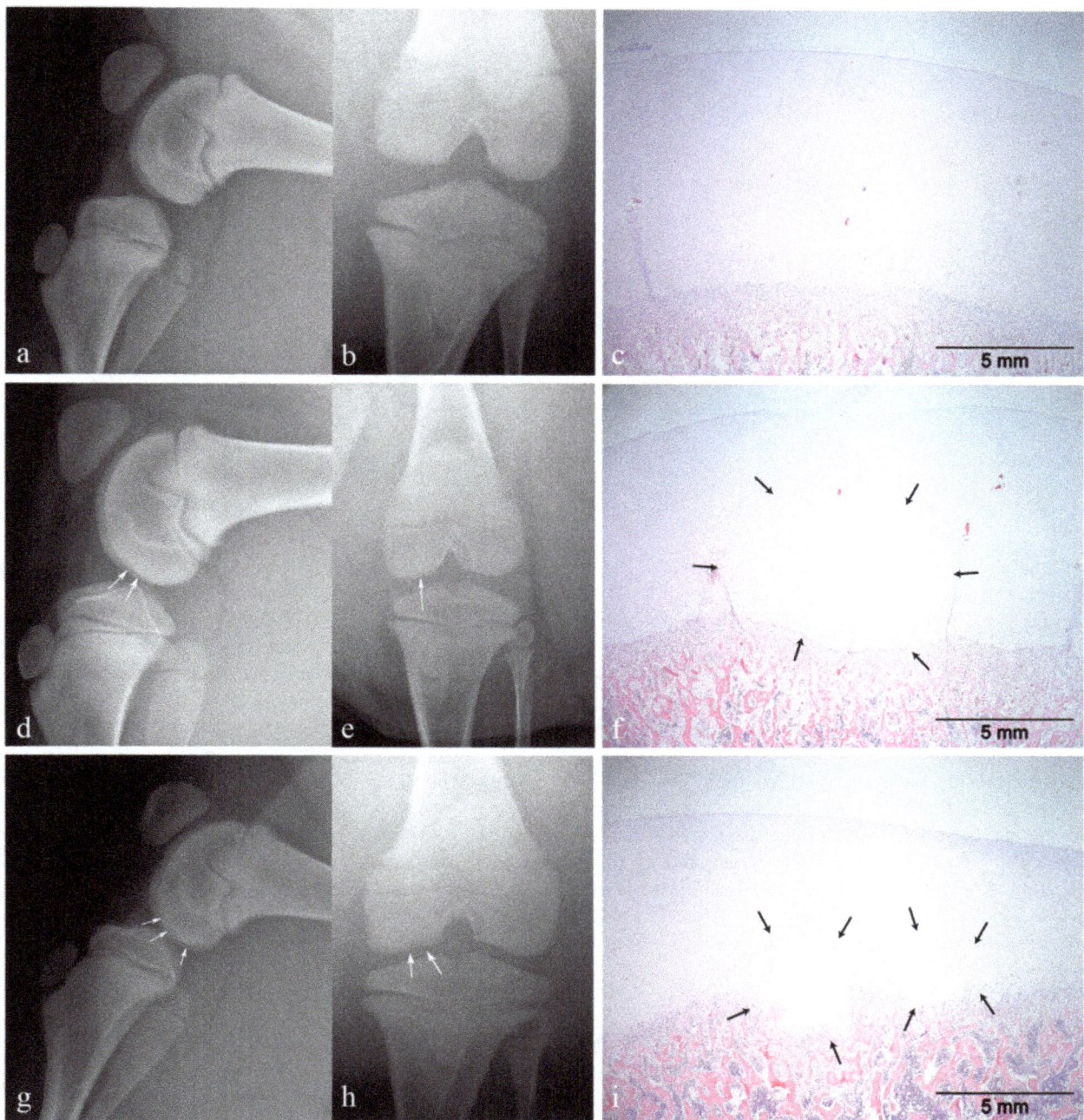

Fig. 3 a–i. Typical images of the histological grades with corresponding radiographic findings. Photo pairs **a**, **b**, **d**, **e** and **g**, **h** are radiographic images taken of 3 different respective pigs at 14 weeks of age. Pictures C, F and I are the corresponding histological slides in chronological order to the radiographic paired images. All stained with hematoxylin and eosin. All pictures depict assessments of the femoropatellar joint with G/H/I being from the left limb only. **a–c** represent the lateral side and **d–f** and **g–i** represent the medial side of the distal femoral condyle. All three pairs represent animals with no OC lesions present (**a–c**), with mild OC lesions ($R = 2$ and $H = 3$, **d–f**) and with severe OC lesions ($R = 4$ and $H = 4$, **g–i**)

are mostly of a low grade nature for radiological standards (R scores <3). Although OC lesions progressed, as expected, between 11 to 14 weeks of age, mostly in the humeroradial and femoropatellar joints, simultaneous repair processes were also taking place. This means that

the PNR, which was anticipated to be before 13 weeks of age, is probably situated somewhat later in time. In horses, OC is known to be a very dynamic disease in which lesions appear, but then may, to a large extent, resolve spontaneously afterwards as a repair process starts

Table 3 Frequencies of lesion development detected radiographically at 7, 11 and 14 weeks of age ($n = 20$, for 10 pigs and 2 sides for each joint)

	Grade	Lesions present	Stationary	Progression	Regression	Resolution
7–11 weeks						
Humeroradial	0	13	13	0	0	0
	1–4	7	4	0	0	3
Femoropatellar	0	11	3	0	0	0
	1–4	9	2	11	0	4
Tarsocrural	0	18	17	0	0	0
	1–4	2	0	1	0	2
11–14 weeks						
Humeroradial	0	16	14	0	0	0
	1–4	4	2	2	0	2
Femoropatellar	0	7	3	0	0	0
	1–4	13	7	5	2	3
Tarsocrural	0	19	19	0	0	0
	1–4	1	0	0	0	1

immediately after occurrence of a lesion [27]. For the pig, we can conclude that a similar dynamic picture emerges, with differences per joint. Unexpectedly, there does not seem be a straightforward relation with the time window until 7–13 weeks of age, when the growth cartilage is still vascularized [11, 12]. The current paradigm is that OC is very likely caused by a failure of the vascular supply of the growth cartilage through cartilage canals in specific focal areas, which then leads to small necrotic areas that are not yet visible radiographically and are hence called osteochondrosis latens. If not resolved, these lesions may become larger and of clinical relevance, at which stage the term osteochondrosis manifesta is used [27]. Whether or not OC latens lesions will turn into OC manifesta lesions depends on both the

initial size of the lesions and the repair capacity of the cartilage, which decreases rapidly with time, as the animal gets older [28]. This latter aspect is influenced by the changes in vascularization of the epiphyseal growth cartilage due to the ongoing process of endochondral ossification that will ultimately lead to the mature state in which there is only avascular articular cartilage left.

Visual scoring and pressure mat reproducibility and comparison with histology

In this study, visual scoring of lameness and the pressure mat showed both a good reproducibility within session measurements (Tables 4 and 5). Only one relationship was found between the hind pasterns and the femoropatellar joint. In the literature, visual assessment as a

Table 4 Percentages of the frequencies of all visual scoring categories scored (VS scores) by both raters for all pigs measured at all time points

Traits	Agreement (%)	Disagreement (%)	Missing scores (%)
O or X shape fore legs	95	4	1
Sickled or Buckled fore legs	46	54	0
Steep/Low angled Pasterns fore	94	6	0
Claw size fore	99	0	1
O or X shape hind legs	88	11	1
Straight or Sickled hind legs	94	5	1
Steep/Low angled Pasterns hind	91	8	1
Claw size hind	99	0	1
Ham	91	8	1
Gait pattern	93	6	1
Twisting hocks	99	0	1
Swaying hind	95	4	1

Table 5 Intra-class correlation data for the pressure mat parameters

	Average FMAX	Average VI	Average CA	Average Pmax
ICC(1)	0.58	0.60	0.56	0.44
F-Test	5.51	5.75	4.98	3.50
P value	<0.05	<0.05	<0.05	<0.05

potential indication of OC has only been compared with macroscopic joint assessments thus far [19, 29]. Some reports claim no relationship [30, 31], or weak associations [13], whereas others claim to have found positive associations when lesions were severe [32], or even strong associations between leg weakness and OC macroscopic scores, at joint level only [33], or animal and joint level [29]. Our results hence seem to reflect the current contradictory position in the literature regarding this relation. Until recently, quantifying lameness was usually performed by a force plate system, measuring the ground reaction forces. However, our more recent pressure mat system is able to provide information on the pressure distribution pattern of the foot or claw, and has already been used successfully in recent studies in horses, cattle, dogs and pigs [16–18, 34], however never in this study format for detecting OC. For the pressure mat, very few associations with histological data were established in this study, mostly with hind limb and animal level parameters. Other more direct comparisons between the two methods were unsuccessful. The reason for this could be the low number of pigs having clinical OC, and the presence of only sub clinical OC. This would be the reason why conformational changes, in terms of lameness, are not occurring due to OC, or that if there are conformational changes taking place, that they are actually occurring but they are not linked to OC. This makes it difficult to appreciate the full potential of the pressure mat in detecting the earliest signs of OC in-vivo, as intended in this study, and requires more testing to be performed on larger populations with more clinical OC present.

Conclusions

Whereas the frequency of clinical OC lesions in this study was much lower than expected, histological prevalence of OC lesions was relatively high, especially in the femoropatellar joint, providing a good opportunity to assess the potential value of the in-vivo measuring systems under scrutiny for the detection of subclinical porcine OC. Moreover, as in other studies, the strong bilateral character of OC was confirmed. From the positive correlations found between radiological and histological findings in all joints and especially in the femoropatellar joint, we can conclude that in-vivo radiography performs well in terms of diagnosing and monitoring OC only at high radiographic scores and works safely in pigs at a young age. Lesions were already radiographically detected at 7 weeks of age and progression and regression/resolution of lesions continued until 14 weeks of age, which suggests that the PNR may occur later in time, than previously expected. Both visual scoring of lameness and pressure mat measurements appeared to be repeatable and well feasible in a logistical sense, but their relationship to the gold standard was weak and they are not apt to detect subclinical OC. Nevertheless, further development of the pressure mat approach seems to show potential for high-throughput assessments of joint disorders in pigs, even at a very young age and warrants further exploration in more heavily affected populations and over a longer timescale.

Methods
Animals and experimental set-up

Nineteen "Topigs 20" piglets at weaning age (4 weeks of age, Van Beek BV, Lelystad, NL) were used for this study. Male-female numbers were close to a 50/50 ratio with 10 males and 9 females to have a good gender distribution. The animals were housed in three groups in relatively equal size pens (153 cm × 256 cm) equipped with toys and smooth wooden plates covering half of the floor to increase slipperiness and the chance of developing OC [20]. An 18 h light-dark cycle was used and stall temperature was set at 24 °C. Pigs were fed ad libitum with a customised diet with a high concentration of proteins (20 % feed) including a high volume of lysine (14.0 g/kg feed) and without cartilage residues (Research Diet Services B.V., Wijk bij Duurstede, NL) and had constant access to water. The pigs were housed for 1 week on arrival for adaptation and training purposes, and were 14 weeks old at the end of the study. The study was divided into four periods of three weeks, separated by weekly in-vivo measuring time-points. Every week, all pigs were weighed using a standard scale (Schippers BV, Bladel, NL). At each period, randomly selected pigs were euthanized and the humeroradial, femoropatellar and tarsocrural joints were examined macroscopically and histologically at specific locations for the presence of OC lesions. All joints were examined

Table 6 Associations between visual scoring and histology (9 joint locations - n = 12)

Visual scoring element	Animal or Joint level	Spearman's rank correlation test	Benjamini and Hochberg correction
Pasterns hind	Femoropatellar	−.786 (P = 0.00)	0.01

Table 7 Associations between pressure mat and histology data (9 joint locations - $n = 12$)

Pressure mat ASIs of kinetic factors		Histology OC score level (Joint or animal level)	Spearman's test	Benjamini and Hochberg correction
Fore/Hind ASIs	ASI CA Right	Animal	-0.772 ($p = 0.00$)	0.05
	ASI Pmax Right	Femoropatellar	0.822 ($p = 0.00$)	0.03
	ASI Pmax Left	Femoropatellar	0.751 ($p = 0.00$)	0.05

specifically at specific sites known to develop osteochondrotic lesions (24 total). The list of sites is based on a previous study for macroscopic evaluations in pigs that originated from horses [19]. Three pigs were euthanized at the first three time points and ten at the last one. The ten pigs, of the last time point, were the only ones to be radiographed. Radiographic monitoring of these pigs took place at the last 3 time points (Additional file 1).

Visual scoring

The method used to score the pigs was derived from a scoring system previously devised by Van Steenbergen [35] and used in a previous study on OC [29]. Briefly, nine anatomical and three mobility categories were scored each with two opposing traits. In this study, due to the small number and young age of the pigs, we regrouped the 1–9 category scale system into a three-point scale (Additional file 2).

Pressure mat analysis

The claw pressures were measured using a Footscan 3D 2 m-system (RsScan International, Olen, Belgium) containing 16 384 force pressure sensors operating at a frequency of 126 Hz. The mat was protected by a 0.5 mm thick mat (shore value $65° ± 5$) and was set into a "runway" system so the pigs could "strike" the mat precisely

(Fig. 4). Pigs were trained in the first week to perform "successful runs", in which they trotted smoothly, in straight line and with their heads staying horizontal. To avoid any influence of velocity and other interfering parameters (e.g. growth) on the results, asymmetry indexes (ASI's) of the Peak Vertical Force (Fmax in N), Peak Vertical Pressure (Pmax in N/cm^2), Contact Area (CA in cm^2) and Vertical Impulse (VI in Ns) were used as outcome parameters, as previously described in Meijer et al. [31]. The ASIs were calculated using the following formula: [(Left parameter – Right parameter)/ 0.5X (Left parameter + Right parameter)]X100.

Radiographical analysis

Radiographs were taken of the ten pigs that were euthanized at the end of the study, at 8, 11 and 14 weeks of age. The images were taken with a two-view setting for all three joints on both sides (Left and Right): a lateromedial view and cranial/dorsal view, using a Philips Optimus full radiography digital system with flat panel detectors (FPDs). Settings for the contrast (mA) and density (kVp) were for pigs at 8 weeks of age: 55 kV and 5.5mAs for all joints (no grid used) using medial/lateral and cranial caudal views. The settings for the larger pigs (11 weeks and 14 weeks old) medial/lateral views (no grid) were used with 60 kV and 8mAs settings for the

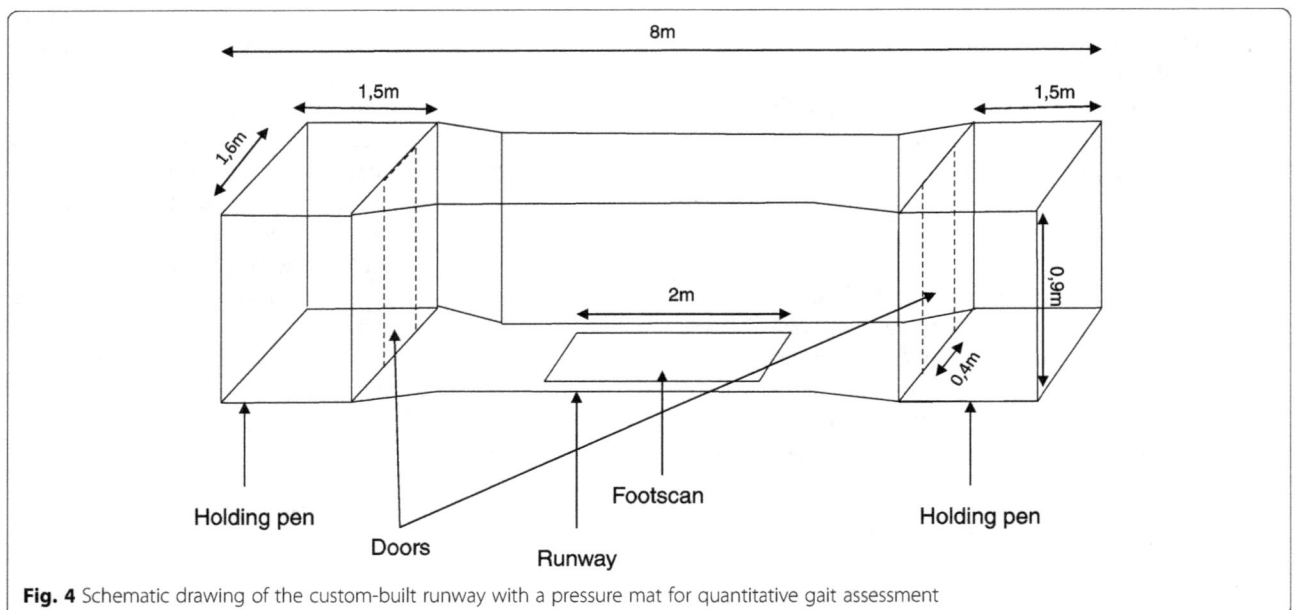

Fig. 4 Schematic drawing of the custom-built runway with a pressure mat for quantitative gait assessment

Table 8 Radiography grading (from Dik et al. [9])

Grade	Classification	Bone contour	Subchondral bone texture	Fragment
0	Normal	Rounded	Diffuse density	Absent
1	Minimal	Smoothly flattened	Obscure lucency	Absent
2	Mild	Irregularly flattened	Obvious, ill-bordered local lucency	Absent
3	Moderate	Small, rounded/irregular concavity	Obvious, well-bordered local lucency	Small fragment(s)
4	Severe	Large, rounded/irregular concavity	Obvious, well-defined extensive lucency	Large fragment(s)

humeroradial and femoropatellar joints. These 2 joints had also a caudal cranial view taken (with grid) at 60–63 kV and 10–12.5mAs. For the tarsocrural joint we used a plantar/dorsal and medial/lateral view (no grid) with a 55–60 kV and 5-10mAs setting. The variations in the settings are due to the fact settings had to be adapted due to the growth of the animals during the study. Prior to radiography, the pigs were sedated using Stresnil (azaperone 40 mg/ml-Janssen Animal Health Benelux) intramuscularly, in a dose of 1 ml/10 kg. X-rays were assessed independently by 2 veterinary radiologists, using a 0–4 scale adapted from a grading system previously used in horses [9]. To allow a better comparison between radiography and histology, the joints were imaged in such a way that they would encompass the same locations that were assessed with histology. The grading is detailed in Table 8.

Macroscopic grading of joints
At 5, 8 and 11 weeks of age, three pigs were euthanized at each time point and at 14 weeks of age for the remaining ten pigs were euthanized. Euthanasia was carried out using pentobarbital (200 mg/kg) injected intracardially (Euthanimal, Alfasan, Woerden, NL) after they had been previously sedated with Stresnil (azaperone 40 mg/ml-Janssen Animal Health Benelux) intramuscularly, in a dose of 1 ml/10 kg. At necropsy, all three joints were scored using a 0–4 scale, as described by van Grevenhof et al. [19] with 0 meaning OC-free, and 1–4 representing minimal, mild, moderate and severe lesions. The scoring was carried out independently by both researchers involved in the study.

Histology
After necropsy, all joints were cut into 0.3 cm thick slices using a band saw, in a sagittal approach in order to visualise cartilage and subchondral bone (except for the right side samples of the pigs from the 2nd time point, which were cut in a frontal plane). Samples were directly processed for histology, except for the left side samples for the 3 pigs of the 3rd time point and the ten pigs from the last time point. Only few samples from these ten pigs were selected for logistical reasons, as it was technically too difficult to be able to analyse all samples. This selection was made using a simple visual examination of the cut tissue and first took place on the cut samples that showed OC lesions on radiographs during any one of the radiograph imaging time points. Two locations of that same joint (medial and lateral) were selected, to obtain an OC positive and negative sample (as normally only either the medial or lateral side is affected). If both sides were OC positive, then control samples were selected on the contralateral joint. Controls were randomized by selecting samples blindly from the list of samples which had corresponding negative radiographs for OC. From all the other radiographically negative joint locations random samples were chosen as control samples. All these pre-selected samples were examined by a veterinary clinician. All samples from this study were fixated in formalin, decalcified in 10 % EDTA and embedded in paraffin before being cut into 3 μm slices and stained with haematoxylin and eosin, and then scored on a 0–4 scale (Table 9).

Table 9 Schematic overview of the histological grading system used in this study

Grade	Classification	Signs
0	Normal	Vessels with erythrocytes, No necrotic areas, No irregularities on ossification front, Normal thickness of cartilage
1	Normal-Mod	Vessels with erythrocytes, Few to no very small necrotic areas, Some irregularities on ossification front, Some mild focal thickening of cartilage
2	Moderate	Vessels with/without erythrocytes, Some mild necrotic areas (not large), Mild irregularities on ossification front, Moderate focal thickening of cartilage
3	Mod-Severe	Vessels mostly without erythrocytes, Some large necrotic areas, Distinct irregularities on ossification front, areas of thickened cartilage
4	Severe	Vessels without or with extremely few erythrocytes, very large necrotic areas, Very severe irregularities of ossification front, Severe thickening of cartilage areas

Statistical analysis

All statistical analyses were performed using the statistical program R (R: A Language and Environment for Statistical Computing, R Core Team, R Foundation for Statistical Computing, Vienna, Austria) and IBM SPSS Statistics 22 (IBM Corp. Released 2013. IBM SPSS Statistics for Windows, Version 22.0. Armonk, NY: IBM Corp).

Bilaterally symmetrical aspect of OC

The degree of bilateral occurrence of presence and absence of lesions was examined by calculating the percentages of occurrences of total, partial or no bilateralism of histological findings between Left and Right joints.

Radiography versus histology

The agreement strength between both methods was assessed using a receiver operating characteristic (ROC) curve that generates an area under the curve (AUC) statistic. This was performed on the histological and radiographic data from all examined joints from the last ten pigs of the study, where results for both techniques were available (Additional file 3).

Visual scoring and pressure mat reproducibility

The frequencies of agreement and disagreement between both raters for all visual scoring aspects were counted and transformed into percentages (Table 4). For the pressure mat, an ICC (inter class correlation) was calculated using a one-way system, to measure the intra-assay variation of all four kinetic parameters from three successful runs each pig performed at all time points. The pressure mat software rating the subjects was considered as the constant and the pigs' runs during one session as the random effects (Table 5). ICC outcomes were deduced using the Landis and Koch's agreement scale [36].

Visual scoring and pressure mat versus histology

Categorical H scores were transformed into continuous ones using the liability model used by van Grevenhof et al. for all joint locations [37, 38] to be able to compare them to the in-vivo method results which measure according to joint or animal level. Joint locations H scores were added up into joint level H scores and these were added up to make animal level H scores. A Spearman's correlation test was used and a Benjamini and Hochberg false discovery rate was applied to $p = 0.05$ to determine significant p-values for individual tests. This study was approved by the Ethical committee of Utrecht University (DEC N. 2011.III.08.092.).

Additional files

Additional file 1: Description: Illustration of the timeline of the analysis.

Additional file 2: Description: Visual scoring characteristics and scoring categories for anatomical and mobility traits

Additional file 3: Description: Final data set obtained for all pigs for histology and radiography Table 4a. Radiographic and histological joint assessments for the right side Table 4b. Radiographic and histological joint assessments for the left side.

Abbreviations
ASI: asymmetry index; AUC: area under the curve; CA: contact area; Fmax: peak vertical force; FPD: flat panel detector; ICC: intra-class correlation; OC: osteochondrosis; Pmax: peak vertical pressure; PNR: point of no return; ROC: receiver operating characteristic; VI: vertical impulse.

Competing interests
There are no present financial, professional, or personal conflicting interests from any of the authors involved in the writing of this article.

Authors' contributions
CPB was involved in devising the aim and the experimental design of this study, in generating and analysing the data and the writing of this manuscript. EM contributed to the planning of the study and the gathering and analysis of the data. AvN and AS assisted in the analysis of the data. RvW and WB provided assistance in designing the study, and assisted in the analysis of the data. All the authors assisted in the reviewing of this manuscript during the writing process and have read and accepted the final version of this manuscript. All authors read and approved the final manuscript.

Acknowledgments
The authors would like to thank the animal caretakers of our department for caring for the animals and for their help designing and building the runway system and the pens. We are also very grateful for the insights and expertise provided by Nancy Rietbroek and Leo van Leengoed, on the practical, scientific and medical aspects of the study. We would like to thank Andrea Gröne for her advice and support to enable the analysis of the histological samples and Guy Grinwis for his assistance with making the pictures of the histological slides. This work is funded by the Technology Foundation STW (grant number 11116), by the Institute for Pig Genetics BV (IPG) and by the Product Board Animal Feed (PDV).

Author details
[1]Department of Farm Animal Health, Faculty of Veterinary Medicine, Utrecht University, Yalelaan 7, NL-3584 CL Utrecht, The Netherlands. [2]Department of Equine Sciences, Faculty of Veterinary Medicine, Utrecht University, Yalelaan 112-114, NL-3584 CM Utrecht, The Netherlands. [3]Department of Surgery and Anaesthesia of Domestic Animals, Ghent University, 17 Salisburylaan 133, B-9820 Merelbeke, Belgium.

References
1. Crenshaw TD. Arthritis or OCD-identification and prevention. Adv Pork Prod. 2006;17:199.
2. Jørgensen B. Osteochondrosis/osteoarthrosis and claw disorders in sows, associated with leg weakness. Acta Vet Scand. 2000;41:123–38.
3. Nakano T, Brennan JJ, Aherne FX. Leg weakness and osteochondrosis in swine: a review. Can J Anim Sci. 1987;67:883–901.
4. Jørgensen B, Sorensen MT. Different rearing intensities of gilts: II. Effects on subsequent leg weakness and longevity. Livest Prod Sci. 1998;54:167–71.
5. Ytrehus B, Carlson CS, Ekman S. Etiology and Pathogenesis of Osteochondrosis. Vet Pathol Online. 2007;44:429–48.
6. Thorp BH, Farquharson C, Kwan APL, Loveridge N. Osteochondrosis/dyschondroplasia: failure of chondrocyte differentiation. Equine Vet J. 1993; 25:13–8.

7. Olstad K, Ytrehus B, Ekman S, Carlson CS, Dolvik NI. Early lesions of articular osteochondrosis in the distal femur of foals. Vet Pathol Online. 2011;48(6):1165–75.

8. Olstad K, Ytrehus B, Ekman S, Carlson CS, Dolvik NI. Epiphyseal cartilage canal blood supply to the tarsus of foals and relationship to osteochondrosis. Equine Vet J. 2008;40(1):30–9.

9. Dik KJ, Enzerink E, Weeren PR. Radiographic development of osteochondral abnormalities, in the hock and stifle of Dutch Warmblood foals, from age 1 to 11 months. Equine Vet J. 1999;31:9–15.

10. Laverty S, Girard C. Pathogenesis of epiphyseal osteochondrosis. Vet J. 2013;197(1):3–12.

11. Ytrehus B, Carlson CS, Lundeheim N, Mathisen L, Reinholt FP, Teige J, Ekman S. Vascularisation and osteochondrosis of the epiphyseal growth cartilage of the distal femur in pigs—development with age, growth rate, weight and joint shape. Bone. 2004;34:454–65.

12. Ytrehus B, Ekman S, Carlson CS, Teige J, Reinholt FP. Focal changes in blood supply during normal epiphyseal growth are central in the pathogenesis of osteochondrosis in pigs. Bone. 2004;35:1294–306.

13. Jørgensen B, Arnbjerg J, Aaslyng M. Pathological and Radiological Investigations on Osteochondrosis in Pigs, Associated with Leg Weakness. J Vet Med Ser A. 1995;42:489–504.

14. Olstad K, Kongsro J, Grindflek E, Dolvik NI. Consequences of the natural course of articular osteochondrosis in pigs for the suitability of computed tomography as a screening tool. BMC Vet Res. 2014;10:212.

15. Olstad K, Cnudde V, Masschaele B, Thomassen R, Dolvik N. Micro-computed tomography of early lesions of osteochondrosis in the tarsus of foals. Bone. 2008;43(3):574–83.

16. Van der Tol PPJ, Metz JHM, Noordhuizen-Stassen EN, Back W, Braam CR, Weijs WA. The vertical ground reaction force and the pressure distribution on the claws of dairy cows while walking on a flat substrate. J Dairy Sci. 2003;86:2875–83.

17. Oosterlinck M, Bosmans T, Gasthuys F, Polis I, Van Ryssen B, Dewulf J, Pille F. Accuracy of pressure plate kinetic asymmetry indices and their correlation with visual gait assessment scores in lame and nonlame dogs. Am J Vet Res. 2011;72:820–5.

18. Oosterlinck M, Pille F, Back W, Dewulf J, Gasthuys F. Use of a stand-alone pressure plate for the objective evaluation of forelimb symmetry in sound ponies at walk and trot. Vet J. 2010;183:305–9.

19. Van Grevenhof EM, Ott S, Hazeleger W, van Weeren PR, Bijma P, Kemp B. The effects of housing system and feeding level on the joint-specific prevalence of osteochondrosis in fattening pigs. Livest Sci. 2011;135:53–61.

20. Dewey CE, Friendship RM, Wilson MR. Clinical and postmortem examination of sows culled for lameness. Can Vet J. 1993;34:555.

21. van Grevenhof EM, Ducro BJ, van Weeren PR, van Tartwijk JM, van Den Belt AJ, Bijma P. Prevalence of various radiographic manifestations of osteochondrosis and their correlations between and within joints in Dutch Warmblood horses. Equine Vet J. 2009;41:11–6.

22. Bourzac C, Alexander K, Rossier Y, Laverty S. Comparison of radiography and ultrasonography for the diagnosis of osteochondritis dissecans in the equine femoropatellar joint. Equine Vet J. 2009;41:685–92.

23. Biezyński J, Skrzypczak P, Piatek A, Kościółek N, Drozdzyńska M. Assessment of treatment of Osteochondrosis dissecans (OCD) of shoulder joint in dogs-the results of two years of experience. Pol J Vet Sci. 2012;15:285–90.

24. Eibed M, Zimmermann W. The radiographic diagnosis of osteochondrosis in pigs: A retrospective study. Swine Health Prod. 1998;6:121–5.

25. Devine DV, VanPelt SR, Boileau MJ. What Is Your Diagnosis? J Am Vet Med Assoc. 2011;238:39–40.

26. Van Weeren PR, van Oldruitenborgh-Oosterbaan MMS, Barneveld A. The influence of birth weight, rate of weight gain and final achieved height and sex on the development of osteochondrotic lesions in a population of genetically predisposed Warmblood foals. Equine Vet J. 1999;31:26–30.

27. Van Weeren PR. Chapter 88 - Osteochondrosis. In: Auer JA, Stick JA, editors. Equine Surg. Fourthth ed. Saint Louis: W.B. Saunders; 2012. p. 1239–55.

28. Ekman S, Carlson CS, van Weeren PR. Third International Workshop on Equine Osteochondrosis, Stockholm, 29-30th May 2008. Equine Vet J. 2009;41:504–7.

29. de Koning DB, van Grevenhof EM, Laurenssen BFA, Ducro BJ, Heuven HCM, de Groot PN, Hazeleger W, Kemp B. Associations between osteochondrosis and conformation and locomotive characteristics in pigs. J Anim Sci. 2012;90:4752–63.

30. Goedegebuure SA, Rothschild MF, Christian LL, Ross RF. Severity of osteochondrosis in three genetic lines of Duroc swine divergently selected for front-leg weakness. Livest Prod Sci. 1988;19:487–98.

31. Carlson CS, Hilley HD, Henrikson CK, Meuten DJ. The ultrastructure of osteochondrosis of the articular-epiphyseal cartilage complex in growing swine. Calcif Tissue Int. 1986;38:44–51.

32. Lundeheim N. Genetic analysis of osteochondrosis and leg weakness in the Swedish pig progeny testing scheme. Acta Agric Scand. 1987;37:159–73.

33. Stern S, Lundeheim N, Johansson K, Andersson K. Osteochondrosis and leg weakness in pigs selected for lean tissue growth rate. Livest Prod Sci. 1995;44:45–52.

34. Meijer E, Bertholle CP, Oosterlinck M, van der Staay FJ, Back W, van Nes A. Pressure mat analysis of the longitudinal development of pig locomotion in growing pigs after weaning. BMC Vet Res. 2014;10.

35. Van Steenbergen EJ. Description and evaluation of a linear scoring system for exterior traits in pigs. Livest Prod Sci. 1989;23:163–81.

36. Landis JR, Koch GG. The measurement of observer agreement for categorical data. Biometrics. 1977;33:159–74.

37. Falconer DS. The inheritance of liability to certain diseases, estimated from the incidence among relatives. Ann Hum Genet. 1965;29:51–76.

38. Van Grevenhof EM, Heuven HCM, van Weeren PR, Bijma P. The relationship between growth and osteochondrosis in specific joints in pigs. Livest Sci. 2012;143:85–90.

2

An investigation into aflatoxin M₁ in slaughtered fattening pigs and awareness of aflatoxins in Vietnam

Hu Suk Lee[1]* ⓘ, Johanna Lindahl[2,3], Hung Nguyen-Viet[1], Nguyen Viet Khong[4], Vuong Bui Nghia[4], Huyen Nguyen Xuan[4] and Delia Grace[2]

Abstract

Background: Aflatoxin M_1 (AFM_1) is a hydroxylated metabolite formed after aflatoxin B_1 (AFB_1) is consumed by humans and animals; it can be detected in urine, milk and blood. It is well recognized that AFB_1 is toxic to humans and other animals. The International Agency for Research on Cancer (IARC) classifies aflatoxins as group 1 carcinogens and AFM_1 as group 2B carcinogen. The main objective of this study was to evaluate the exposure of pigs to aflatoxins as well as to assess the public awareness of aflatoxins among people in five provinces in Vietnam.

Results: A total of 1920 urine samples were collected from slaughterhouses located in five provinces. Overall, the positive rate of AFM_1 was 53.90% (95% confidence interval 51.64–56.15) using a cut-off of 0.15 μg/kg (range: limit of detection to 13.66 μg/kg, median: 0.2 μg/kg and mean: 0.63 μg/kg). A total of 252 people from the general population were interviewed from 5 provinces, and overall 67.86% reported being aware of aflatoxins. We also found that men and more highly educated had significantly increased awareness of aflatoxins compared to the females and primary/secondary school group. The respective odds ratios (ORs) were as follows: "male" group (OR: 2.64), "high school educated" group (OR: 3.40) and "college/university or more educated" group (OR: 10.20).

Conclusions: We can conclude that pigs in Vietnam are exposed to aflatoxins to varying degrees, and there may be a risk that pork products could contain AFM_1. Further investigation is needed into the possible health impacts as well as to aid in establishing regulations for animal feed to reduce the health impacts in humans and animals.

Keywords: Vietnam, ELISA, Pig, Urine, Aflatoxins, Survey, Perception/knowledge

Background

Aflatoxins are natural toxic metabolites of *Aspergillus* spp. (*A. flavus* and *A. parasiticus*) [1–3]; they may occur in a wide range of food commodities, and some, such as peanuts, maize and nuts are especially prone to contamination [4–6]. Aflatoxin M_1 (AFM_1) is a hydroxylated metabolite of aflatoxin B_1 (AFB_1) produced in humans and other animals that consume contaminated food, and can be detected in urine, milk and blood [7–10]. It is well recognized that AFB_1 is toxic to humans and other animals. The International Agency for Research on Cancer (IARC) classifies aflatoxins as group 1 carcinogens and AFM_1 as group 2B carcinogen [11–14]. Human exposure to aflatoxins can occur via consumption of agricultural products (such as maize, rice, peanuts and nuts etc.) or following consumption of dairy products (such as milk, cheese and yoghurt), meat and eggs produced by livestock exposed to aflatoxins [15–17]. Long-term exposure to aflatoxins is a major risk factor for liver cancer [18]. In animals, chronic exposure to aflatoxins is associated with weight loss and reproductive problems [19–21]. Particularly in pigs, the associated clinical signs are lethargy, hypothermia and icterus [22–24]. The main effect of aflatoxin exposure in pigs, however, is reduction in feed intake and average daily weight gain [19, 25, 26].

Vietnam is a tropical country which is hot and humid, providing favorable conditions for fungal growth [27, 28]. Some studies have been conducted to assess

* Correspondence: H.S.Lee@cgiar.org
[1]International Livestock Research Institute, Regional Office for East and Southeast Asia, Room 301-302, B1 Building, Van Phuc Diplomatic Compound, 298 Kim Ma Street, Ba Dinh District, Hanoi, Vietnam

AFB_1 contamination in agricultural products in Vietnam. One study found AFB_1 in 83.3–100% of pig feed products [29]. Other studies reported AFB_1 in rice, cassava and maize [4, 28, 30].

However, to our knowledge, few studies have been conducted to evaluate the concentrations of AFM_1 in pig urine in Vietnam as well as to assess the perception and knowledge of aflatoxins. One study in two pigs found only up to 16% of a dose of AFB_1 fed to the animals could be detected in the urine [31]. In another pig trial in Vietnam, approximately 23% of ingested AFB_1 was converted to AFM_1 and excreted in the urine [32]. Another older study found that the average AFM_1 concentration in pig urine was 2.29 ng/ml after feeding 12.7 µg/kg AFB_1 over 12 weeks [33]. Therefore, the main objective of this study was to evaluate the concentrations of AFM_1 in the urine of pigs slaughtered for human consumption and to assess public awareness of aflatoxins among the general population in five provinces in Vietnam.

Methods

Study locations and data collection

Vietnam's climate shows much variation because of its geography (Fig. 1). According to the Köppen-Geiger classification, the climate of southern Vietnam is mainly 'tropical wet dry', northern Vietnam has a "humid sub-tropical" climate and most of the middle and the extreme south are "tropical monsoonal" (Table 1) [34]. Vietnam is commonly divided into eight ecological zones based on geographical features and climatic conditions [35]. Here, the provinces were selected based on high maize production and to represent different ecological and climatic zones.

Urine samples were collected from slaughtered fattening pigs (mostly 6–9 months old and weighing 60–120 kg; 11 pigs were out of range) in five provinces (Son La, Hanoi, Nghe An, Dak Lak and An Giang) between January and early June 2016. The sample size was calculated based on 50% prevalence, a precision level of 5% and 95% confidence interval. At least 385 samples

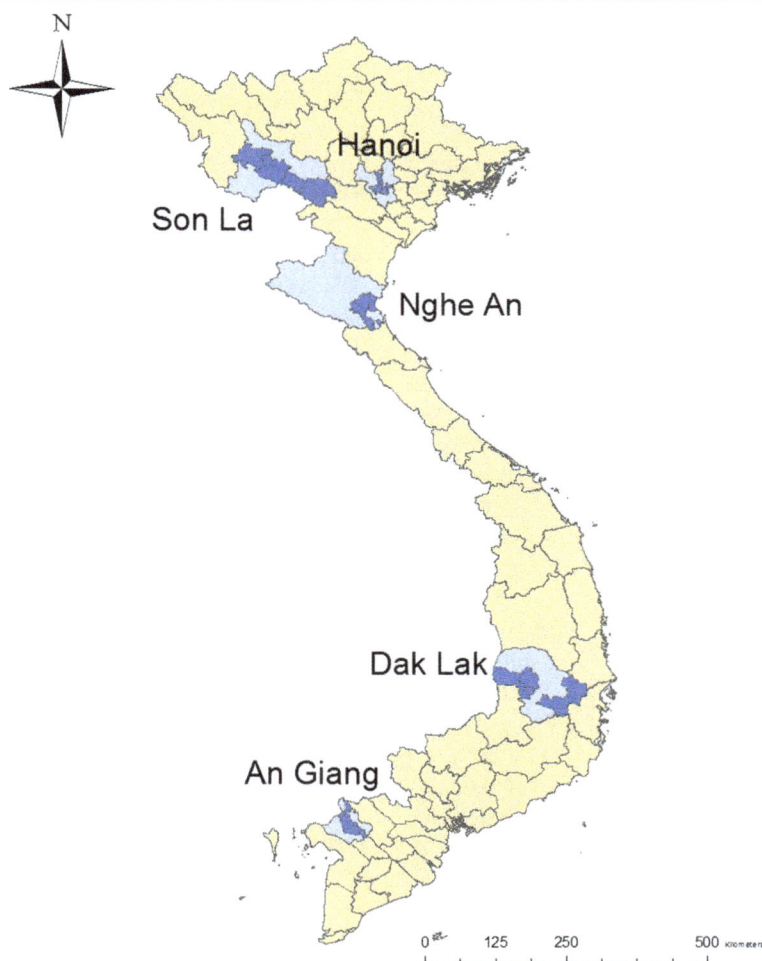

Fig. 1 Selected sampling districts (dark blue) within each province (light blue)

Table 1 Climate classification and ecological region of each province

Province	Köppen climate classification	Ecological region
Hanoi	Cw: Humid subtropical	Red River Delta
Son La	Cw: Humid subtropical	North West
Nghe An	Aw: Tropical wet dry, and Am: tropical monsoonal	North Central Coast
Dak Lak	Aw: tropical wet dry	Central Highlands
An Giang	Aw: Tropical wet dry	Mekong River Delta

per province were collected using multi-stage sampling (province-district-commune). For each province, a total of 25 communes (5 communes per district) were selected from 5 districts based on the availability of pig slaughterhouses. Within the commune, 15–16 samples were randomly collected from more than one slaughterhouse (Fig. 1). Before collecting the samples, it was confirmed that the pigs had been raised in the selected province only. Data were collected on the sex and breed of the pigs. In communes where pigs were sampled, households and pedestrians close to the slaughterhouses were selected by convenience sampling and interviewed in Vietnamese to assess their awareness of aflatoxins.

Laboratory analysis

All urine samples were immediately placed in cool boxes at the slaughterhouses and stored at −20 °C at local laboratories until transportation to the National Institute of Veterinary Research (NIVR) in Hanoi where they were analyzed. Prior to the analysis, all samples were thawed and centrifuged at 3000 g for 5 min to eliminate debris and then supernatant was used for the determination of AFM_1 levels. The concentration of AFM_1 was determined using a commercially available enzyme-linked immunosorbent assay (ELISA) (Helica Biosystems Inc., Santa Ana, CA. USA). This commercial ELISA has been specifically developed and validated for urine testing, and has been used in many previous studies. We followed all the procedures based on the protocol provided by the manufacturer. Finally, the optical density (OD) of the sample was compared to a standard curve, and then each sample level (µg/kg) was determined based on the regression equation. The standard curve covers 150–4000 ppt. A cut-off level of 0.15 µg/kg [(limit of detection (LOD) determined by the manufacturer] was used for calculating the proportion of positive samples. In addition, the mean, median and range were investigated for each province based on samples with AFM_1 concentrations above 0.15 µg/kg.

Data analysis

A logistic regression model was developed to assess the association between the demographic variables (sex and

breed) and positive status (≥LOD) while province was included as a random effect in the model.

For the awareness study, demographic information was collected via questionnaires [variables: age group (<20 years, 20–29 years, 30–39 years, 40–49 years, 50–59 years and ≥60 years), gender (male and female), education level (none, primary/secondary school, high school and college/university or more) and occupation (farmers, retailers, feed manufactures and others)]. A multivariable logistic regression model was used to evaluate the association between the demographic variables and awareness of aflatoxins as the dependent variable (question: *Have you heard about aflatoxins?*). For variable screening, chi-square tests were conducted for each variable, with only significant variables included in the final model. In addition, province was treated as a random effect to account for clustering. The final model fit was assessed using the Hosmer-Lemeshow test [36]. Variables with *p*-values <0.05 were set for statistical significance in the model. Odds ratio (OR) and 95% confidence interval (CI) were calculated by exponentiation of the coefficients from the regression model. All data were entered into Microsoft Excel 2013 and analyzed using STATA (version 14.0, StataCorp, College Station, TX, USA). ArcGIS version 10.3.1 ArcMap (ESRI, Redlands, CA, USA) was used to generate the map.

Results

A total of 1920 urine samples were collected from slaughterhouses located in five provinces (Hanoi: n = 385, Son La: n = 383, Nghe An: n = 375, Dak Lak: n = 384 and An Giang: n = 393). Overall, the positive rate of AFM_1 was 53.90% (95% CI 51.64–56.15) using a cut-off of 0.15 µg/kg (range: LOD to 13.66 µg/kg, median: 0.2 µg/kg and mean: 0.63 µg/kg) (Table 2). Son La and Hanoi had significantly higher positive rates, whereas An Giang had a significantly lower positive rate compared to other provinces.

From collected demographic information (breed and sex), we found that female pigs were significantly more likely to have AFM_1 (OR: 1.45, 95% CI: 1.08–1.96) in their urine as opposed to male pigs. Also, "indigenous breed" pigs (OR: 12.40, 95% CI: 5.15–29.82) pigs were significantly more likely to have AFM_1 in their urine compared to "exotic breed" pigs (Table 3).

A total of 252 people were interviewed from five provinces (Hanoi: n = 49, Son La: n = 50, Nghe An: n = 50, Dak Lak: n = 53 and An Giang: n = 50) to assess their awareness of aflatoxins. Overall, we found that 67.86% (95% CI: 61.71–73.58) of people were aware of aflatoxins. In addition, age groups 21–29 and 30–39 had relatively high awareness of aflatoxins whereas those aged under 20 and those between 50 and 59 years of age had the lowest awareness of aflatoxins (Table 4).

Table 2 Distribution of aflatoxin M_1 levels in pigs from five provinces in Vietnam

Province (No.)	Samples above LOD (% with 95% CI)[a]	Mean (μg/kg)[a]	Median (μg/kg)[a]	Range (μg/kg)[a]
Hanoi (385)	292 (75.84, 95% CI 71.25–80.04)	0.41	0.19	<LOD - 8.05
Son La (383)	316 (82.51, 95% CI 78.32–86.18)	1.23	0.32	<LOD - 7.35
Nghe An (375)	245 (65.33, 95% CI 60.28–70.15)	0.24	0.18	<LOD - 1.42
Dak Lak (384)	167 (43.49, 95% CI 38.47–48.61)	0.50	0.18	<LOD - 13.66
An Giang (393)	15 (3.82, 95% CI 2.15–6.22)	0.19	0.17	<LOD - 0.30
Total (1920)	1035 (53.90, 95% CI 51.64–56.15)	0.63	0.20	<LOD - 13.66

[a]Mean and median were calculated from the samples above limit of detection (LOD $\geq$ 0.15 μg/kg)

Gender and education were significantly associated with awareness in the univariable analysis and were therefore included in the model. Our final model showed that male (OR: 2.64, 95% CI: 1.15–6.09), "High school" group (OR: 3.40, 95% CI: 1.69–6.86) and "College/university or more" group (OR: 10.20, 95% CI: 4.54–22.89) had significantly increased awareness of aflatoxins compared to the reference group (female and primary/secondary school group) (Table 5). The Hosmer-Lemeshow goodness of fit test showed that there was no evidence of poor fit (p-value = 0.27).

Discussion

This was the first national study to systematically evaluate the concentrations of AFM_1 in the urine of pigs in Vietnam. We found that samples from Son La had the highest positive rate; this was consistent with previous research on maize samples which found that Son La had a higher proportion of samples with AFB_1 above 2 and 5 μg/kg, compared to five other provinces [30]. Son La is one of the most important maize production areas in Vietnam, mainly due to its suitable agro-climatic conditions and high altitude [37]. Maize is an important source of income for ethnic minorities as well as being one of their staple foods. Moreover, it is the main source feed for livestock (such as pigs and cattle) in Vietnam. In humans and animals, aflatoxins in urine have been used as a marker of exposure, and it has been shown that there is a correlation between aflatoxin intake and serum concentrations, as well as serum concentrations and urine AFM_1 concentrations [32, 38]. Therefore, a rigorous investigation is necessary to assess the full impact of AFB_1 and AFM_1 on plant, animal and human health in Son La province.

Overall, significantly higher positive rates in the northern region (Hanoi and Son La) were observed compared to the southern region (Dak Lak and An Giang), which may be attributed to differences in climatic conditions. In Vietnam, the northern and southern regions are classified into subtropical and tropical conditions, respectively. Some studies have suggested that climatic conditions, particularly temperature and humidity, affect *Aspergillus* growth and aflatoxin production [39–42], and it is possible that this also had an influence on the regional differences observed in this study. Processing

Table 3 Logistic regression model of AFM_1 (positive was considered if above the limit of detection, $\geq$ 0.15 μg/kg) for each category of pigs

Variable	Category	Adjusted Odds ratio	95% CI	P-value
Sex				
	Male	Reference	N/A	N/A
	Female	1.45	1.08–1.96	0.014*
Breed				
	Exotic	Reference	N/A	N/A
	Indigenous	12.40	5.15–29.82	<0.001*
	Mixed	1.51	0.36–6.32	0.576

CI confidence interval, *NA* not applicable as reference group
*= statistically significant at p < 0.05

Table 4 Demographic characteristics of survey respondents from "*Have you heard about aflatoxins?*"

Category	Characteristic (n)	Have you heard about aflatoxins?
Age (year)	< 20 (n = 3)	1 (33.33%)
	21–29 ($n = 21$)	16 (76.19%)
	30–39 ($n = 65$)	46 (70.77%)
	40–49 ($n = 89$)	62 (69.66%)
	50–59 ($n = 54$)	32 (59.26%)
	$\geq$ 60 ($n = 20$)	14 (70.0%)
Gender	Male ($n = 154$)	114 (74.03%)
	Female ($n = 98$)	57 (58.16%)
Education	None (n = 3)	1 (33.33%)
	Primary & Middle school ($n = 97$)	49 (50.52%)
	High school ($n = 115$)	88 (76.52%)
	College/University or more ($n = 37$)	33 (89.19%)
Occupation	Farmers ($n = 141$)	90 (63.83%)
	Retailers ($n = 36$)	27 (75.0%)
	Feed manufacturers ($n = 10$)	9 (90.0%)
	Others (office workers and businessmen) ($n = 65$)	45 (69.23%)

Table 5 Final multivariable logistic regression model of awareness of aflatoxins in five provinces in Vietnam

Variable	Category	Odds ratio	95% CI	P-value
Gender				
	Female	Reference	N/A	N/A
	Male	2.64	1.15–6.09	0.022*
Education				
	None	0.34	0.03–3.91	0.389
	Primary & Secondary school	Reference	N/A	N/A
	High school	3.40	1.69–6.86	0.001*
	College/university or more	10.20	4.54–22.89	<0.001*

CI confidence interval, NA not applicable as reference group
*= statistically significant at p < 0.05

also affects aflatoxin production, with insect damaged and high moisture corn being major predisposing factors for contamination. Based on our work, livestock in Vietnam are exposed to AFB_1, and there is a risk that meat, eggs and dairy products contain AFM_1. This is because the feed used in these livestock systems is very similar and contains a large proportion of maize. Although few studies have been conducted on dairy products, one study in Ho Chi Minh City found that 32.6% milk samples contained AFM_1 while one of 46 samples exceeded the limit (0.05 μg/L) under Vietnamese regulation [43]. However, there is currently no guidance on AFB_1 levels in animal feed under the Vietnamese regulations, while 5 μg/kg is commonly used as the tolerated level for feedstuffs for dairy cattle in European Union countries [44]. The United States Food and Drug Administration (FDA) guidelines for total aflatoxin levels are between 20 μg/kg and 300 μg/kg, depending on the commodities and intended species [45]. For corn, less than 20 μg/kg is considered safe for use in all animal feed. Exposure to aflatoxins has adverse effects on human and animal health. Therefore, future research is needed to assess the potential adverse effects of AFM_1 residues in meat and dairy products in Vietnam as well as to establish regulations for animal feed to reduce the negative health impacts in humans and animals.

The main limitation of this study was that most of our samples were collected during the dry season, yet aflatoxin levels are seasonally heterogeneous. In a study in Sierra Leone, AFM_1 levels from urine samples in humans were higher during the rainy season than the dry season [46]. Another potential bias is that clinically healthy pigs are over-represented as they are more likely to be slaughtered. Although neither slaughterhouses nor participants in the survey were selected probabilistically, we did not expect much bias to have been introduced here, and these samples to be representative.

We found that female pigs had significantly higher exposure to AFM_1, although fattening pigs may be expected to be fed the same type of feed independent of gender. However, male animals are known to be more susceptible to aflatoxins than females [47, 48] and it is possible that females could have a higher clearance of aflatoxins through urine, and thus be more protected against the harmful effects. Even among the two sows studied by Lüthy et al. [31], there was a difference in the proportion of fed AFB_1 that was excreted as AFM_1, 9.6% and 15.7% respectively, so difference in excretion among pigs may be possible. If pregnant and suckling sows are exposed to aflatoxins, this could negatively impact productivity as some studies have shown that chronic exposure to aflatoxins led to lower growth rate in piglets [49, 50]. Urine concentrations of AFM_1 were higher in indigenous pigs but the reason for this is unclear. Indigenous pigs are more likely to be raised by ethnic minorities who are less educated and thus have low levels of awareness of aflatoxins, increasing the chances of the animals consuming feed with high levels of AFB_1. Moreover, susceptibility to aflatoxins varies by breed and this may play a role.

Aflatoxin exposure in pigs has multiple implications. From the food safety point of view, aflatoxins carried over to pork are a food safety hazard. Given that pork accounts for about 70% of livestock production in Vietnam [51], this could be a significant source of human exposure. Pork is of concern as a possible exposure route as it is consumed by more than 95% of the population in Vietnam, with an annual consumption of approximately 24.7 kg per capita [52]. From the farm management point of view, the control of aflatoxin contamination needs to be considered for animal health and economic benefits. In addition further studies are needed to confirm if differences in urine aflatoxin levels are due to differences in exposure or clearance, although human studies have used it as an exposure assessment [38].

Our survey showed that men and more educated groups were more aware of aflatoxins. It is well recognized that Vietnamese women are undervalued and it is not uncommon for women to have limited access to higher education and suffers from lower pay in occupational sectors [53, 54]. The age groups less than 20 and 5−59 showed that awareness of aflatoxins relatively lower while it is worthwhile to conduct further investigation why awareness was low for these groups. It is recommended that women and less well educated groups are targeted for raising the public awareness of aflatoxin risks as well as introduction of control and prevention strategies. The intervention programs may include timing of planting, avoiding drought and rodent/insect control for field management, washing, rapid and proper drying and cleaning for post-harvest [55].

Conclusion

We can conclude that pigs in Vietnam are exposed to aflatoxins to varying degrees, and there may be a risk that pork products could contain AFM_1. Further investigation is needed into the possible health impacts as well as to aid in establishing regulations for animal feed to reduce the health impacts in humans and animals.

Abbreviations
AFB_1: Aflatoxin B_1; AFM_1: Aflatoxin M_1; CI: Confidence interval; ELISA: Enzyme-linked immunosorbent assay; FDA: Food and Drug Administration; IARC: International Agency for Research on Cancer; LOD: Limit of detection; NIVR: National Institute of Veterinary Research; OD: Optical density; OR: Odds ratio

Acknowledgments
We would like to thank the National Institute of Veterinary Research (NIVR) under the Ministry of Agriculture and Rural Development in Vietnam for supporting sample collection and analysis of AFM_1. We would also like to thank Tezira for editing.

Funding
We acknowledge the CGIAR Fund Council, the Australian Centre for International Agricultural Research, Irish Aid, the European Union, the International Fund for Agricultural Development, the United States Agency for International Development and governments of the Netherlands, New Zealand, Switzerland, Thailand and the United Kingdom for funding to the CGIAR Research Program on Climate Change, Agriculture and Food Security.

Authors' contributions
Conceived and designed the experiments: HSL Performed the experiments: HSL, NVK, HNX and VBN Analyzed the data: HSL, HNX and VBN. Wrote the paper: HSL, JL, HNV and DG. All authors read and approved the final manuscript.

Competing interests
The authors declare that they have no competing interests.

Author details
[1]International Livestock Research Institute, Regional Office for East and Southeast Asia, Room 301-302, B1 Building, Van Phuc Diplomatic Compound, 298 Kim Ma Street, Ba Dinh District, Hanoi, Vietnam. [2]International Livestock Research Institute, Nairobi, Kenya. [3]Swedish University of Agricultural Sciences, Uppsala, Sweden. [4]National Institute of Veterinary Research, 86 Truong Chinh, Phuong Mai, Dong Da, Hanoi, Vietnam.

References
1. Gourama H, Bullerman LB. Aspergillus flavus and Aspergillus parasiticus: Aflatoxigenic fungi of concern in foods and feeds: a review. J Food Prot. 1995;58:1395–404.
2. Richard J, Payne G, Desjardins A, Maragos C, Norred W, Pestka J. Mycotoxins: risks in plant, animal and human systems. CAST Task Force Rep. 2003;139:101–3.
3. Reddy K, Farhana NI, Salleh B. Occurrence of Aspergillus spp. and aflatoxin B1 in Malaysian foods used for human consumption. J Food Sci. 2011;76:T99–104.
4. Nguyen MT, Tozlovanu M, Tran TL, Pfohl-Leszkowicz A. Occurrence of aflatoxin B1, citrinin and ochratoxin a in rice in five provinces of the central region of Vietnam. Food Chem. 2007;105:42–7.
5. Sangare-Tigori B, Moukha S, Kouadio HJ, Betbeder AM, Dano DS, Creppy EE. Co-occurrence of aflatoxin B1, fumonisin B1, ochratoxin a and zearalenone in cereals and peanuts from Côte d'Ivoire. Food Addit Contam. 2006;23:1000–7.
6. Sapsford KE, Taitt CR, Fertig S, Moore MH, Lassman ME, Maragos CM, Shriver-Lake LC. Indirect competitive immunoassay for detection of aflatoxin B 1 in corn and nut products using the array biosensor. Biosens Bioelectron. 2006;21:2298–305.
7. Gacem MA, El Hadj-Khelil AO. Toxicology, biosynthesis, bio-control of aflatoxin and new methods of detection. Asian Pac J Trop Biomed. 2016;6:808–4.
8. Prandini A, Tansini G, Sigolo S, Filippi L, Laporta M, Piva G. On the occurrence of aflatoxin M1 in milk and dairy products. Food Chem Toxicol. 2009;47:984–91.
9. Redzwan SM, Rosita J, Sokhini AMM, Aqilah ARN. Association between aflatoxin M1 excreted in human urine samples with the consumption of milk and dairy products. Bull Environ Contam Toxicol. 2012;89:1115–9.
10. Sabbioni G, Sepai O. Determination of human exposure to aflatoxins. Mycotoxins in Agriculture and Food Safety. New York: Marcel Dekker Inc; 1998. p. 183–226.
11. IARC, 2002. Summaries & Evaluations AFLATOXINS (Group1) volumes 82. Avilable from: http://www.inchem.org/documents/iarc/vol82/82-04.html
12. Bennett JW, Klich M. Mycotoxins. Clin Microbiol Rev. 2003;16:497–516.
13. Robens J, Richard J. Aflatoxins in animal and human health. Rev Environ Contam Toxicol. 1992;127:69–94.
14. Williams JH, Phillips TD, Jolly PE, Stiles JK, Jolly CM, Aggarwal D. Human aflatoxicosis in developing countries: a review of toxicology, exposure, potential health consequences, and interventions. Am J Clin Nutr. 2004;80:1106–22.
15. Bintvihok A. Toe liver WLW. Controlling aflatoxin danger to duck and duck meat. Control. 2001;17:1–2.
16. Chun HS, Kim HJ, Ok HE, Hwang JB, Chung DH. Determination of aflatoxin levels in nuts and their products consumed in South Korea. Food Chem. 2007;102:385–91.
17. Zaghini A, Martelli G, Roncada P, Simioli M, Rizzi L. Mannanoligosaccharides and aflatoxin B1 in feed for laying hens: effects on egg quality, aflatoxins B1 and M1 residues in eggs, and aflatoxin B1 levels in liver. Poult Sci. 2005;84: 825–32.
18. Henry SH, Bosch FX, Troxell TC, Bolger PM. Reducing liver cancer–global control of aflatoxin. Science. 1999;286:2453–4.
19. Atherstone C, Grace D, Lindahl JF, Kang'ethe EK, Nelson F. Assessing the impact of aflatoxin consumption on animal health and productivity. Afr J Food Agric Nutr Dev. 2016;16:10949–66.
20. Fink-Grernmels J. Mycotoxins: their implications for human and animal health. Vet Q. 1999;21:115–20.
21. Cheeke PR. Endogenous toxins and mycotoxins in forage grasses and their effects on livestock. J Anim Sci. 1995;73:909–18.
22. Cook W, Van Alstine W, Osweiler G. Aflatoxicosis in Iowa swine: eight cases (1983-1985). J Am Vet Med Assoc. 1989;19:554–8.
23. Miller D, Stuart B, Crowell W. Experimental aflatoxicosis in swine: morphological and clinical pathological results. Can J Comp Med. 1981;45:343–51.
24. Panangala V, Giambrone J, Diener U, Davis N, Hoerr F, Mitra A, Schultz R, Wilt G. Effects of aflatoxin on the growth performance and immune responses of weanling swine. Am J Vet Res. 1986;47:2062–7.
25. Andretta I, Kipper M, Lehnen CR, Hauschild L, Vale MM, Lovatto PA. Meta-analytical study of productive and nutritional interactions of mycotoxins in growing pigs. Animal. 2012;6:1476–82.
26. Dersjant-Li Y, Verstegen MW, Gerrits WJ. The impact of low concentrations of aflatoxin, deoxynivalenol or fumonisin in diets on growing pigs and poultry. Nutr Res Rev. 2003;16:223–39.
27. Trung T, Bailly J, Querin A, Le Bars P, Guerre P. Fungal contamination of rice from south Vietnam, mycotoxinogenesis of selected strains and residues in rice. Rev Med Vet. 2001;152:555–60.
28. Wang DS, Liang YX, Chau NT, Dien LD, Tanaka T, Ueno Y. Natural co-occurrence of Fusarium toxins and aflatoxin B1 in com for feed in North Vietnam. Nat Toxins. 1995;3:445–9.

29. Thieu NQ, Ogle B, Pettersson H. Screening of Aflatoxins and Zearalenone in feedstuffs and complete feeds for pigs in southern Vietnam. Trop Anim Health Prod. 2008;40:77–83.

30. Lee HS, Nguyen-Viet H, Lindahl J, Thanh H, Khanh T, Hien L, Grace D. A survey of aflatoxin B1 in maize and awareness of aflatoxins in Vietnam. World Mycotoxin J. in press;2017

31. Lüthy J, Zweifel U, Schlatter C. Metabolism and tissue distribution of [14 C] aflatoxin B 1 in pigs. Food Cosmet Toxicol. 1980;18:253–6.

32. Thieu NQ, Pettersson H. Zearalenone, deoxynivalenol and aflatoxin B1 and their metabolites in pig urine as biomarkers for mycotoxin exposure. Mycotoxin Res. 2009;25:59–66.

33. Ho C. The metabolites of aflatoxins B1 and G1 in the pig urine. J Chin Soc of Vet Sic. 1987;13:197–201.

34. Kottek M, Grieser J, Beck C, Rudolf B, Rubel F. World Map of the Köppen-Geiger climate classification updated. Meteorol Z. 2016;15:259–63. doi:10.1127/0941-2948/2006/0130.

35. Sawada H, Araki M, Chappell NA, LaFrankie JV, Shimizu A. (Eds.). Forest environments in the Mekong River basin. Springer. 2007;169.

36. Hosmer DW, Hosmer T, Le Cessie S, Lemeshow S. A comparison of goodness-of-fit tests for the logistic regression model. Stat Med. 1997;16: 965–80.

37. Karimov AA, Thinh NT, Cadilhon JJ, Khanh TT, Van Thuy T, Long CTM, Truc DT N. Value chain assessment report for avocado, cattle, pepper and cassava in Dak Lak province of Central Highlands of Vietnam. ILRI (aka ILCA and ILRAD); 2016.

38. Gan LS, Skipper PL, Peng X, Groopman JD, Chen JS, Wogan GN, Tannenbaum SR. Serum albumin adducts in the molecular epidemiology of aflatoxin carcinogenesis: correlation with aflatoxin B1 intake and urinary excretion of aflatoxin M1. Carcinogenesis. 1988;9:1323–5.

39. Magan N, Aldred D. Post-harvest control strategies: minimizing mycotoxins in the food chain. Int J Food Microbiol. 2007;119:131–9.

40. Magan N, Aldred D, Hope R, Mitchell D. Environmental factors and interactions with mycobiota of grain and grapes: effects on growth, deoxynivalenol and ochratoxin production by Fusarium culmorum and Aspergillus carbonarius. Toxins. 2010;2:353–66.

41. Medina A, Rodríguez A, Sultan Y, Magan N. Climate change factors and Aspergillus flavus: effects on gene expression, growth and aflatoxin production. World Mycotoxin J. 2014;8:171–9.

42. Wu F, Bhatnagar D, Bui-Klimke T, Carbone I, Hellmich R, Munkvold G, Paul P, Payne G, Takle E. Climate change impacts on mycotoxin risks in US maize. World Mycotoxin J. 2011;4:79–93.

43. VTH T. Detection and qualification of aflatoxin in raw materials, cattle feeds and fresh milk. PhD Thesis. Vietnam: International University HCMC; 2014.

44. FAO. 2013 Worldwide regulations for mycotoxins in food and feed. Available from: http://www.fao.org/docrep/007/y5499e/y5499e00.htm.

45. FDA, 2016. Action Levels for Aflatoxins in Animal Feeds. Available from http://www.fda.gov/ICECI/ComplianceManuals/ CompliancePolicyGuidanceManual/ucm074703.htm.

46. Jonsyn-Ellis FE. Seasonal variation in exposure frequency and concentration levels of aflatoxins and ochratoxins in urine samples of boys and girls. Mycopathologia. 2001;152:35–40.

47. Cote LM, Beasley VR, Bratich PM, Swanson SP, Shivaprasad HL, Buck WB. Sex-related reduced weight gains in growing swine fed diets containing deoxynivalenol. J Anim Sci. 1985;61:942–50.

48. Gurtoo HL, Motycka L. Effect of sex difference on the in vitro and in vivo metabolism of aflatoxin B1 by the rat. Cancer Res. 1976;36:4663–71.

49. Dilkin P, Zorzete P, Mallmann C, Gomes J, Utiyama C, Oetting L, Correa B. Toxicological effects of chronic low doses of aflatoxin B 1 and fumonisin B 1-containing Fusarium moniliforme culture material in weaned piglets. Food Chem Toxicol. 2003;41:1345–53.

50. Marin D, Taranu I, Bunaciu R, Pascale F, Tudor D, Avram N, Sarca M, Cureu I, Criste R, Suta V. Changes in performance, blood parameters, humoral and cellular immune responses in weanling piglets exposed to low doses of aflatoxin. J Anim Sci. 2002;80:1250–7.

51. Lemke U, Mergenthaler M, Roßler R, Huyen L, Herold P, Kaufmann B, Zarate AV. Pig production in Vietnam–a review. CAB Rev: Pers Agri Vet Sci Nutri Nat Res. 2008;23:1–15.

52. Nga NTD, Lapar L, Unger F, Van Hung P, Ha DN, Huyen NTT, Van Long T, Be D T. Conference on International Research on Food Security. Tropentag, Berlin, Germany September 16-18, 2015.

53. Schuler SR, Anh HT, Ha VS, Minh TH, Mai BTT, Thien PV. Constructions of gender in Vietnam: in pursuit of the 'three criteria'. Cult Health Sex. 2006;8: 383–94.

54. UN, 2016. Cross-cutting Themes: Gender. Available from: http://www.un.org. vn/en/component/content/article.html?Itemid=&id=1081:cross-cutting-themes-gender.

55. Hell K, Mutegi C, Fandohan P. Aflatoxin control and prevention strategies in maize for sub-Saharan Africa. Julius-Kühn-Archiv. 2010;425:534–41.

3

Cross-sectional survey of selected enteric viruses in Polish turkey flocks between 2008 and 2011

K. Domańska-Blicharz[1*], Ł. Bocian[2], A. Lisowska[1], A. Jacukowicz[1], A. Pikuła[1] and Z. Minta[1]

Abstract

Background: Enteric diseases are an important health problem for the intensive poultry industry, resulting in considerable economic losses. Apart from such microbiological agents associated with enteritis as bacteria and parasites, a lot of research has been recently conducted on viral origin of enteric diseases. However, enteric viruses have been identified in intestinal tract of not only diseased but also healthy poultry, so their role in enteritis is still unclear. The present study aimed at determination of the prevalence of four enteric viruses, namely astrovirus, coronavirus, parvovirus and rotavirus in meat-type turkey flocks in Poland as well as at statistical evaluation of the occurrence of the studied viruses and their relationships with the health status and the age of birds. Two hundred and seven flocks of birds aged 1-20 weeks originating from different regions of the country were investigated between 2008 and 2011. Clinical samples (10 individual faecal swabs/flock) were duly processed and examined using molecular methods targeting the conservative regions of viral genomes: RNA-dependent RNA polymerase gene of astrovirus, non-structural 1 gene of parvovirus, non-structural protein 4 gene of rotavirus, and 5′ untranslated region fragment of turkey coronavirus. Different statistical methods (i.e. the independence chi-square test, the correspondence analysis and the logistic regression model) were used to establish any relationships between the analyzed data.

Results: Overall, 137 (66.2%, 95% CI: 59.3-72.6) of the 207 turkey flocks sampled were infected with one or more enteric viruses. Among the 137 flocks, 74 (54%, 95% CI: 45.3-62.6) were positive for one virus, whereas 54 (39.4%, 9 5% CI: 31.2-48.1) and 9 (6.6%, 95% CI: 3.1-12.1) were co-infected with two or three different enteric viruses, respectively. No flock was simultaneously infected with all four viruses studied. The prevalence of astrovirus infection was 44.9% (95% CI: 38.0-52.0), parvovirus 27.5% (95% CI: 21.6-34.2), rotavirus 18.8% (95% CI: 13.8-24.8), and coronavirus 9.7% (95% CI: 6.0-14.5). Young turkeys aged 1-4 weeks old had the highest (82.1%, 95% CI:71.7-89.8) prevalence of viral infection. Applied statistical methods have indicated the dependence of rotavirus infection as well as the co-infection with multiple viruses and the health status of turkeys. Furthermore, our results statistically confirm that especially young birds are susceptible to infection with rotavirus and astrovirus.

Conclusions: The study demonstrated the presence of astrovirus, coronavirus, parvovirus and rotavirus infections in Polish turkey farms. These viruses were detected in both healthy and diseased birds. However, the presented results provide valuable feedback which could help to evaluate the role of some enteric viruses in the etiology of enteritis in turkey.

Keywords: Enteritis, Turkey, Astrovirus, Coronavirus, Rotavirus, Parvovirus, Statistics, Poland

* Correspondence: domanska@piwet.pulawy.pl
[1]Department of Poultry Diseases, National Veterinary Research Institute, Al. Partyzantów 57, 24-100 Puławy, Poland
Full list of author information is available at the end of the article

Background

Enteric diseases are an important health problem for the intensive poultry industry and can result in considerable economic losses. The most characteristic of these conditions are diarrhea, stunting, huddling, increased feed conversion and extended time-to-market. In the more severe forms, immune dysfunction and increased mortality may occur. Different terms have been used to describe enteric disease syndrome in turkey: poult enteritis complex (PEC) and poult enteritis mortality syndrome (PEMS) or light turkey syndrome (LTS) [1–3]. However, none of these descriptions relate to any specific agents, and numerous factors have been associated with them, including environmental, such as housing, ventilation, temperature and humidity; management, such as biosecurity programmes; and microbiological agents (viruses, bacteria and parasites). In recent years there has been a lot of research regarding the viral origin of enteric diseases, and different viruses, such as astrovirus, coronavirus, reovirus, rotavirus, parvovirus, and adenovirus have been identified in intestinal tract of not only diseased but also healthy poultry. The role of some viruses in enteritis is well-defined, i.e. coronavirus is a known factor of mud fever/bluecomb disease of turkeys [4]. However, the clinical and epidemiological significance of turkey infection with most of them still remains unclear. It seems that viruses are present in large amounts and different combinations in bird intestines, but in optimal nutritional and environmental conditions they do not cause clinical disease. In favorable conditions, such as feed change, sudden temperature shift or other stress factors, the viruses present in the gut might damage its walls making birds more susceptible to secondary bacterial infection. Furthermore, it is also possible that enteric viruses could contribute to the development of symptoms only by co-infections with multiple viruses. For a better understanding and control of enteric disease in turkeys, more studies on enteric viruses are needed.

In Polish commercial turkey flocks, despite the high level of hygiene/biosecurity implemented, the clinical signs of enteritis are often observed. In such situations, bacteria and parasites are most often investigated microbiological agents. The aim of the study was to examine the prevalence of four enteric viruses, namely astrovirus (AstV), coronavirus (CoV), rotavirus (RoV) and parvovirus (PV) in Polish meat-type turkey flocks. Recently statistical methods have been used more often to explain which factors contribute to disease characteristics, thus we also attempted to determine whether a statistical correlation occurs between the presence of the viruses studied and the health condition and age of turkeys.

Methods

Sample collection

Between 2008 and 2011, a total of 2070 fecal swabs were collected from 207 turkey flocks (10 individual fecal swabs per flock) located in two regions of Poland (north-eastern, and western part). A lot of turkey farms, which account for about 70% of total production in Poland are located in these regions (Warmia-Mazury, Wielkopolska and Lubuskie voivodeships) [5]. The samples originated from both healthy and enteric turkeys, aged 1 to 140 days. To standardize sample collection, as well as to acquire additional information about flocks, turkey farm owners or veterinarians therein were provided with instructions for samples collection and questionnaires inquiring about such data as flock location, age of birds, number of birds in the flock, state of health, possible treatments. All samples were stored in -20 °C until processing.

The information about the health status of studied turkeys was obtained for 187 flocks. The health status was a subjective criterion, assessed in the descriptive form by individual veterinarians taking care of the studied turkey flocks. Based on this description the flocks were divided into three categories, i.e. healthy when the behavior and weight gain of birds were normal; PEC when any symptoms of enteritis, such as prolonged diarrhorea, low body gain and feed intake, or uneven growth occurred; and PEMS when more serious clinical signs of disease, including increased mortality, were observed.

The information about birds age was obtained for 149 flocks. Depending on the age, turkey flocks were separated into three following groups reflecting the production cycles: the growing phase of 1 to 4-week-old, the fattening phase of 5 to 12-week-old and the finishing phase of birds older than 13 weeks.

Sample processing

After slow thawing, each individual swab (of 10 swabs sampled from flock) was hydrated in phosphate-buffered saline supplemented with antibiotics (100 U of penicillin with 100 mg of streptomycin/ml), incubated for 1 h at room temperature and clarified by centrifugation at 1500 g for 20 min. Five such obtained supernatants were combined into one pool so that two pooled samples were then tested per flock. Total RNA and DNA were then extracted from 250 µl of each two pools into 50 µl RNase-free water using commercial kits (RNeasy and DNeasy Mini Kits, respectively, Qiagen, Germany) according to manufacturer's instruction. Extracted nucleic acids were stored in -70 °C for further molecular analysis.

Molecular methods for enteric viruses

Four different assays were used for detection of astrovirus, coronavirus, parvovirus and rotavirus targeting the conservative regions of their genomes. In spite these viruses were detected both in diseased and healthy birds, they were recently regarded as those which are mainly responsible for enteritis in turkeys. Presence of astroviruses, irrespective of their affiliation to virus type was identified by using RT-PCR aimed at RNA dependent RNA polymerase gene fragment according to the protocol described by Tang et al. [6]. This method enables detection of both turkey astrovirus type 1 (TAstV-1) and type 2 (TAstV-2) as well as avian nephritis virus (ANV). Detection of CoV was performed in real-time RT-PCR aimed at 5′ untranslated region (5′UTR) fragment present in gammacoronaviruses of poultry i.e. turkey coronavirus and infectious bronchitis virus [7]. For parvovirus detection, primers which enable the amplification of non structural 1 (NS1) of *Aveparvovirus* gene were used [8]. For RoV identification the method according to Day et al. was applied [9]. This method enable the amplification of highly conserved non-structural protein 4 (NSP4) gene region which was believed not group specific. But next studies revealed that NSP4 gene of avian D, F and G rotavirus groups are only distantly related to that of avian group A so it seems that applied method is useful only for this group of rotavirus [10]. Details on primers sequences, annealing temperatures and sizes of expected amplicons are presented in the Table 1. Tests were performed using Qiagen One-Step RT-PCR kit (Qiagen, Germany) for RT-PCR, OptiTaq DNA Polymerase Kit (EURx, Poland) for PCR and QuantiTect Probe RT-PCR Kit (Qiagen, Germany) for real time RT-PCR. The above tests were conducted in a 2720 Thermal Cycler or 7500 Real Time PCR System for conventional or real time amplifications, respectively (both Applied Biosystems, USA). The details on reaction mixtures and the protocols of applied assays are available on request.

Statistical analysis

The data obtained from completed questionnaires were analyzed. To assess the normal distribution, categorized graphs of normality and the Shapiro-Wilk test were used. To compare the quantitative correlation between age, health status as well as the number of identified viruses in all groups of birds, the Kruskal-Wallis test was used. In turn, the Mann-Whitney test was used to correlate the age of infected and uninfected birds. The interactions between qualitative variables, i.e. the age (categorized in groups) or health condition of the tested flocks and the prevalence of the studied viruses, as well as correlations between the incidences of the investigated pathogens, were also examined. These variables were summarized in the contingency tables which were examined by the independence chi-square (χ^2) statistical test. The statistical methods which inform about the strength of the relationship between qualitative variables were then applied. Consequently, we calculated the Yule coefficients. However, these tests do not describe the nature of the relationships, and thus we applied the correspondence analysis (CA) to this end. The CA is the method of displaying the relationships between categorical variables as scatterplot diagrams; they are, however, not shown here, and only the conclusions from the obtained diagrams are discussed. We also implemented logistic regression (LR) model, which is also helpful in description of the strength and nature of the relationship between the variables. In the LR model, a dichotomous scale is used, so the health status of flocks has been rearranged as healthy or sick (PEC/PEMS) animals. Odds ratio (OR) and 95% confidence interval (CI) were calculated and the differences with $P < 0.05$ were considered as significant. Data analysis was performed using STATISTICA ver. 10 (StatSoft, Inc., 2011 Dell Software, Aliso Viejo, CA, USA).

Ethics statement

Sampling was conducted under the permission of the owners or other responsible persons. The birds in the farms were under the supervision of veterinarians, who took different samples as part of their routine work (i.e. as screening flocks for efficacy evaluation of applied vaccines or presence of any infections) and thus part of them were used in this study. For this reason, sampling did not require the approval of the Ethics Committee.

Table 1 Primers and probe used in the study

Primer name	Target virus	Target gene	Sequence	Annealing temperature (°C)	Amplicon (bp)	Ref
TAPG-L1	AstV	ORF1b	TGGTGGTGYTTYCTCAARA	50	601	[6]
TAPG-R1			GYCKGTCATCMCCRTARCA			
F30	RoV	NSP4	GGGCGTGCGGAAAGATGGAGAAC	50	630	[9]
R660			GGGGTTGGGGTACCAGGGATTAA			
PVF-1	PV	NS1	TTCTAATAACGATATCACTCAAGTTTC	55	500	[8]
PVR-1			TTTGCGCTTGCGGTGAAGTCTGGCTCG			
GU391			GCTTTTGAGCCTAGCGTT			
GL533	CoV	5′UTR	GCCATGTTGTCACTGTCTATTG	60	143	[7]
G probe			CACCACCAGAACCTGTCACCTC			

Results

Enteric viruses occurrence

Out of the 207 flocks tested, 137 (66.2%, 95% CI: 59.3-72.6) were positive for one or more enteric viruses. Among the positive flocks, 74 (54%, 95% CI: 45.3-62.6) were infected with one virus, 54 (39.4%, 95% CI: 31.2-48.1) with two viruses, and 9 (6.6%, 95% CI: 3.1-12.1) with three viruses. No flock was simultaneously infected with four studied viruses. AstV was the most commonly detected virus in the samples and its prevalence was 44.9% (95% CI: 38.0-52.0), followed by 27.5% prevalence of parvovirus (95% CI: 21.6-34.2), 18.8% of rotavirus (95% CI: 13.8-24.8), and 9.7% occurrence of coronavirus (95% CI: 6.0-14.5), as presented in Table 2.

The majority of the samples were collected in 2008 (from 80 flocks) and the least flocks were sampled in 2010 (32 flocks). The prevalence of viruses varied depending on the year of the study. Generally, the frequencies of astrovirus, rotavirus and parvovirus occurrence during the four-year study revealed an increasing tendency. The most visible increase was observed in the case of RoV the occurrence of this virus was 12.5% in 2008 and rose up to 34.2% in 2011. Adversely, the occurrence of TCoV initially increased from 5% in 2008 up to 33.3% in 2010 and then decreased to a zero level of virus detection in 2011.

Health status

Birds in 45 flocks were described as healthy, whereas in 94 flocks PEC and in 48 flocks PEMS symptoms were observed. The information on the health status of turkeys in 20 flocks was not available. Of the 45 flocks that were in good health condition, 25 were negative, 11 were positive for one virus and 9 were infected with two or three viruses. The most frequent viruses detected among healthy flocks was astrovirus (28.9%) and parvovirus (26.7%). RoV and TCoV were detected only in 4.4% and 6.7% of flocks without any disease signs, respectively. The frequencies of astrovirus, parvovirus, rotavirus and coronavirus detection in enteric birds were 44.9%, 27.5%, 18.8% and 9.7%, respectively. In 40 flocks

showing clinical signs of enteritis no virus was detected. However, among the rest 102 diseased flocks 56.9% were infected with single virus. The most frequently identified in this group was AstV present in 51.7% of singly infected flocks. Thirty percent of flocks with enteritis were positive for more than one virus. Among enteric flocks co-infected with multiple viruses 50.8% included RoV. The results regarding correlations between virus prevalence and the health status of the flock are shown in Table 3.

Flock age

Seventy flocks consisted of birds of 1 to 4 weeks age group (growing phase), 63 flocks comprised birds of 5 to 12 weeks age group (fattening phase), and the least numerous age group of over 13 weeks old (finishing phase) constituted 16 flocks. Table 4 presents the data about the numbers of diseased or healthy flocks and detected viruses according to the age of the flocks. In the case of 49 flocks, the information about the age of turkeys was not provided. Only 9 samples (12.9%) collected from turkeys in the growing phase came from flocks in good health condition although seven of them were infected with 1 or 2 viruses (Table 5). All the remaining samples originated from flocks that showed PEC or PEMS symptoms and 82% of them harbored 1 to 3 viruses per flock. The viruses most frequently detected in diseased turkeys in growing phase were TAstV (82%), RoV (38%) and PV (34%). Among 11 (18%) enteric flocks none of the 4 studied viruses were found.

No enteric viruses were detected in either 16 flocks of fattening turkeys which showed clinical signs of intestinal disease or in 16 healthy flocks. In the remaining 6 healthy flocks one or more viruses were detected. Among 25 enteric flocks most of the samples (80%) were infected with one virus and two viruses were only present in 5 flocks. The PV (56%) and TAstV (36%) were

Table 2 Frequencies of individual and multiple enteric virus infections detected in swab samples from all 207 commercial turkey flocks collected between 2008 and 2011

Item	TAstV	TCoV	RoV	TuPV
Number of viruses detected				
1 virus	39	7	7	21
2 viruses	45	9	26	28
3 viruses	9	4	6	8
Number of positive samples (%)	93 (44.9)	20 (9.7)	39 (18.8)	57 (27.5)

Table 3 Distribution of viruses in studied samples from 187 commercial turkey flocks depending on health status of birds (healthy/PEC/PEMS)

Virus	Healthy (%)	PEC (%)	PEMS (%)	Total (%)
TAstV	6 (3.2)	18 (9.6)	12 (6.4)	36 (19.3)
TCoV	1 (0.5)	5 (2.7)	1 (0.5)	7 (3.7)
RoV	0	5 (2.7)	0	5 (2.7)
TuPV	4 (2.2)	10 (5.3)	7 (3.8)	21 (11.2)
Total one virus	11 (5.9)	38 (20.3)	20 (10.7)	69 (36.9)
Two viruses	8 (4.3)	24 (12.8)	16 (8.6)	48 (25.7)
Three viruses	1 (0.5)	3 (1.6)	1 (0.5)	5 (2.7)
Total coinfections	9 (4.8)	27 (14.4)	17 (9.1)	53 (28.3)
No virus	25 (13.4)	29 (15.5)	11 (5.9)	65 (34.8)
Total	45 (24.1)	94 (50.3)	48 (25.7)	187

Table 4 The numbers of diseased (PEC/PEMS) or healthy flocks and associated viruses according to the age of the flocks

Age	Flocks No.	PEC	PEMS	healthy	TAstV	TCoV	RoV	PV
1 to 4 wks	70	38	23	9	50	4	19	20
5 to 12 wks	63	26	15	22	13	6	3	18
> = 13 wks	16	12	1	3	2	3	0	0

the most frequently detected virus families in the samples from fattening flocks.

As regards turkeys in the finishing phase, most of them (8/16) showed mild clinical signs of enteritis independently of infection with any of the studied viruses. Among the remaining 5 diseased turkey flocks, three were infected with CoV and two with AstV.

Statistical analysis

Only statistically significant results are presented; insignificant relationships of variables are not included in this section. Since the normality was not fulfilled, non-parametric Kruskal-Wallis and Mann-Whitney tests were selected to assess the differences in the average age between groups of healthy, PEC and PEMS birds. Additionally, due to the relatively restricted amount of data (especially for multifactorial models), two separate logistic regression (LR) models were used. The first model investigated the impact of virus infections on the variable 'health condition', while the second model explored the impact of the number of viruses and turkey's age on the same variable, i.e. 'health condition'.

Age and health status of turkeys

The results of applied statistical analysis revealed the existence of significant differences between the age of

Table 5 Distribution of virus number in diseased (PEC/PEMS) or healthy flocks according to the age of the flocks

Age	Number of				
	viruses	flocks	healthy	PEC	PEMS
1 to 4 wks	0	13	2	8	3
	1	24	4	10	10
	2	31	3	19	9
	3	2	0	1	1
5 to 12 wks	0	32	16	10	6
	1	23	3	14	6
	2	7	2	2	3
	3	1	1	0	0
> = 13 wks	0	11	3	8	0
	1	5	0	4	1
	2	0	0	0	0
	3	0	0	0	0

birds depending on their health status (healthy, PEC and PEMS). The Kruskal-Wallis test showed that the average age of healthy, PEC and PEMS turkeys differs significantly ($P = 0.036$). Multiple comparison test showed that this difference was only between healthy and PEMS turkeys ($P = 0.03$); healthy turkeys were about 7 weeks old and PEMS about 4 weeks old. Generally the older the turkeys were, the healthier they were. Such correlation was also implied by the independence chi-square test ($P = 0.007$, $\phi = 0.31$). The calculated ORs of =3.57 and 3.75 indicated that the chance of PEC and PEMS symptoms in turkeys aged 1-4 weeks are above 3.5 times higher than the chance of such disease symptoms in the older group of 5-12-week-old animals. The OR =6.92 indicates that the possibility of PEMS relative to PEC symptoms in turkeys in the fattening phase is almost sevenfold higher than in turkeys over 13 weeks of age.

Age and astrovirus infection

The correlation between the age of turkeys and AstV infection was established by three statistical methods. The results of the Mann-Whitney test revealed that the difference between the age of uninfected and astrovirus-infected birds was significant ($P = 0.0000$). The median age of turkeys infected with AstVs was 3 weeks and uninfected turkeys were about 6 weeks old. The graphs of correspondence analysis also indicated relatively more astrovirus infections among the youngest individuals aged 1-4 weeks. The CA diagrams confirmed the correlation between the age and AstV infections. Such correlations were additionally confirmed in the independence chi-square test ($P = 0.0000$, $\phi = 0.50$). The lowest prevalence of AstVs was in turkeys older than 13 weeks (12%, 95% CI: 1.5-36.4), slightly higher in turkeys aged 5-12 weeks (21%; 95% CI: 11.5-32.7), and the highest in young individuals aged 1-4 weeks (68%, 95% CI: 56.4%-78.1%). The calculated OR =0.12 indicates that the chance of astrovirus infection among turkeys aged 5-12 weeks is above eight-fold lower than the chance of this infection in the youngest group of 1-4-week-olds. Furthermore, it is almost twofold lower (OR =0.51) in the oldest group (> = 13 weeks) than among turkeys aged 5-12 weeks.

Age and parvovirus infection

The correlation between turkey age and PV infection was only found by the independence chi-square statistical test ($P = 0.04$, $\phi = 0.20$). No parvovirus infection was observed in turkeys older than 13 weeks while its prevalence in turkeys aged 5-12 weeks and 1-4 weeks was similar (29%, 95% CI:17.9-41.4 and 28%, 95% CI: 18.6-39.5, respectively). The CA diagrams confirmed this observation; the finishing group corresponded with PV-free birds while the growing and fattening phase groups corresponded with the infection.

Age and rotavirus infection

The correlation between the age of turkeys and RoV infection was also found by three applied methods. The results of the Mann-Whitney test revealed that the difference between the age of uninfected and rotavirus-infected birds was significant (P = 0.0000). The median age of turkeys infected with rotaviruses was 2 weeks and uninfected turkeys were about 5 weeks old. The independence chi-square test indicated a significant correlation between turkey age and RoV infection (P = 0.00007, φ = 0.35). There were no rotavirus infections among turkeys older than 13 weeks. The prevalence of RoVs in turkeys aged 5-12 weeks was about 5% (95% CI: 1.0-13.3) and in birds aged 1-4 weeks it reached the highest level (29%, 95% CI: 19.7-40.9), which means that the incidence of rotavirus infection diminished with age. The calculated OR =0.12 indicates that the chance of RoV infection among turkeys aged 5-12 weeks is above eightfold lower than the chance of this infection in the youngest group of 1-4-week-olds. The CA graphs confirmed this relationship.

Health status and rotavirus infection

The independence chi-square test revealed a significant relationship between RoV occurrence and the health status of turkeys (P = 0.01, φ = 0.22). The OR =6.96 indicates that the odds of disease symptoms in turkeys infected with rotaviruses are almost sevenfold higher than the odds of occurrence of these symptoms in the uninfected group. The CA plots also revealed that rotavirus infection is most strongly correlated with PEC symptoms in turkeys, and PEMS symptoms as well as healthy status correspond with turkeys uninfected with rotaviruses. The first LR model showed that only RoV infection has a significant impact on health status (diseased turkeys), with P = 0.017. The OR =5.06 indicates that the odds of disease symptoms of turkeys infected with rotaviruses are over fivefold higher than the odds of occurrence of these symptoms in the uninfected turkeys.

Age and number of viruses

The correlation between turkey age and the number of viruses was found in the Kruskal-Wallis test (P = 0.0000). Generally, the older the birds, the number of infecting viruses decreases. The median age of turkeys infected with one virus was 5 weeks, those infected with 2 viruses - 3 weeks, and in the case of birds uninfected with any virus it was 7 weeks- the differences between the age of differently infected birds were significant. Such correlation was also found in the independence chi-square test (P = 0.0000, φ = 0.49). The OR =2.58 indicates that the odds of infection with one virus in turkeys up to 4 weeks of age are more than 2.5-fold higher than the odds for the occurrence of such

infections in turkeys aged 5 to 12 weeks and 1.7-fold higher in 5 to 12-week-old turkeys than in the oldest ones (≥13 weeks) (OR =1.73). Furthermore, the OR =4.3 indicates that the odds of infection with two viruses (as opposed to infection with one virus) in turkeys up to 4 weeks of age is more than fourfold higher than the odds for the occurrence of such infections at the age of 5 to 12 weeks. The CA plots also suggested the lowest number of infecting viruses/no infection in the oldest group, and the highest (2-3 viruses) in the youngest group, whereas turkeys aged 5 to 12 weeks were mostly associated with single virus infection.

Health status and number of viruses

The relationship between co-infection with two or three viruses and the health status of turkeys was identified in χ^2 statistics (P = 0.046, φ = 0.26). The health condition of turkeys deteriorated with an increasing number of viruses; the OR =2.98 indicates that the odds of PEC symptoms in turkeys infected with one virus are almost threefold higher (and above fourfold higher for PEMS symptoms; OR =4.13) than the odds of occurrence of these symptoms in the uninfected turkeys. Additionally, the odds of PEMS symptoms (in relation to PEC symptoms) in turkeys infected with two viruses are about 1.27-fold higher than the odds of occurrence of these symptoms in birds infected with one virus and they are about 1.39-fold higher in turkeys infected with one virus than in the uninfected turkeys. Similarly, the CA diagrams also indicated that healthy turkeys had no virus infection, and infections with two or three different viruses most commonly resulted in disease symptoms. The second logistic regression model also indicated that the odds of disease symptoms in turkeys infected with one virus are more than 3.5-fold higher than in turkeys free from these infections (OR =3.61, P = 0.01) and that younger age of turkeys is a factor stimulating illness. The OR =2.75 (P = 0.04) indicated that the odds of PEC or PEMS symptoms in turkeys up to 4 weeks of life are more than 2.5-fold higher than in turkeys 5 to 12 weeks of age.

Correlations between investigated pathogens

The χ^2 statistics revealed that there is a significant relationship between RoV and AstV infections (P = 0.0002; φ = 0.26). Among turkeys infected with rotavirus, about 72% were also infected with astroviruses. Contrastingly, only about 30% of AstV-infected birds were concomitantly infected with RoVs. However, among rotavirus-free turkeys, above 60% were also astrovirus-free and inversely, among AstV-free birds about 90% were also RoV-free. The OR =4.03 indicates that the odds of astrovirus (rotavirus) infection in turkeys infected with rotavirus (astrovirus) are about fourfold higher than the odds of occurrence of this infection in turkeys uninfected with rotavirus (astrovirus).

Discussion

The present study was undertaken to estimate the prevalence of four enteric viruses, namely astrovirus, coronavirus, parvovirus and rotavirus in commercial meat-type turkey farms in Poland. As the role of these viruses in the development of enteritis is still unclear, statistical analysis was also applied in an attempt to better understand the influence of these viruses on the health status of studied turkey flocks.

All applied statistical analyses confirmed the common knowledge that the older the turkeys are, the healthier they are. The intestines during the first weeks of bird's life (up to 2-3 week) are in a phase of morphological, biochemical, immunological and molecular maturation and during this time they are concomitantly exposed to a variety of factors [11]. They contain their own microbial communities which are initially influenced by the breeder, but later also by environment [12]. The health status of intestines depends on maintaining the balance of all components of the turkey gut and environmental factors. Any disturbances in this balance may cause enteritis. Clinical signs of enteritis in turkey aged 1-4 weeks are most probably due to the disturbance of this delicate balance between various factors in immature intestines.

Enteric viruses were detected in 66.2% of Polish commercial turkey flocks. Among the studied viruses, the most frequently detected in both healthy as well as in enteric turkeys was AstV, accounting for 28.9% and 47.2%, respectively. Our previous studies on Polish turkey flocks also revealed the presence of turkey astroviruses, independently of detected types (TAstV-1, TAstV-2 or ANV), in both turkeys experiencing enteritis as well as in clinically healthy birds [13]. Previously applied statistical methods indicated a weak correlation between astrovirus incidence and the health status, with calculated coefficients between combined TAstV-1 and TAstV-2 presence or TAstV-2 presence and the health status equal to 0.14 and 0.1, respectively [13]. However, rank Spearman correlation used at that time seems to be not fully appropriate for qualitative variables. Thus in this study we have applied other methods, more useful for such kind of analysis. None of them indicated any correlation between AstVs prevalence and turkey health status. There are many reports of astrovirus presence in turkeys, with the prevalence varying from about 24 to 100% depending on the methods used for virus detection and the country; however, according to our knowledge, none of them specifically explored the impact of astroviruses on the health of turkeys. The surveys of commercial turkeys conducted in the USA revealed

that 47.2 and 100% of the analyzed samples were astrovirus-positive [14]. The majority of detected astroviruses belonged to TAstV-2 and their level ranged from 69.6 to 71.8%. The occurrence of TAstV-1 and ANV was less frequent, accounting for 9.8-28.1% and 2.7 – 12.5%, respectively, but all of these turkey astrovirus types were detected irrespectively of the turkey health status [14]. Brazilian studies also revealed a high prevalence of astroviruses in turkey flocks ranging from 44% to 81.6% [15, 16]. The share of individual astrovirus types was different than in the USA, but no clear influence on the health status was suggested. Cattoli et al. reported TAstV in Italian and Spanish turkey farms with enteritis [17]. On the other hand, Jindal et al. reported TAstV-2 in apparently healthy breeder turkey poults by RT-PCR; 47.2% of pooled fecal samples were positive [18]. Recently, TAstV-2, TAstV-1 and ANV were detected in 50%, 20% and 13.8% of market-age turkeys with lower weight but also in 33%, 20% and 12.5% of turkeys with standard breed character, respectively [3, 19]. Moreover, the same author found TAstV-2 in fecal swabs of both experimentally infected and PBS-inoculated turkeys [3]. In several reports the existence of two turkey astrovirus pathotypes were suspected: pathogenic and non-pathogenic [19, 20]. Presented results of applied statistical methods do not exclude the existence of such pathotypes, but it seems that the frequency of pathogenic astrovirus strains occurrence is rather low.

PVs were detected in 26.7% and 27.5% of healthy and enteric birds, respectively. Slightly higher rates of parvovirus infection in Polish turkey farms were demonstrated previously [21]. However, the occurrence of parvoviruses in turkey flocks in Poland was lower when compared to the 71-78% prevalence in commercial turkey flocks reported in a survey in the USA between 2003 and 2008 [8, 22]. The presence of PV infections in 46.9% of Hungarian turkey flocks experiencing enteric disease syndrome was also reported recently [23]. However, it should be noted, that most of the information about the PV came from studies of birds with enteritis problems. On the other hand, enteritis was experimentally reproduced in chickens infected with intestinal content containing parvoviral particles; however, the used inoculum might have contained other unknown viruses (apart from astrovirus, rotavirus and reovirus tested) which might have caused disease symptoms [24]. Our results identified the presence of PV with the same intensity in healthy individuals as in diseased ones. Moreover, the applied statistical investigation did not indicate any correlation between parvoviruses and turkey health status.

The overall observed frequency of RoV detection was 20.8%, but its prevalence was different in regards to health status of studied turkeys: in enteric flocks (21.1%)

it was above fourfold higher than in healthy ones (4.7%). All three applied different statistical methods have indicated the correlation between rotavirus prevalence and turkey health status. Calculated odds of disease symptoms in turkeys infected with RoV, depending on the methods used, were from five- to almost sevenfold higher than the occurrence of these symptoms in the uninfected group. Rotaviruses are a major cause of acute enteritis in young children and in several mammalian and avian species [25–27]. However, the impact of RoV on the health of turkeys is ambiguous, as its presence was reported in healthy as well as in enteric birds. Periodic monitoring of eight turkey flocks in North Carolina revealed the presence of rotaviruses in 69.7% of all tested samples during the whole production cycle. Moreover, RoVs were also detected in poults at the hatchery and vertical transmission was suggested as the reason of such detection. Interestingly, all studied flocks were described as healthy and normally performing by field personnel, but compared to the turkeys raised in experimental conditions they had significantly lower body weight and worse feed conversion rate [1]. The monitoring of enteric turkeys in Minnesota showed the presence of RoV in 48.3-93% of tested flocks [28, 29]. In subsequent studies by the above authors, the monitoring of five flocks of apparently healthy breeder turkeys over the period of 9 weeks revealed RoV presence with maximum occurrence until 5 weeks of age [18]. In a Brazilian study, rotavirus infections were identified in 52.6% of surveyed turkey farms; its prevalence in enteric and healthy flocks was 57.8% and 39.3%, respectively [16]. The presence of RoV in 18.8% of Polish diseased and healthy turkey flocks has also been previously reported [30]. It should be stressed that in all abovementioned studies the same protocol was used for RV detection as in present study [9]. The possible reason for the observed ambiguous impact of rotavirus infection on enteritis is the circulation of virus strains that differ in pathogenicity. Although our previous studies reported only several amino acid changes in NSP4 both in strains identified in enteric and healthy turkeys, the biological characteristics of RoVs, such as the replication ability, virulence and pathogenicity, are determined by a combination of many genes, such as VP4, VP7, VP3, NSP1, NSP2, and NSP4 gene [31]. Thus in order to get more insight into the pathogenicity as well as to establish a definite classification of detected rotavirus strain, the complete genome sequence should be characterized.

Turkey coronaviruses were found in 9.7% of the Polish flocks analyzed between 2008 and 2010. They were detected in 6.7% of healthy flocks and 9.2% of flocks with clinical signs of enteritis. Generally, TCoV is directly connected with enteritis [4]. The virus was responsible for enormous losses in turkey production in Minnesota

in the early 1970s and then for several outbreaks in multiple states in the United States since the 1990s [32, 33]. Maurel et al. reported that 37% of intestinal samples from diseased turkey flocks in France were TCoV-positive [34]. Villareal et al. demonstrated the presence of TCoV in 82.4% of studied diseased turkey flocks, but also in one apparently normal [15]. In another study from Brazil, TCoVs were detected in 71.1% and 28.6% of enteric and healthy flocks, respectively [16]. The lack of clinical signs of enteritis in spite of detected infection with turkey coronavirus may be interpreted in a few ways. Firstly, the sampling time – compilation of the results when the sampling was performed in the early stages of infection before the onset of clinical symptoms of the disease or at a later stage when clinical symptoms can be observed in the presence of the virus below its detection limit will lead to doubtfulness. Secondly, the existence of turkey coronaviruses with various pathogenicity. In addition, the severity of the disease could be influenced by other zoohygenic or infectious factors.

The results of most of the applied statistical methods have indicated clear correlations between astrovirus and rotavirus incidences, and this corresponds with the results of metagenomic studies which revealed that the most abundant RNA virus families in the gut of clinically normal 5-week old turkeys were *Astroviridae* and *Reoviridae*, and of these families, viruses of *Rotavirus* and *Astrovirus* genera [35]. We also found the correlation between AstV and RoV and the age of turkeys in studied flocks pointing at young birds as more susceptible to infection with these viruses. Our observations statistically support the view that astroviruses and rotaviruses are rarely detected after 4 weeks or below 6 weeks of age, respectively [36].

Conclusions

Our results revealed the prevalence of astroviruses, coronaviruses, rotaviruses and parvoviruses in Polish turkey flocks. The presence of detected viruses was commonly identified in flocks of both clinically normal and enteric birds. Moreover, we have tried to statistically analyze the obtained results for any correlations between such variables as health status/virus species/number of detected viruses/age groups. The dependence between rotavirus infection and the health status of turkeys was found. The occurrence of TAstV, TuPV and TCoV had no statistical effect on the development of clinical signs of enteritis. Additionally, our results suggest that infection with two or three viruses also has an effect on the health of turkeys, which is in line with recent observations, so far statistically unsupported, suggesting that viral co-infections may be relevant in the onset and the severity of enteric disease [16, 18]. Final determination of the role of these viruses could be clarified in experimental conditions using SPF or commercial turkeys. However, in order to do this, the virus should be available as propagated on embryos/cell lines, which however currently has not been achieved.

Abbreviations

AstV: Astrovirus; CoV: Coronavirus; PV: Parvovirus; RoV: Rotavirus

Acknowledgements

Not applicable.

Funding

This work was partially supported by the Polish Ministry of Science and Higher Education (Grant No. N308 578,040). Publication costs were covered by the Polish Ministry of Science and Higher Education (Grant No. 05-1/KNOW2/2015 within Scientific Consortium "Healthy Animal - Safe Food"). The funding body has no effect on design of the study and collection, analysis, and interpretation of data as well as writing the manuscript. The purchase of reagents and equipment was only possible.

Authors' contributions

KDB designed the study, carried out the molecular genetic studies, drew up a questionnaire, analyzed fulfilled questionnaires and obtained results, drafted the manuscript. LB performed the statistical analysis. AL, AJ, and AP contributed to laboratory work and results analysis. ZM conceived of the study, and helped to draft the manuscript. All authors read and approved the final manuscript.

Competing interests

The authors declare that they have no competing interests.

Consent for publication

Not applicable.

Author details

[1]Department of Poultry Diseases, National Veterinary Research Institute, Al. Partyzantów 57, 24-100 Puławy, Poland. [2]Department of Epidemiology and Risk Assessment, National Veterinary Research Institute, Al. Partyzantów 57, 24-100 Puławy, Poland.

References

1. Pantin-Jackwood MJ, Spackman E, Day JM, Rives D. Periodic monitoring of commercial turkeys for enteric viruses indicates continuous presence of astrovirus and rotavirus on the farms. Avian Dis. 2007;51(3):674–80.
2. Barnes HJ, Guy JS, Vaillancourt JP. Poult enteritis complex. Rev Sci Tech. 2000;19(2):565–88.
3. Mor SK, Sharafeldin TA, Abin M, Kromm M, Porter RE, Goyal SM, Patnayak DP. The occurrence of enteric viruses in light Turkey syndrome. Avian Pathol. 2013;42(5):497–501.
4. Ismail MM, Tang AY, Saif YM. Pathogenicity of turkey coronavirus in turkeys and chickens. Avian Dis. 2003;47(3):515–22.
5. Dybowski G. Polska liderem w produkcji mięsa drobiowego. In: Biuletyn informacyjny ARR. 00-400 Warsaw, ul. Nowy Świat 6/12, Agencja Rynku Rolnego. 2015;2:10–13.
6. Tang Y, Ismail MM, Saif YM. Development of antigen-capture enzyme-linked immunosorbent assay and RT-PCR for detection of turkey astroviruses. Avian Dis. 2005;49(2):182–8.
7. Callison SA, Hilt DA, Boynton TO, Sample BF, Robison R, Swayne DE, Jackwood MW. Development and evaluation of a real-time Taqman RT-PCR assay for the detection of infectious bronchitis virus from infected chickens. J Virol Methods 2006; 138(1-2):60-65.
8. Zsak L, Strother KO, Day JM. Development of a polymerase chain reaction procedure for detection of chicken and turkey parvovirus. Avian Dis. 2009; 53:83–8.
9. Day JM, Spackman E, Pantin-Jackwood M. A multiplex RT-PCR test for the differential identification of turkey astrovirus type 1, turkey astrovirus type 2, chicken astrovirus, avian nephritis virus, and avian rotavirus. Avian Dis. 2007; 51(3):681–4.
10. Kindler E, Trojnar E, Heckel G, Otto PH, Johne R. Analysis of rotavirus species diversity and evolution including the newly determined full-length genome sequences of rotavirus F and G. Infection Genetics and Evolution. 2013;14:58–67.
11. Yegani M, Korver DR. Factors affecting intestinal health in poultry. Poult Sci. 2008;87(10):2052–63.
12. Shah JD, Desai PT, Zhang Y, Scharber SK, Baller J, Xing ZS, Cardona CJ. Development of the intestinal RNA virus Community of Healthy Broiler Chickens. PLoS One. 2016;11:2.
13. Domanska-Blicharz K, Jacukowicz A, Bocian L, Minta Z. Astroviruses in Polish commercial Turkey farms in 2009–2012. Avian Dis. 2014;58:158–64.
14. Pantin-Jackwood MJ, Day JM, Jackwood MW, Spackman E. Enteric viruses detected by molecular methods in commercial chicken and turkey flocks in the United States between 2005 and 2006. Avian Dis. 2008;52(2):235–44.
15. Villarreal LYB, Assayag MS, Braqndao PE, Chacon JLV, Bunger ND, Astolfi-Ferreira CS, Gomes CR, Jones RC, Ferreira AJP. Identification of turkey astrovirus and turkey coronavirus in an outbreak of poult enteritis and mortality syndrome. Brazilian Journal of Poultry Science. 2006;8:131–5.
16. Moura-Alvarez J, Chacon J, Scanavini L, Nunez L, Astolfi-Ferreira C, Jones R, Piantino-Ferreira A. Enteric viruses in Brazilian turkey flocks: single and multiple virus infection frequency according to age and clinical signs of intestinal disease. Poult Sci. 2013;92:945–55.
17. Cattoli G, De Battisti C, Toffan A, Salviato A, Lavazza A, Cerioli M, Capua I. Co-circulation of distinct genetic lineages of astroviruses in turkeys and guinea fowl. Arch Virol. 2007;152(3):595–602.
18. Jindal N, Patnayak DP, Chander Y, Ziegler AF, Goyal SM. Detection and molecular characterization of enteric viruses in breeder turkeys. Avian Pathol. 2010;39(1):53–61.
19. Singh A, Mor SK, Jindal N, Patnayak D, Sobhy NM, Luong NT, Goyal SM. Detection and molecular characterization of astroviruses in turkeys. Arch Virol. 2016;161(4):939–46.
20. Jindal N, Patnayak DP, Chander Y, Ziegler AF, Goyal SM. Comparison of capsid gene sequences of turkey astrovirus-2 from poult-enteritis-syndrome-affected and apparently healthy turkeys. Arch Virol. 2011;156(6):969–77.
21. Domanska-Blicharz K, Jacukowicz A, Lisowska A, Minta Z. Genetic characterization of parvoviruses circulating in turkey and chicken flocks in Poland. Arch Virol. 2012;157:2425–30.
22. Murgia MV, Rauf A, Tang Y, Gingerich E, Lee CW, Saif YM. Prevalence of Parvoviruses in commercial Turkey flocks. Avian Dis. 2012;56(4):744–9.
23. Palade EA, Demeter Z, Hornyak A, Nemes C, Kisary J, Rusvai M. High prevalence of turkey parvovirus in turkey flocks from Hungary experiencing enteric disease syndromes. Avian Dis. 2011;55(3):468–75.
24. Zsak L, Cha RM, Day JM. Chicken parvovirus-induced runting-stunting syndrome in young broilers. Avian Dis. 2013;57(1):123–7.
25. Parashar UD, Gibson CJ, Bresee JS, Glass RI. Rotavirus and severe childhood diarrhea. Emerging Infectious Dis. 2006;12(2):304–6.
26. Otto PH, Ahmed MU, Hotzel H, Machnowska P, Reetz J, Roth B, Trojnar E, Johne R. Detection of avian rotaviruses of groups a, D, F and G in diseased chickens and turkeys from Europe and Bangladesh. Vet Microbiol. 2012;156(1-2):8–15.
27. Otto P, Liebler-Tenorio EM, Elschner M, Reetz J, Lohren U, Diller R. Detection of rotaviruses and intestinal lesions in broiler chicks from flocks with runting and stunting syndrome (RSS). Avian Dis. 2006;50(3):411–8.
28. Jindal N, Patnayak DP, Chander Y, Ziegler AF, Goyal SM. Detection and molecular characterization of enteric viruses from poult enteritis syndrome in turkeys. Poult Sci. 2010;89(2):217–26.
29. Jindal N, Patnayak DP, Ziegler AF, Lago A, Goyal SM. A retrospective study on poult enteritis syndrome in Minnesota. Avian Dis. 2009;53(2):268–75.
30. Domańska-Blicharz K, Jacukowicz A, Minta Z. Prevalence and molecular characteristics of rotaviruses from Polish turkey flocks between 2008 and 2011. Bull Vet Inst Pulawy. 2013;57:461–5.

31. Park JG, Kim DS, Matthijnssens J, Kwon HJ, Zeller M, Alfajaro MM, Son KY, Hosmillo M, Ryu EH, Kim JY, et al. Comparison of pathogenicities and nucleotide changes between porcine and bovine reassortant rotavirus strains possessing the same genotype constellation in piglets and calves. Vet Microbiol. 2014;172(1-2):51–62.

32. Day JM, Gonder E, Jennings S, Rives D, Robbins K, Tilley B, Wooming B. Investigating Turkey enteric Coronavirus circulating in the southeastern United States and Arkansas during 2012 and 2013. Avian Dis. 2014;58(2):313–7.

33. Chen YN, Loa CC, Ababneh MMK, Wu CC, Lin TL. Genotyping of turkey coronavirus field isolates from various geographic locations in the unites states based on the spike gene. Arch Virol. 2015;160(11):2719–26.

34. Maurel S, Toquin D, Briand FX, Queguiner M, Allee C, Bertin J, Ravillion L, Retaux C, Turblin V, Morvan H, et al. First full-length sequences of the S gene of European isolates reveal further diversity among turkey coronaviruses. Avian Pathol. 2011;40(2):179–89.

35. Shah JD, Baller J, Zhang Y, Silverstein K, Xing Z, Cardona CJ. Comparison of tissue sample processing methods for harvesting the viral metagenome and a snapshot of the RNA viral community in a turkey gut. J Virol Methods. 2014;209:15–24.

36. Reynolds DL, Saif YM, Theil KW. Enteric viral infections of turkey poults: incidence of infection. Avian Dis. 1987;31(2):272–6.

4

Development and validation of a house finch interleukin-1β (HfIL-1β) ELISA system

Sungwon Kim[2], Myeongseon Park[1], Ariel E. Leon[3], James S. Adelman[4], Dana M. Hawley[3] and Rami A. Dalloul[1*]

Abstract

Background: A unique clade of the bacterium *Mycoplasma gallisepticum* (MG), which causes chronic respiratory disease in poultry, has resulted in annual epidemics of conjunctivitis in North American house finches since the 1990s. Currently, few immunological tools have been validated for this songbird species. Interleukin-1β (IL-1β) is a prototypic multifunctional cytokine and can affect almost every cell type during *Mycoplasma* infection. The overall goal of this study was to develop and validate a direct ELISA assay for house finch IL-1β (HfIL-1β) using a cross-reactive chicken antibody.

Methods: A direct ELISA approach was used to develop this system using two different coating methods, carbonate and dehydration. In both methods, antigens (recombinant HfIL-1b or house finch plasma) were serially diluted in carbonate-bicarbonate coating buffer and either incubated at 4 °C overnight or at 60 °C on a heating block for 2 hr. To generate the standard curve, rHfIL-1b protein was serially diluted at 0, 3, 6, 9, 12, 15, 18, 21, and 24 ng/mL. Following blocking and washing, anti-chicken IL-1b polyclonal antibody was added, plates were later incubated with detecting antibodies, and reactions developed with tetramethylbenzidine solution.

Results: A commercially available anti-chicken IL-1β (ChIL-1β) polyclonal antibody (pAb) cross-reacted with house finch plasma IL-1β as well as bacterially expressed recombinant house finch IL-1β (rHfIL-1β) in immunoblotting assays. In a direct ELISA system, rHfIL-1β could not be detected by an anti-ChIL-1β pAb when the antigen was coated with carbonate-bicarbonate buffer at 4°C overnight. However, rHfIL-1β was detected by the anti-ChIL-1β pAb when the antigen was coated using a dehydration method by heat (60°C). Using the developed direct ELISA for HfIL-1β with commercial anti-ChIL-1β pAb, we were able to measure plasma IL-1β levels from house finches.

Conclusions: Based on high amino acid sequence homology, we hypothesized and demonstrated cross-reactivity of anti-ChIL-1β pAb and HfIL-1β. Then, we developed and validated a direct ELISA system for HfIL-1β using a commercial anti-ChIL-1β pAb by measuring plasma HfIL-1β in house finches.

Keywords: Interleukin-1beta, ELISA, House finch, *Mycoplasma gallisepticum*

Background

A member of the interleukin (IL)-1 family, IL-1beta (IL-1β) is a pivotal pro-inflammatory cytokine for host-defense responses to infection and injury [1–3]. It is a major mediator of innate immunity, as well as adaptive immune responses. The guardian cells of the innate immune system – macrophages and monocytes – are a major source of IL-1β [4, 5], but many other cell types including epithelial cells [6], endothelial cells [7], and fibroblasts [8] can also produce this cytokine. IL-1β is produced as an inactive 31 kDa precursor – termed pro-IL-1β, which is proteolytically processed to its active form by cytosolic caspase-1, followed by secretion via an unconventional protein secretion pathway [9]. Activated IL-1β affects diverse major innate immune processes including immune cell recruitment, cell proliferation, tissue destruction, bone resorption, vascular smooth muscle cell contraction, blood pressure and central nervous system cell function [10, 11].

A chicken homolog of mammalian IL-1β was first identified and characterized with CXCLi1 (K60)-inducing activity in 1998 [12], and some research has focused on its biological roles in avian species. Expression of ChIL-1β typically increases in response to both bacterial

* Correspondence: RDalloul@vt.edu
[1]Avian Immunobiology Laboratory, Department of Animal and Poultry Sciences, Virginia Tech, Blacksburg, VA 24061, USA

and viral infections, consistent with its role as a rapidly induced pro-inflammatory mediator. For example, IL-1β expression in bursal cells increases in chickens infected with infectious bursal disease virus [13], and IFN-γ-primed heterophils stimulated with *Salmonella* Enteritidis show increased IL-1β expression [14]. Infection of chicken kidney cells (CKCs) with *Escherichia coli* caused a reduction of IL-1β compared to non-infected control, but infection with *S. typhimurium* or *S. dublin* led to significantly increased IL-1β transcripts [15]. Chicken macrophage cell line (HD11) and CKCs induced significant IL-1β mRNA expression during stimulation with *Campylobacter jejuni* [16]. Finally, a study of *Mycoplasma gallisepticum* (MG) infection in chickens revealed the down-regulation of IL-1β at 1-day post-inoculation (dpi), and then a three-fold increase in expression at 4-dpi [17].

The bacterium MG is a common cause of chronic respiratory disease of poultry, but a unique clade of this pathogen emerged in a common North American backyard songbird species, house finch (*Haemorhous mexicanus*), in the mid-1990s [18, 19]. MG causes severe conjunctivitis in finches and significantly reduces survival in free-living birds [20, 21]. This pathogen spreads by either direct contact or short-term indirect contact on bird feeders, and MG is now endemic throughout most of the house finch range in North America [21, 22]. MG in house finches induces a series of local and systemic inflammatory responses, including severe conjunctivitis and rhinitis [21], local infiltration of lymphocytes and heterophils [23], and systemic responses such as fever, sickness behaviors, and expression of pro-inflammatory cytokines [24, 25]. A recent microarray study reported a different gene expression profile during innate and adaptive immune responses between MG-resistant (Alabama population) and MG-susceptible (Arizona population) finches [26]. MG-susceptible finches exhibited significant downregulation of gene expression patterns at 3-dpi (innate response) and 14-dpi (adaptive response) compared to MG-resistant finches. These two populations showed distinct transcriptional responses in the early stages of infection. Additionally, while gene expression profiles were similar on 3- and 14-dpi in MG-susceptible finches, MG-resistant finches had significantly different gene expression profile between 3-dpi and 14-dpi, suggesting genes associated with the adaptive immune response are only upregulated after population differences in transcription are first observed [26]. We recently documented population differences in relative IL-1β mRNA expression early in experimental MG infection, and these expression differences correspond to population differences in the severity of conjunctivitis [25]. However, a full understanding of the role of IL-1β in MG infection requires tools to measure and potentially manipulate the IL-1β protein in house finches.

Previously, our team identified and characterized HfIL-1β including successfully expressing rHfIL-1β in both prokaryotic and eukaryotic systems [27]. Recombinant HfIL-1β induced cell proliferation, as well as nitric oxide production in house finch splenocytes. Additionally, rHfIL-1β resulted in increased mRNA levels of Th1/Th2 cytokines in splenocytes, as well as an acute phase protein and an antimicrobial peptide in hepatocytes [27]. The goal of this study was to develop and validate a direct ELISA system for house finch IL-1β (HfIL-1β) using a commercially available anti-chicken IL-1β (ChIL-1β). Based on the high amino acid sequence homology of IL-1β between chicken and house finch, our previous work demonstrated the cross-reactivity of anti-ChIL-1β polyclonal antibody (pAb) to recombinant HfIL-1β [27]. In this study, we validated the cross-reactivity of anti-ChIL-1β pAb to nature form of serum HfIL-1β by immunoblotting. Then, a direct ELISA system using anti-ChIL-1β pAb was developed and test-validated with plasma samples collected from house finches.

Methods
Blood samples
Blood was collected from house finches via wing vein puncture using heparinized microcapillary blood collection tubes (approximately 100 μL per bird) and plasma was separated via centrifugation and frozen at −20°C. Once thawed, all blood samples were further diluted with PBS for the immunoblotting and ELISA analyses. Plasma samples from four captive house finches (non-infected with MG) were randomly selected for assay development (Table 1). Plasma samples from four free-living house finches were used to test whether HfIL-1β levels are elevated during MG infection: two of these individuals were clinically healthy at capture and were thus considered non-infected, and the other individuals had severe clinical signs of mycoplasmal conjunctivitis and thus were considered MG-infected. All housing and animal procedures were approved by the Institutional Animal Care and Use Committee of Virginia Tech.

Immunoblotting
Concentration of the purified recombinant HfIL-1β (rHfIL-1β) was measured using a BCA Protein Assay Kit

Table 1 Bird identification for blood collection

Bird ID	Infection	Capture Date	Status
380	No	07/13/2012	Had been captive for >1 year, but always control bird (non-infected)
412	No	16/01/2012	
1401	Yes	24/07/2013	Captured in the field without pathology, broke with MG while housed in captivity prior to time of sampling
1410	Yes	26/07/2013	

(Pierce, IL) following the manufacturer's instructions. For SDS-PAGE, 1 µg of the purified protein or 1 µL of plasma samples was mixed with 10 µL of SDS loading buffer (New England Biolabs, MA) containing DTT. The samples were then boiled at 97°C on a hot plate for 7 min. The prepared samples were electrophoresed on a 12% SDS-polyacrylamide gel at 90 V for 140 min. Proteins were transferred to PVDF membranes (Millipore, MA) by the submarine method at 90 V for 1.5 h. The membranes were incubated with anti-ChIL-1β pAb (Thermo Scientific, IL; 1:1000) overnight, followed by incubation with goat anti-rabbit antibody conjugated with HRP (Santa Cruz Biotechnology, CA, 1:2000) for 45 min. Using the SuperSignal West Pico Chemiluminescent Substrate (Pierce, IL), HRP signal was enhanced, followed by exposure and development on CL-XPosure film (Thermo Scientific, IL).

Enzyme-linked Immunosorbent assay (ELISA)

To develop the ELISA system for HfIL-1β, we adopted a direct ELISA approach using two different coating methods, carbonate and dehydration, on Nunc MaxiSorp® flat-bottom 96-well plates (Thermo Scientific, IL). In both methods, antigens – either rHfIL-1β or plasma – were diluted in carbonate-bicarbonate coating buffer (0.05 M; pH 9.5). For the standard curve, rHfIL-1β protein was serially diluted with 0.05 M carbonate-bicarbonate buffer at the following concentrations: 0, 3, 6, 9, 12, 15, 18, 21, and 24 ng/mL. The plasma samples were diluted with carbonate-bicarbonate coating buffer (0.05 M; pH 9.5) using 10-fold dilutions. An aliquot of 100 µL of each diluted antigen was added to the assigned well of the 96-well plates. For the carbonate method, the plates were then incubated at 4°C overnight, while for the dehydration method, the plates were incubated at 60°C on a heating block for 2 h for complete dehydration. Then, each plate was incubated with a blocking buffer (PBS [pH 7.4] containing 0.05% Tween-20 and 1% BSA) at room temperature (RT) for 45 min.

The primary (rabbit anti-ChIL-1β pAb) and the secondary (goat anti-rabbit) antibodies were prepared by dilution with the blocking buffer at 1:1000 and 1:2000, respectively. Fifty microliters of anti-ChIL-1β pAb were added to each assigned well, followed by incubating on a microplate shaker for 1 h at RT and washing 3 times (washing buffer as PBS [pH 7.4] containing 0.05% Tween-20). The plates were then incubated with 50 µL of the goat anti-rabbit antibody for 1 h at RT with continuous shaking. To develop HRP signal, a 3.3′,5.5′-tetramethylbenzidine (TMB, Sigma-Aldrich, MO) solution was prepared in 1 mL DMSO, 9 mL of 0.05 M phosphate-citrate buffer (pH 5.0) and 2 µL of 30% hydrogen peroxide (0.03% as final concentration, Sigma-Aldrich) per 10 mL total volume. One hundred

microliters of the prepared TMB solution were added to each well, followed by incubation for 30–45 min at RT. Colorimetric development was stopped by adding 100 µL of 2 N sulfuric acid (H_2SO_4) then absorbance was quantified spectrophotometrically at 450 nm with a microplate reader. The raw OD values were normalized by subtracting OD value of buffer. The standard curve was drawn and calculated with Excel (Microsoft Corp., WA) and elisaanalysis.com (www.elisaanalysis.com) using a logarithmic equation (i.e. Y = aln (x) + b), which was applied to calculate the concentration of HfIL-1β. The data were analyzed by independent two-sample t-test using program R (https://www.r-project.org/).

Results

Cross-reactivity of commercial anti-ChIL-1β antibody to HfIL-1β

To validate cross-reactivity of commercially available anti-ChIL-1β pAb with HfIL-1β, immunoblotting was performed with rHfIL-1β (Fig. 1); we also included plasma samples from non-infected and MG-infected house finches to verify such cross-reactivity. Approximately 25 kDa of rHfIL-1β was clearly detected by anti-ChIL-1β pAb. Plasma from non-infected birds showed only approximately 60 kDa protein band, whereas MG-infected birds showed 25 kDa and 60 kDa protein bands. The immunoblot results verified that commercial anti-

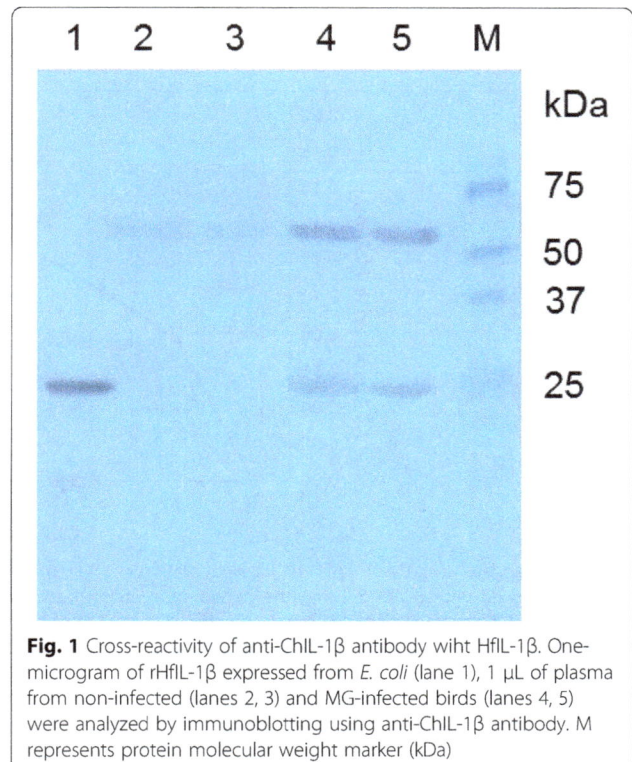

Fig. 1 Cross-reactivity of anti-ChIL-1β antibody wiht HfIL-1β. One-microgram of rHfIL-1β expressed from *E. coli* (lane 1), 1 µL of plasma from non-infected (lanes 2, 3) and MG-infected birds (lanes 4, 5) were analyzed by immunoblotting using anti-ChIL-1β antibody. M represents protein molecular weight marker (kDa)

ChIL-1β pAb can cross-react with both recombinant and natural forms of HfIL-1β.

Development of direct ELISA assay using anti-ChIL-1β antibody

To develop a direct ELISA system for HfIL-1β, we first tested the carbonate method for antigen coating. However, there was no signal detected even when coating with 2000 ng/mL rHfIL-1β (data not shown). To improve coating the antigen on the plate, the dehydration method was adopted. The ELISA results showed increased optical density (OD450) in a dose-dependent fashion (Fig. 2). The negative controls (coating buffer and the goat anti-rabbit antibody) were only in the 0.034–0.044 range. Using the dehydration coating method, we were able to detect as low as 2 ng/mL of rHfIL-1β protein. The standard curve showed a logarithmic scale (Y = a – b*ln (x + c)) with 0.94–0.95 adjusted R-square value.

Validation of the developed ELISA system using house finch plasma

To validate the developed ELISA system, we measured HfIL-1β plasma level from randomly selected captive house finches. The samples were diluted with coating buffer at 1:10 to 1:160 dilution factors (Fig. 3a), followed by coating using the dehydration method. Based on the OD450 values, the plasma samples diluted at 1:20 showed a similar range as the standard curve. Therefore, we used a 1:20 dilution of plasma samples to measure circulating HfIL-1β levels in house finches. The standard curve showed "Y = a - b*ln (x + c)" equation with:

a = 0.08494 (± 0.01225), b = −0.01948 (± 0.00456), and c = −2.35041 (± 0.52948). This translates into 2.92 ng/mL, 2.77 ng/mL, 2.41 ng/mL and 3.58 ng/mL of plasma HfIL-1β in each house finch, respectively (Fig. 3b).

We also measured HfIL-1β plasma levels in non-infected and MG-infected birds with a 1:20 dilution factor (Fig. 4a), followed by coating onto Nunc MaxSorp® flat-bottom plates using the dehydration method. Two non-infected birds revealed 2.54 ng/mL and 3.23 ng/mL plasma HfIL-1β, whereas MG-infected house finches had 19.20 ng/mL and 8.67 ng/mL HfIL-1β levels (Fig. 4b).

Discussion

Unlike mammalian research, few immune reagents including specific antibodies are available for immunological studies in wild birds. Although the infection of house finches by the bacterial pathogen MG is among the best-studied wildlife disease systems [22], there is still limited knowledge of the host's immune response especially with respect to cytokine levels. Recently, our lab identified and characterized HfIL-1β, which modulates the expression of Th1/Th2 cytokines as well as enhances the expression of acute phase protein by activated immune cells [27]. We also reported increased plasma levels of HfIL-1β in MG-infected house finches using the immunoblotting method [27], which could not quantify circulating HfIL-1β levels. In this study, we developed and validated a direct ELISA system using commercially available anti-ChIL-1β antibody to quantify plasma levels of HfIL-1β. The immunoblotting assays validated the cross-reactivity of commercial anti-ChIL-1β pAb with the recombinant form and plasma HfIL-

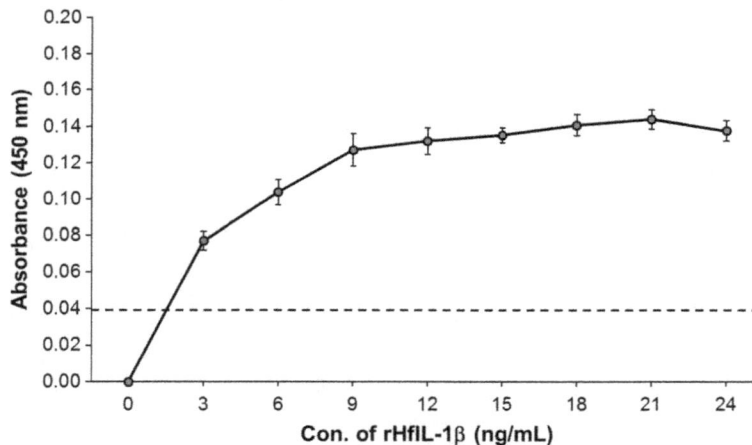

Fig. 2 A standard curve with rHfIL-1β by dehydration coating method. To establish an ELISA system for HfIL-1β, first a standard curve was established with rHfIL-1β using the dehydration coating method. The purified rHfIL-1β was diluted with the coating buffer with the following final concentrations: 0, 3, 6, 9, 12, 15, 15, 18, 21, and 24 ng/mL. Fifty-microliter of serially diluted rHfIL-1β were added to each well of Nunc MaxiSorp® flat-bottom 96-well plate, followed by incubation at 60°C for 2 h. Then, the plate was sequentially incubated with anti-ChIL-1β pAb (1:1000) and goat anti-rabbit antibody (1: 2000). The HRP signal was developed with TMB solution for 30 min. The coating buffer itself was used as negative control. Values represent the mean of three independent experiments. Error bars represent standard error of the mean. The dashed line indicates the threshold line, representing the value of negative control and limitation of the developed ELISA system (OD450 = 0.038)

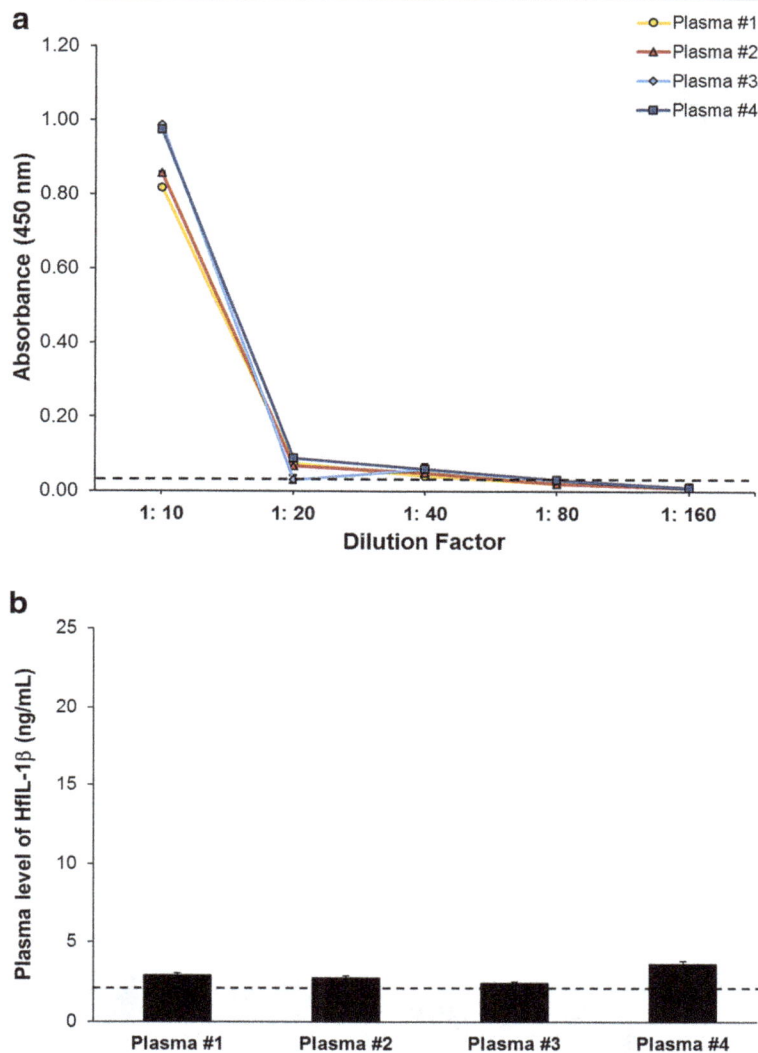

Fig. 3 Quantification of HfIL-1β in the plasma of non-infected house finches using the developed ELISA system. House finch plasma from randomly-selected non-infected birds were serially diluted with PBS as follows: 1:10, 1:20, 1:40, 1:80, and 1:160 (**a**). The diluted samples were coated with the dehydration method (60°C for 2 h), followed by sequential incubation with anti-ChIL-1β antibody (1:1000) as the primary antibody and the secondary antibody goat anti-rabbit antibody (1:2000). The HRP signal was developed with TMB solution for 45 min. The concentration of HfIL-1β plasma levels was calculated using Excel (Microsoft Corp) (**b**). The values represent the mean of triplicate wells with standard deviation bars. The dashed line indicates the threshold line, representing the value of negative control and limitation of the developed ELISA system (OD450 = 0.034)

1β. Similar to previously reported (27), MG-infected house finches showed two prominent bands observed at approximately 25 kDa and 60 kDa, whereas non-infected birds showed very weak 60 kDa protein band [27], implying the 25 kDa may be the more active secreted form of HfIL-1β.

Coating, which is the process where a suitably diluted antigen or antibody is incubated until adsorbed to the surface of the well, is the first step in any ELISA. Adsorption occurs passively as the result of hydrophobic interactions between the amino acid side chains on the antibody or antigen used for coating, and the plastic surface. It is dependent upon time, temperature and the pH

of the coating buffer, as well as the concentration of the coating agent. Bicarbonate buffer (pH 9.6) is the most common coating buffer, and typical coating condition involves 50–100 μL of coating buffer containing 1–10 μg/mL of either antigen or antibody incubated overnight at 4°C or for 1–3 h at 37°C. However, optimal coating conditions should be tested during the development of a new ELISA system. In this study, the typical coating condition of incubation at 4°C overnight did not work in detecting HfIL-1β in plasma samples. To increase adsorption to the plate, we applied a dehydration method for antigen coating, although we

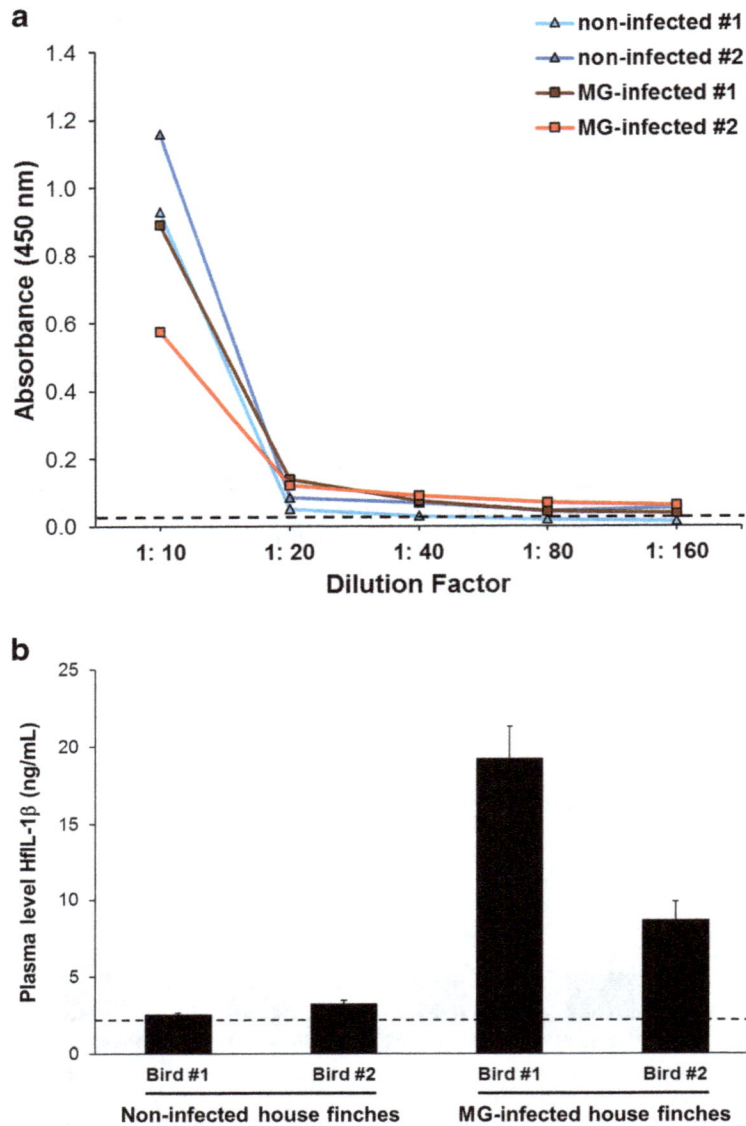

Fig. 4 Quantification of plasma HfIL-1β in non-infected and MG-infected house finches using the developed ELISA system. Plasma samples were collected from two individual birds from non-infected and MG-infected groups. Then, plasma samples were serially diluted with PBS as follows: 1:10, 1:20, 1:40, 1:80, and 1:160 (**a**). The diluted samples were coated with the dehydration method (60°C for 2 h), followed by sequential incubation with anti-ChIL-1β antibody (1:1000) as the primary antibody and the secondary antibody goat anti-rabbit antibody (1:2000). The HRP signal was developed with TMB solution for 30 min. The concentration of HfIL-1β plasma level was calculated using Excel (Microsoft Corp) (**b**). The values represent the mean of triplicate wells with standard deviation bars. The dashed line indicates the threshold line, representing the value of negative control and limitation of the developed ELISA system (OD450 = 0.042)

were concerned with the background level of spectrometric readings. To eliminate such background, we used the coating buffer itself as a negative control. We also incubated rHfIL-1β and random plasma samples with goat anti-rabbit antibody alone. In both cases, the absorbance at 450 nm was in the 0.034–0.044 range, which was significantly low compared to the 2 ng/mL rHfIL-1β, the minimum concentration that the established ELISA system can detect. It is currently unclear why the dehydration method rather than a typical

coating method is superior for detecting HfIL-1β in house finch plasma. One potential explanation is the denaturation of HfIL-1β during the heat-based dehydration, resulting in the increased binding affinity between HfIL-1β and rabbit anti-ChIL-1β, as well as adsorption to the plate.

Unlike commercial systems, which mostly use a sandwich ELISA with capture and detection antibodies, we developed a direct ELISA by coating the antigens on the plate using dehydration. Therefore, the sensitivity of the developed HfIL-1β ELISA system is low, and the assay can only detect a limited concentration range (2–20 ng/ mL). In our trials with random plasma samples and non-infected and MG-infected plasma samples, a 1:20 dilution with PBS fell within the measurable standard curve range. However, it is necessary to serially dilute plasma samples for every ELISA assay to obtain the best fit within the standard curve range. For the standard curve with rHfIL-1β, first we attempted to use a simple logarithmic equation (i.e. $Y = a*\ln(x) + c$); however, the adjusted R-square value was 0.89, resulting in high variation among samples. Thus, we modified the logarithmic equation to "$Y = a - b*\ln(x + c)$", which yielded an adjusted R-square value of 0.94–0.95.

The ELISA results showed the non-infected house finches had an average of 2.89 ng/mL of plasma HfIL-1β, while the MG-infected birds had on average 13.93 ng/mL of plasma HfIL-1β. Two-sample t-test showed no statistical significance between non-infected and MG-infected birds ($P = 0.28$), likely due to the low sample size tested. However, the MG-infected birds showed average plasma HfIL-1β levels almost 5-fold higher than those of non-infected birds. Further studies should use a larger number of birds to examine how plasma levels of HfIL-1β change during MG infection.

Conclusions
We validated the cross-reactivity of anti-ChIL-1β pAb with HfIL-1β using immunoblotting. Then, a direct ELISA system was developed using anti-ChIL-1β antibody to quantify HfIL-1β plasma levels using the dehydration coating method. The developed ELISA system can quantify HfIL-1β in plasma of test birds including non-infected and MG-infected house finches. This developed direct ELISA system using commercially available chicken antibody will provide valuable research and diagnostic tools for house finch research.

Acknowledgments
We thank Laila Kirkpatrick for her technical assistance.

Funding
This work was partially supported by National Science Foundation NSF grant IOS-1054675 (to DMH), and by the Virginia Agricultural Experiment Station and the Hatch Program of the National Institute of Food and Agriculture, U.S. Department of Agriculture (to RAD).

Authors' contributions
SK drafted the manuscript and conducted the experimental work along with MP and AEL. JSA, DMH and RAD provided substantial contributions to the design, and acquisition and interpretation of data, as well as preparing the manuscript. All authors read and approved the final manuscript.

Consent for publication
Not applicable.

Competing interests
The authors declare that they have no competing interests.

Author details
[1]Avian Immunobiology Laboratory, Department of Animal and Poultry Sciences, Virginia Tech, Blacksburg, VA 24061, USA. [2]The Roslin Institute and R(D)SVS, University of Edinburgh, Easter Bush, Midlothian EH25 9RG, UK. [3]Department of Biological Sciences, Virginia Tech, Blacksburg, VA 24061, USA. [4]Department of Natural Resource Ecology and Management, Iowa State University, Ames, IA 50011, USA.

References
1. Gaestel M, Kotlyarov A, Kracht M. Targeting innate immunity protein kinase signalling in inflammation. Nat Rev Drug Discov. 2009;8:480–99.
2. Dinarello CA. Immunological and inflammatory functions of the interleukin-1 family. Annu Rev Immunol. 2009;27:519–50.
3. Martinon F, Mayor A, Tschopp J. The inflammasomes: guardians of the body. Annu Rev Immunol. 2009;27:229–65.
4. Granowitz EV, Clark BD, Vannier E, Callahan MV, Dinarello CA. Effect of interleukin-1 (IL-1) blockade on cytokine synthesis: I. IL-1 receptor antagonist inhibits IL-1-induced cytokine synthesis and blocks the binding of IL-1 to its type II receptor on human monocytes. Blood. 1992;79:2356–63.
5. Netea MG, Nold-Petry CA, Nold MF, Joosten LA, Opitz B, van der Meer JH, van de Veerdonk FL, Ferwerda G, Heinhuis B, Devesa I, et al. Differential requirement for the activation of the inflammasome for processing and release of IL-1beta in monocytes and macrophages. Blood. 2009;113:2324–35.
6. Hoffmann E, Thiefes A, Buhrow D, Dittrich-Breiholz O, Schneider H, Resch K, Kracht M. MEK1-dependent delayed expression of Fos-related antigen-1 counteracts c-Fos and p65 NF-kappaB-mediated interleukin-8 transcription in response to cytokines or growth factors. J Biol Chem. 2005;280:9706–18.
7. Bandman O, Coleman RT, Loring JF, Seilhamer JJ, Cocks BG. Complexity of inflammatory responses in endothelial cells and vascular smooth muscle cells determined by microarray analysis. Ann N Y Acad Sci. 2002;975:77–90.
8. Holzberg D, Knight CG, Dittrich-Breiholz O, Schneider H, Dorrie A, Hoffmann E, Resch K, Kracht M. Disruption of the c-JUN-JNK complex by a cell-permeable peptide containing the c-JUN delta domain induces apoptosis and affects a distinct set of interleukin-1-induced inflammatory genes. J Biol Chem. 2003;278:40213–23.
9. Keller M, Ruegg A, Werner S, Beer HD. Active caspase-1 is a regulator of unconventional protein secretion. Cell. 2008;132:818–31.
10. Dinarello CA. Proinflammatory cytokines. Chest. 2000;118:503–8.
11. Dinarello CA, Pomerantz BJ. Proinflammatory cytokines in heart disease. Blood Purif. 2001;19:314–21.
12. Weining KC, Sick C, Kaspers B, Staeheli P. A chicken homolog of mammalian interleukin-1 beta: cDNA cloning and purification of active recombinant protein. Eur J Biochem. 1998;258:994–1000.
13. Eldaghayes I, Rothwell L, Williams A, Withers D, Balu S, Davison F, Kaiser P. Infectious bursal disease virus: strains that differ in virulence differentially modulate the innate immune response to infection in the chicken bursa. Viral Immunol. 2006;19:83–91.
14. Kogut MH, Rothwell L, Kaiser P. IFN-gamma priming of chicken heterophils upregulates the expression of proinflammatory and Th1 cytokine mRNA following receptor-mediated phagocytosis of Salmonella enterica serovar enteritidis. J Interf Cytokine Res. 2005;25:73–81.
15. Kaiser P, Rothwell L, Galyov EE, Barrow PA, Burnside J, Wigley P. Differential cytokine expression in avian cells in response to invasion by Salmonella typhimurium, Salmonella enteritidis and Salmonella gallinarum. Microbiology. 2000;146:3217–26.
16. Smith CK, Kaiser P, Rothwell L, Humphrey T, Barrow PA, Jones MA. Campylobacter jejuni-induced cytokine responses in avian cells. Infect Immun. 2005;73:2094–100.
17. Mohammed J, Frasca S Jr, Cecchini K, Rood D, Nyaoke AC, Geary SJ, Silbart LK. Chemokine and cytokine gene expression profiles in chickens inoculated with Mycoplasma gallisepticum strains Rlow or GT5. Vaccine. 2007;25:8611–21.
18. Ley DH, Berkhoff JE, McLaren JM. Mycoplasma gallisepticum isolated from house finches (Carpodacus Mexicanus) with conjunctivitis. Avian Dis. 1996;40:480–3.
19. Hochachka WM, Dhondt AA, Dobson A, Hawley DM, Ley DH, Lovette IJ. Multiple host transfers, but only one successful lineage in a continent-spanning emergent pathogen. Proc Biol Sci. 2013;280:1–7.
20. Faustino CR, Jennelle CS, Connolly V, Davis AK, Swarthout EC, Dhondt AA, Cooch EG. Mycoplasma gallisepticum infection dynamics in a house finch

population: seasonal variation in survival, encounter and transmission rate. J Anim Ecol. 2004;73:651–69.

21. Kollias GV, Sydenstricker KV, Kollias HW, Ley DH, Hosseini PR, Connolly V, Dhondt AA. Experimental infection of house finches with *Mycoplasma gallisepticum*. J Wildl Dis. 2004;40:79–86.

22. Dhondt AA, Altizer S, Cooch EG, Davis AK, Dobson A, Driscoll MJ, Hartup BK, Hawley DM, Hochachka WM, Hosseini PR, et al. Dynamics of a novel pathogen in an avian host: Mycoplasmal conjunctivitis in house finches. Acta Trop. 2005;94:77–93.

23. Hawley DM, Grodio J, Frasca S, Kirkpatrick L, Ley DH. Experimental infection of domestic canaries (Serinus Canaria Domestica) with *Mycoplasma gallisepticum*: a new model system for a wildlife disease. Avian Pathol. 2011;40:321–7.

24. Luttrell MP, Stallknecht DE, Fischer JR, Sewell CT, Kleven SH. Natural *Mycoplasma gallisepticum* infection in a captive flock of house finches. J Wildl Dis. 1998;34:289–96.

25. Adelman JS, Kirkpatrick L, Grodio JL, Hawley DM. House finch populations differ in early inflammatory signaling and pathogen tolerance at the peak of *Mycoplasma gallisepticum* infection. Am Nat. 2013;181:674–89.

26. Bonneaud C, Balenger SL, Zhang J, Edwards SV, Hill GE. Innate immunity and the evolution of resistance to an emerging infectious disease in a wild bird. Mol Ecol. 2012;21:2628–39.

27. Park M, Kim S, Adelman JS, Leon AE, Hawley DM, Dalloul RA. Identification and functional characterization of the house finch interleukin-1beta. Dev Comp Immunol. 2017;69:41–50.

5

Effects of intramuscularly administered enrofloxacin on the susceptibility of commensal intestinal *Escherichia coli* in pigs (sus scrofa domestica)

Antje Römer[1*†] (iD), Gesine Scherz[2†], Saskia Reupke[2], Jessica Meißner[2], Jürgen Wallmann[1], Manfred Kietzmann[2] and Heike Kaspar[1]

Abstract

Background: In the European Union, various fluoroquinolones are authorised for the treatment of food producing animals. Each administration poses an increased risk of development and spread of antimicrobial resistance. The aim of this study was to investigate the impact of parenteral administration of enrofloxacin on the prevalence of enrofloxacin and ciprofloxacin susceptibilities in the commensal intestinal *E. coli* population.

Methods: *E. coli* isolates from faeces of twelve healthy pigs were included. Six pigs were administered enrofloxacin on day 1 to 3 and after two weeks for further three days. The other pigs formed the control group. MIC values were determined. Virulence and resistance genes were detected by PCR. Phylogenetic grouping was performed by PCR. Enrofloxacin and ciprofloxacin were analysed in sedimentation samples by HPLC.

Results: Susceptibility shifts in commensal *E. coli* isolates were determined in both groups. Non-wildtype *E. coli* could be cultivated from two animals of the experimental group for the first time one week after the first administration and from one animal of the control group on day 28. The environmental load with enrofloxacin in sedimentation samples showed the highest amount between days one and five. The repeated parenteral administration of enrofloxacin to pigs resulted in rapidly increased MIC values (day 28: MIC up to 4 mg/L, day 35: MIC ≥ 32mg/L). *E. coli* populations of the control group in the same stable without direct contact to the experimental group were affected.

Conclusion: The parenteral administration of enrofloxacin to piglets considerably reduced the number of the susceptible intestinal *E. coli* population which was replaced by *E. coli* strains with increased MIC values against enrofloxacin. Subsequently also pigs of the control were affected suggesting a transferability of strains from the experimental group through the environment to the control group especially as we could isolate the same PFGE strains from both pig groups and the environment.

Keywords: Parenteral administration, Enrofloxacin, Resistance, PFGE, *E. coli*, Macrorestriction

* Correspondence: antje.roemer@bvl.bund.de
†Equal contributors
[1]Federal Office of Consumer Protection and Food Safety, Berlin, Germany
Full list of author information is available at the end of the article

Background

Each administration of antibiotics in human and in veterinary medicine exerts a selective pressure and poses an increased risk of development and spread of antimicrobial resistance (AMR) [1]. This may affect not only zoonotic bacteria but also commensal bacteria in the intestine of food producing animals which are of special concern under public health aspects [2–4].

Fluoroquinolones are important antibiotics for the treatment of various bacterial infections in both humans and in animals [5], which have been categorised as "highest priority critically important antimicrobials" (HPCIA) for human and animal health [6, 7].

Fluoroquinolones exhibit a broad spectrum of antimicrobial activity against Gram-negative and Gram-positive bacteria. Their pharmacokinetic properties are characterised by a good bioavailability after oral as well as after parenteral application, an adequate distribution in tissue in association with high plasma levels and adequate renal clearance. Elimination is rapid via both urine and faeces [8–11]. In previous kinetic studies bioavailability and plasma concentrations were higher and absorption faster after intramuscular administration of fluoroquinolones [9, 12]. Wiuff et al. showed that i.m. administration of enrofloxacin in pigs resulted in a faster absorption and a more efficient distribution of enrofloxacin to plasma, lymph nodes and tissues of the intestine than p.o. administration [13]. But no significant differences in intestinal content concentrations between the administration routes were found except one measurement after 24 hours. The concentration dependent bactericidal activity of fluoroquinolones is based on inhibition of the target enzymes gyrase and topoisomerase IV [14, 15].

Diverse mechanisms may cause resistance to fluoroquinolones. Chromosomal mediated resistance develops by step-wise occurring point mutations in the target genes encoding for DNA gyrase (*gyrA* and *gyrB*) and topoisomerase IV (*parC* and *parE*). Increased number of mutations lead to an accumulating reduction of susceptibilities against fluoroquinolones [5, 16]. Other mechanisms are related to decreased intracellular accumulations of fluoroquinolones, e.g. those affecting the expression of outer membrane proteins or efflux pumps [17]. Plasmid mediated quinolone resistance (PMQR) in Enterobacteriaceae has been reported for 17 years [18–20]. But without additional chromosomally located mutations in the target genes plasmid mediated fluoroquinolone resistance results only in slight shifts of the particular minimal inhibitory concentration (MIC) [19, 21].

In the European Union, various fluoroquinolones are authorised for the treatment of food producing (cattle, pigs, poultry and rabbits) and companion animals [22]. Enrofloxacin was the first fluoroquinolone developed exclusively for the use in veterinary medicine. In food producing animals, main indications are the therapy of respiratory and gastrointestinal diseases. Enrofloxacin can be administered by subcutaneous injection to cattle and intramuscular injection to pigs or by preparations for oral use for cattle, pigs, turkeys and chicken. For piglets, numerous solutions for injections but only few preparations for oral administration are available. Therefore, administration of enrofloxacin to pigs seems mainly done by injection [23].

Resistance to enrofloxacin and other fluoroquinolones in intestinal commensal as well as pathogenic *E. coli* strains is low to moderate in European pig production systems although there are differences among countries and production types [4, 24]. As high-level resistance to fluoroquinolones commonly requires a mutational component, occurrence of fluoroquinolone-resistant *E. coli* strains is mainly influenced by the degree of its use [25] whereby some resistant *E. coli* strains may persist for at least two weeks after three-day treatment [26]. Resistance to one fluoroquinolone often result in resistance to all fluoroquinolones [27].

Via faeces and/or urine depending on the particular active ingredients, enrofloxacin and its antibiotic active metabolite ciprofloxacin are released into the surrounding. This can lead to carry-over or reconsumption of subtherapeutic concentrations by the animals or to the exposure of environmental bacteria to the residues [28, 29].

The objective of this study was to investigate the impact of parenteral administration of the fluoroquinolone enrofloxacin and its metabolite ciprofloxacin on the prevalence of non-wild type (N-WT)-*E. coli* isolates in the commensal intestinal *E. coli* population of pigs with and without direct administration of enrofloxacin. Additionally, the distribution of enro- and ciprofloxacin in sedimentation samples of dust from the stable was measured to determine the release of the active ingredients into the direct surrounding of the animals.

Methods

Animals

Prior to placement of the pigs onto the holding several pigs from different farms were screened for N-WT-*E. coli* isolates. Rectal swabs of 30 pigs per farm were enriched in 3 mL lysogeny broth (LB) [30] overnight at 37°C; 10 μL of the cell suspension were streaked onto Endo agar [31] supplemented with 0.125 mg/L, 0.25 mg/L and 4 mg/L enrofloxacin and onto Endo agar without supplementation as a control. Only pigs from farms without any coliform growth onto Endo agar with 0.125 mg/L, 0.25 mg/L and 4 mg/L enrofloxacin were assumed to be free of N-WT-*E. coli* and included in this study.

Twelve five to six weeks old pigs from a single farm were randomly divided into two groups of six pigs. Both pig groups were housed in straw bedded pens of 1.5 X 3 metres with a distance of 3 metres between both pens. Temperature was between 23°C and 24°C and the relative humidity was between 50% and 60%. Ventilation was done by a positive pressure system. In addition to daylight a lighting program of 10 hours of light was used simultaneously. The pigs were fed a commercial pig feed once daily and had free access to tap water.

Before the animals were placed in the pens, the whole stable was cleaned and disinfected according to good manufacturing practices (dry cleaning first, followed by high-pressure cleaning, afterwards disinfection with Venno® FF super (Menno Chemie Vertrieb GmbH, Norderstedt, Germany)). Environmental samples were taken by wiping with sterile boot swabs soaked in hypotonic sodium chloride. Thereafter each boot swab was enriched in 250 mL LB and streaked onto Endo agar supplemented with enrofloxacin (0.125 mg/L, 0.25 mg/L, 0.5 mg/L, 0.75 mg/L, 1 mg/L, 1.5mg/L and 4 mg/L) to screen for contamination with N-WT-*E. coli*. If, after cleaning and disinfection, no lactose-fermenting coliform growth occurred on these plates, the stable was assumed to be free of N-WT *E. coli* and pigs were placed in. All persons handling animals wore disposable protective gloves and appropriate protective clothing. All animals were clinically healthy during the entire experiment. Experimental procedures started after one week of acclimatisation to avoid excessive stress to the pigs.

The study was authorised by the Lower Saxony State Office for Consumer Protection and Food Safety, Niedersachsen, Germany (reference number 33.9-42502-04-11/0338).

Study protocol

Six pigs (experimental group) were administered an intramuscular injection in the neck behind the ear with the recommended dosage of 2.5 mg/kg bodyweight (bw) enrofloxacin (Baytril® 10%) on day 1 to 3 and two weeks later (day 18–20). The other six pigs housed in a second pen in the same stable were used as control animals. Animals of both groups had no direct contact but the transmission of airborne particles or via vectors could not be excluded. To reduce carryover into the control group, feeding, administration of enrofloxacin and sampling of pigs from the control group were always done first. Rectal faecal samples were taken one day before the administration started and on the days 4, 11, 18, 21, 28, 35, 42 and 54. Additionally, Endo agar plates without supplementation of enrofloxacin were placed directly in front of the boxes close to the feeding trough of each group, between both boxes on the stall gangway and in

the working area for animal keepers to isolate *E. coli* from the direct environment (Fig. 1).

Isolation of *E. coli* from faeces and environment

Faecal samples were serially diluted in sodium chloride solution (0.9%), and 100 μL of each dilution were streaked onto Endo agar plates without supplementation of enrofloxacin and onto Endo agar plates supplemented with 0.125 mg/L enrofloxacin, a concentration equivalent to the epidemiological cut-off value [32]. Plates were incubated at 37°C for 20–24 hours. Ten lactose-fermenting coliform colonies per animal and plate were picked by a self-produced template, which localised ten fix points on every plate to avoid subjective selections. Isolates were confirmed as *E. coli* using LMX broth modified by Manafi and Ossmer [33] and indole reaction.

The animals were also screened for *E. coli* isolates with MIC values > 4 mg/L for enrofloxacin. After enrichment of each faecal sample in 9 mL LB-Bouillon at 37°C for 20–24 hours, 100 μL of the enriched cell suspension were streaked onto Endo agar plates supplemented with 4 mg/L enrofloxacin and incubated at 37°C for 20–24 hours as described by Scherz et al. [34].

During environmental sampling four Endo agar plates without supplementation of enrofloxacin were placed uncovered for 1.5 h at four defined locations of the stable.

If fewer than ten colonies were observed on a plate, all suitable colonies were used for further analysis. According to a previous study, the probability to isolate at least one colony of the most common strain in a faecal sample was more than 99% if five bacterial isolates were tested [35]. Another study showed that it would be sufficient to type

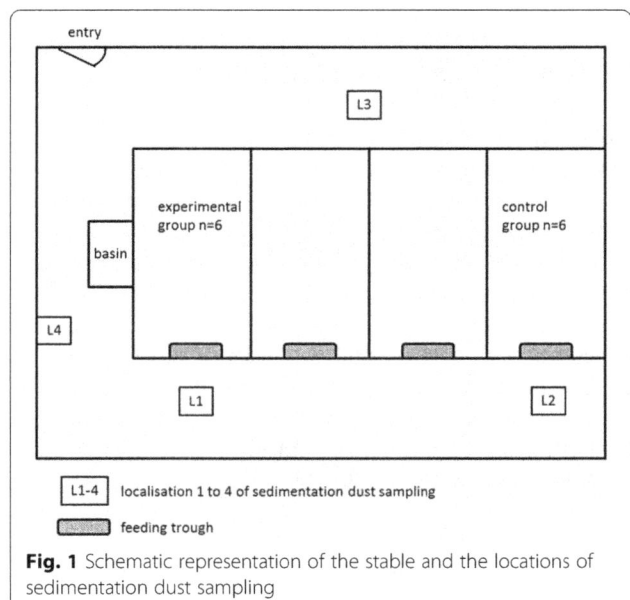

Fig. 1 Schematic representation of the stable and the locations of sedimentation dust sampling

ten isolates per sample to find all strains of verotoxigenic and non-type-specific *E. coli* in faeces samples with an 85% probability [36].

Antimicrobial susceptibility testing

MIC values of ten *E. coli* colonies from each Endo agar plate supplemented with 0.125 mg/L enrofloxacin, one colony from each Endo agar plate with 4 mg/L enrofloxacin and maximum ten *E. coli* colonies from the environment were determined by using Etest® for enrofloxacin (Biomèrieux) according to the manufacturer's instructions.

After assigning individual *E. coli* isolates to strains by macrorestriction analysis, broth microdilution method in accordance to CLSI was used to determine MIC values against enrofloxacin and nalidixic acid. Selected *E. coli* strains, which occurred repeatedly at various points in time over the duration of the experiments in the same animal, or which were found in both groups and/or in the environment, were monitored. Therefore, MIC_{90}-values were used for evaluation.

Macrorestriction analysis

To monitor the spread and the distribution of individual *E. coli* strains pulsed-field gel electrophoresis (PFGE) was used. *E. coli* isolated from Endo agar without supplementation of enrofloxacin from two randomly chosen pigs per group as well as selected isolates from Endo agar with supplementation of 0.125 mg/L and 4 mg/L enrofloxacin were chosen for macrorestriction analysis as previously described [37]. Bacterial DNA was digested with the restriction enzyme XbaI. 100 µM thiourea were added to the running buffer (0.5XTBE) to avoid DNA degradation by Tris radicals [38]. DNA fragment patterns were examined by ethidium bromide staining and compared using the software platform BioNumerics (version 7.1, Applied Maths, Sint-Martens-Latem, Belgium).

Molecular characterisation by PCR

One representative isolate of each *E. coli* strain was tested for the occurrence of virulence genes, adhesion gene and genes for iron acquisition. The selected genes are shown in Table 3. PCRs were performed as described previously [39, 40]. Classification according to the *E. coli* Reference Collection (EcoR) system was based on the rapid phylogenetic grouping PCR technique described by Clermont et al. [41]. Isolates were assigned to one of four groups (A, B1, B2 or D). Genes for plasmid-mediated fluoroquinolone resistance (*qnrA*, *qnrB* and *qnrS*) were detected as described by Robicsek et al. [19].

Enro- and ciprofloxacin in sedimentation samples

Sedimentation samples of dust from the stable were collected at four different localisations in the stable before (Fig. 1), during and after both administration periods, respectively. For that, two Makrolon® panels with 0.16 m² surface each were respectively placed on the alley in front of each feeding trough of both groups, on the opposite alley between both groups and apart from the animals in the working area of the stable, to serve as surface for the sedimentation of dust. The Makrolon® panels were placed five days before administration started to ensure a sufficient amount of sedimented dust to collect for analysis. The extraction of enrofloxacin and its metabolite ciprofloxacin as well as the analysis performed by using high-performance liquid chromatography with fluorescence detection was carried out as described by Scherz [42].

Definitions and statistical analysis

Each coliform colony picked from a plate and identified as *E. coli* was defined as an *E. coli* isolate. *E. coli* isolates with MIC values against enrofloxacin above the epidemiological cut-off value (ECOFF) of 0.125 mg/L were defined as N-WT isolates.

Isolates with the same specific macrorestriction enzyme pattern were assigned to one *E. coli* strain. The diversity of *E. coli* strains was measured as defined by Katouli et al. [43] with Simpson's index of diversity (Di) [44]. *E. coli* strains detected from at least two of three potential sources (experimental group, control group, environment) were described as transferable strains.

Statistical evaluations were performed by the software GraphPad Prism software (GraphPad Prism®5.01, Graph Pad Software, San Diego, CA, USA) and SAS (SAS®9.3 Software, SAS, Cary, NC, USA). Medians of MIC values from *E. coli* isolates of experimental and control group on each trial day were compared with the non-parametric Wilcoxon test for independent samples. Fisher's exact test was used to calculate significant differences between the occurrence of single genes and the phylogenetic affiliation of individual *E. coli* strains. Correlation between frequency and transferability of detected *E. coli* strains and single genes were tested by Spearman correlation coefficient. Results were considered statistically significant at a significance level of $\alpha < 0.05$.

Results

Number of *E. coli* isolates

Throughout the sampling period, a total number of 1444 *E. coli* isolates were obtained from the experimental group (EG) (748 isolates), the control group (CG) (508 isolates) and the environment (En) (188 isolates). Further information is shown in Table 1.

On Endo agar plates without supplementation of enrofloxacin, coliform growth occurred with each sampling in the control group and in the environment. Within the experimental group limited or no coliform growth was

Table 1 Number of *E. coli* isolates cultured included in further analyses

	Enrofloxacin concentration of Endo agar plates			Σ
	0 mg/L	0.125 mg/L	4 mg/L	
EG	361	267	120	748
CG	423	55	30	508
En	188	n.t.	n.t.	188
Σ	972	322	150	1444

EG experimental group, *CG* control group, *En* environment

shown always one day after administration of enrofloxacin. Therefore, it was not possible to obtain ten isolates from all pigs one day after administration.

On Endo agar plates supplemented with enrofloxacin, coliform growth was found earlier and more often in the experimental group, where first isolates above the

ECOFF could be detected on day 11 (Endo agar with 0.125 mg/L enrofloxacin) and on day 35 (Endo agar with 4 mg/L enrofloxacin). Within the control group first coliform growth occurred on day 28 (Endo agar with 0.125 mg/L enrofloxacin) and on day 42 (Endo agar with 4 mg/L enrofloxacin).

Enrofloxacin susceptibility of isolated *E. coli*
Susceptibility shifts of commensal *E. coli* isolates were determined in both groups at various points in time of the trial as shown in Fig. 2. Only susceptible wild type (WT)-*E. coli* isolates with MIC values below the ECOFF were found before starting the experimental trial. One week after the first administration period (day 11), N-WT-*E. coli* with MIC values above the ECOFF could be cultivated from faecal samples of two animals from the experimental group for the first time. After the second

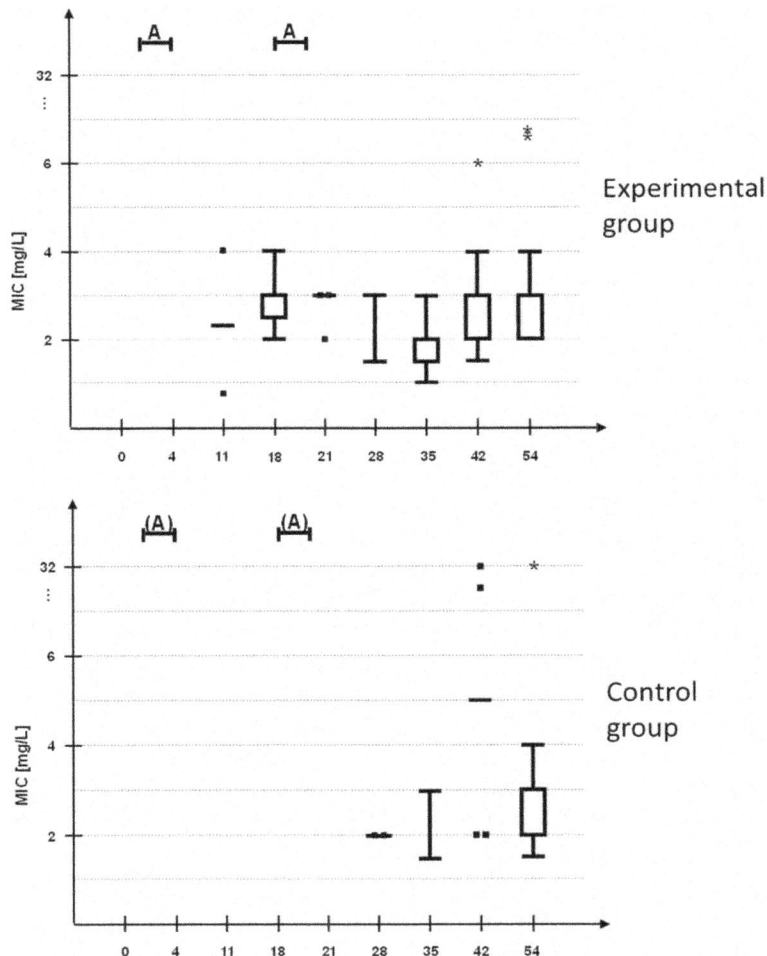

Fig. 2 Distribution of MIC values of *E. coli* isolates. MIC values were determined by Etest® during the experimental trial in the experimental group (EG) (n=6 pigs) compared to the control group (CG) (n=6 pigs) housed in the same stable. MIC values are represented in Tukey boxplots, points represent single MIC-values, asterisks outliers. Up to 10 *E. coli* colonies per animal and day were isolated from Endo agar plates supplemented with 0.125 mg/L enrofloxacin to screen for MIC-shifts (total number of isolated *E. coli* colonies per day represented in brackets). A= intramuscular administration of EG with 2.5 mg/kg bw for three days

administration period faecal samples from all pigs of the experimental group harboured N-WT-*E. coli*. N-WT-*E. coli* in faecal samples of the control group could be detected for the first time seven days after the second administration period on day 28. Only one out of six animals was affected. At the end of the trial *E. coli* with MIC values above the ECOFF were determined in faecal samples from all animals of the control group. *E. coli* isolates with MIC values > 4 mg/L were determined in both groups about three weeks after the second administration period. Environmental N-WT-*E. coli* isolates could be isolated for the first time on day 32, ten days later one *E. coli* isolate with a MIC value above 4 mg/L was detected.

Significant differences between MIC values from *E. coli* isolates of the experimental and the control group could not be detected at no trial day.

Time-dependent occurrence of *E. coli* strain pattern

The total number of *E. coli* isolates selected for macrorestriction analysis was 429. Among these isolates, we identified 64 different pulsed-Field Gel Electrophoresis (PFGE) strains with a mean occurrence of 6.7 times (range 1 to 64). More than 50% of the strains could not be detected more than twice, whereas 23.4% occurred more than ten times. Further information on the distribution of *E. coli* strains can be found in Table 2.

27 of the detected strains (42.2%) were transferred between at least two of the three potential sources (experimental group, control group, environment). Few strains could be detected throughout the entire experimental period; most strains were seen once or for up to two weeks.

Changes of the composition of *E. coli* population of two pigs, which were randomly chosen from each group, were determined. DNA fragment patterns of fifty *E. coli* isolates from each pig (10 isolates per trial day) were analysed. Before administration, diversity of *E. coli* strains varied strongly regardless of pig's group affiliation (diversity indices (Di) between 0.6 and 0.93). Directly after each administration, almost no *E. coli* colony could be isolated from the two pigs of the experimental group, resulting in a diversity index close to zero. One week after each administration period diversity indices of *E. coli* isolates from both pigs of the experimental group

varied strongly (0.2 vs 0.76), whereas the diversity of *E. coli* from the control group was still high (Di ≥ 0.84). On day 54 diversity of *E. coli* strains from both pigs of the experimental group and one pig of the control group was high (Di ≥ 0.83). However, *E. coli* from the other pig of the control group showed lower diversity (Di = 0.38). Overall diversity of *E. coli* strains from each individual pig varied uniquely during the whole trial regardless of administration or no administration.

Changes of individual *E. coli* strains regarding their susceptibilities during the experimental trial

In total, 25 *E. coli* strains could be detected repeatedly during the whole trial isolated from at least two of three potential sources. MIC determination was done for at least two isolates of the same strain. MIC changes of more than two titre steps were assumed as a relevant shift in susceptibility behaviour.

An increase in MIC values specifically against fluoroquinolones was found in four strains. These results are presented in Fig. 3.

Molecular characterisation of *E. coli* strains by PCR

Phylogenetic affiliation and presence of virulence genes, adhesion genes and genes for iron acquisition of 64 *E. coli* strains are listed in Table 3. Most strains belonged to EcoR group A (71.9% of isolates) followed by groups B1 (21.9%), D (4.7%) and B2 (1.5%). Virulence genes were not detected in any strain. The genes *crlA* and *fimC* were present in more than three-quarters of all isolated *E. coli* strains. Other genes related with adhesion or iron acquisition occurred only sporadically. No strain harboured the adhesion related gene *afa/draB*. We also screened the strains for plasmid-mediated horizontally transferable genes encoding quinolone resistance: 12 *E. coli* strains were at least temporarily positive for *qnrS*. Further information is given in Table 4. *QnrA* and *qnrB* could not be detected in any strain.

Significant associations between occurrence, frequency and transferability of detected *E. coli* strains and the occurrence of single genes as well as their phylogenetic affiliation were not detected.

Environmental load with enrofloxacin and ciprofloxacin

As shown in Fig. 4 the highest amounts of enrofloxacin were found in front of the feeding trough of the

Table 2 Occurrence of *E. coli* PFGE strains during the whole trial

Total number of different PFGE type from this source:	Distribution within all sources				
	Only in this group	Coexistent in En	Coexistent in CG	Coexistent in EG	In all groups
En: *n*=20	7		3	3	7
CG: *n*=44	20	3		14	
EG: *n*=34	10	3	14		

EG experimental group, *CG* control group, *En* environment, *PFGE* Pulsed-Field Gel Electrophoresis

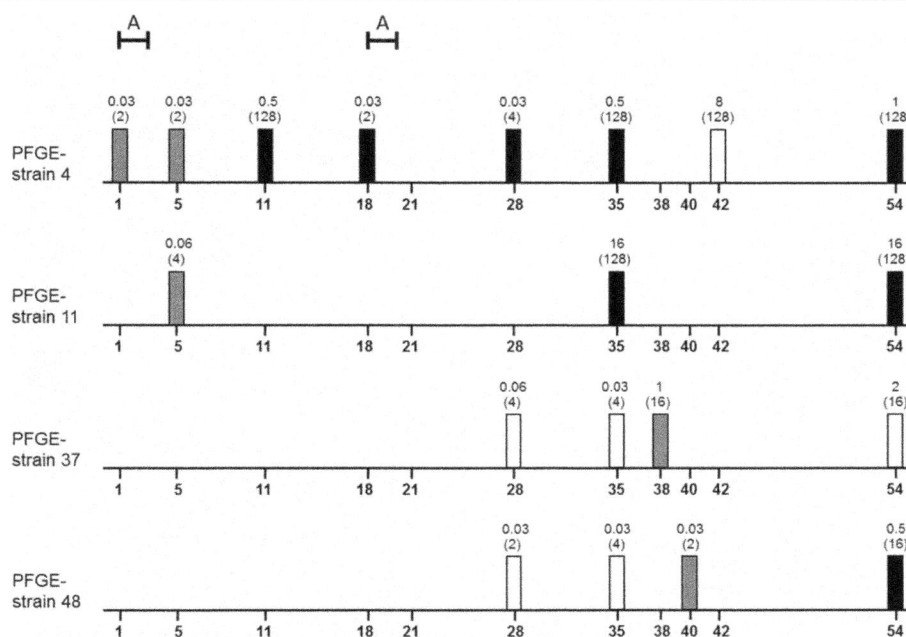

Fig. 3 Distribution of four *E. coli* PFGE strains with susceptibilities shifts against fluoroquinolones. The X-axis represents the trial period (in days). The columns show the origin of *E. coli* strains (white: CG, grey: environment, black: EG). The data above the columns are MIC values against enrofloxacin and nalidixic acid (in brackets). A=administration period

Table 3 Prevalence of virulence, adhesion, and iron acquisition genes in 64 intestinal *E. coli* strains

Gene or operon	Description	Total Prevalence [%]
Virulence gene		
stx1	shigatoxin 1	0
stx2	shigatoxin 1	0
eae	intimin	0
sta	heat-stable enterotoxin	0
F41	fimbriae F41	0
K99	fimbriae K99	0
Adhesins		
afa/draB	Afimbrial/Dr antigen-specific adhesion	0
crlA	curli fibre gene	92.2
fimC	Type 1 fimbriae	78.1
hra	Heat-resistant agglutinin	6.3
papC	Pilus associated with pyelonephritis	1.6
sfa/foc	S fimbriae and F1C fimbriae	1.6
tsh	Temperature-sensitive haemagglutinin	3.1
Iron acquisition		
chuA	Heme receptor gene	6.3
fyuA	Ferric yersinia uptake	10.9
ireA	Iron-responsive element	1.6
iroN	Catecholate siderphore	3.1
iucD	Aerobactin synthesis	4.7
sitD chr.	*Salmonella* iron transport system gene	1.6
sitD ep.	*Salmonella* iron transport system gene	4.7

experimental group. The environmental load with enrofloxacin in sedimentation dust increased during the first administration period and showed the highest amount of the antibiotic during the first administration and two days afterwards. At that time the concentration ranged between 5 and 10 ng enrofloxacin/mg sedimentation dust at the different locations. The more time passed, the more sedimentation dust accumulated and minimal quantities between 0 and 2 ng enrofloxacin/mg sedimentation dust were detectable.

Only traces of the active metabolite ciprofloxacin could be found during the experimental trial.

Discussion

Each use of antibiotics increases the risk of antimicrobial resistance [45, 46]. Besides, several studies have shown that antimicrobial use in food producing animals leads to increasing resistance levels in both pathogenic and commensal bacteria [47]. Many recent studies also demonstrated the increased risk of AMR in commensal *E. coli* after oral administration of antibiotics compared to untreated pigs [48–50]. However, there is a lack of studies comparing the impact of the route of administration on the development of resistance.

In this study we focused as a first step on changes in the intestinal commensal *E. coli* population after parenteral administration of pigs with fluoroquinolones as a potential and approved alternative to oral medication.

Table 4 Occurrence of *E. coli* PFGE strains with *qnrS*

PFGE strain	Also detected without *qnrS*	First detection		MIC ENRO [mg/L]	ECOR group
		Source	Day		
I-12	yes	EG (CG)[a]	35 (11)	0.03 (0.06)	A
I-21	no	EG	54	1	A
I-25	yes	CG (En)	42 (5)	8 (0.03)	A
I-28	no	EG	54	1	A
I-31	no	CG	54	1	B1
I-34	no	EG	21	1	B1
I-37	yes	EG (EG)	54 (54)	1 (0.03)	A
I-38	yes	CG (CG)	54 (18)	2 (n.t.)	A
I-41	yes	EG (CG)	54 (21)	1 (n.t.)	A
I-45	no	EG	54	1	A
I-51	yes	EG (CG)	54 (11)	1 (n.t.)	A
I-56	no	CG	54	1	A

EG experimental group, *CG* control group, *En* environment, *PFGE* Pulsed-Field Gel Electrophoresis, *n.t.* not tested, *ENRO* Enrofloxacin
[a]Number in brackets apply to the same PFGE strain without *qnrS*

Since we could not detect any *E. coli* strain during the whole experimental trial showing virulence genes typical for porcine enteropathogenic *E. coli*, it is likely that only commensal intestinal *E. coli* were included in this study.

The parenteral administration of enrofloxacin at the recommended dose (2.5 mg/kg bw) to piglets considerably reduced the number of the susceptible intestinal *E. coli* population. After a few days a rising number of *E. coli* with increased MIC values against enrofloxacin could be detected. These findings are in accordance with

other studies regarding the administration of fluoroquinolones [27, 51, 52].

There are two different types of criteria to interpret MIC values: clinical breakpoints and ECOFFs [53]. For *E. coli* from chicken and turkeys the Clinical Laboratory Standard Institute (CLSI) has determined a veterinary specific breakpoint of ≥ 2 mg/L enrofloxacin [54]. A veterinary specific breakpoint for enrofloxacin is not available for Enterobacteriaceae in pigs. By ECOFFs bacterial populations are divided into a wild type (WT) population without any acquired resistance mechanisms and a N-WT population with higher MIC values or smaller zone diameters [53]. To evaluate enrofloxacin susceptibilities of isolated *E. coli* we used the ECOFF (0.125 mg/L) from the European Committee on Antimicrobial Susceptibility Testing [33].

E. coli isolates with MIC values above the enrofloxacin ECOFF were observed one week after the first administration period for the first time. This result was in agreement with those of Wiuff et al. [26] who described a very rapid occurrence of fluoroquinolone resistance among the porcine coliform flora during administration with enrofloxacin. As expected, increased MIC values above 0.125 mg/L enrofloxacin occurred initially in *E. coli* isolates from the experimental group with increasing detection rates until the end of the trial. Subsequently and to a lesser extent also pigs of the control group in the same pen exhibited N-WT-*E. coli*. In similar studies, it has been shown that ceftiofur resistant *E. coli* occurred in untreated pigs which were housed in the same stable like a treated pig group [55]. These findings may be due to the transferability of

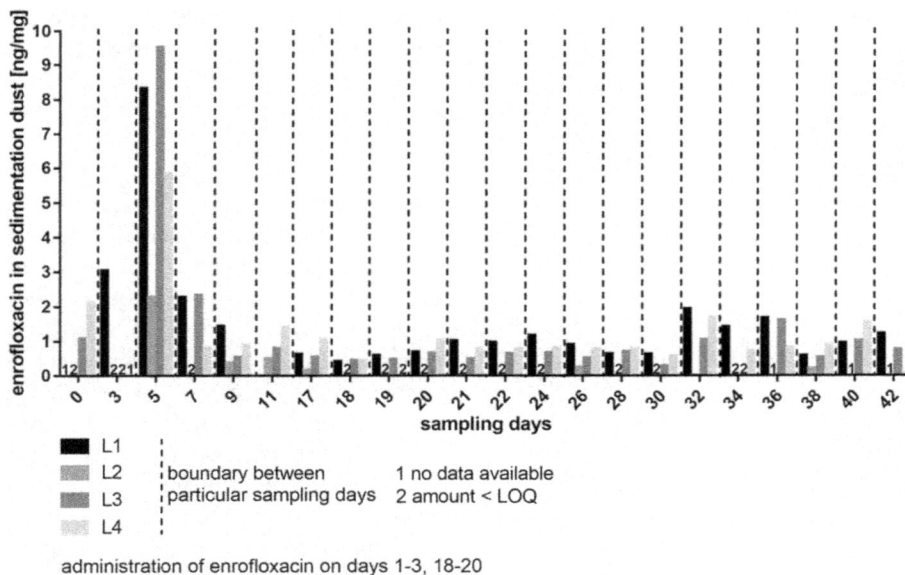

Fig. 4 Distribution of enrofloxacin in sedimentation dust in dependence of the location of sampling. L1-4: Location 1 to 4 of sedimentation dust sampling. For orientation make use of the schematic representation of the stable (Fig. 1)

N-WT strains from the experimental group through the environment to the control group especially as we could isolate the same PFGE strains from both pig groups and the environment. However, a causal link between the prevalence and transferability of single *E. coli* strains and detected adhesion genes, genes for iron acquisition or phylogenetic affiliation could not be identified.

Another possible root for the detection of N-WT strains within pigs of the control group might be the exposure of WT *E. coli* strains to subtherapeutic doses of enrofloxacin. Enrofloxacin and its main and active metabolite ciprofloxacin, which is known as a broad spectrum antimicrobial in human medicine, were found in minimal concentrations in sedimentation dust in the stable. These findings indicate that *E. coli* from untreated pigs could be exposed to enrofloxacin concentrations in lower than therapeutic doses from the environment resulting in an effective selective pressure. Scherz et al. [34] showed that a long-time exposure of 21 days of the commensal flora of poultry to subtherapeutic doses of enrofloxacin leads to an amplification and selection of N-WT-*E. coli* strains, which persist in the commensal microbiota. In the present study, the concentrations of enrofloxacin vary in the lower nanogram range in sedimented dust while only traces of ciprofloxacin are detectable. Gullberg et al. [56] demonstrated for ciprofloxacin *in vitro* that amounts of 1/10 and 1/230 of the MIC of a susceptible strain can select for N-WT strains by reducing the growth rates of the susceptible ones and balancing the fitness costs of the resistant bacteria. It is unknown to what extent the minimal amounts of enro- and ciprofloxacin in sedimented dust in the present case affect the commensal flora. However, in case of carry-over *in vivo* environmental bacteria are exposed to these contaminations.

In addition, the cross contamination between both groups by veterinarians and animal care staff during feeding, administration and sampling measures cannot be completely ruled-out in spite of all preventive measures taken.

In this study, 12 of 64 PFGE strains were at least temporarily positive for *qnrS*. As we did not analyse mutations within the gyrase and topoisomerase we cannot determine whether the observed reduced fluoroquinolone susceptibilities is caused by chromosomal mutations, by PMQR or a combination of both. But it is assumed that especial *qnr* mediated resistance results in fluoroquinolone resistance to a limited extent and leads to slightly elevated MIC values [19, 57]. Nevertheless it is horizontally transferable in Enterobacteriaceae and thereby contributes to the wide spread of reduced fluoroquinolone susceptibility.

To prevent selection of chromosomal mutations resulting in reduced susceptibility to fluoroquinolones, the dose regimen should be based on pharmacokinetics parameters in both plasma and target tissues. The mutant prevention concentration (MPC) is the minimum concentration preventing growth at a high inoculum ($\geq 10^9$ CFU/mL) using agar dilution methodology [58]. Drug levels in target tissues above the MPC are necessary to severely restrict a first step mutation as far as concentration dependent antibiotics like fluoroquinolones are concerned [59]. However, further studies are required to assess the clinical significance of the MPC.

N-WT strains may also be already present in small amounts within the coliform microflora or in the environment and could have used the selective advantage during antimicrobial administration. In our study, exclusively pigs from farms without any coliform growth on Endo agar supplemented with 0.125 mg/L enrofloxacin were included. No N-WT strain could be detected in the stable and all animals within the first five days after the beginning of administration. In addition, we were able to detect the occurrence of *E. coli* strains with the same PFGE patterns showing different susceptibilities against the tested fluoroquinolones. Therefore we assume that the observed N-WT strains are based at least partially on the acquisition of resistance traits by previously completely susceptible strains and the following spread of these N-WT strains. However, our screening of the stable prior to the arrival of the pigs could not include the whole area of all pens and our sampling design only included the examination of a small amount of faeces. Therefore, we cannot totally exclude the presence of N-WT strains prior to the first administration. *E. coli* strains isolated from faeces may not be representative of *E. coli* from other regions of the gastrointestinal tract [60]. Therefore, further studies are necessary which include the coliform flora colonizing the lower intestine (jejunum, ileum and colon).

Conclusions

The parenteral administration of enrofloxacin at the recommended dose to piglets considerably reduced the number of the susceptible intestinal *E. coli* population which was replaced by a rising number of *E. coli* with increased MIC values. Also, pigs of the control group in the same pen without direct contact exhibited N-WT-*E. coli* suggesting the transferability of single *E. coli* strains.

Resistance in particular to critically important antimicrobials is a significant public health threat as it limits the number of effective antimicrobial agents available for therapy. The constant excretion of N-WT strains within more than four weeks after the second administration and the putative transmission of N-WT *E. coli* strains to pigs of the control group or to the environment may contribute directly to the spread of resistant bacteria to farm workers or slaughterhouse staff. Furthermore, the introduction of N-WT bacteria and resistance determinants into the food chain or the spreading of animal-by-products such as manure and slurry harbouring high levels of potentially N-WT bacteria may facilitate the exchange of resistance determinants between livestock and humans.

Abbreviation

°C: degree Celsius; µL: microliter; AMR: Antimicrobial resistance; bw: bodyweight; CFU: Colony-forming units; CG: Control group; CLSI: Clinical and Laboratory Standards Institute; Di: Simpson's index of diversity; DNA: Deoxyribonucleic acid; *E. coli*: *Escherichia coli*; e.g.: exempli gratia; ECOFF: Epidemiological cut-off value; EcoR: *E. coli* reference collection; EG: Experimental group; En: environment; EUCAST: European Committee on Antimicrobial Susceptibility Testing; HPCIA: Highest priority critically important antimicrobials; HPLC: High performance liquid chromatography; i.m.: intramuscular; kg: kilogram; LB: Luria-Bertani Bouillon; mg/L: milligrams per Litre; MIC: Minimum inhibitory concentration; mL: millilitre; MPC: Mutant prevention concentration; ng: nanogram; N-WT: Non-Wild Type, in this context: bacterial populations with acquired resistance mechanisms resulting in decreased susceptibilities to fluoroquinolones; OIE: World Organisation for Animal Health; PCR: Polymerase chain reaction; PFGE: Pulsed-field gel electrophoresis; PMQR: Plasmid mediated Quinolone Resistance; WHO: World Health Organization; WT: Wild Type, in this context: bacterial populations without any acquired resistance mechanisms against fluoroquinolones

Acknowledgements
We thank Hannelore Willeken for excellent technical assistance.

Funding
Not applicable.

Authors' contributions

Design of the study: JW, MK, HK, GS. Executing the animal experiments: SR, GS. Executing the microbiology: SR, AR. Executing the HPLC-analysis: SR, JM. Analysis of the data: SR, AR, GS. Preparation of the manuscript: AR, GS. All authors read and approved the final manuscript.

Consent for publication

Not applicable.

Competing interests

The authors declare that they have no competing interests.

Author details
[1]Federal Office of Consumer Protection and Food Safety, Berlin, Germany. [2]University of Veterinary Medicine Hannover, Foundation, Institute of Pharmacology, Toxicology and Pharmacy, Hanover, Germany.

References

1. Lee CR, Cho IH, Jeong BC, Lee SH. Strategies to minimize antibiotic resistance. Int J Environ Res Public Health. 2013;10:4274–305.
2. Mazurek J, Pusz P, Bok E, et al. The phenotypic and genotypic characteristics of antibiotic resistance in *Escherichia coli* populations isolated from farm animals with different exposure to antimicrobial agents. Pol J Microbiol. 2013;62:173–9.
3. Volkova VV, Lanzas C, Lu Z, Gröhn YT. Mathematical model of plasmid-mediated resistance to ceftiofur in commensal enteric *Escherichia coli* of cattle. PLoS One. 2012;7:e36738.
4. Chantziaras I, Boyen F, Callens B, Dewulf J. Correlation between veterinary antimicrobial use and antimicrobial resistance in food-producing animals: a report on seven countries. J Antimicrob Chemother. 2014;69(3):827–34.
5. Rodríguez-Martínez JM, Cano ME, Velasco C. Plasmid-mediated quinolone resistance: an update. J Infect Chemother. 2011;17:149–82.
6. World Health Organization (WHO), WHO Advisory Group on Integrated Surveillance of Antimicrobial Resistance (AGISAR): Critically important antimicrobials for human medicine – 3rd rev. http://www.who.int (2011). Accessed 18 Nov 2016.
7. World Organization for Animal Health (OIE): OIE List of antimicrobials of veterinary importance. http://www.oie.int (2007). Accessed 18 Nov 2016
8. Appelbaum PC, Hunter PA. The fluoroquinolone antibacterials: past, present and future perspectives. Int J Antimicrob Agents. 2000;16:5–15.
9. Bugyei K, Black WD, Ewen S. Pharmacokinetics of enrofloxacin given by the oral, intravenous and intramuscular routes in broiler chickens. Can J Vet Res. 1999;63:193–200.
10. Brown SA. Fluoroquinolones in animal health. J Vet Pharmacol Ther. 1996; 19:1–14.
11. Scheer M. Concentrations of active ingredient in serum and in tissues after oral and parenteral administration of Baytril. Vet. Med. Rev. 1987;2:104–18.
12. Wiuff C, Lykkesfeldt J, Aarestrup FM, Svendsen O. Distribution of enrofloxacin in intestinal tissue and contents of healthy pigs after oral and intramuscular administrations. J Vet Pharmacol Ther. 2002;25(5):335–42.
13. Lode H, Borner K, Koeppe P. Pharmacodynamics of fluoroquinolones. Clin Infect Dis. 1998;27(1):33–9.
14. Drlica K, Zhao XDNA. gyrase, topoisomerase IV, and the 4-quinolones. Microbiol Mol Biol Rev. 1997;61:377–92.
15. Hopkins KL, Davies RH, Threlfall EJ. Mechanisms of quinolone resistance in *Escherichia coli* and *Salmonella*: recent developments. Int J Antimicrob Agents. 2005;25:358–73.
16. Chenia HY, Pillay B, Pillay D. Analysis of the mechanisms of fluoroquinolone resistance in urinary tract pathogens. J Antimicrob Chemother. 2006;58:1274–8.
17. Veldman K, Cavaco LM, Mevius D, et al. International collaborative study on the occurrence of PMQR in *Salmonella enterica* and *Escherichia coli* isolated from animals, humans, food and the environment in 13 European countries. J Antimicrob Chemother. 2011;66:1278–86.
18. Robicsek A, Jacoby GA, Hooper DC. The worldwide emergence of plasmid-mediated quinolone resistance. Lancet Infect Dis. 2006;6:629–40.
19. Martínez-Martínez L, Pascual A, Jacoby GA. Quinolone resistance from a transferable plasmid. Lancet. 1998;351:797–9.
20. Jacoby GA. Mechanisms of resistance to quinolones. Clin Infect Dis. 2005; 41(Suppl 2):S120–6.
21. Devreese M, Antonissen G, De Baere S, De Backer P, Croubels S. Effect of administration route and dose escalation on plasma and intestinal concentrations of enrofloxacin and ciprofloxacin in broiler chickens. BMC Vet Res. 2014;10:289.
22. European Medicines Agency (EMA): Opinion following an Article 35 referral for Baytril 2.5% injectable, Baytril 5% injectable and Baytril 10% injectable and their associated names, and related veterinary medicinal products - Background information (EMEA/V/A/097). http://www.ema.europa.eu (2014). Accessed 18 Nov 2016.
23. Robanus M, Hegger-Gravenhorst C, Mollenhauer Y, Hajek P, Käsbohrer A, Honscha W, Kreienbrock L. Feasibility study of veterinary antibiotic consumption in Germany - comparison of ADDs and UDDs by animal production type, antimicrobial class and indication. BMC Vet Res. 2014;10:7.
24. European Food and Safety Agency and. (EFSA) and European Food Safety Authority and European Centre for Disease Prevention and Control (ECDC): The Community Summary Report on antimicrobial resistance in zoonotic and indicator bacteria from humans, animals and food in 2012. EFSA Journal. 2014;12(3):3590–926.
25. Pereira RV, Siler JD, Ng JC, Davis MA, Grohn YT, Warnick LD. Effect of on-farm use of antimicrobial drugs on resistance in fecal *Escherichia coli* of preweaned dairy calves. J Dairy Sci. 2014;97(12):7644–54.
26. Wiuff C, Lykkesfeldt J, Svendsen O, Aarestrup FM. The effects of oral and intramuscular administration and dose escalation of enrofloxacin on the selection of quinolone resistance among *Salmonella* and coliforms in pigs. Res Vet Sci. 2003;75:185–93.
27. Walker RD. Fluoroquinolones. In: Prescott JF, Baggot JD, Walker RD, editors. Antimicrobial Therapy in Veterinary Medicine. 3. Ames, Iowa: Iowa State Univ Pr; 2000. p. 315–38.
28. Fadário Frade VM, Dias M, Costa Teixeira ACS, Alves Palma MS. Environmental contamination by fluoroquinolones. Braz. J Pharm Sci. 2014;50:1.
29. Douglas R. Call, Louise Matthews, Murugan Subbiah, Jinxin Liu, (2013) Do antibiotic residues in soils play a role in amplification and transmission of antibiotic resistant bacteria in cattle populations?. Frontiers in Microbiology 4.

30. Bertani G. Studies on lysogenesis. I. The mode of phage liberation by lysogenic *Escherichia coli*. J. Bacteriol. 1951;62:293–300.

31. Atlas RM, Snyder JW. Handbook of Media for Clinical Microbiology. Boca Raton, Fla: CRC Press; 2006. p. 190–1.

32. The European Committee on Antimicrobial Susceptibility Testing (EUCAST): MIC distributions and ECOFFs. http://www.eucast.org/mic_distributions_and_ecoffs (2016). Accessed 18 Nov 2016.

33. Corry JEL, Curtis GDW, Baird RM. Handbook of culture media for food and water Microbiology. Cambridge: Royal Society of Chemistry; 2012. p. 801.

34. Scherz G, Stahl J, Glünder G, Kietzmann M. Effects of carry-over of fluoroquinolones on the susceptibility of commensal *Escherichia coli* in the intestinal microbiota of poultry. Berl Munch Tierarztl Wochenschr. 2014;127:478–85.

35. Edén CS, Eriksson B, Hanson LA. Adhesion to normal human uroepithelial cells of *Escherichia coli* from children with various forms of urinary tract infection. J Pediatr. 1978;93:398–403.

36. Döpfer D, Buist W, Soyer Y, et al. Assessing Genetic Heterogeneity within Bacterial Species Isolated from Gastrointestinal and Environmental Samples: How Many Isolates Does It Take? Appl Environ Microbiol. 2008;74(11):3490–6.

37. Schierack P, Roemer A, Jores J, et al. Isolation and characterization of intestinal *Escherichia coli* clones from wild boars in Germany. Appl Environ Microbiol. 2009;75:695–702.

38. Liesegang A, Tschäpe H. Modified pulsed-field gel electrophoresis method for DNA degradation-sensitive *Salmonella enterica* and *Escherichia coli* strains. Int J Med Microbiol. 2002;291:645–8.

39. Ewers C, Li G, Wilking H, et al. Avian pathogenic, uropathogenic, and newborn meningitis-causing *Escherichia coli*: how closely related are they? Int J Med Microbiol. 2007;297:163–6.

40. Franck SM, Bosworth BT, Moon HW. Multiplex PCR for enterotoxigenic, attaching and effacing, and Shiga toxin-producing *Escherichia coli* strains from calves. J Clin Microbiol. 1998;36:1795–7.

41. Clermont O, Bonacorsi S, Bingen E. Rapid and simple determination of the *Escherichia coli* phylogenetic group. Appl Environ Microbiol. 2000;66:4555–8.

42. Scherz G. Carryover of subtherapeutic antimicrobial dosages of enrofloxacin and the influence on the development of antibiotic resistance of commensal *Escherichia coli* in the intestine of poultry (Ph.D. Thesis). University of Veterinary Medicine Hannover, Foundation; 2013.

43. Katouli M, Lund A, Wallgren P, et al. Phenotypic characterization of intestinal *Escherichia coli* of pigs during suckling, postweaning, and fattening periods. Appl Environ Microbiol. 1995;61:778–83.

44. Simpson EH. Measurement of diversity. Nature. 1949;163:688.

45. Davies J, Davies D. Origins and Evolution of Antibiotic Resistance. Microbiol Mol Biol Rev. 2010;74:417–33.

46. Dunlop RH, McEwen SA, Meek AH. Associations among antimicrobial drug treatments and antimicrobial resistance of fecal *Escherichia coli* of swine on 34 farrow-to-finish farms in Ontario, Canada. Prev Vet Med. 1998;34:283–305.

47. Van den Bogaard AE, Stobberingh EE. Epidemiology of resistance to antibiotics. Links between animals and humans. Int J Antimicrob Agents. 2000;14:327–35.

48. Lutz EA, McCarty MJ, Mollenkopf DF, et al. Ceftiofur use in finishing swine barns and the recovery of fecal *Escherichia coli* or *Salmonella* spp. resistant to ceftriaxone. Foodborne Pathog Dis. 2011;8:1229–34.

49. Varga C, Rajić A, McFall ME, et al. Associations among antimicrobial use and antimicrobial resistance of *Salmonella* spp. isolates from 60 Alberta finishing swine farms. Foodborne Pathog Dis. 2009;6:23–31.

50. Wagner BA, Straw BE, Fedorka-Cray PJ, Dargatz DA. Effect of antimicrobial dosage regimen on *Salmonella* and *Escherichia coli* isolates from feeder swine. Appl Environ Microbiol. 2008;74:1731–9.

51. Lastours de V, Cambau E, Guillard T, et al. Diversity of Individual Dynamic Patterns of Emergence of Resistance to Quinolones in Escherichia coli From the Fecal Flora of Healthy Volunteers Exposed to Ciprofloxacin. J Infect Dis. 2012;206(9):1399–406.

52. Fantin B, Duval X, Massias L, et al. Ciprofloxacin Dosage and Emergence of Resistance in Human Commensal Bacteria. J Infect Dis. 2009;200(3):390–8.

53. Schwarz S, Kadlec K, Silley P. Enteric infection. In: Antimicrobial resistance in bacteria of animal origin. 1st ed. Steinen, Germany: Zett-Verlag; 2013. p. 70–8.

54. Clinical and Laboratory Standards Institute (CLSI). Performance Standards for Antimicrobial Disk and Dilution Susceptibility Tests for Bacteria Isolated From Animals, Third Informational Supplement VET01S. Wayne, PA; 2015.

55. Beyer A, Baumann S, Scherz G, et al. Effects of ceftiofur treatment on the susceptibility of commensal porcine *E. coli* – comparison between treated and untreated animals housed in the same stable. BMC Vet Res. 2015;11:265.

56. Gullberg E, Cao S, Berg OG, et al. Selection of resistant bacteria at very low antibiotic concentrations. PLoS Pathog. 2011;7:e1002158.

57. Yue L, Jiang HX, Liao XP, Liu JH, Li SJ, Chen XY, Chen CX, Lü DH, Liu YH. Prevalence of plasmid-mediated quinolone resistance *qnr* genes in poultry and swine clinical isolates of *Escherichia coli*. Vet Microbiol. 2008;132(3–4):414–20.

58. Mouton JW, Vinks AA. PK-PD modelling of antibiotics in vitro and in vivo using bacterial growth and kill kinetics: the MIC vs stationary concentrations. Clin Pharmacokinet. 2005;44:201–10.

59. Wang J, Hao H, Huang L, et al. Pharmacokinetic and Pharmacodynamic Integration and Modeling of Enrofloxacin in Swine for *Escherichia coli*. Front Microbiol. 2016;7:36.

60. Dixit SM, Gordon DM, Wu X, et al. Diversity analysis of commensal porcine *Escherichia coli* – associations between genotypes and habitat in the porcine gastrointestinal tract. Microbiol. 2004;150:1735–40.

Phenotypic and genotypic characterization of *Enterococcus cecorum* strains associated with infections in poultry

Beata Dolka[1*], Dorota Chrobak-Chmiel[2], László Makrai[3] and Piotr Szeleszczuk[1]

Abstract

Background: From the beginning of the 21st century *Enterococcus cecorum* has emerged as a significant health problem for poultry raised under intensive production systems. To obtain new insights into this bacterial species, we investigated 82 clinical isolates originating from different poultry flocks in Poland between 2011 and 2014.

Results: Phenotypically, isolates from clinical cases showed ability to growth at low temperatures (4 °C, 10 °C), and differences in growth at 45 °C (74.4 %). Survival at high temperatures (60 °C, 70 °C) was observed for 15, 30 min. More than half of strains survived at 60 °C even after prolonged incubation (1 h), but none survived after 1 h at 70 °C. Total growth inhibition was observed on agar supplemented with tergitol or potassium tellurite. Relatively high number of isolates gave positive reactions for β-galactosidase (βGAL 80 %), Voges Proskauer test (60 %), less for β-mannosidase (17 %), glycogen and mannitol (12 %). The metabolic fingerprinting for *E. cecorum* obtained in Biolog system revealed ability to metabolise 22 carbon sources. Only 27/82 strains contained ≥ 1 virulence genes of tested 7, however 2.4 % isolates carried 6. Increased antimicrobial resistance was observed to enrofloxacin (87 %), teicoplanin (85 %), doxycycline (83 %), erythromycin (46 %). Most strains (75/82) showed multidrug resistance. The single isolate was resistant to vancomycin (VRE) and high level gentamicin (HLGR). Linezolid resistance among clinical isolates was not found. PFGE revealed diversity of *E. cecorum* from cases. It could be assumed that transmission of pathogenic strains between flocks regardless of type of production or geographical region may be possible.

Conclusions: Clinical infections in poultry caused by *E. cecorum* may indicated on new properties of this bacterial species, previously known as a commensal. Despite many common phenotypic features, differences were found among clinical isolates. Several, widely distributed pathogenic *E. cecorum* strains seemed to be responsible for infection cases found in different poultry types.

Keywords: *Enterococcus cecorum*, Phenotyping, Genotyping, PFGE, Enterococcal spondylitis, Chicken

Background

First time *Enterococcus cecorum* was isolated from cecal flora of chickens and described as *Streptococcus cecorum* in 1983, thereafter well known as commensal in gastrointestinal tract of various mammals and birds [1]. On the other hand, *Enterococcus cecorum* belongs to opportunistic pathogens and may also play a role as etiological agent of diseases in humans (nosocomial infections) [2, 3], chickens [4], and racing pigeons [5]. Recently, this bacteria appears to be a new threat ("emerging pathogen") to poultry industry worldwide [6–15]. *E. cecorum* has been increasingly recognized as a cause of enterococcal spondylitis (ES), previously called enterococcal vertebral osteoarthritis (EVOA) in chickens [12]. Disease outbreaks were diagnosed mostly in broiler chicken flocks raised under an intensive production system. Clinically affected birds suffered from locomotor problems due to compression of the spinal cord at the thoracic vertebrae resulting from *E. cecorum* induced osteomyelitis and due to femoral head necrosis (FHN) [6, 7, 9, 12, 13]. Disease outbreaks can lead to high morbidity, mortality, culling, carcass condemnations, and may result in severe economic losses within a short time [9].

* Correspondence: beata_dolka@sggw.pl
[1]Department of Pathology and Veterinary Diagnostics, Faculty of Veterinary Medicine, Warsaw University of Life Sciences-SGGW, Nowoursynowska 159c St., Warsaw 02-776, Poland
Full list of author information is available at the end of the article

Recently, poultry or domestic animals (cats, dogs) are thought to be a possible source for transmission leading to *E. cecorum*–associated septicaemia in humans [2, 3].

Various methods using conventional biochemical tests and molecular techniques have been commonly used for identification and typing enterococci [16–18]. Pulsed field gel electrophoresis (PFGE) is considered to be the "gold standard" for subtyping enterococci and has been used extensively for molecular epidemiological characterization of enterococcal outbreaks [19, 20]. The PCR assay based on specific amplification followed by sequencing and nucleotide sequence comparison of target genes (such as 16S ribosomal RNA, *sod*A, *ddl*, *tuf*, *gro*ESL) or tDNA-PCR have served for the genotypic identification of enterococci [21–23].

Despite of available literature biochemical and molecular analysis of *E. cecorum* strains with poultry origin isolated in Europe are still limited. Moreover, there is not enough data regarding the properties of isolates, usually referred as pathogenic for poultry [1, 7, 8, 10]. The purpose of this study was phenotypic characterization of clinical *E. cecorum* isolates associated with infections in poultry and investigation their genetic relatedness.

Methods
Bacterial isolates
Eighty two *E. cecorum* isolates of poultry-origin used in this study were obtained from archival bacterial collection deposited at Department of Pathology and Veterinary Diagnostics, or were obtained from clinical specimens submitted by veterinarians for routine diagnostic work to the Diagnostic Laboratory in Division of Avian Diseases, Faculty of Veterinary Medicine at the Warsaw University of Life Sciences-SGGW (Poland). Authors ensure that the ARRIVE guidelines were followed. Among 82 clinical strains collected between 2011 and 2014, 49 came from broiler chicken flocks (CB), 20 from broiler breeder flocks (BB), 10 from commercial layer flocks (CL), 2 from geese flocks (G) and 1 from turkey flock (T). According to adopted criteria in this study, one *E. cecorum* isolate represented one different flock in which clinical problems due to *E. cecorum* infection were reported by veterinarians on farms. Affected birds displayed a variety of clinical signs, however in all types of flocks the lameness, paralysis, hock sitting, weakness, pododermatitis, decreased water and food intake were usually noted. Subsequently, disease caused lower results of production, increased losses due to mortality and culling. Necropsies and pathological examinations revealed usually femoral head necrosis, (purulent) arthritis, fibrinous pericarditis, endocarditis, hepatitis and congested lungs. Characteristic osteomyelitis lesions at caudal thoracic vertebrae we found only in chicken flocks (mainly in CB). Isolates were recovered from tissue samples such as vertebral column, femoral heads, heart, liver, lungs or yolk sac, which were collected during necropsy.

Bacterial analysis
The tissue samples were inoculated onto Columbia agar with 5 % sheep blood (CA) (Graso, Poland) and agar plates with esculin (KAA, Biocorp, Poland; Enterococcosel Agar, Graso, Poland), then incubated at 37 °C for 24 h in a CO_2-enriched atmosphere. Bacteria were identified as *Enterococcus* based on their phenotypic properties such as colonial morphology, hemolysis (on CA), Gram-staining, catalase production (using a 3 % H_2O_2), cytochrome oxidase production (OXItest, Erba Lachema s.r.o., Czech Republik), and esculin hydrolysis (Enterococcosel Agar, KAA). Pigment production was visually assayed by growing the bacteria on CA for 24 h and scraping off the growth with a white cotton swab. Motility was examined using Motility Test Agar (Graso, Poland). The ability to growth was estimated in 6.5 % NaCl (salt tolerance test) after 48 h at 37 °C, and on different media (Graso, Poland) (Table 2). Serological identification of Lancefield group was conducted by rapid latex agglutination method using Slidex Strepto Plus D (bioMérieux, France). Tests for *E. cecorum* growth were performed in BHI broth (Brain-heart infusion; bioMérieux, France) tubes preincubated at 4 °C, 10 °C, 45 °C for 24 h. Then cultures in BHI broth were spread onto CA and incubated at 37 °C. The growth response was assessed after 24 h and 48 h. The ability to survive at 60 °C, 70 °C was estimated for 15 min, 30 min, 1 h in BHI broth tubes, followed by incubation of inoculated CA plates. The results were recorded after 24 h and 48 h.

Biochemical tests
Identification to the species level based on biochemical characterization was performed by API rapid ID 32 STREP (bioMérieux, France) and on the basis of carbon source utilisation using Biolog system (Biolog Inc., Hayward, USA). Isolates ($n = 13$) were determined according to Biolog GP2 MicroPlates, which performed 95 discrete tests simultaneously and gave a characteristic reaction pattern (metabolic fingerprint). The MicroPlates were incubated at 37 °C and read visually after 4 h and 24 h. The metabolic fingerprint patterns were compared and identified using the MicroLog™ 4.20.05 database software.

Virulence factors
All 82 isolates were tested for the presence of seven virulence factors: *asa*1 (aggregation substance), *gelE* (gelatinase), *hyl* (hyaluronidase), *esp* (enterococcal surface protein), *cylA* (cytolisin), *efaA* (endocarditis antigen), *ace* (collagen-binding protein) according to

Martín-Platero et al. [24], Jung et al. [5] using duplex PCRs (*asa1/gelE, cylA/esp, efaA/ace*) and single PCR (*hyl*). PCR reaction mix contained 12.5 µl DreamTaq PCR Master Mix (Thermo Fisher Scientific Inc., USA) 0.3 µl of each primer (50 pmol/µl), 4 µl DNA and PCR-clean water (added up to a volume of 25 µl). Thermocycler conditions were as follows: initial denaturation at 94 °C for 5 min, followed by 30 cycles: denaturation at 94 °C for 1 min, annealing at 56 °C for 1 min (55 °C for *efaA/ace*), extension at 72 °C for 1 min, followed by final extension step 72 °C for 10 min and a 4 °C hold. Amplification products (10 µl) were analyzed by 1.2 % agarose gel electrophoresis after ethidium bromide staining and visualized under UV light (UVP, USA). A 100-bp DNA ladder (Thermo Fisher Scientific Inc., USA) was used as a molecular size marker.

Production of gelatinase was additionally determined using Difco Nutrient Gelatin (BD, USA) according to the manufacturer's recommendations. The tubes inoculated with *E. cecorum* ATCC 43198, *S. aureus* ATCC 25923 (gelatinase positive), *E. coli* ATCC 25922 (gelatinase negative) and an uninoculated tube were used for quality control testing.

Antibiotic susceptibility

Susceptibility for 13 antimicrobial agents: amoxicillin/clavulanic acid (AUG 20/10 µg), ampicillin (AP 10 µg), penicillin (PG 10 µg), enrofloxacin (ENF 5 µg) tetracycline (TEC 30 µg), nitrofurantoin (NI 300 µg), doxycycline (DXT 30 µg), chloramphenicol (C 30 µg), erythromycin (E 15 µg), teicoplanin (T 30 µg), vancomycin (VA 30 µg), high level gentamicin (GM 120 µg) and linezolid (LZD 30 µg) was tested by Kirby-Bauer disk diffusion method and the results were interpreted according to Clinical and Laboratory Standards Institute guidelines [25]. The criteria for selection of antibiotics based on CLSI guidelines for *Enterococcus* spp. and on their practical significance for the clinical use. Among tested antibiotics, tetracycline, doxycycline, amoxicillin, enrofloxacin have been actually approved for use in poultry (erythromycin until 2014) and have practical relevance. Vancomycin resistance genes (*van*A, *van*B) were tested by PCR using primers and conditions previously reported [24]. *Staphylococcus aureus* ATCC 25923 (vancomycin susceptible), *E. faecalis* ATCC 51299 (vancomycin resistant), *E. cecorum* ATCC 43198 were used as controls.

Molecular identification

Rapid extraction of bacterial genomic DNA was carried out by using boiling method. PCR assay targeting *sodA* gene was performed for identification and determination the diversity of 82 *E. cecorum* strains [22]. PCR products were visualized after electrophoresis on agarose gel (2 %) by staining with ethidium bromide, then purified using GeneMATRIX PCR/DNA Clean-Up Purification Kit (EURx, Poland) and submitted for sequencing to commercial services (IBB PAN, Genomed, Poland). The *sodA* gene sequences were analyzed with NCBI BLAST. The genetic distances based on the partial sequences of *sodA* was calculated by the two-parameter method of Kimura by using the MEGA6, and the phylogenetic tree was constructed using the Neighbor-Joining method (NJ) with 1000 bootstrap replicates.

PFGE

The standard PFGE procedure was adapted from previously published studies with minor modifications [18, 26, 27]. The 82 *E. cecorum* strains were cultured overnight on CA and then suspended in sterile saline to obtain the density of 3.5 on McFarland scale and centrifuged 10 min. at 4000 rpm/min. The bacterial pellets were mixed with 150 µl Tris-EDTA buffer solution (10 mM Tris-HCl, 1 mM disodium EDTA, pH 8.0) and 150 µl liquid 2 % agarose (InCert Agarose, Lonza, Rockland, USA) and small discs were formed (20 µl). The solidified discs were incubated at 37 °C for 18 h in 1 ml of EC buffer (6 mM Tris-HCl pH 8.0, 1 M NaCl, 0.1 M EDTA, 0.2 % deoxycholate, 0.2 % sarkosyl) containing 10 mg lysozyme (A&A Biotechnology, Poland), and 0.02 mg RNase A (Thermo Fisher Scientific Inc., USA). DNA discs were washed 3 times in 5 ml EBS solution (0.5 M EDTA pH 9.0, 1 % sarkosyl) and incubated overnight at 50 °C in 1 ml EBS solution containing 1 mg of proteinase K (ESP buffer) (A&A Biotechnology, Poland). Then the discs were washed 4 times (each time upside down for 30 times at room temperature) with 10 ml TE buffer (10 mM Tris, 1 mM EDTA, pH 8.0) and stored in 1 ml TE buffer at 4 °C. Subsequently, each disc was pre-incubated in 100 µl restriction buffer for 30 min at room temperature. The agarose discs were digested with *Sma*I (20 U/µl; Fermentas, Lithuania) overnight (at 37 °C). The restriction fragments were separated by clamped homogenous electric field (CHEF) electrophoresis with a CHEF-DR II System (Bio-Rad Laboratories, USA) in a 1.2 % (w/v) agarose gel using pulse time at 0.5 s followed by 35 s at 6 V/cm and temperature 14 °C for 24 h [17]. Afterwards the gel was stained with ethidium bromide for 30 min, then washed in distilled water for 30 min, photographed under UV light and documented in the system VersaDoc (Bio-Rad Laboratories, USA). Lambda Ladder PFG marker (New England Biolabs Inc., USA) was used as molecular size marker. Gel images were analyzed by Gel Compar II version 6.6 (Applied Maths, Belgium) and cluster analysis was performed by UPGMA using dice similarity coefficient with optimization set at 1 % and position tolerance at 1 %. Isolates were clustered using an 80 % homology cut-off, above which strains were considered to be closely related and assigned to the same PFGE type [19].

Results

Phenotypic characterization

Table 1 shows results of conventional tests and effects of different temperatures on the growth and survival of *E. cecorum* strains. Bacterial growth was characterized on 7 different microbiological media (Table 2).

Biochemical tests

The strains were identified as *E. cecorum* with the API rapid ID 32 STREP and Biolog system. API revealed perfect identification profile (ID 99.9 %, T 0.83) for 40 (49 %) *E. cecorum* strains, very good identification (ID 99.9 %, T 0.67) for 21 (26 %) strains, good identification (ID 99.8 %, T 0.38) for 2 (2 %) strains, doubtful profile (99.9 %, T 0.4) for 16 (20 %) strains, and unacceptable profile for 3 (4 %) strains. Among perfect identification profiles for *E. cecorum*, the code 6717–4607–131 was recorded the most often. Based on the analysis of 82 obtained profiles in API (each with 32 tests), we defined one common code 2317–4607–111 for clinical strains which gives perfect identification as *E. cecorum* with the API database. Biochemical results obtained in API were presented in Table 3. The vast majority of isolates was

positive in tests for βGLU, RAF, SAC, MβDG, CDEX (100 %), αGAL, RIB, TRE (99 %), MAL, MEL (98 %). All isolates were completely negative for ADH, APPA, HIP, PYRA, LARA. The discrepancies among tested and control isolates or recommendations for *E. cecorum* were noted in 6 tests: βGAR, MAN, VP, βGAL, GLYG, βMAN.

All of examined isolates were identified as *E. cecorum* in Biolog system (index: probability 91.7 %, similarity 0.806). The metabolic fingerprinting for *E. cecorum* was showed in Fig. 1. All of the examined isolates were able to metabolise 22 carbon sources (α-cyclodextrin, dextrin, N-acetyl-D-glucosamine, N-acetyl-D-mannosamine, arbutin, D-cellobiose, D-fructose, D-galactose, gentiobiose, α-D-glucose, maltose, maltotriose, D-mannose, D-melibiose, D-psicose, D-raffinose, salicin, stachyose, sucrose, D-trehalose, pyruvic acid methyl ester, adenosine). Not all of examined isolates were able to metabolise 14 carbon sources: amygdalin, D-melezitose, β-methyl-D-glucoside, inosine, thymidine, uridine (metabolised by 92.3 % strains), α-D-lactose (84.6 %), D-ribose (76.9 %), lactulose, palatinose (69.2 %), 2'-deoxy-adenosine (61.5 %), adenosine-5'-monophosphate, uridine-5'-monophosphate (53.8 %), β-methyl-D-galactoside (15.4 %). Further 59 carbon sources present in the GP2 microplate were not utilised by *E. cecorum*.

Virulence factors

Of all 82 *E. cecorum* strains, 22 (26.8 %) were positive for *asa1*, 21 (25.6 %) for *gelE*, 12 (14.6 %) for *ace*, 11 (13.4 %) for *efaA*. The *cylA* and *esp* PCR amplification yielded positive results in 4 (4.9 %) and 2 (2.4 %) *E. cecorum* strains. The *hyl* gene was not detected in any strain. The isolates from CB were positive for *asa1* (24.5 %), *gelE* (22.4 %), *ace* (14.3 %), *efaA* (14.3 %), *cylA* (2.1 %). The isolates from BB were positive for *asa1* (20 %), *gelE* (20 %), *ace* (15 %), *esp* (10 %), *cylA* (10 %). The isolates from CL were positive for *asa1* (60 %), *gelE* (60 %), *efaA* (20 %), ace (20 %). None of 7 virulence factors was found in isolates from G and T flocks. Most of virulence-gene positive isolates (11; 13.4 %) contained 2 of tested 7 virulence genes, then 6 (7.3 %) *E. cecorum* contained 4 virulence genes, 5 (6.1 %) harbored 1 virulence gene, while 3 (3.7 %) carried 3 virulence genes. In two isolates (2.4 %) 6 virulence genes were identified. None of isolates carried 5 or 7 virulence genes. Phenotypically, non of isolates produced gelatinase despite being *gelE*-positive in PCR.

Antibiotic susceptibility

One (0.82 %) out of the 82 clinical *E. cecorum* was susceptible to 13 antibiotics tested, the rest were resistant to one or more antibiotics (Table 4). All isolates were susceptible to amoxicillin/clavulanic acid (AUG) and

Table 1 Test or characteristic for *E. cecorum* isolates (*n* = 82)

Test or characteristic	*E. cecorum* isolates from clinical cases
Hemolysis	α (strong)
Gram-staining	Gram-positive
Cell morphology	ovoid cocci (single, double or short chains)
Catalase-production	negative
Oxidase-production	negative
Yellow pigment-production	negative
Lancefield group D	negative
Motility	negative
Halotolerance (6.5 % NaCl)	limited growth
Growth at:	% positive (n)
4 °C	100 % (82)
10 °C	98.8 % (81)
45 °C	74.4 % (61)
Survival at 60 °C for:	% positive (n)
15 min	76.8 % (63)
30 min	64.6 % (53)
1 h	54.9 % (45)
Survival at 70 °C for:	% positive (n)
15 min	36.6 % (30)
30 min	15.9 % (13)
1 h	0 % (0)

Table 2 Results of *E. cecorum* (*n* = 82) growth on different media

Medium	Observed growth (YES/NO)	Description of colonies of *Enterococcus cecorum*
Columbia Agar with 5 % Sheep Blood (CA)	YES	Small, round, white-grey colonies with α-hemolysis
Columbia CNA Agar with 5 % Sheep Blood	YES	Small, grayish colonies with α-hemolysis, resistant to two antibiotics colistin and nalidixic acid
Edwards Agar with 5 % sheep blood	YES	Blue-grayish coloured colonies with α-hemolysis
Bile Esculin Azide Agar (Enterococcosel Agar)	YES	Colonies beige with strong black halos
KAA agar (Kanamycine Esculin Azide Agar)	YES (weak)	Brown to black colonies and blackening zones around the colonies
Slanetz and Bartley Agar (with tetrazolium chloride)	NO or poor	Red, maroon colonies
TCC agar with tergitol	NO	Total inhibition
Tellurite Agar (potassium tellurite)	NO	Total inhibition

penicillin (PG), nitrofurantoin (NI), and linezolid (LZD). The majority of isolates were susceptible to ampicillin (AP), and high level gentamicin (GM) (*n* = 81; 99 %), chloramphenicol (C) (*n* = 79; 96 %), vancomycin (VA) (*n* = 75; 91 %). The lower level susceptibility was to erythromycin (E) (*n* = 42; 51 %), tetracycline (TEC) (*n* = 24; 29 %), teicoplanin (T) and doxycycline (DXT) (*n* = 11; 13 %). Most isolates noted intermediate susceptibility to TEC (*n* = 53; 65 %) with 29 % susceptible and 6 % resistant. None of clinical *E. cecorum* isolates was susceptible to enrofloxacin (71 resistant isolates). A high percentage of antimicrobial resistance was also observed to teicoplanin (T) (*n* = 70; 85 %), and doxycycline (DXT) (*n* = 68; 83 %). Linezolid resistance among *E. cecorum* isolates was not found. Of the vancomycin resistance genes tested by PCR, *vanA* gene was present in one strain, *vanB* gene was not detected.

Molecular identification

The obtained sequences *sodA* gene fragment showed similarity to *E. cecorum* (BLAST database) and allowed for identification strains. Dendrogram showed the genetic similarity between reference strain of *E. cecorum* and clinical isolates based on the *sodA* gene sequences (Fig. 2). Phylogenetic analysis supported the separation of clinical isolates into three main groups (A, B, C). Genetic distances between groups ranged from 0.00 to 0.04 (Table 5). The group A comprised 69 strains (CB *n* = 43, BB *n* = 15, CL *n* = 10, G *n* = 1) and had one subgroup (A') with 5 strains (BB *n* = 4, CB *n* = 1). Five CB isolates were clustered together in the group B, and three isolates (BB, G, T) in the group C (all isolates from 2014). Among all groups, the group C revealed the highest values of genetic distance with B group (0.04) and with reference strain (0.03).

PFGE

The PFGE analysis (based on >80 % similarity index) of 82 clinical *E. cecorum* isolates exhibited 21 pulsotypes (A-U) with 60 strains (41 CB, 10 BB, 8 CL, 1 G) (Table 6,

Fig. 3). The highest degree of band similarity (>90 %) was demonstrated in pulsotype B (with two CB isolates) and in pulsotype S (with G and CB isolate). Pulsotype M was the predominant type, and included 8 isolates (8/60, 13.3 %), then E, L, T pulsotypes (each included 4 isolates). However, 11 of the 21 pulsotypes included only 2 isolates. Twenty isolates (20/60, 33.3 %) representing CB flocks (20/41; 48.8 %), were distributed among 8 pulsotypes: A, B, C, D, P (each 3.33 %), F, K (each 5 %), L (6.6 %). The majority of BB isolates (7/10, 70 %) were clustered with CB isolates (13) in distinct 6 pulsotypes (E, G, I, M, Q, R). Among isolates representing CL flocks (8/60, 13.3 %), three of these (3/8, 37.5 %) were clustered in one profile (L). Three pulsotypes (H, N, O) were created by clustering both CB (5) and CL (3) isolates. Generally, no clear temporal and geographical clustering was visible, but with the exceptions of 7 pulsotypes (A, D, I, L, M, P, U).

Discussion

In order to characterize clinical *E. cecorum*, we investigated 82 strains isolated from clinical samples originated from different poultry flocks (1 isolate per flock). Our observations were consistent with reports on a succession of disease outbreaks in broiler flocks raised in the intensive farming systems [9]. Previously, clinical *E. cecorum* was not described in commercial chicken layers or geese flocks. We found that the problem may affect hens or other bird species than chicken. Our results were consistent with the literature in regarding on certain characteristics traditionally considered to be typical for the genus *Enterococcus* or *E. cecorum* including intestinal isolates of poultry origin [28, 29]. According to the literature, *E. cecorum* are often NaCl sensitive [17, 30], and intestinal *E. cecorum* of poultry may be also NaCl-resistant [28]. In our study, clinical isolates appeared to be less salt-tolerant, however no complete inhibition of growth was observed. Authors suggested possible higher ability to survive clinical *E. cecorum* in saline environment or even higher resistance to chlorine-based disinfectants.

Table 3 Percent of positive reactions (%) in rapid ID 32 STREP (bioMérieux, France) for clinical *E. cecorum* isolates in this study (*n* = 82) compared with standard isolates (manufacturers recommendations) and control strain (*E. cecorum* ATCC 43198)

Parameter	% Positive reactions in rapid ID 32 STREP for *Enterococcus cecorum*		
	Clinical isolates (this study) % (n)	Manufacturers recommendations %	Reference strain *E. cecorum* ATCC 43198 (+/-)
ADH	0 (0)	0	-
βGLU	100 (82)	100	+
βGAR	73 (60)	11	-
βGUR	94 (77)	88	+
αGAL	99 (81)	100	+
PAL	71 (58)	94	+
RIB	99 (81)	98	+
MAN	12 (10)	38	-
SOR	10 (8)	11	-
LAC	88 (72)	100	+
TRE	99 (81)	100	+
RAF	100 (82)	88	+
VP	60 (49)	66	-
APPA	0 (0)	0	-
βGAL	80 (66)	33	-
PYRA	0 (0)	0	-
βNAG	82 (67)	88	+
GTA	89 (73)	94	+
HIP	0 (0)	1	-
GLYG	12 (10)	27	-
PUL	4 (3)	0	-
MAL	98 (80)	100	+
MEL	98 (80)	98	+
MLZ	88 (72)	55	+
SAC	100 (82)	100	+
LARA	0 (0)	0	-
DARL	1 (1)	0	-
MβDG	100 (82)	98	+
TAG	65 (53)	64	+
βMAN	17 (14)	41	-
CDEX	100 (82)	66	+
URE	6 (5)	0	-

ADH (arginine dihydrolase), βGLU (β-glucosidase), βGAR (β-galactosidase), βGUR (β-glucuronidase), αGAL (α-galactosidase), PAL (alkaline phosphatase), RIB (ribose), MAN (mannitol), SOR (sorbitol), LAC (lactose), TRE (trehalose), RAF (rafinose), VP (Voges Proskauer, aceton production), APPA (alanyl-phenylalanyl-proline arylamidase), βGAL (β-galactosidase), PYRA (pyroglutamic acid arylamidase), βNAG (N-acetyl-β-glucosaminidase), GTA (glycyl-tryptophan arylamidase), HIP (hydrolysis of hipurate), GLYG (glycogen), PUL (pullulane), MAL (maltose), MEL (melibiose), MLZ (melezitose), SAC (saccharose), LARA (L-arabinose), DARL (D-arabitol), CDEX (cyclodextrin), MβDG (methyl-βD-glucopyranoside), TAG (tagatose), βMAN (β-mannosidase), URE (urease)

Previous research demonstrated no growth of poultry cecal *E. cecorum* on Slanetz medium, and on KAA agar [1], while clinical strains showed variable growth on these media. The growth was clearly more abundant on bile esculine azide agar than on esculin azide agar with kanamycin. Based on results, we suggested that complete growth inhibition on a solid medium supplemented with tergitol or with potassium tellurite may be used in identification of this enterococcal species. According to the literature, *Enterococcus* species are able to survive a range of stresses and hostile environments [31], but *E. cecorum* was described as unable to grow at 10 °C or survive 30 min at 60 °C [5, 29]. In contrast to above authors, clinical isolates were able to grow at low temperatures (4 °C, 10 °C) and some of them might survive even longer heating at 60 °C for 1 h and even 70 °C for

	1	2	3	4	5	6	7	8	9	10	11	12
A	A1 Water (Control)	A2 α-Cyclodextrin (100%)	A3 β-Cyclodextrin (0%)	A4 Dextrin (100%)	A5 Glycogen (0%)	A6 Inulin (0%)	A7 Mannan (0%)	A8 Tween 40 (0%)	A9 Tween 80 (0%)	A10 N-Acetyl-D-Glucosamine (100%)	A11 N-Acetyl-D-Mannosamine (100%)	A12 Amygdalin (92.3%)
B	B1 L-Arabinose (0%)	B2 D-Arabitol (0%)	B3 Arbutin (100%)	B4 D-Cellobiose (100%)	B5 D-Fructose (100%)	B6 L-Fucose (0%)	B7 D-Galactose (100%)	B8 D-Galacturonic Acid (0%)	B9 Gentiobiose (100%)	B10 D-Gluconic Acid (0%)	B11 α-D-Glucose (100%)	B12 m-Inositol (0%)
C	C1 α-D-Lactose (84.6%)	C2 Lactulose (69.2%)	C3 Maltose (100%)	C4 Maltotriose (100%)	C5 D-Mannitol (0%)	C6 D-Mannose (100%)	C7 D-Melezitose (92.3%)	C8 D-Melibiose (100%)	C9 α-Methyl-D-Galactoside (0%)	C10 β-Methyl-D-Galactoside (15.4%)	C11 3-Methyl-D-Glucose (0%)	C12 α-Methyl-D-Glucoside (0%)
D	D1 β-Methyl-D-Glucoside (92.3%)	D2 α-Methyl-D-Mannoside (0%)	D3 Palatinose (69.2%)	D4 D-Psicose (100%)	D5 D-Raffinose (100%)	D6 L-Rhamnose (0%)	D7 D-Ribose (76.9%)	D8 Salicin (100%)	D9 Sedoheptulosan (0%)	D10 D-Sorbitol (0%)	D11 Stachyose (100%)	D12 Sucrose (100%)
E	E1 D-Tagatose (0%)	E2 D-Trehalose (100%)	E3 Turanose (0%)	E4 Xylitol (0%)	E5 D-Xylose (0%)	E6 Acetic Acid (0%)	E7 α-Hydroxybutyric Acid (0%)	E8 β-Hydroxybutyric Acid (0%)	E9 γ-Hydroxybutyric Acid (0%)	E10 p-Hydroxy Phenyl Acetic Acid (0%)	E11 α-Ketoglutaric Acid (0%)	E12 α-Ketovaleric Acid (0%)
F	F1 Lactamide (0%)	F2 D-Lactic Acid Methyl Ester (0%)	F3 L-Lactic Acid (0%)	F4 D-Malic Acid (0%)	F5 L-Malic Acid (0%)	F6 Pyruvic Acid Methyl Ester (100%)	F7 Succinic Acid Mono-Methyl Ester (0%)	F8 Propionic Acid (0%)	F9 Pyruvic Acid (0%)	F10 Succinamic Acid (0%)	F11 Succinic Acid (0%)	F12 N-Acetyl-L-Glutamic Acid (0%)
G	G1 L-Alaninamide (0%)	G2 D-Alanine (0%)	G3 L-Alanine (0%)	G4 L-Alanyl-glycine (0%)	G5 L-Asparagine (0%)	G6 L-Glutamic Acid (0%)	G7 Glycyl-L-Glutamic Acid (0%)	G8 L-Pyroglutamic Acid (0%)	G9 L-Serine (0%)	G10 Putrescine (0%)	G11 2,3-Butanediol (0%)	G12 Glycerol (0%)
H	H1 Adenosine (100%)	H2 2'-Deoxy Adenosine (61.5%)	H3 Inosine (92.3%)	H4 Thymidine (92.3%)	H5 Uridine (92.3%)	H6 Adenosine-5'-Monophosphate (53.8%)	H7 Thymidine-5'-Monophosphate (0%)	H8 Uridine-5'-Monophosphate (53.8%)	H9 D-Fructose-6-Phosphate (0%)	H10 α-D-Glucose-1-Phosphate (0%)	H11 D-Glucose-6-Phosphate (0%)	H12 D-L-α-Glycerol Phosphate (0%)

Fig. 1 Percent of positive profiles for *Enterococcus cecorum* in Biolog GP2 MicroPlate™

Table 4 Antibiotics resistance patterns of *E. cecorum* strains isolated from clinical cases of different bird species

Antibiotics (n, number of antibiotics)	% (n) of resistant isolates
ENF/E/T/DXT/TEC (5)	2.5 (2)
ENF/E/T/DXT (4)	33.3 (27)
ENF/T/DXT/TEC (4)	1.2 (1)
ENF/T/DXT/AP (4)	1.2 (1)
ENF/T/DXT (3)	33.3 (27)
T/E/DXT (3)	4.9 (4)
ENF/E/T (3)	2.5 (2)
ENF/E/GM (3)	1.2 (1)
ENF/DXT/VA (3)	1.2 (1)
T/DXT (2)	6.2 (5)
ENF/E (2)	2.5 (2)
ENF/TEC (2)	1.2 (1)
T/TEC (2)	1.2 (1)
ENF (1)	7.4 (6)
0	1.2 (1)

Ampicillin (AP 10 μg), enrofloxacin (ENF 5 μg), tetracycline (TEC 30 μg), Doxycycline (DXT 30 μg), erythromycin (E 15 μg), teicoplanin (T 30 μg), Vancomycin (VA 30 μg), high level gentamicin (GM 120 μg)

30 min. The results may indicate to the possibly longer survival *E. cecorum* at more extreme temperatures in the poultry house environment.

We confirmed the efficacy of two biochemical systems for identification poultry-origin *E. cecorum* strains. Instead of doubtful or unacceptable profile in API, all strains were properly recognized by *sod*A gene sequencing. We found, that almost all clinical strains gave positive reactions in 10 biochemical tests, and negative in 5 tests (API). Similar results were reported for other *E. cecorum* including commensal or reference strains with some exceptions [1, 17, 28, 32]. We observed that all of strains were able to metabolise α-cyclodextrin. Makrai et al. [10] observed differences among clinical isolates in metabolism of both α- and β-cyclodextrin. We noted relatively high positive reactions for βGAL, βGAR, VP, opposed to reference *E. cecorum* strain and despite the discrepancies in the literature [17]. In contrast to other studies [2, 17], some clinical *E. cecorum* showed ability to produce urease, β-mannosidase, and metabolize glycogen. On the other hand, results for β-mannosidase, glycogen, mannitol were lower for clinical isolates than reported for standard strains. Our results were consistent with Borst et al. [33] who noted that pathogenic *E. cecorum* isolates are more deficient in mannitol metabolism. Recently molecular aspects for the defect mannitol metabolism in pathogenic strains were investigated [34].

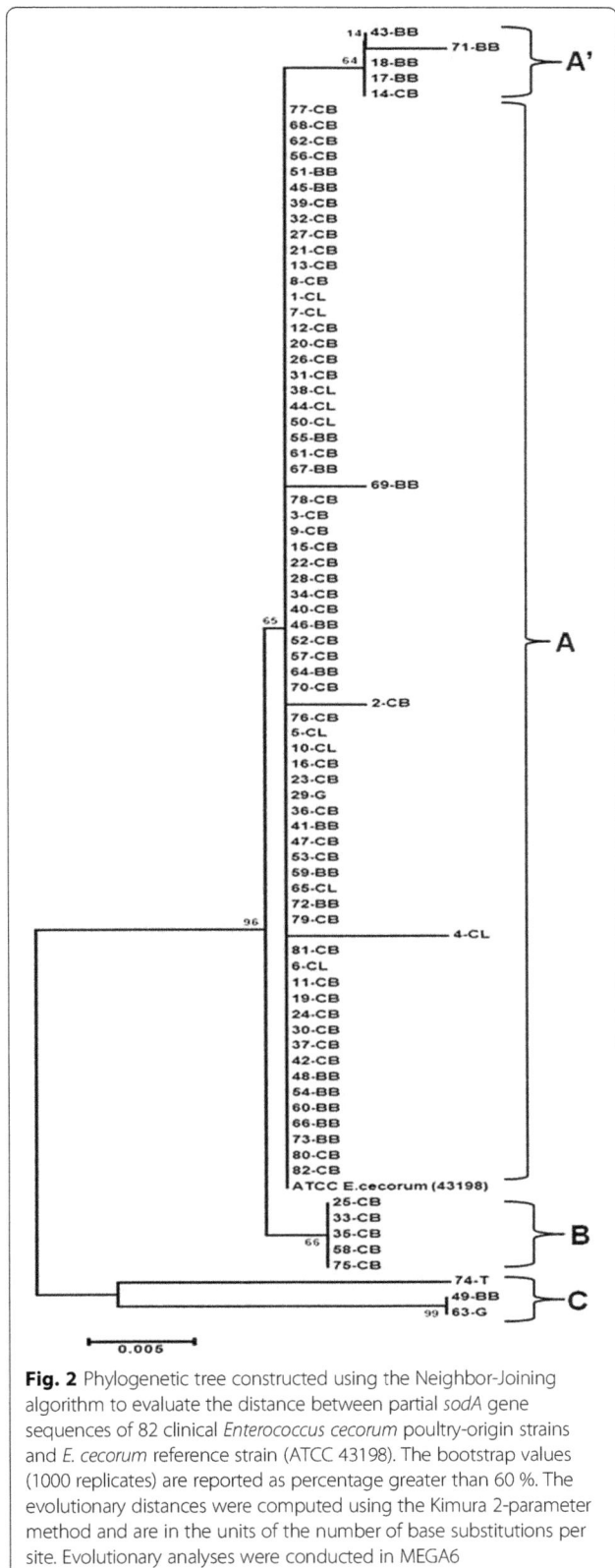

Fig. 2 Phylogenetic tree constructed using the Neighbor-Joining algorithm to evaluate the distance between partial *sodA* gene sequences of 82 clinical *Enterococcus cecorum* poultry-origin strains and *E. cecorum* reference strain (ATCC 43198). The bootstrap values (1000 replicates) are reported as percentage greater than 60 %. The evolutionary distances were computed using the Kimura 2-parameter method and are in the units of the number of base substitutions per site. Evolutionary analyses were conducted in MEGA6

Table 5 Kimura 2-parameter genetic distances between groups of clinical *E. cecorum* (A, A', B, C) and reference strain (ATCC 43198)

	1	2	3	4	5
1. A					
2.A'	0.00				
3.B	0.00	0.01			
4.C	0.03	0.04	0.03		
5.ATCC (43198)	0.00	0.00	0.00	0.03	

Based on comparative analysis of our results with study of Makrai et al. [10], it could be assumed, that all clinical *E. cecorum* may metabolise 18 carbon sources (adenosine, arbutin, D-cellobiose, dextrin, D-fructose, D-mannose, D-psicose, D-raffinose, D-trehalose, gentiobiose, maltose, maltotriose, N-acetyl-D-glucosamine, N-acetyl-β-D-mannosamine, pyruvic acid methyl ester, salicin, sucrose, α-D-glucose). Similarly to above mentioned authors, clinical isolates may give differences in 3 tests: α-D-lactose, D-ribose, 2-deoxy adenosine.

Recently several potential mediators of virulence were found in pathogenic *E. cecorum* isolated from chickens in the southeast US. These virulence determinants conserved in pathogenic EC were found to be similar to those utilized by other medically important enterococci [33]. In the present study, only 32.9 % clinical *E. cecorum* strains contained one or more virulence genes. *E. cecorum* from chicken flocks contained mainly *asa1/gelE/ace* genes. The pathogenicity of *E. cecorum* may be associated with other species-specific virulence factors. Similar observations were presented by Jackson et al. [35] who detected only few virulence genes among US *E. cecorum* isolates, and the incidences of virulence determinants tested were lower than ours. In our study the most of positive isolates contained two *asa1/gelE* or four *asa1/gelE/efaA/ace* virulence genes. We speculated about possible linkage between *asa1/gelE* (74 % of all virulence positive isolates) or *efaA/ace* (33.3 % of all virulence positive isolates) in clinical *E. cecorum*. It may have impact on pathogenesis and clinical course of infection.

Because none of the investigated strains harbored *hyl* gene, we suggest that this virulence determinant may be not widespread among clinical isolates. Our results were consistent with other authors who described the lack of *hyl* in *E. cecorum* from poultry carcass rinsates, diseased chickens [35] and pigeons [5]. According to the literature, hyaluronidase is a degradative enzyme that is associated with tissue damage. Among *Enteroccocus* species the *hyl* gene has been reported more often in ampicillin-resistant VRE *E. faecium* isolates [36]. We suggest that

Table 6 Twenty one PFGE profiles (A-U) of clinical isolates *E. cecorum* derived from poultry in Poland between 2011-2014

Pulsotype	Poultry type	No. of strain	Year	Poland's voivodeship	Number of isolates of each pulsotype	% Similarity (>80 %)
A	CB	76	2014	Greater Poland	2	84.2
	CB	39	2014	Greater Poland		
B	CB	32	2014	Greater Poland	2	92.3
	CB	31	2013	Świętokrzyskie		
C	CB	3	2011	Greater Poland	2	84.2
	CB	23	2012	Silesian		
D	CB	80	2014	Masovian	2	81.8
	CB	57	2014	Masovian		
E	BB	60	2014	Warmian-Masurian	4	85.7
	CB	52	2014	Pomeranian		
	CB	47	2014	Greater Poland		
	CB	27	2013	Greater Poland		
F	CB	81	2014	Masovian	3	84.2
	CB	53	2014	Pomeranian		
	CB	34	2014	Greater Poland		
G	BB	46	2014	Masovian	2	87.0
	CB	2	2011	Greater Poland		
H	CL	44	2014	Podlaskie	2	84.6
	CB	28	2013	Greater Poland		
I	CB	82	2014	Masovian	2	81.2
	BB	66	2014	Opolskie		
J	BB	67	2014	Masovian	3	80.1
	BB	48	2014	Masovian		
	CL	4	2011	Greater Poland		
K	CB	42	2014	Pomeranian	3	80.8
	CB	40	2014	Pomeranian		
	CB	21	2012	Warmian-Masurian		
L	CB	70	2014	Masovian	4	84.3
	CB	62	2014	Masovian		
	CB	61	2014	Masovian		
	CB	20	2014	Masovian		
M	CB	77	2014	Kuyavian-Pomeranian	8	82.1
	CB	56	2014	Greater Poland		
	CB	35	2014	Greater Poland		
	CB	25	2014	Pomeranian		
	CB	58	2014	Pomeranian		
	CB	75	2014	Lodzkie		
	BB	51	2014	West Pomeranian		
	BB	55	2014	West Pomeranian		
N	CB	68	2014	Masovian	3	84.0
	CL	65	2014	Masovian		
	CB	13	2011	Greater Poland		
O	CB	30	2013	Masovian	3	81.8
	CB	19	2011	Świętokrzyskie		

Table 6 Twenty one PFGE profiles (A-U) of clinical isolates *E. cecorum* derived from poultry in Poland between 2011-2014 *(Continued)*

	CL	10	2011	Greater Poland		
P	CB	78	2014	Greater Poland	2	86.7
	CB	37	2014	Greater Poland		
Q	BB	45	2014	West Pomeranian	2	81.3
	CB	11	2011	Greater Poland		
R	BB	71	2014	Masovian	2	81.5
	CB	14	2011	Greater Poland		
S	G	29	2013	Greater Poland	2	90.3
	CB	22	2012	Pomeranian		
T	BB	59	2014	Warmian-Masurian	4	84.1
	CL	6	2011	Greater Poland		
	CB	16	2011	Greater Poland		
	CB	24	2012	Masovian		
U	CL	1	2012	Greater Poland	3	87.0
	CL	5	2012	Greater Poland		
	CL	7	2011	Greater Poland		

hyl is not specific for *E. cecorum* and could has minor role in pathogenicity of *E. cecorum*, however more studies are needed to elucidate this aspect.

The present study showed lack of correlation between the presence of *gelE* gene and its expression. The literature provide no data in regard this aspect on *E. cecorum*, however similar observations are available for *E. faecalis* [37].

Generally, pathogenic isolates from poultry were found to be significantly more drug resistant than commensal strains [33]. In the present study almost all of clinical isolates showed high level of antibiotic resistance and 91.5 % of them showed multidrug resistance (resistance to ≥ 2 antimicrobials). Other authors identified lower multidrug resistance in *E. cecorum* from carcass rinsates and diseased poultry, however the panel of used antibicrobials were not completely the same [35]. Affected flocks were treated against *E. cecorum* usually with amoxicillin, doxycycline or enrofloxacin. All of the above antibiotics were tested in this study. Similarly to other authors, the overwhelming majority of the isolates were susceptible to penicillin, which appear to be drug of choice [4, 5, 7, 10, 14, 15]. However, the majority of *E. cecorum* was resistant to enrofloxacin > teicoplanin > doxycycline > erythromycin. Our results were opposed to clinical *E. cecorum* from other countries, in which sensitivity to enrofloxacin (in Germany, Holland, Hungary, South Africa), doxycycline (in Germany, Hungary) and macrolides (in Belgium, Germany) were identified [4, 5, 7, 10, 14, 15]. Similarly to the isolates from Canada, USA, Holland and Belgium, clinical *E. cecorum* from Poland showed the increased resistance to tetracycline or erythromycin (macrolides) [4, 7, 27, 33, 35].

This antimicrobial resistance pattern may be common and characteristic for pathogenic *E. cecorum*. The presented study indicated on the presence clinical *E. cecorum* (1.2 %) with the resistance to vancomycin (VRE) and to high level gentamicin (HLGR). Similarly to Jackson et al. [35], we found out that none of the isolates were resistant to linezolid. According to the literature, enterococci have both an intrinsic and acquired resistance to antibiotics which complicate treatment of infections. The acquired resistance includes resistance to i.a. chloramphenicol, tetracyclines, fluoroquinolones, aminoglycosides (high levels), and vancomycin. Enterococci have demonstrated a huge potential for acquiring and disseminating resistant genes. We found, that the high level of the resistance to enrofloxacin, doxycycline, tetracycline in *E. cecorum* isolates is probably related to the wide use of these antibiotics in poultry production. In previous years erythromycin was also commonly applied in the therapy of poultry. Other authors confirmed the presence of resistance genes (including *van* genes) among *E. cecorum* from broilers or retail chicken meat [38, 39]. We suggest, that poultry may play an important role as reservoirs of antibiotic resistant *E. cecorum* in the environment. However, further studies are needed to investigate the resistance genes in clinical isolates.

In the present study *sodA* gene fragment was successfully used to confirm phenotypic identification of *E. cecorum*, however it was not sufficiently discriminative to differentiate them from each other. In the collection it was possible to distinguish for three phylogenetic groups and one subgroup. The strains from group B showed the same type of production (CB), year of

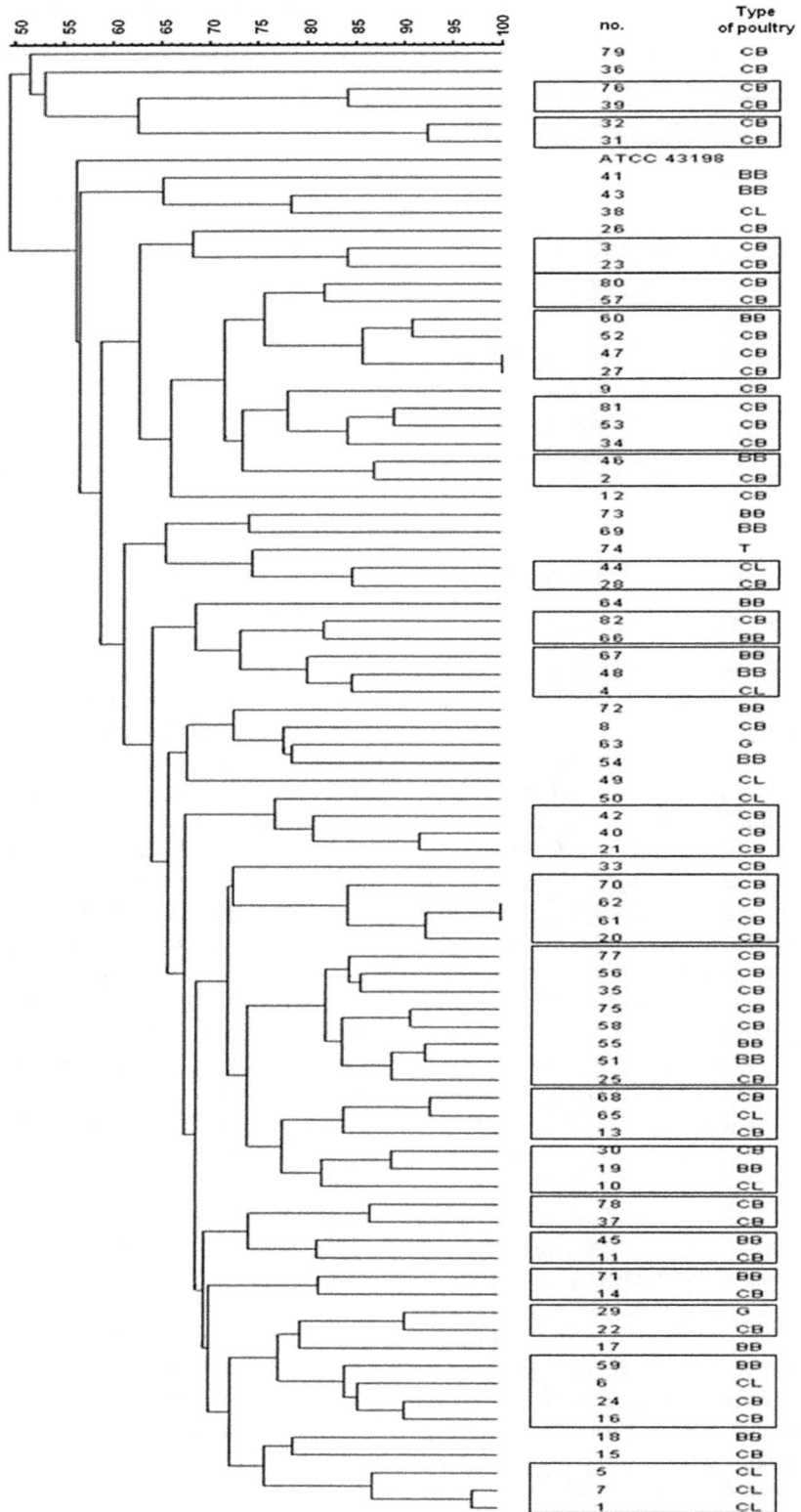

Fig. 3 Results of pulsed field gel electrophoresis (PFGE) examination of *Enterococcus cecorum* clinical isolates. Dendrogram based on Dice coefficient with 1 % position tolerance. Cut-off value of 80 % similarity was used to assign the pulsotypes

isolation, virulence determinants and multidrug resistance pattern, but different geographical origin; 80 % of them belong to pulsotype M. The strains of group C shared only the same year of isolation, virulence and multidrug resistance patterns. The low genetic distance (based on *sod*A gene sequences) indicated on the very close genetic relationships between clinical *E. cecorum*. No clear genetic differences were observed between clinical strains and reference strain.

Recent data indicated that pathogenic *E. cecorum* from the southeast US were clonal, however comparative genomic analysis revealed fundamental differences in their genomes [34]. According to the previous report, isolates recovered from spinal abscesses were highly similar and could be detected by using PFGE [33]. In our study, PFGE results showed the genetic heterogeneity between clinical *E. cecorum* isolates, that is consistent with the other studies [18]. Therefore, the usage of PFGE in distinguishing pathogenic strains may be difficult and limited. This genetic diversity was seen between poultry flocks, however some clustering was visible in relation of type of production (CB, CL). Moreover, some temporal and geographical clustering was visible. Many CB isolates from the same year and geographical origin were clustered together (pulsotype A, D, L, P) indicating their close genetic relationship. Some CL isolates from the same location but different years were grouped into a single pulsotype (U) indicating on the possible horizontal transmission among CL flocks in this area. We found that CB and BB isolates from the same year which were clustered together into separate pulsotypes (I, M). Based on the relatively close relationship between isolates from geese and chicken flocks, it could be assumed that isolates from the single clonal lineage may cause outbreaks in different bird species. The results may suggest the transmission of potential disease-causing *E. cecorum* between flocks.

Conclusions

These data indicate that several, widely distributed pathogenic *E. cecorum* clones seemed to be responsible for infection cases found in different poultry types. The isolates causing infection in different CB flock in the same year and region may be somewhat genetically distinct from each other and from those that cause disease in CL or BB flocks in the same year and region. Phenotypically, clinical isolates were generally found to be very similar, however some properties or characteristics described in some isolates were not found in others. The study presented here is the first in Poland as well as one of the few in Europe which provides phenotypic and genotypic characterization of *E. cecorum* isolates associated with disease outbreaks in poultry flocks. Further research needs to focus on finding new virulence determinants of *E. cecorum* and recognition of transmission routes.

Additional files

Additional file 1: Alignment of partial *sod*A gene sequences from *E. cecorum* isolates and reference strain (ATCC 43198). Nucleotide differences are specified by the nucleotide, while dot represented no nucleotide changing.

Additional file 2: Alignment of partial 16S rRNA gene sequences from *E. cecorum* isolates and reference strain (ATCC 43198). Nucleotide differences are specified by the nucleotide, while dot represented no nucleotide changing.

Abbreviations
BB, broiler breeder flocks; CA, Columbia agar with 5 % sheep blood; CB, chicken broiler flocks (commercial broilers); CL, commercial layer flocks; *E. cecorum, Enterococcus cecorum*; ES, enterococcal spondylitis; G, geese flocks; HLGR, high level gentamicin resistance; PFGE, Pulsed Field Gel Electrophoresis; T, turkey flock; VRE Vancomycin-Resistant Enterococcus

Acknowledgements
The authors thank mgr Beata Sienkiewicz for excellent technical assistance. The authors are very grateful veterinary laboratories and private practice veterinarians for help in collecting strains: Microbiological Laboratory SGGW, SLW Biolab Ostróda, Lab-Vet Sp. z o.o. Tarnowo Podgórne, Vet-Lab Brudzew, Vetdiagnostica Solec Kujawski, Animal Pharma, Wet-Net s.c. Giżycko, Gabinet Weterynaryjny "Gallus" Sylwester Barabasz, "Spec-Drób" Mariusz Lorek, and DVM: Anna Biegańska, Magdalena Wiczk, Tomasz Nowak, Ismaila Massaly, Mirosław Berezowski, Paweł Tubielewicz, Natalia Bednarz.

Funding
This work was financially supported by grant no. 505-10-023700-L00183-99 at Department of Pathology and Veterinary Diagnostics, Faculty of Veterinary Medicine, Warsaw University of Life Sciences-SGGW, Poland. The analysis by Biolog system was supported by Department of Microbiology and Infectious Diseases, Faculty of Veterinary Science, Szent István University in Budapest, Hungary.

Authors' contributions
BD conceived and designed the study, collected the strains, performed all works, analysis and interpretation of data, wrote manuscript. DCC input in PFGE. LM contributed to perform analysis in Biolog system. PS gave conceptual advice and additional inputs in study project, contributed materials. All authors read and approved the final manuscript.

Competing interests
The authors declare that they have no competing interests.

Consent to publication
Not applicable.

Ethics approval and consent to participate
Poultry samples were collected for laboratory diagnosis by as part of the usual clinical practice on farms, and Polish ethical guidelines (Dz.U. 2015 poz. 266) and animal welfare regulations were strictly respected. Veterinarians who provided healthcare for poultry flocks were in contact with the owners, and gave an oral informed consent and acceptance for using obtained isolates for research studies. As this study was focused on bacterial isolates collected from routine samples, approval from Ethics Committee at the Warsaw University of Life Sciences University was not necessary.

Author details
[1]Department of Pathology and Veterinary Diagnostics, Faculty of Veterinary Medicine, Warsaw University of Life Sciences-SGGW, Nowoursynowska 159c St., Warsaw 02-776, Poland. [2]Department of Preclinical Sciences, Faculty of Veterinary Medicine, Warsaw University of Life Sciences-SGGW, Ciszewskiego 8 St., Warsaw 02-786, Poland. [3]Department of Microbiology and Infectious Diseases, Faculty of Veterinary Science, Szent István University, Hungária krt. 23-25, Budapest H-1143, Hungary.

References

1. Devriese LA, Dutta GN, Farrow JAE, van de Kerckhove A, Phillips BA. *Streptococcus cecorum*, a new species isolated from chickens. Int J Syst Bacteriol. 1983;33:772–6.
2. Greub G, Devriese LA, Pot B, Dominguez J, Bille J. *Enterococcus cecorum* septicemia in a malnourished adult patient. Eur J Clin Microbiol Infect Dis. 1997;16:594–8.
3. Warnke P, Köller T, Stoll P, Podbielski A. Nosocomial infection due to *Enterococcus cecorum* identified by MALDI-TOF MS and Vitek 2 from a blood culture of a septic patient. Eur J Microbiol Immunol (Bp). 2015;5:177–9.
4. Devriese LA, Cauwerts K, Hermans K, Wood AM. *Enterococcus cecorum* septicaemia as a cause of bone and joint lesions resulting in lameness in broiler chickens. Flemish Vet J. 2002;71:219–21.
5. Jung A, Teske L, Rautenschlein S. *Enterococcus cecorum* infection in a racing pigeon. Avian Dis. 2014;58:654–8.
6. Wood AM, MacKenzie G, McGiliveray NC, Brown L, Devriese LA, Baele M. Isolation of *Enterococcus cecorum* from bone lesions in broiler chickens. Vet Rec. 2002;150:27.
7. De Herdt P, Defoort P, Van Steelant J, Swam H, Tanghe L, Van Goethem S, Vanrobaeys M. *Enterococcus cecorum* osteomyelitis and arthritis in broiler chickens. Vlaams Diergen Tijds. 2008;78:44–8.
8. Stalker MJ, Brash ML, Weisz A, Ouckama RM, Slavic D. Arthritis and osteomyelitis associated with *Enterococcus cecorum* infection in broiler and broiler breeder chickens in Ontario, Canada. J Vet Diagn Invest. 2010;22: 643–5.
9. Armour NK, Collett SR, Williams SM. *Enterococcus cecorum* – related arthritis and osteomyelitis in broilers and broiler breeders. Poult Inf Profess. 2011;17:1–7.
10. Makrai L, Nemes C, Simon A, Ivanics E, Dudás Z, Fodor L, Glávits R. Association of *Enterococcus cecorum* with vertebral osteomyelitis and spondylolisthesis in broiler parent chicks. Acta Vet Hung. 2011;59:11–21.
11. Dolka B, Szeleszczuk P. Enterococcal vertebral osteoarthritis in chickens. Med Weter. 2012;68:157–62.
12. Robbins KM, Suyemoto MM, Lyman RL, Martin MP, Barnes HJ, Borst LB. An outbreak and source investigation of enterococcal spondylitis in broilers caused by *Enterococcus cecorum*. Avian Dis. 2012;56:768–73.
13. Szeleszczuk P, Dolka B, Żbikowski A, Dolka I, Peryga M. First case of enterococcal spondylitis in broiler chickens in Poland. Med Weter. 2013;69: 298–303.
14. Aitchison H, Poolman P, Coetzer M, Griffiths C, Jacobs J, Meyer M, Bisschop S. Enterococcal-related vertebral osteoarthritis in South African broiler breeders: A case report. J S Afr Vet Assoc. 2014;85:1077.
15. Jung A, Rautenschlein S. Comprehensive report of an *Enterococcus cecorum* infection in a broiler flock in Northern Germany. BMC Vet Res. 2014;10:311.
16. Murray BE. The life and times of the *Enterococcus*. Clin Microbiol Rev. 1990;3: 46–65.
17. Manero A, Blanch AR. Identification of *Enterococcus* spp. with a biochemical key. Appl Environ Microbiol. 1999;65:4425–30.
18. Wijetunge DS, Dunn P, Wallner-Pendleton E, Lintner V, Lu H, Kariyawasam S. Fingerprinting of poultry isolates of *Enterococcus cecorum* using three molecular typing methods. J Vet Diagn Invest. 2012;24:1166–71.
19. Tenover FC, Arbeit RD, Goering RV, Mickelsen PA, Murray BE, Persing DH, Swaminathan B. Interpreting chromosomal DNA restriction patterns produced by pulsed-field gel electrophoresis: criteria for bacterial strain typing. J Clin Microbiol. 1995;33:2233–9.
20. Kense MJ, Landman WJ. *Enterococcus cecorum* infections in broiler breeders and their offspring: molecular epidemiology. Avian Pathol. 2011;40:603–12.
21. Ke D, Picard FJ, Martineau F, Menard C, Roy PH, Ouellette M, Bergeron MG. Development of a PCR assay for rapid detection of enterococci. J Clin Microbiol. 1999;37:3497–503.
22. Jackson CR, Fedoka – Cray PJ, Barett JB. Use of a genus- and species-specific multiplex PCR for identication of enterococci. J Clin Microbiol. 2004;42: 3558–65.
23. Tsai JC, Hsueh PR, Lin HM, Chang HJ, Ho SW, Teng LJ. Identification of clinically relevant enterococcus species by direct sequencing of groES and spacer region. J Clin Microbiol. 2005;43:235–41.
24. Martín-Platero AM, Valdivia E, Maqueda M, Martínez-Bueno M. Characterization and safety evaluation of enterococci isolated from Spanish goats' milk cheeses. Int J Food Microbiol. 2009;132:24–32.
25. CLSI. Performance Standards for Antimicrobial Disk and Dilution Susceptibility Tests for Bacteria Isolated from Animals; Approved Standard –

Fourth Edition. Wayne: Clinical and Laboratory Standards Institute; 2013. PA CLSI document VET01-A4, supplement VET01-S2.
26. Van den Braak N, van Belkum A, van Keulen M, Vliegenthart J, Verbrugh HA, Endtz HP. Molecular characterization of vancomycin-resistant enterococci from hospitalized patients and poultry products in The Netherlands. J Clin Microbiol. 1998;36:1927–32.
27. Boerlin P, Nicholson V, Brash M, Slavic D, Boyen F, Sanei B, Butaye P. Diversity of *Enterococcus cecorum* from chickens. Vet Microbiol. 2012;157: 405–11.
28. Devriese LA, Hommez J, Wijfels R, Haesebrouck F. Composition of the enterococcal and streptococcal intestinal flora of poultry. J Appl Bacteriol. 1991;71:46–50.
29. Devriese LA, Pot B, Collins MD. Phenotypic identification of the genus *Enterococcus* and differentiation of phylogenetically distinct enterococcal species and species groups. J Appl Bacteriol. 1993;75:399–408.
30. Domig KJ, Mayer HK, Kneifel W. Methods used for the isolation, enumeration, characterisation and identification of *Enterococcus* spp. 2. Pheno- and genotypic criteria. Int J Food Microbiol. 2003;55:165–88.
31. Fisher K, Phillips C. The ecology, epidemiology and virulence of *Enterococcus*. Microbiology. 2009;155:1749–57.
32. Devriese LA, Ceyssens K, Haesebrouck F. Characteristics of *Enterococcus cecorum* strains from the intestines of different animal species. Lett Appl Microbiol. 1991;12:137–9.
33. Borst LB, Suyemoto MM, Robbins KM, Lyman RL, Martin MP, Barnes HJ. Molecular epidemiology of *Enterococcus cecorum* isolates recovered from enterococcal spondylitis outbreaks in the southeastern United States. Avian Pathol. 2012;41:479–85.
34. Borst LB, Suyemoto MM, Scholl EH, Fuller FJ, Barnes HJ. Comparative genomic analysis identifies divergent genomic features of pathogenic *Enterococcus cecorum* including a type IC CRISPR-Cas system, a capsule locus, an epa-like locus, and putative host tissue binding proteins. PLoS One. 2015;10:e0121294.
35. Jackson CR, Kariyawasam S, Borst LB, Frye JG, Barrett JB, Hiott LM, Woodley TA. Antimicrobial resistance, virulence determinants and genetic profiles of clinical and nonclinical *Enterococcus cecorum* from poultry. Lett Appl Microbiol. 2015;60:111–9.
36. Vankerckhoven V, Van Autgaerden T, Vael C, Lammens C, Chapelle S, Rossi R, Jabes D, Goossens H. Development of a multiplex PCR for the detection of asa1, gelE, cylA, esp, and hyl genes in enterococci and survey for virulence determinants among European hospital isolates of *Enterococcus faecium*. J Clin Microbiol. 2004;42:4473–9.
37. Olsen RH, Schønheyder HC, Christensen H, Bisgaard M. *Enterococcus faecalis* of human and poultry origin share virulence genes supporting the zoonotic potential of *E. faecalis*. Zoonoses Public Health. 2012;59:256–63.
38. Cauwerts K, Decostere A, De Graef EM, Haesebrouck F, Pasmans F. High prevalence of tetracycline resistance in *Enterococcus* isolates from broilers carrying the erm(B) gene. Avian Pathol. 2007;36:395–9.
39. Harada T, Kawahara R, Kanki M, Taguchi M, Kumeda Y. Isolation and characterization of vanA genotype vancomycin-resistant *Enterococcus cecorum* from retail poultry in Japan. Int J Food Microbiol. 2012;153:372–7.

7

Sero-prevalence, risk factors and distribution of sheep and goat pox in Amhara Region, Ethiopia

Tsegaw Fentie[1], Nigusie Fenta[2], Samson Leta[3]* ⓘ, Wassie Molla[1], Birhanu Ayele[1], Yechale Teshome[4], Seleshe Nigatu[1] and Ashenafi Assefa[1]

Abstract

Background: Sheep pox and goat pox are contagious viral diseases of sheep and goats, respectively. The diseases result in substantial economic losses due to decreased milk and meat production, damage to hides and wool, and possible trade restriction. A study was undertaken in Amhara region of Ethiopia. A cross-sectional study design was used to estimate the sero-prevalence and identify associated risk factors, while retrospective study design was used to assess the temporal and spatial distribution of the disease. A total of 672 serum samples were collected from 30 Kebeles and tested using virus neutralization test.

Results: From a total of 672 sera tested, 104 (15.5%) were positive for sheep and goat pox virus antibody; from which 56 (17%) were sheep and 48 (14%) were goats. The diseases were prevalent in all study zones, the highest sero-prevalence was observed in South Gondar (20.9%) and the lowest in North Gondar and West Gojjam zones (11.9% each). From the potential risk factors considered (species, sex, age, agro-ecology and location); only sex and age were significantly associated ($p < 0.05$) with the diseases in multivariable logistic regression. Female and young animals were at higher risk than their counterparts. From January 2010 to December 2014, a total of 366 outbreaks, 12,822 cases and 1480 deaths due to SP and 182 outbreaks, 10,066 cases and 997 deaths due to GP were recorded in Amhara National Regional State.

Conclusion: Both the serological and the outbreak data revealed that sheep and goat pox is one of the most prevalent and widespread diseases of sheep and goats in the study area. Hence, annual mass vaccination program must be implemented for economic and viable control of sheep and goat pox diseases in the Amhara region in particular and at a national level in general.

Keywords: Amhara region, Sero-prevalence, Sheep and goat pox, Ethiopia

Background

Estimates indicate that about 25.5 million sheep and 24.1 million goats are reared in the sedentary areas of Ethiopia excluding the non-sedentary population of three zones of Afar and six zones of Somali region. This makes Ethiopia be home to one of the largest heads of small ruminants in Africa after Nigeria [1, 2]. However, studies indicate that the current contributions of the livestock subsector which includes small ruminant production to the national economy at either the macro- or micro-level to be limited and below the potential [3].

Infectious diseases are amongst the major factors which limit the production and productivity of small ruminant; sheep pox (SP) and goat pox (GP) being topping the list [4]. Sheep pox and GP are viral diseases of sheep and goats characterized by fever, pyrexia and generalized skin lesions [5]. In susceptible herds, morbidity is 75-100% and case fatality, depending on the virulence of the virus is between 10 and 85% (19). The viruses causing these diseases are members of the genus *Capripoxvirus*, subfamily *Chordopoxvirinae* and family *Poxviridae*. There is close genetic relatedness of these *Capripoxvirus* isolates. The

* Correspondence: samiwude@gmail.com; samson.leta@aau.edu.et
[3]College of Veterinary Medicine and Agriculture, Addis Ababa University, P. O. Box 34, Bishoftu, Ethiopia

diseases induced by strains of SP virus, GP virus and lumpy skin disease virus cannot be differentiated clinically and serologically, including virus neutralization test. However, distinct host preferences exist with most strains of SP virus and GP virus causing more severe disease in the homologous host [6, 7].

The geographical distribution of SP and GP have been shown to extends from Africa, north of the equator to the Middle East and Asia including the former Soviet Union, India and China [4]. Many authors have reported the occurrence of SP and GP from east African countries namely Sudan and Kenya [8–11]. SP and GP are among the most important diseases of sheep and goats in Ethiopia following Peste des petits ruminants (PPR) and contagious caprine pleuropneumonia (CCPP). Questionnaire surveys and case reports indicate an occurrence of the diseases in Ethiopia including Amhara National Regional State (ANRS) [12, 13]. A recent study by Gelaye et al., [14], indicated GP and lumpy skin disease virus to be responsible for the Capripox outbreaks in small ruminants and cattle in different parts of Ethiopia.

Sheep pox and GP are major constraints to trade and introduction of exotic breeds of sheep and goats; hindering efforts to improve local sheep and goats through importation of improved breeds [15]. The diseases are associated with significant production losses because of reduced milk yield, decreased weight gain, increased abortion rates, damage to wool and hides, and increased susceptibility to pneumonia and fly strike, while also being a direct cause of mortality [16].

Despite its considerable economic importance and threats to trade, information on the sero-prevalence and distribution of the diseases in Ethiopia in general and Amhara region, in particular, is absent. A better understanding of its prevalence and distribution would lead to improved disease control measures. Therefore, this study was aimed to estimate the sero-prevalence, assess risk factors and distribution of sheep and goat pox in ANRS.

Methods
Study area
The study has two components: sero-surveillance and retrospective studies. The sero-surveillance study was conducted in western part of the Amhara Regional National State (ANRS) whereas the retrospective study was undertaken in whole ANRS. The ANRS is located in the north western part of Ethiopia between 9°20′ and 14°20′ North latitude and 36° 20′ and 40° 20′ East longitude. The region borders with Tigray in the North, Afar in the East, Oromia in the South and Benishangul-Gumuz in the Southwest and the country of Sudan in the West (Fig. 1). The study area has diverse agro-climatic conditions; ranging from hot lowlands to cold highlands. Areas less than 1500 m.a.s.l. were considered as lowland,

areas ranging from 1500 to 2500 m.a.s.l. were considered as midland and areas greater than 2500 m.a.s.l. were considered as highland. Western Amhara region has five zones (North Gondar, South Gondar, East Gojjam, West Gojjam and Awi). There are 66 districts in the five zones.

Study population and sampling strategy
An estimated 6.6 million sheep and goats are reared in western Amhara region [1]. Sheep and goats that were kept under the extensive farming system, aged greater than 5 months were considered for the sero-surveillance study. Information about Capripox vaccination practice was collected from district animal husbandry departments and livestock owners. Kebeles/localities free of capripox vaccination in the past 1 year were included in the study. From 5 zones, districts and Kebeles were selected based on their agro-climatic zones and history of Capripox vaccination. A total of 10 districts, proportionally from each administrative zone (3 from North Gondar, 2 from South Gondar, 2 from West Gojjam, 2 from East Gojjam and 1 from Awi zone) were selected. Two districts were selected from highland, three districts from midland and five districts from lowland agro-ecology. Thirty Kebeles from ten selected districts; five Kebeles from highland, nine Kebeles from midland and 16 Kebeles from lowland were considered for the cross-sectional study. Systematic random sampling technique was used to sample individual animals in selected Kebeles.

Study design
Cross-sectional and retrospective study designs were employed in this study. Cross-sectional study design was used to estimate the sero-prevalence of SP and GP and retrospective study design was used to assess their temporal and spatial distribution.

Sample size determination
The sample size was calculated according to the formula given by Thrusfield [17], using 50% expected prevalence (since there is no previous prevalence report from the study area), 5% desired absolute precision and 95% confidence level.

The formula used:

$$n = \frac{z^2 * pq}{d^2}$$

Where: z = 1.96, p = 0.5, q = 1-p and d = 0.05 (the desired level of precession). The calculated sample size became 384. To increase the precision, the sample size was almost doubled to a total of 672 (329 sheep and 343 goats).

Due to the difference in population size in the 5 zones, 10 districts and even within Kebeles in the same district,

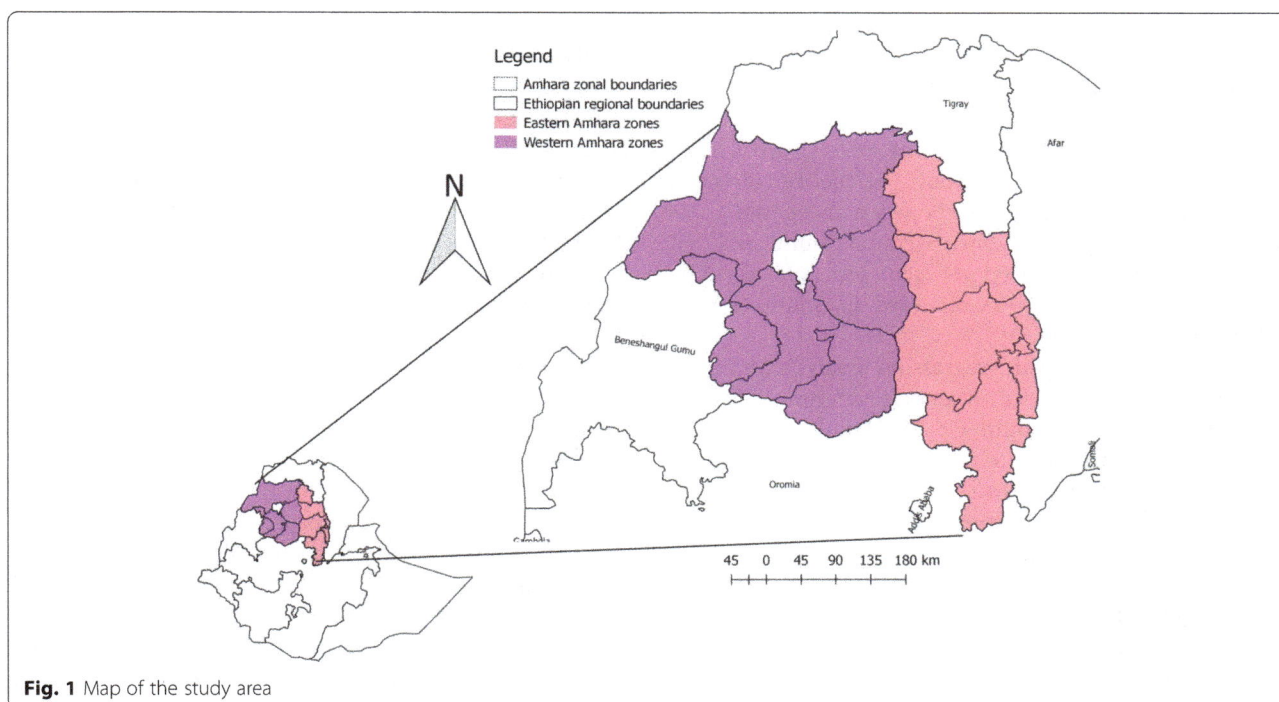

Fig. 1 Map of the study area

sample size was allocated proportionally based on the existing sheep and goat population per zones/districts. Hence 22.47% ($n = 151$) of the sample was drawn from North Gondar zone, 22.77% ($n = 153$) from South Gondar zone, 22.47% ($n = 151$) from West Gojjam zone, 16.96% ($n = 114$) from East Gojjam zone and 15.33% ($n = 103$) from Awi zone.

Data collection and laboratory analysis

Whole blood was collected from the jugular vein of sheep and goats using plain 10 ml vacutainer tubes and 19 gauge sterile needles. The samples were labelled to allow identification of each animal. Additional file 1 on potential risk factors (such as species, age, and sex) was recorded during sampling. The samples were kept in slant position overnight to allow serum separation. Serum was decanted and aliquoted into cryovials and stored in a freezer (-20 °C) at University of Gondar laboratory before transported to National Veterinary Institute (NVI). The serum samples were tested for the presence of sheep and goat pox antibodies using virus neutralization test (VNT) in NVI virology laboratory following the procedures described by Boshra et al. [18]. Sheep pox and GP antibodies cannot be distinguished by VNT, thus, it is called 'sheep and goat pox antibody' when referring to the serological result. Information regarding the age of the animal was obtained from the owner of the animal and classified as young (5 months to ≤1.5 years) and adults (> 1.5 years).

Five years (January 2010 to December 2014) retrospective data on SP and GP outbreak was also compiled. The data was obtained from the Epidemiology Unit of Federal Ministry of Livestock and Fisheries (MoLR). The retrospective epidemiological data contain the number of SP and GP outbreaks, sick and deaths, the population at risk and the species involved. The data were extracted from district monthly reports. Sheep pox and GP cases were identified by veterinary clinical diagnosis.

Data management and analysis

The collected data was classified; filtered, coded using Microsoft Excel spreadsheet (Additional file 1) and analyzed by using STATA version 12. Descriptive statistics were used to present the results. Univariable and multivariable logistic regression analyses were used to identify the risk factors associated with Sheep and goat pox. Likelihood ratio test was used to determine the variables in the final multivariable logistic regression model. Backward variable illumination procedure was employed by using $p = 0.05$ as a cut-off value. In all analysis, the result is considered statistical significance if $p < 0.05$. QGIS v 2.18.0 was used to map the distribution of SP and GP in ANRS.

Results

Sero-prevalence and risk factors

Serum samples of both sheep and goat from all the five zones of western Amhara region were evaluated for previous exposure to sheep and goat poxviruses using virus neutralization test (VNT). Of the total 672 (329 sheep and 343 goats) blood sera tested, 104 (56 sheep and 48 goats) samples were positive for sheep and goat pox antibody. The overall sero-prevalence of sheep and goat pox was 15.5% (95% CI: 12.7, 18.2) in which 17% (95% CI: 12.9, 21.1) in sheep and 14% (95% CI: 10.3, 17.7) in goats. However, the antibody prevalence variation was not statistically significant between sheep and goats ($P > 0.05$). All the studied zones had small ruminants that were sero-positive for sheep and goat pox antibody. The highest sero-prevalence was observed in South Gondar (20.9%) followed by Awi (19.4%) (Fig. 2).

The univariable logistic regression analysis indicates that the sero-positivity varies among different localities, the sero-positivity in South Gondar being significantly higher than North Gondar ($p < 0.05$). Animals from all the study districts were positive for sheep and goat pox viruses antibody; the highest sero-prevalence was recorded in Fogera (27.3%), followed by Guangua (19.4%) and Debre-Elias (18.3%). The sero-prevalence of sheep and goat pox in Fogera is significant ($p < 0.05$) higher than Gonder zuria, West Belesa, Mecha, Javitehinan, and Gozamin. The antibody prevalence was not varied significantly among the agro-ecologies (Table 1).

The host related risk factors; species, sex, and age were also analysed using univariable logistic regression. As shown in Table 2, the sero-prevalence of sheep and goat pox varies significantly between male and female animals as well as among different age categories. The odds of sero-positivity in female animals and young age animals were 1.75 and 2.2 times higher than male and adult animals, respectively.

Among host-related factors, only sex and age were significant in the univariable analysis and were fitted to the final multivariable logistic regression model (Table 3) and both sex and age of animals were identified as risk factors for the occurrence of sheep and goat pox.

Retrospective study

Analysis of the retrospective data indicates that SP and GP outbreaks occur in all zones of the region, both in the western and eastern Amhara regions (Fig. 3). From January 2010 to December 2014, a total of 548 SP and GP outbreaks with 22, 888 cases and 2477 deaths were recorded in the region. Species-wise, 366 outbreaks, 12, 822 cases and 1480 deaths were in sheep, and 182 outbreaks, 10,066 cases and 997 deaths in goats. The highest outbreak was recorded in North Shoa (154) followed by South Wollo (115) and the least was recorded in Waghemra (12) within the last 5 years period (Fig. 4).

As to the temporal pattern of outbreaks, the highest numbers of outbreak were reported in 2012 (155 outbreaks)

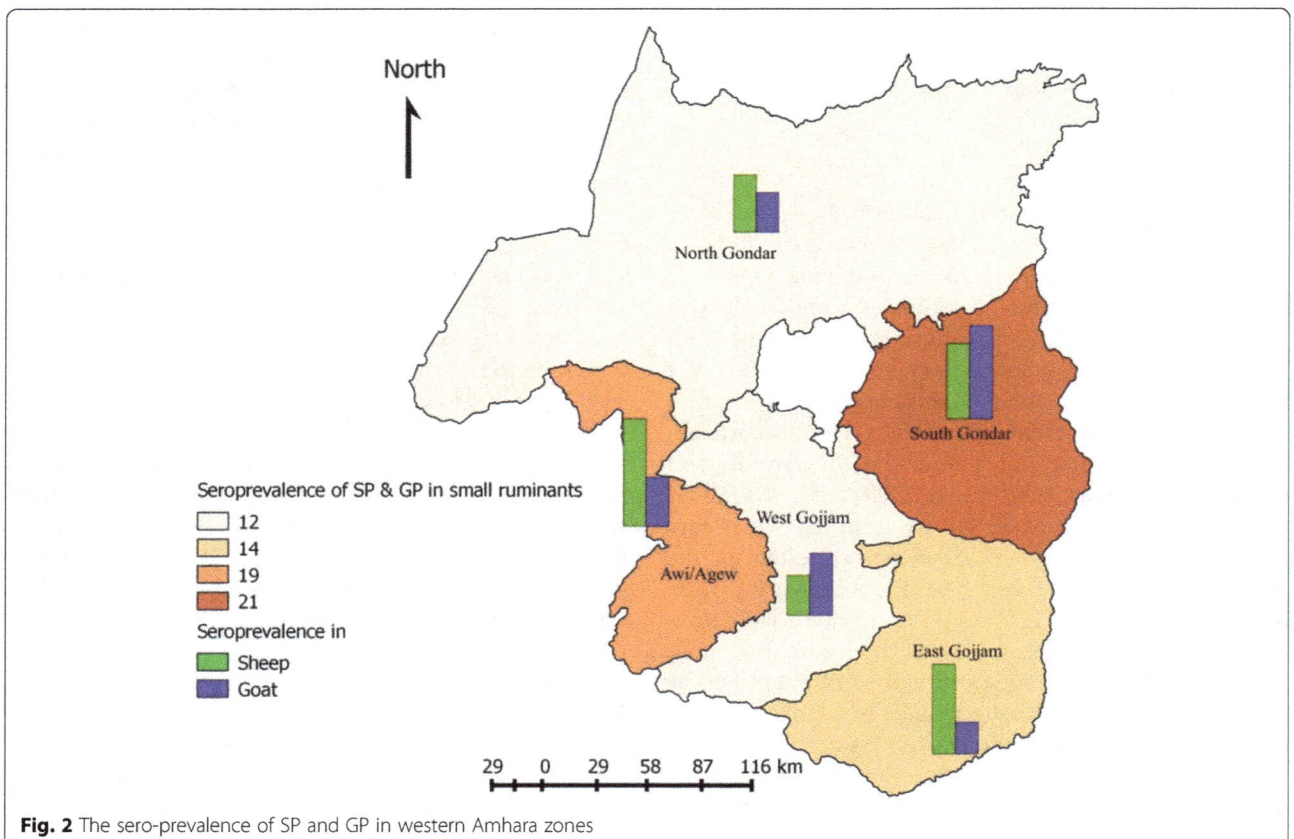

Fig. 2 The sero-prevalence of SP and GP in western Amhara zones

Table 1 Univariable analysis of risk factors for sheep and goat pox antibodies categorized by location

Variables		Number sampled	Prevalence (%)	Odds ratio	Confidence interval	P value
Zone	North Gondar	151	18(11.9)	Ref	–	–
	West Gojjam	151	18(11.9)	1.0	0.49-2.0	1.0
	South Gondar	153	32(20.9)	1.9	1.04-3.66	0.036
	Awi	103	20(19.4)	1.8	0.89-3.56	0.103
	East Gojjam	114	16(14.03)	1.2	0.59-2.48	0.611
District	Fogera	66	18(27.3)	Ref	–	–
	Simada	87	14(16.1)	0.51	0.23- 1.12	0.095
	Gondar zuria	69	7(10.1)	0.30	0.12-0.78	0.013
	West Belesa	40	4(10)	0.30	0.09-0.95	0.041
	Dabat	42	7(16.6)	0.53	0.20-1.41	0.207
	Mecha	81	10(12.4)	0.38	0.16-0.88	0.025
	Javitehinan	70	8(13.4)	0.34	0.14-0.86	0.022
	Gozamin	54	5(9.3)	0.27	0.09-0.79	0.017
	Debre Elias	60	11(18.3)	0.60	0.26-1.39	0.236
	Guangua	103	20(19.4)	0.64	0.31-1.33	0.235
Agro-ecology	Highland	96	12(12.5)	Ref	–	–
	Midland	199	33(16.6)	1.4	0.68-2.83	0.362
	Lowland	377	59(15.7)	1.3	0.67-2.53	0.441

followed by 2013 (115 outbreaks), and 2011 (108 outbreaks), whereas the lowest number of outbreaks (12 outbreaks) were recorded in 2014. The 5 years outbreak record indicates that the diseases occurred in all months of the year, but the highest number of outbreaks has occurred in May and August, and the lowest in February and October (Fig. 5).

Figure 4 shows the distribution of SP and GP in Amhara National Regional State. The map is based on the retrospective data obtained from MoLF database. The distribution of SP and GP in the region could be much wider, as many of the districts in the region haven't reported the status.

Discussion

Occurrences of sheep and goat pox in Ethiopia have been documented in World Organization for Animal Health Information Database since 1996. In the present study, we used VNT to assess whether sheep and goat in Ethiopia are exposed to sheep and goat poxviruses. Using VNT as a measure of sero-prevalence, it was found that 17% of sheep and 14% of goats were exposed to SP and/or GP viruses. The sero-prevalence report showed that both sheep and goat were equally exposed to sheep and goat pox in the study area.

The distribution of sheep and goat pox were widespread, all the study zones and districts had animals that were sero-positive for sheep and goat poxviruses. However, the sero-positivity was varied among districts. The sero-prevalence finding in this study was quite lower than the reports of Elshafie and Ali [10] and Masoud et al. [19] in Sudan and Pakistan. The difference in sero-prevalence by location could be attributed to the difference in animal movement and introduction of new animals (possibly infected). The district which showed higher sero-positivity (Fogera) is associated with extensive livestock movements

Table 2 Univariable analysis of risk factors for sheep and goat pox categorized by animal factor

Variables		Number sampled	Prevalence (%)	Odds ratio	Confidence interval	P value
Species	Goat	343	48(14.0)	Ref	–	–
	Sheep	329	56(17.0)	1.26	0.83-1.92	0.279
Sex	Male	251	28(11.2)	Ref	–	–
	Female	421	76(18.1)	1.75	1.10-2.79	0.018
Age	Adult	444	52(11.7)	Ref	–	–
	Young	228	52(22.8)	2.2	1.46-3.40	<0.001

Table 3 Multivariable logistic regression analysis of risk factors of sheep and goat pox by animal factor

Variables		Number sampled	Prevalence	Odds ratio	Confidence interval (95%CI)	P value
Sex						
	Male	251	28(11.15%)	Ref	–	–
	Female	421	76(18.05%)	1.9	1.18 3.05	0.008
Age						
	Adult	444	52(11.7%)	Ref	–	–
	Young	228	52(22.8%)	2.2	1.46 3.40	<0.001

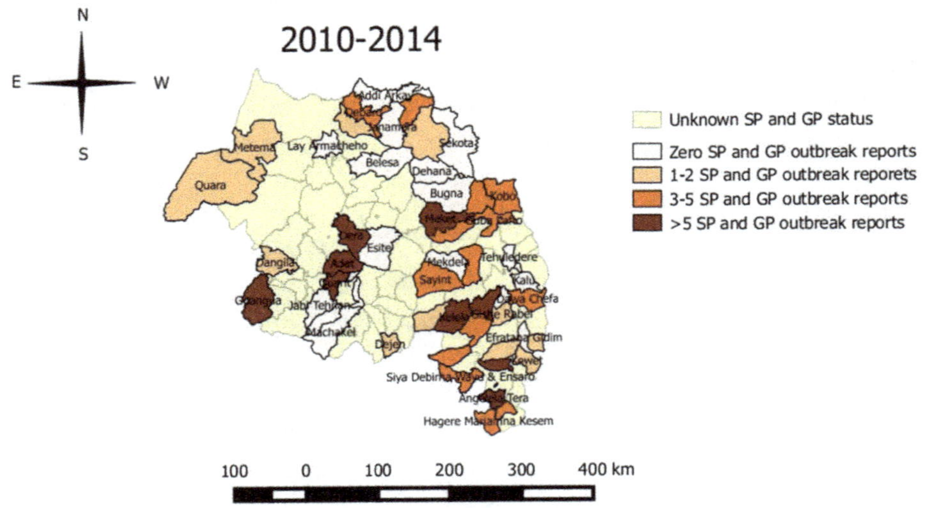

Fig. 3 Map showing the number of SP and GP outbreak and their spatial distribution in various parts of Amhara region (sum of 2010-2014 outbreak reports)

for grazing and marketing activities and due to recent occurrence of the disease as an outbreak. Outbreaks of SP and GP have reported from all administrative zones of the Amhara region between 2010 and 2014 and the sero-prevalence findings in the study districts confirm the retrospective study result.

Comparison of the sero-prevalence of sheep and goat pox in different age groups of sheep and goat showed a higher seropositivity in young animals than adult animals (OR: 2.2; 95%OR: 1.46, 3.40) and the difference was highly significant ($p < 0.001$). The odds of seropositivity also showed that female (OR: 1.99; 95%CI: 1.18, 3.05; $P = 0.008$) animals were more affected than male animals. This result is in agreement with previous works reported by Elshafie and Ali [10] in Sudan and Masoud et al. [19] in Pakistan. The higher sero-prevalence in young and female animals

could be explained by the low level of immunity in young and female animals associated with lambing season or poor physiological condition. In lambs and kids, the malignant form of SP and GP has been recorded as the most common type and maternal immunity provides protection only for up to 3 months (19). Here it is important to note that, due to relatively small sample size, the risk factor analyses may not have adequate power to detect differences between risk factors. Thus, care should be taken when interpreting these results.

The retrospective data indicated a difference in SP and GP cases between years and even months. Seasonality of the diseases is also reported (6, 19) that could be explained by the capability of the viruses to survive for several months in wet and cold weather, by association with lambing season and transportation of animals for

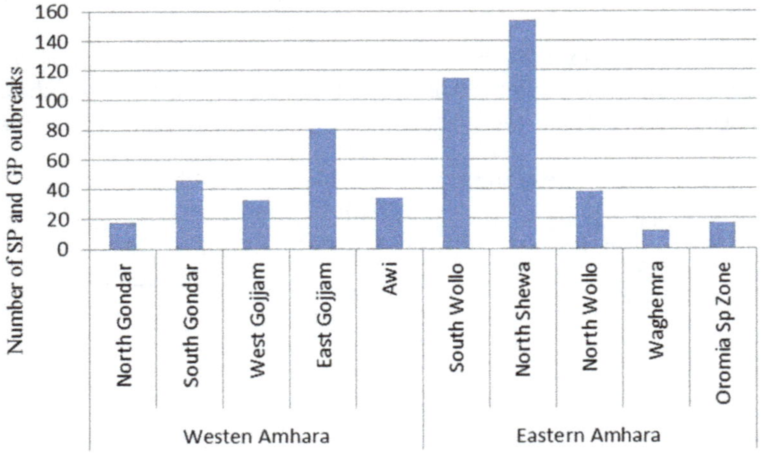

Fig. 4 SP and GP outbreaks in Amhara Regional State from January 2010 to December 2014

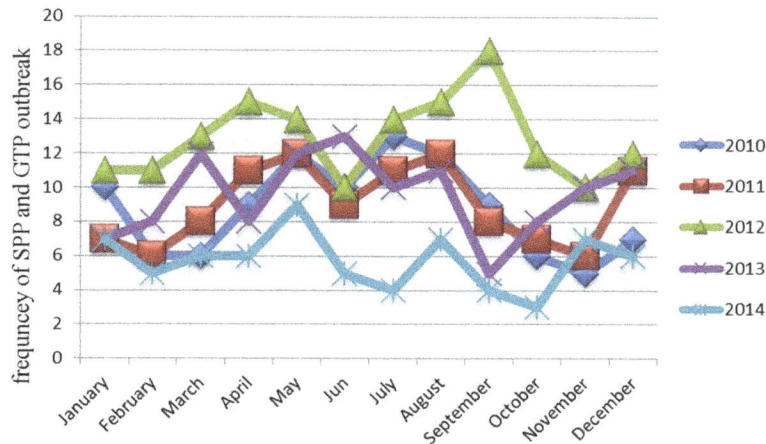

Fig. 5 Seasonal pattern of SP and GP disease outbreaks from 2010 to 2014

marketing, Besides, the absence of regular, obligatory and constant disease reporting or underreporting is a problem in some areas of the study region. A study by Bayissa and Bereda [20] indicated, transport and communication to be the two limiting factors in disease reporting in Ethiopia. These problems could result in irregular or absence of outbreak reports for some remote health posts and district veterinary offices. If not for the low level of disease reporting, the distribution and burden of SP and GP in Ethiopia could be much more than what is documented here. The previous study by Gelaye et al. [14] indicated GP virus to be responsible for the widespread outbreaks of SP and GP in different parts of Ethiopia. In Ethiopia, a live attenuated SPPV Kenya O-180 vaccine strain (KS1-O180) is used to immunize small ruminants against sheep and goat pox and cattle against lumpy skin disease. It's believed that sheep and goat naturally infected with sheep and goat pox viruses and recovered from the disease remain protected for lifelong (19). However, there have been repeated reports on the insufficient protection provided by the vaccine against lumpy skin disease virus in Ethiopia. This vaccination failure would likely be caused by poor vaccine handling where the availability of electricity is limited for keeping cold chain.

Conclusion

To our knowledge, this is the first serological evidence of sheep and goat pox presence in Ethiopia. The study indicated a wide spread distribution of sheep and goat pox in Amhara region. The widespread occurrence and the high sero-positivity of sheep and goat pox observed in this study is alarming; thus, extensive studies on epidemiology, transmission, and economic impact will be necessary. Although direct and indirect animal

contacts during grazing, watering, and trading are the most important transmission means in extensive production system, it is also important to explore whether there is involvement of vector in the transmission of SP and GP or not; since genetically related viruses in the genus Capripoxvirus such as Lumpy Skin Disease Virus (LSDV) are arthropod-borne [21, 22]. Considering the role of vectors in the transmission of LSDV, the possible role of vectors in the transmission of SP and GP virus should be investigated. In the context of Ethiopia where livestock movement is uncontrolled, compulsory annual mass vaccination is recommended as the best feasible, economic and viable method for the control of sheep and goat pox.

Abbreviations
CCPP: Contagious caprine pleuropneumonia; CI: Confidence interval; GP: Goat pox; LSDV: Lumpy skin disease virus; MoLF: Federal ministry of livestock and fisheries; PPR: Peste des petits ruminants; SP: Sheep pox; VNT: Virus neutralization test

Acknowledgements
We would like to acknowledge animal owners, district and Kebele officials for their corporations during sample collection. We would also like to acknowledge MoLF for providing the 5 year SP and GP outbreak records.

Funding
University of Gondar is the funding organization. The funding organization didn't participate in the design of the study and collection, analysis, and interpretation of data and in writing the manuscript.

Authors' contributions
TF was involved in initiation and design of the study. NF was involved in the laboratory work and manuscript preparation. SL was involved in analyzing

the data and manuscript preparation. WM, YT, BA, AA and SN were involved in the design and implementation of the study. All authors read and approved the final manuscript.

Consent for publication
Not applicable

Competing interests
The authors declare that they have no competing interests.

Author details
[1]College of Veterinary Medicine and Animal Sciences, University of Gondar, P. O. Box 196, Gondar, Ethiopia. [2]Livestock and Fisheries Development Office, Dembia District, North Gondar, Ethiopia. [3]College of Veterinary Medicine and Agriculture, Addis Ababa University, P. O. Box 34, Bishoftu, Ethiopia. [4]Faculty of Agriculture, Debre Markos University, Debre Markos, Ethiopia.

References
1. CSA (2013): Central statistical agency Federal Democratic Republic of Ethiopia agricultural sample survey report on livestock and livestock characteristics 2012-2013, vol. (2), Addis Ababa, Ethiopia.
2. Behnke R, Metaferia F. The contribution of livestock to the Ethiopian economy – part II. IGAD LPI working paper 02-11. Addis Ababa, Ethiopia: IGAD Livestock Policy Initiative; 2011.
3. Asresie A, Zemedu L. Contribution of livestock sector in Ethiopian economy: a review. Adv Life Sci Technol. 2015;29:79–90.
4. Babiuk S, Bowden TR, Boyle DB, Wallance DB, Kitching RP. Capripoxviruses: an emerging worldwide threat to sheep, goats and cattle. Transbound Emerg Dis. 2008;55:263–72.
5. Rao TVS, Bandyopadhyay SK. A comprehensive review of goat pox and sheep pox and their diagnosis. Anim Health Res Rev. 2000;1(2):127–36.
6. Kitching P. Sheep pox and goat pox. In: Coetzer JAW, Tustin RC, editors. Infectious diseases of livestock. 2nd ed. Capetown: Oxford University Press Southern Africa; 2004. p. 1277–81.
7. Bowden T, Babiuk S, Parkyn G, Copps J, Boyle D. Capripox virus tissue tropism and shedding: a quantitative study in experimentally infected sheep and goats. Virology. 2008;371(2):380–93.
8. Ahmed MM. (2012): A study on prevalence, risk factors and economic impact of sheep pox in North and South Kordofan States of the Sudan. International Symposia on Veterinary Epidemiology and Economics proceedings, ISVEE13: Proceedings of the 13th International Symposium on Veterinary Epidemiology and Economics, Belgium, Netherlands, Poster topic 9 - Surveillance and diagnostic test evaluation, p 524.
9. Enan KA, Intisar KS, Haj MA, Hussien MO, Taha KM, Elfahal AM, Ali YH, El Hussein M. Seroprevalence of two important viral diseases in small ruminants in Marawi Province Northern State, Sudan. Int J Livest Prod. 2013;4(2):18–21.
10. Elshafie EI, Ali AS. Participatory epidemiological approaches and Seroprevalence of sheep pox in selected localities in Kassala State, Sudan. Sudan J Vetterinary Res. 2008;23:47–58.
11. Davies FG, Mbugwa G. The alterations in pathogenicity and immunogenicity of a Kenya sheep and goat pox virus on serial passage in bovine foetal muscle cell cultures. J Comp Pathol. 1985;95(4):565–72.
12. Moges N, Bogale B. Assessment of major animal production and health problems of livestock development in lay-Armacheho District, Northwestern Ethiopia. American-Eurasian J Sci Res. 2012;7(3):136–41.
13. Teshome D, Derso S. Prevalence of major skin diseases in ruminants and its associated risk factors at University of Gondar Veterinary Clinic, North West Ethiopia. J Vet Sci Technol. 2015;S13:002.
14. Gelaye E, Belay A, Ayelet G, Jenberie S, Yami M, Loitsch A, Tuppurainen E, Grabherr R, Diallo A, Lamien CE. Capripox disease in Ethiopia: genetic differences between field isolates and vaccine strain, and implications for vaccination failure. Antivir Res. 2015;119:28–35.
15. ESGPIP (Ethiopian sheep and goat productivity improvement program). technical bulletin no.29 sheep and goat pox: causes, prevention and treatment. Addis Ababa, Ethiopia; 2009.
16. Yeruham I, Yadin H, Van Ham M, Bumbarov V, Soham A, Perl S. Economic and epidemiological aspects of an outbreak of sheeppox in a dairy sheep flock. Vet Rec. 2007;160(7):236–7.
17. Thrusfield M. Veterinary epidemiology. 3rd ed. UK: Blackwell science Ltd; 2007. p. 233–50.
18. Boshra H, Truong T, Babiuk S, Hemida MG. Seroprevalence of sheep and goat pox, Peste des Petits ruminants and Rift Valley fever in Saudi Arabia. PLoS One. 2015;10(10):e0140328.
19. Masoud F, Mahmood M, Hussain I. Seroepidemiology of goat pox disease in district Layyah, Punjab, Pakistan. J Vet Med Res. 2016;3(1):1043.
20. Bayissa B, Bereda A. Assessment of veterinary service delivery, livestock disease reporting, surveillance systems and prevention and control measures across Ethiopia/Kenya border enhanced livelihoods in southern ethiopia (ELSE) project: CIFA ethiopia/CARE ethiopia; 2009.
21. Chihota CM, Rennie LF, Kitching RP, Mellor PS. Mechanical transmission of lumpy skin disease virus by Aedes Aegypti (Diptera: Culicidae). Epidemiol Infect. 2001;126(2):317–21.
22. Tuppurainen ES, Stoltsz WH, Troskie M, Wallace DB, Oura CA, Mellor PS, Coetzer JA, Venter EH. A potential role for Ixodid (hard) tick vectors in the transmission of lumpy skin disease virus in cattle. Transbound Emerg Dis. 2011;58(2):93–104.

8

Important knowledge gaps among pastoralists on causes and treatment of udder health problems in livestock in southern Ethiopia: results of qualitative investigation

Kebede Amenu[1,2,4*] ⓘ, Barbara Szonyi[2], Delia Grace[3] and Barbara Wieland[2]

Abstract

Background: Ethiopia has high prevalences of udder health problems including clinical and subclinical mastitis across production systems in different livestock species. Previous studies on udder health problems have largely focused on identification of mastitis causing microbial pathogens and associated risk factors. However, relatively little is known about the knowledge and beliefs of livestock keepers regarding udder health problems. An understanding of the beliefs on the other hand would facilitate effective communication between livestock keepers and animal health professionals. Therefore, this study aimed at exploring the knowledge and belief surrounding the causes, clinical signs and treatments for udder health problems in (agro-) pastoral communities in southern Ethiopia using qualitative investigation.

Results: The result showed that udder health problem, locally known as '*dhukkuba muchaa*', which translates to 'disease of teats', was classified into three main types: (1) tick infestation (*dirandisa*), (2) swelling of udder often with pus discharge (*nyaqarsa*) and (3) acute mastitis caused by evil eye (*buda*) with 'bloody milk'. Tick infestation was perceived to directly cause mechanical damage to udder tissue or to resulting in swelling leading to *nyaqarsa*. Our analysis also revealed the strong misperception that acute and severe swelling of udder was caused by evil eye. According to the pastoralists, cows with large udders in the late pregnancy are prone to evil eye infliction upon giving birth. The pastoralists often treat udder health problems by combining both modern and traditional methods. Removal of ticks by hand and acarcide application were the preferred methods for limiting tick infestation while swelling and evil eye cases were treated with antibiotics (e.g. oxytetracycline).

The study also revealed that specific herbs, only known by the herbalists, were used for traditional treatment of udder health. Although this information could not be divulged at the time, it should form the subject of further investigation. Traditional treatment for evil eye was often administered through nostrils, raising questions about its effectiveness.

(Continued on next page)

* Correspondence: kamenu@gmail.com
[1]School of Veterinary Medicine, Hawassa University, P.O. Box 5, Hawassa, Ethiopia
[2]International Livestock Research Institute, P. O. Box 5689, Addis Ababa, Ethiopia
Full list of author information is available at the end of the article

(Continued from previous page)

Conclusion: The narration given by the pastoralists in associating tick infestation with udder health problems was compatible with existing scientific evidences. In this respect, such local knowledge can be better utilized for the educational messages targeting control and management of tick infestation in livestock. However, the misperception of causes for acute udder swelling as evil eye can be problematic as far as the application of appropriate treatment and management of the problem is concerned. The misperception can significantly impact the welfare of animals and highlights the need for capacity building of the pastoralists on the causes and treatment of udder health problems.

Keywords: Tick infestation, Udder swelling, Evil eye, (mis)-perception, Traditional medicine, Udder health management, Dairy hygiene, Qualitative study, Animal welfare,

Background

Livestock production is the cornerstone for pastoral and agro-pastoral livelihoods for provision of foods and plays an important economic, social and cultural role [1]. However, significant seasonal fluctuations in feed and water availability, ecosystem changes, extreme weather conditions and presence of animal diseases [2] threaten this important income source. Diseases result in poor livestock health, low productivity, mortality, reduced livestock products and increased risk of disease transmission to humans. When looking at the dairy sector specifically, milk is important in the diet of the pastoralists and can contribute to more than 50% of the energy intake of families, especially in children [1, 3]. In Borana pastoral areas (where the present study was carried out), animals kept primarily for milk production include cattle, camels and goats [4]. In this respect, maintaining the health and productivity of milk producing animals is crucial to ensure good health and nutrition of the livestock keepers.

Udder health problem is one of the major diseases of dairy animals with negative impacts on milk production, milk safety and animal welfare [5]. Mastitis constitutes one of the main udder problems and by definition refers to an inflammation of the mammary glands characterized by changes in the physical and chemical features of milk and pathological changes in the glandular tissue often with non-specific and complex factors responsible for its occurrence. Mastitis occurs in two forms: clinical mastitis with evidence of milk and udder changes upon physical examination and subclinical mastitis only detected by indirect tests such as milk somatic cell count and pathogen isolation [6]. In Ethiopia, a number of epidemiological studies showed high prevalence of mastitis in different livestock species and production systems. For example, animal level mastitis prevalence in dairy cattle was 32.6% in central Ethiopia [7], 64.6% in southeastern Ethiopia [8] and 59.1% in Borana pastoral areas in southern Ethiopia [9]. Studies in the pastoral areas of Ethiopia in camels revealed animal level prevalence of 29.0% [10] and 44.8% [11]. Based on indicator paper test, 15.5% subclinical mastitis prevalence in goats has been reported in Borana pastoral areas [12]. Etiologically,

Gram positive cocci are the most common bacterial agents isolated from the milk of animals affected by clinical or subclinical mastitis. Mastitis, especially the subclinical form, causes significant economic losses in dairy cattle. In Ethiopia, the total loss caused by subclinical mastitis associated to *Staphylococcus aureus* was estimated at USD 78.65 per cow per lactation in small-size farms (<5 heads of cattle) and USD 150.35 in large-size farms (≥50 heads of cattle) [13].

Past studies mainly focused on identification of microbial causative agents and associated risk factors. However, research focusing on social and cultural aspects of how livestock keepers perceive and manage generally udder health problem (e.g. mastitis) in their livestock is lacking. This information and involvement of livestock keepers is needed in order to implement suitable and effective treatments, control and management of mastitis. Therefore, there is a need to assess knowledge and beliefs of livestock keepers regarding prevailing livestock health problems before designing possible disease control programs through awareness creation and education of livestock keepers. In spite of the importance of milk and milk products in in Borana, detail studies on milk handling and processing are limited. The present study was implemented as part of a research project aiming to improve the handling practices and microbiological safety of milk and traditionally produced dairy products in the area. The study aimed at exploring the knowledge and belief of communities in the area about the causes, clinical signs and treatment of different udder health problems. In pastoral livestock keeping communities women are mostly involved in routine management of milking animals [14] and are supposed to be more familiar with udder health problems and related diseases of dairy animals than men. In line with this, we involved largely women in our study.

Methods

Study area

The present study was carried out in four village administrations (Kebeles) of the Yabello district in the Borana

Zone, southern Ethiopia. Borana is located in southern part of Ethiopia bordering Kenya (Fig. 1). The zone has a semi-arid to arid climate with high variability of rainfall resulting in seasonality in the off-take of livestock and livestock products. The rainfall distribution of the area is bimodal with the long rainy season extending from March to May and the short rainy season from September to November. From June to August is the cool dry season while December to February is the warm dry season [15].

The zone has an estimated population of 962,489 (487,024 male and 475,465 female) with 91.2% of the population living in rural areas [16]. The same information source also indicated that 90.9% of the population speaks Afan Oromo as mother tongue and the population follows different religions with 47.3% Protestant Christian, 35.0% traditional, 9.6% Muslim and the remaining other religions. Historically, Borana people are nomadic people residing in southern parts of Ethiopia and northern Kenya, keeping cattle as their main livelihood strategies and seasonally moving in search of feed and water or to escape the risk of potential diseases. Associated with various environmental, social, political and cultural changes, Borana people are diversifying their livelihoods strategies. The livelihood diversification of traditionally cattle keeping

Borana families is often evident in terms of growing shift towards crop farming, keeping of other types of livestock such as camels and small ruminants and involvement in non-farming activities such as petty trading, employment as daily labourer and charcoal production [17, 18]. Based on discussion made with people working in Yabello District Pastoral Development Office, animal health delivery systems in Borana pastoral areas include government veterinary clinics/ animal health posts and private drug shops. Community-based animal health workers are also active in the pastoral areas of Borana in providing animal health services. The education levels of people involved in animal health delivery vary which include few weeks of training for the community-based animal health workers and formal training for veterinary doctors and animal health assistants. Livestock keepers themselves also buy and administer drugs especially in remote villages. Use of traditional veterinary medicine is also common in the area.

For this present study, four village administrations (Dharito, Elweya, Surupha and Dida Yabello) were selected based on ongoing research projects aiming to improve the small ruminant value chain and the high milk production potential of the villages. The qualitative data collection was carried out in July 2015 with subsequent

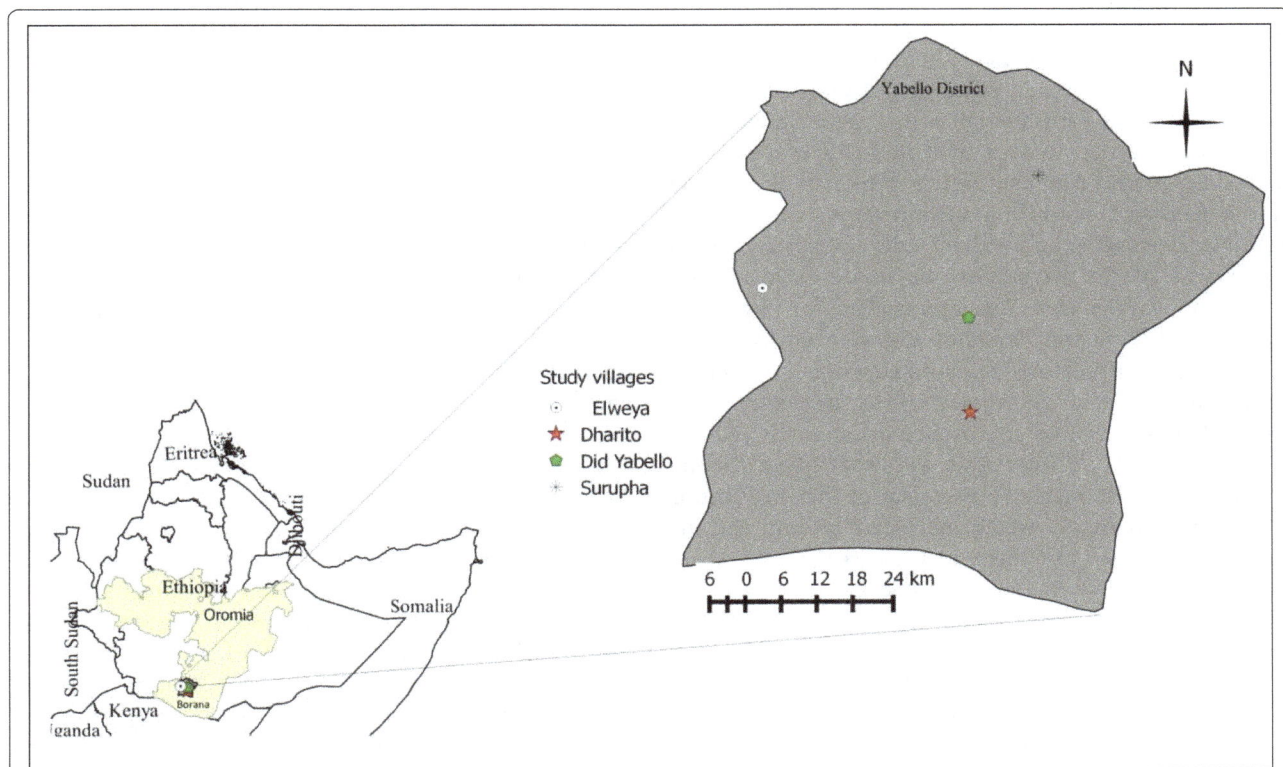

Fig. 1 Map of study area (four villages in Yabello District, Borana Zone, Oromia Regional State, Southern Ethiopia) (Note: The boundaries are unofficial)

substantiation of unclear concepts through informal discussions with the pastoralists in December 2015.

Methodological approaches

Different participatory data collection methodologies which included individual semi-structured in-depth interviews (IDI), focus group discussions (FGD), informal discussions with key informants and observations were used in the present study. Individual interviews were held with 40 women (10 in each village) using a semi-structured questionnaire which was initially developed in English and then translated to Afan Oromo (Borana dialect), a language widely spoken in the study area. As background information, age of the respondents and types of livestock they were keeping was first asked in the IDIs. For the main study, the questions mainly focused on the knowledge and perceptions of pastoralists about causes, clinical signs and treatments of udder health problems in different livestock species. The questions guide was pretested for clarity by interviewing three women before the actual study started.

In Borana, several households, mostly belonging to same lineage, reside in clusters or neighbourhoods (called *olla*). The respondents for individual interviews were identified by driving through the villages and stopping in different *ollas*. The selection of the respondents was made with assistance of development agents in each village facilitated by a female field assistant from Yabello District Pastoral Development Office with a background in animal production and more than 4 years of working experience in dairy production in the area. The respondents were married women (having major domestic responsibilities) and who own lactating animals at the time of the fieldwork or during the previous rainy season. The interviews were held either under the shade near to the houses of the pastoralists or inside their houses. When the women were first approached for interviewing, most of them were hesitant and even not happy to exchange greetings. However, upon explaining the objectives of the study, all agreed to be interviewed and consequently none of those approached declined the request for participation. All interviews were carried out by the female field assistant in the presence of the first author except in few cases in which men were around and informal discussion with them was necessitated to divert their attention and to minimize distraction during the interviews with women. The interviews were documented using audio recording and field notes.

Following the IDIs, FGD with women in each village on similar topics was held to verifying already obtained information and to obtain further information about udder health problems. To select participants for the FGD, the development agents identified 6–8 women, preferably, from different *ollas* and representing different age groups. The FGD was then held at a central place agreed upon in advance. The group discussion was moderated by the first author with help of the female field assistant. After the interviews and FGDs, informal discussions with the participants or people encountered along the way were held on general aspects of milk handling practices and udder health problems to triangulate and complete information obtained.

Before the interviews and FGDs, verbal consent was obtained from each of the respondents after explaining the objectives of the study and assuring anonymous use of the information. Verbal consent was considered enough given that the information collected in the present study was assumed to be what is shared commonly among the pastoral communities and no personally sensitive information was collected. As part of good practices in undertaking the research project, the respondents were given chances to ask the interviewer on any of the topics discussed. After that brief feedback on the general milk hygiene and causes of mastitis were given to the respondents.

Data management and analysis

The audio recordings of the interviews and discussions were transcribed and translated into English. The data was analysed qualitatively by repeatedly reading the transcript to identify different themes, which were coded using the free software QDA Miner Lite v1.4.3 Provalis Research [19]. Essentially, the themes were identified from the questions guide and further emerging themes were identified iteratively starting from data collection throughout the analysis process. Content analysis was carried out for clinical signs of udder health problem described by the pastoralists to identify commonly used terminologies used by the pastoralists in describing the problem. To portray the qualitative data different quotes in the words of the respondents were highlighted.

Results

Background information (age of respondents and livestock production practices)

The reported age of respondents involved in the individual interviews varied from 17 to 50 years (average of 32 years, median = 30 years). Cattle, camels and goats were the common livestock species kept for the purpose of milk production in Borana. Out of 40 women interviewed individually 39, 33, 13 and 4 women reported keeping of cattle, goats, camels and sheep respectively. Similarly, the milk from the different livestock was not equally preferred by the people. The widely produced and consumed milk in the area was cow milk. Borana pastoralists highly value cattle both for economic and cultural reasons. The following quotes in the words of the pastoralists strengthen above statements.

"Cattle are the foundation of the people here; milk is also mainly obtained from cattle" (IDI 7).

"Borana started rearing camels very recently. What we know more is about cattle" (FGD 3).

Keeping of camels for milk production was preferred by the pastoralists due to the large volume of milk camels produce especially during dry season.

"Good milk is from cow (tasty) but large volume [is] from camels" (IDI 8).

Goat milk was consumed especially by mixing with tea and it was perceived to have better nutritional value as indicated below.

"Though goat milk is small in volume, when we want to dilute with water, goat milk has more vitamins [to indicate nutritional value]" (IDI 3).

Themes identified under udder health problems and treatments

The major themes identified in the present study are indicated in Fig. 2. The concepts of causes, risk factors and clinical signs were intermingled and as a result the identified themes were grouped together in the three types of udder health problems reported.

Knowledge about causes and signs of clinical udder health problems

Locally, udder health problem is known as *dhukkuba muchaa*, which literally means 'disease of teats'. Though the name implies "disease of teat", the term is understood to be a general udder health problems. Pastoralists associated the problems of udder health with different factors and grouped based on the perceived causes and clinical signs into different categories. The main categories identified were: *dirandisa* (tick infestation), *nyaqarsa* (chronic swelling in the form of a boil) and *buda* (which means evil eye and is characterized by bloody milk). The qualitative investigation revealed that there was no clear differentiation between the different types of udder health problems and the multifactorial causes was well appreciated by the pastoralists as is illustrated by the following quote:

"Udder health problem results in swelling of udder, blood can come with milk, ticks can infest the udder and the teat; it can cause swelling. Udder health problem is caused by different factors" (IDI 8).

In some cases differences in the predisposition of the different livestock species was noted. For example, it was stated that udder health problems are common in cows compared with goats. It was described that udder health problem is severe in camels compared with cows.

"Udder health problem occurs in both cattle and goats, udder becomes ill when trauma due to thorny plant (goraa), painful when calf suckle and to prevent the painful situation, smeared with butter. Udder health problem is common in cows compared with other species" (IDI 24).

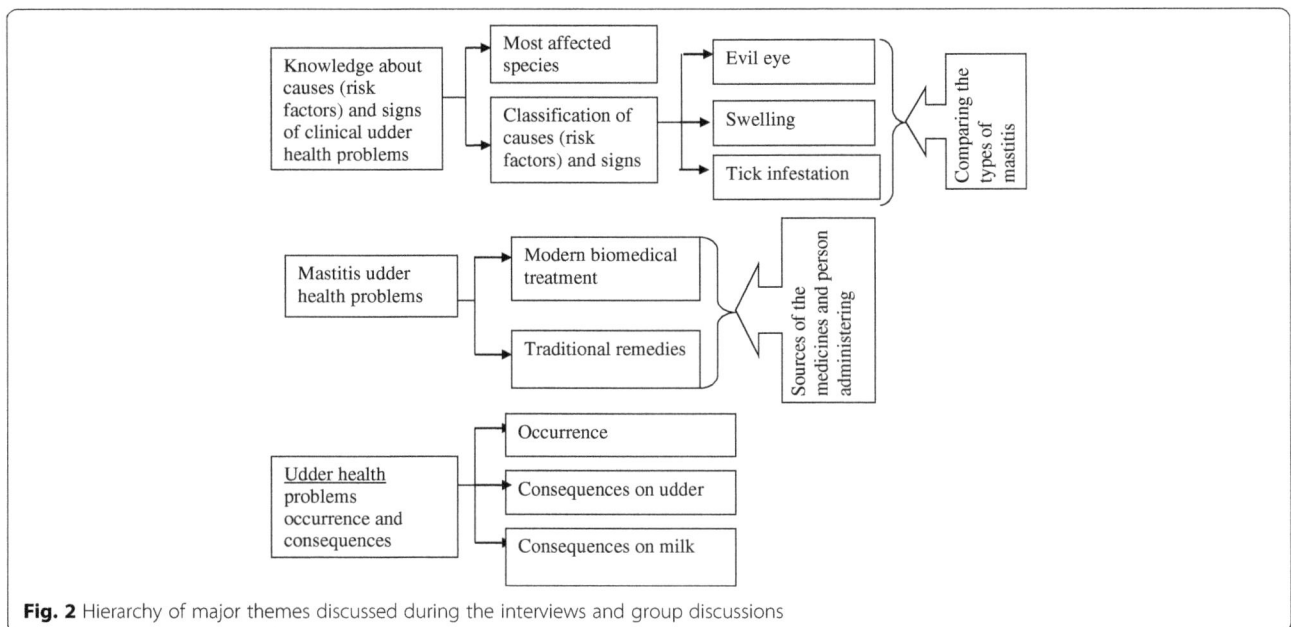

Fig. 2 Hierarchy of major themes discussed during the interviews and group discussions

*"Camels are highly affected by udder health problem.
It is camel which is highly affected by buda or
swelling. When camels are affected by mastitis, no
survival of the calf. It is not like cows. All [teats]
affected once. Regarding mastitis, camels are highly
harmed by mastitis" (FGD 2).*

Content analysis of the interviews and the discussions
about clinical signs indicated that the most common
term used to characterize udder health problem was
'swelling of udder' followed by 'bloody milk' (Table 1).

Tick infestation (diraandisa)

The pastoralist women mentioned the predisposition of
domestic animals for udder health problems due to tick
infestation and they explained well the biology of ticks
during the interviews. They mentioned that tick infestation causes swelling of udder, wounds on the skin of
udder at the attachment site and leads to blind teats.

*"...tick, small creature, moves on the ground and
can infest cattle. Causes wounds on the animal"
(Informal 2).*
*"The main cause is tick infestation. Whether cattle,
goats or camels, tick infestation is the one which
causes major problems" (Informal 1).*
*"Tick infestation can cause udder health problem.
Even if you remove [the tick], the mouth can remain
[in attachment site] and lead to ill health. Also cows
can acquire and experience chronic swelling of udder"
(IDI 14).*
*"There is also tick infestation which causes swelling
and blockage of teat" (IDI 40)*

Table 1 Clinical description of udder health problems by
pastoral people: results of content analysis of individual
interviews and focus group discussions

Clinical description of mastitis	Number of times mentioned[a]
Swelling of udder	33
Bloody (red) milk	16
Pus discharge upon milking	9
Blind teat	8
Painful udder and teat	6
Difficulty in milk letdown	3
Teat perforation	3
Milk with stingy odour	2
Cracks on teat	2
Abnormal discharge from teats	1
Skin sloughing of teats/udder	1

[a]Number of the issue was mentioned (can be multiple times in the IDI, FGD or
informal discussion)

Swelling (nyaqarsa)

Udder health problem characterized by swelling of udder
was referred to as *nyaqarsa* which literally translated
means a boil. As per the description by the pastoralists,
nyaqarsa does not involve diffuse inflammation of the
udder potentially as a result of ascending bacterial infection through teat canal. It was often recognized as a distinct disease of udder but the cause of *nyaqarsa* was
associated with different factors including tick infestation and trauma by thorny plants. The respondents
both in IDI and FGD described that swelling (nyaqarsa)
bursts leading to wound and teat loss.

*"Nyaqarsa is changed to ilke [swelling with pointed
tip]. It becomes hard. If blood and pus is not coming
out (discharged),, the teat opening becomes hard and
blocked" (FGD 1).*
*"Nyaqarsa, is like a boil, causes swelling. Whether
medicine is applied or not, nyaqarsa can burst.
Wound is formed and when washed with medicine can
be healed. It is also this one which causes discharge
from teat" (FGD 4).*
*"Udder becomes ill because of trauma due to thorny
plant" (IDI 24)*

Evil eye affliction (buda)

Pastoralists perceived evil eye affliction of the udder as
one of the main causes of udder health problem. The belief that there are people whose eyes can change 'white
milk' into 'red milk' when they stare at the udder of animals, especially on animals with good milk production,
is very common. According to the respondents cows
with large udder during late pregnancy are highly vulnerable to evil eye infliction after giving birth during
early lactation. Most interviewees expressed the idea that
acute udder health problem (more referring to acute
mastitis) is caused by evil eye and they believe that it is
caused by looking intently at animals. In the words of
one informant:

*"When cows are pregnant and ready to give birth,
there is something in the udder which gives high milk
yield. There is 'evil eye' which sees the milk in udder
and kajeellaa [envy], causes swelling of udder. There is
milking of blood. There is milking of blood. You see
blood like when slaughtering [animals]. There is
discharge of blood from cow teats" (Informal 2).*

*"Evil eye afflicts when the udder of pregnant cows
increases in size" (IDI 38)*

According to the pastoralists, people with evil eye have
super ability to see milk inside the udder of animals. As
per the descriptions of the pastoralists, "seeing the milk
by the people possessing evil eye" was not the

observation obvious to anybody (e.g. differentiating lactating and dry cow) rather it was more referring as if one observes milk in a bucket after milking including the colour.

"There are people which are by nature buda. It is said that there is a person who can see milk inside udder" (IDI 28)

In addition to causing udder health problem in livestock, evil eye was widely believed to cause many health problems in animals or humans and crop failure. It is believed that some people naturally possess evil eye but such information is shared secretly among the community. According to the respondents, saying to somebody "you have *buda*" is an insult and can lead to a disagreement to the extent that who said so can be legally held accountable for defaming the person.

Another related cause of udder health problem was envy (*kajeellaa*), literally referring to an intense desire or need for other people's livestock, crop or property. *Kajeellaa* was believed to aggravate the affliction by evil eye. For example, when someone's belongings with superior quality (could be livestock or crop) are widely praised among the community, this may lead to evil eye among the people getting the information (hearing) and cause the damage. According to the description given in group discussion, the one who is praising about the animals can be innocent (without evil eye) but a person possessing evil eye may hear the information and subsequently inflicting the damage. If information about good milk production of animals is shared and reached 3 or 4 people, all of the people may not be free of evil eye ability. The problem of *kajeellaa* on milk quality was described by a participant in one FGD as follows:

"Three of our cows became pregnant and people started talking about their pregnancy. The whole neighbourhood started talking about by saying: Oh! 'Your cows are pregnant and how big their udders are!' The cows gave birth and after that their milk stayed for 9 days without forming curd [i.e. without solidification]. Thereafter, I requested the neighbours 'to spit for me' (tufaa). By putting the medicine for buda [referring to spitting] in the milk, it started to curd" (FGD 1).

The pastoralists also believe that evil eye is responsible for reduced milk production of animals and calf rejection by cows after birth (*finqilchaa*).

Treatment approaches of udder health problems
Various treatment approaches were used for the treatment of the different types of udder health problems,

including modern and traditional treatment approaches depending on the types of the problems. According to most of the interviews, people tend to use modern medicine for *nyaqarsa* and traditional medicine for evil eye affliction. As revealed during the FGDs, there was no consistency in the order of the use of modern or traditional medicine for treating udder health problem. The overall idea during the FGDs showed that people tend to resort to either traditional or modern medicine when the animal being treated is not responding to the first treatment regime they are using. This was clearly evident from the following quotes:

"When Borana [traditional] medicine is not effective (does not cure), we try black medicine (taramishi) [oxytetracyline]" (FGD 2)
"We try with black medicine, if not successful revert to Borana medicine" (FGD 3)
"When Borana medicine is found not effective (unable to cure), the thing [milk] is not coming out and the milk becomes watery like whey; we injected black medicine [oxytetracyline]; also the white one [penicillin and streptomycin]. We use combinations of the medicines" (FGD 2).

Traditional treatment
The traditional treatment termed as *qorsa* Borana (means Borana medicine) varied from the use of herbs to procedures involving magical practices. Specific plants, only known to the herbalists and not revealed to the users, were used for the treatment of udder health problems. Especially for acute udder health problem caused by evil eye with discharge of blood mixed in milk, traditional treatments were used. The term *walda* was often mentioned by the pastoralists referring to a specific plant used in the traditional treatment of udder health problem and other livestock diseases. However, it was evident from the interviews and discussions that other types of plants were also used.

"Walda is used. We don't know the plant name and we get from healer" (IDI 5).

The pastoralists described that different modes of application of the plants materials were practiced which include dropping the juice of the plant after chewing into nostril or mouth, smearing of the plant on udder or teat after burning and mixing with butter, fumigation of udder with the smoke of burning plant, rubbing with the dying burning wood (*barbadaa*). The intention of applying herbal medicine was either curative or preventive. It was mentioned that traditional medicine can be applied before animals giving birth to prevent occurrence of udder health problem or after delivery to treat clinical cases.

> "After chewing the medicine, the juice is dropped into the nostril. After chewing it is also spitted on the udder. The udder of a pregnant animal is also smeared. It is also smeared after burning of the plant. There is a difference when applied before birth and after birth" (FGD 2).

> "There is medicine which is dropped into the nostrils of lactating or late pregnant cows. There is also medicine smeared on the udder before delivery. To prevent blockage of teats at delivery, traditional medicine is smeared on [udder] in animals having large udders during late pregnancy; also after birth" (IDI 36).

The common magical practice employed to cure animals attacked with evil eye was called *tufaa* (literally means spitting). *Tufaa* was used for different health problems (e.g. mastitis, calf rejection or low milk yield or failure of milk to curdle) if evil eye was thought to be the cause. Spitting as countermeasure against evil eye affliction is done by requesting the person suspected of possessing evil spirit and who might have caused the animal health problem to spit on the animal. If the suspected person agrees, he/she will spit on the animal by reciting the phrase: 'if I have eaten (inflicted) you [referring to the animal], may God recover you from the problem". According to the interviews and discussions, the practice of spitting for the treatment of udder health problem was found to be declining compared to past times due to fear of people to ask the suspected person for spitting. Nowadays, referring somebody to possess evil spirit (*buda*) is becoming an offence for people and in some cases can lead to disagreement. Due to this, alternative means of the procedure of spitting was practiced without pinpointing to a specific person. The person whose animal has been diseased calls all people in the neighbourhood to spit and all people state publicly that they have not attacked the animal. In the words of a discussant and an informant in the informal discussion, respectively:

> "That is in the very past. Currently this practice does not exist. If you now say to somebody, "you are buda and it is you who ate my animal", you may face serious legal consequences. You can be imprisoned. Even if when the heifer is affected by buda and you know very well and [the heifer] entered to the house of the person [indicating that the animal was inflicted by the person], you cannot claim anything. You cannot say to the person you ate my heifer and please come

> and spit for me. In that case you can be imprisoned and no way to come out of prison. No way to ask. You simply leave when the cow is diseased, teats blocked or calf died [due to buda]. [Nowadays], you cannot say that 'you have buda' [to an individual]. In that case, the person can kill you" (FGD 1).

> "In the past (long time) Borana would call the person who was known to be buda and requested from him/her by saying: 'it is you who ate my livestock and now spit for me [laugh...]. It is you who ate and now come and spit for me'. The person is obliged to spit. Nowadays, that is not the case. You are scared to ask the person. That can have legal consequences. It is not said anymore, spit for me. When you suspect somebody among community members, you call all people and request to spit. You say, we don't know who ate, so all come and spit. There can be the suspect among the community. The people spit by saying: 'if it is me who ate you, get well'. Thereafter, the milk changes from blood to milk. Nowadays, this practice [requesting a person for spitting] is not common. If you say that the person can go to court and can lead you to serious legal consequence. As a result, the owner of the affected animal lost his/her animal and simply decline to complain" (Informal 2).

An alternative practice to spitting to counteract the effect of evil eye and simultaneously minimize the possible disagreement was to give milk to the suspected person without his/her knowledge. In an interview it was stated that:

> "When you suspect somebody, you don't complain and not let the person know you are suspecting [him/her]. You simply invite [him/her] for milk and you just pour milk into a cup and give to the person to drink. After that it is considered that the person has spitted for you" (IDI 15).

Traditionally *nyaqarsaa* was treated by incising the swelling and removing the pus. Instead of using traditional medicine, nowadays people tend to resort to modern medicine for the treatment of *nyaqarsa*.

Traditionally, tick infestation was treated by manual removal. The qualitative analysis showed that the different treatment and control strategies of tick infestation are well known among pastoralists and that there was agreement that tick associated problems have been reduced compared to earlier times. A typical example illustrating this is the exact quote below from informal discussion with a pastoralist in the area:

> "Now ticks are not affecting udder. People are wise and use acarcides (skin medicine). This medicine

eliminates ticks. There had been changes with regard to this. When you inject medicine, ticks fall down" (Informal 1)

Biomedical treatments

Borana pastoralists use different modern drugs for the treatment of udder health problems as mentioned during the FGDs. The two common drugs mentioned by the pastoralists were 'black' and 'white' medicines, literally referring to oxytetracycline and penicillin-streptomycin (PenStrep), respectively. Normally, the pastoralists use the medicines not only for udder health problems but also for other livestock diseases. It means that intramuscular was the commonest route of administration. The animals were either taken to a veterinary clinic or the pastoralists buy drugs and treat by themselves. In the words of the respondents:

"We buy from market and have at home. Black medicine is for buda. Black medicine is for many [health] ailments. When udder swells, injection is given on two sides [of rump]. It regresses the swelling" (FGD 3).

"The udder swells. When udder swells, black or white medicine is injected by ourselves or the animal taken to clinic ..." (FGD 4)
"Before causing swelling it can be removed using medicine for scabies [acarcides]. With the injection of the medicine, ticks disappear" (FGD 3).

As the following quotes of the group discussions in different villages show, the participants indicated that people have reverted to modern medicine from traditional medicine for treatment of udder health problem:

"Mostly people use black medicine. It is not like earlier times. Except [newborn] rejection [by dam], swelling of udder is treated with black medicine" (FGD 2).

"Nowadays we are using modern medicine. We use black medicine. When udder swells, we inject with medicine" (FGD 1).
"Nowadays after healthcare came, we take [livestock] to clinic like human and let them be treated using various drugs and injections" (FGD 4).

The acaricide mentioned by the pastoralists and used for treatment of tick infestation was termed as *cululuqaa* (shining medicine) which apparently refers to subcutaneous injectable ivermectin.

Occurrences and consequences of udder health problems

Most informants indicated that mastitis leads to various problems of the udder and teats, weak newborns and milk quality or safety problems. The most frequent recurring theme regarding the consequences of udder health problem was loss or blockage of teats. According to the informants, loss of teats leads to early elimination of animals by selling.

"When teats blocked, the animals is sold and changed" (IDI 17).

The other consequence of udder health problem stated was discarding of the milk of the affected animals due to aesthetic reason. Especially, the 'bloody milk' of acutely affected animal was not consumed by the pastoralists and either left not milked, milked on the ground or the milk given to dogs.

"It is not milked, unless milked on the ground. It has problem for calf and humans. Milked on the ground. Nobody drinks until becomes normal. Milked and given to dogs. Dogs drink" (FGD 1)

However, in the FGDs there was no consensus among the pastoralists regarding the use or discarding of the milk from animals with udder problems. It was revealed that some pastoralists use the milk after boiling or by adding into boiling tea. The pastoralists also explained the effect of udder health problem on milk quality and difficulty of processing milk into different products. The pastoralists traditionally process milk into different products such as yoghurt (*ititu*), butter, ghee and butter milk. *Ititu* is traditional milk product widely consumed in the area and prepared by natural fermentation through accumulating whole milk for several days to weeks and regularly removing the whey (the fluid part). The effects of udder health problems on milk quality and processing were explained by the pastoral women as follows.

"The milk smells extremely stingy even after returning to white. Not good for animals as well as humans. Irrespective of this, it is produced by people and you cannot discard. This thing is consumed. There is nothing, Borana cannot consume" (FGD 2)

"[Milk from animal with udder health problem is] not for drinking. It is boiled in a pot and added to tea. It is not consumed fresh. This milk is not even suitable for churning. Even if you churn it does not yield butter. If you also make ititu (yoghurt), you cannot drink. It has stingy odour. You boil like water and add to tea. We don't discard. We prepare (boil) like that and drink" (FGD 2).
"Now there is a problem with the milk. When milked, the milk is not used fresh. It clots/precipitates. It can be yellowish if it stays for some time. It cannot form

[proper] curd easily, if you want to change to ititu (yoghurt)" (FGD 2).

The other consequence of udder health problem perceived by the pastoralists was the negative effect on health of the newborn animals as a result of consuming the milk of the affected dams.

"The calf can be affected by disease characterized by hair falling when suckling the dam affected with evil eye. ... No appetite for grazing and then death. When the calf suckles, diarrhea, extreme loss of body condition and death" (FGD 1)

Discussion

The present qualitative study of the knowledge and beliefs of pastoralists about the cause, clinical signs and treatment of udder health problem is part of a research project aiming to improve the handling practices and microbiological safety of milk and traditionally produced dairy products in Borana pastoral area in Ethiopia. Employing of different participatory qualitative tools in the present study enabled us to describe the details of the knowledge of pastoralists about udder health problem and current treatment practices in different livestock species. Thanks to qualitative research methods much more detailed information on attitudes and beliefs could be obtained compared to quantitative approaches.

In Borana, the term "dhukkuba muchaa" used by the pastoralists to refer to udder health problem in general and specifically mastitis was not exactly matching the western disease nomenclature. In Oromo language, 'dhukkuba' means disease or illness and 'muchaa' means teat (so literally means 'disease of teats'). The name for udder in Afan Oromo is 'gurruu'. Similar studies also indicated such differences in disease naming between African traditional medical practices and western sciences. On the other hand, udder health problem specifically mastitis was understood by Borana pastoralists to be caused by various factors; often the factors acting collectively in causing udder health problems. For example, the pastoralists had understanding that injuries by thorny plants cause udder problems. Such understanding is compatible with western scientific explanation of the multifactorial causation of mastitis [20]. On the other hand, it should be noted that the description given for mastitis by the pastoralists can be potentially periparturient edema and physiological discoloration of milk as result of colostrums which should be clarified through creating close collaboration with the pastoralists.

Quality and effectiveness of animal health management in small-scale livestock production system is affected by farmer and farm characteristics, and economic, institutional and biophysical factors. The education level and experiences of livestock keepers largely influences animal health management [21]. In this respect, pastoralists have good knowledge about the health of their animals with a huge potential for integrating their knowledge with the veterinary service delivery system to solve prevailing livestock health problems [22–24]. This is due to the fact pastoral people usually live in close proximity to their livestock and do have lifelong experiences about livestock production and diseases management. Therefore it is no surprise that sometimes pastoralists can have animal health knowledge and skills comparable to trained animal health professionals. For example, in a study by Catley [23] there was a good agreement between pastoralists' and veterinarians' disease names and diagnostic criteria.

On the other hand, even if detailed studies are lacking for animal health, studies in human health showed that there can be strong cultural beliefs about diseases which can subsequently be a barrier for seeking treatment and provision of effective health services [25–28]. For example, Masika et al. [29] in their studies about the local tick control strategies by cattle farmers in South Africa indicated that very few farms associated ticks with tick-borne disease such redwater and Gall sickness (anaplasmosis). The authors [29] further noted that wrong understanding of the causes and transmission of diseases can lead to ill-directed treatments and widespread deviation from the recommended directions of use when administering conventional medicines. It was recommended that this should be addressed by farmer training and supplying appropriate information [29]. Similarly, the present study revealed that the pastoral women blame acute udder health problem on evil eye since there is a lack of knowledge on causative agents and how the causes (pathogens) are transmitted.

Scientifically, a number of microbial agents are responsible in causing mastitis in livestock with presence of multitude of predisposing factors which include: animal risk factors (age, udder/teat conformation), environmental risk factors (poor housing condition) and pathogen risk factors [6]. The causes of udder health problems in Ethiopia may be very different from Western developed countries. In the case of Ethiopia in which livestock are largely kept under extensive production system, local damage and infection from superficial skin wounds seems to be a large contributing factor [10]. In the case of Western countries (largely intensive management system), ascending bacterial infection through the teat canal can be a driving factor for udder infection [6]. This would have implications for prevention and treatment strategies under Ethiopian production systems in which greater emphasis should be put on skin inspection and wound care. Prevalence of contagious pathogens such as *Staphylococcus aureus* and *Streptoccus agalactiae* have been significantly reduced in importance in developed countries [30] while still a big challenge in the production systems of developing countries like Ethiopia [8–10, 12].

In Ethiopia, linking evil eye and health problems is a widespread belief [31, 32]. Our findings that pastoralists link evil eye with acute udder health problem characterized by inflammation of udder is consistent with the findings of a study conducted by Mesfin and Shiferaw [33] on the indigenous veterinary practices of South Omo (agro)-pastoralists of southern Ethiopia. Similar to Borana people, the different ethnic groups in South Omo link evil eye with mastitis and apply various traditional treatment approaches such as spitting and smearing of traditional medicine on the udder. A study by Teferra and Shibire [34] in Borana also showed linking of severe human mental illness with heterogeneous supernatural causes such as evil spirit, curse and bewitchment. In a study in southwestern Ethiopia among people suffering from tuberculosis, it was found that 50% of tuberculosis suspects (with signs of cough for at least 2 weeks) linked the cause of tuberculosis with evil eye [31]. Similarly, in other African countries the belief in evil eye as cause of various health ailments is also a common situation. In this respect, Comoro et al. [35] reported strong beliefs of mothers in linking severe malaria among children with supernatural causes such as evil spirit and bewitchment, and mild malaria with biological cause (mosquito bites).

In our present study, the misperception of acute inflammation of udder as evil eye can be problematic in the management of mastitis and lead to poor welfare of the animals. Traditional treatment for evil eye is often administered through nostrils and this procedure may not be effective against the disease. Moreover, resorting of the pastoralists to traditional magical treatment (e.g. spitting) can have negative impacts on welfare of the animals and cause further economic losses. The delay in appropriate treatment of mastitis can lead to loss of teats due to progression of mastitis to terminal stage and leading to early elimination of animals [6, 20]. The practice of spitting for the treatment of udder health problem was found to be declining compared to past times due to fear of people to ask the suspected person for spitting. This can be a good strategy to teach the pastoralists to follow proper management of udder health problems.

Though traditional medicine can complement modern medicine, antibiotic treatment remains necessary to achieve bacteriological cure of infection [35]. Cure rate of mastitis treatment depends on a number factors related to host, pathogens and treatment regimen. For example, it is not recommended to treat mastitis caused by *Staphylococcus aureus* infection in older animals, chronic infections, or penicillin-resistant isolates [36]

and such possibilities should be also considered when the udder health management programs are going to be designed in the pastoral area. There is now a move in developed countries to leave some forms of mastitis untreated (e.g. those caused by *Escherichia coli*) due to the pressure to reduce antimicrobial use in livestock production. This may not be applicable in Ethiopia, where most mastitis is associated with gram-positive species [8–10, 12].

Important to note, though, is the apparent lack of topical intra-mammary antibiotics in these regions and instead injections are used which may not be optimal as a treatment option. In the present study we could not find evidences indicating availability of drugs specifically intended of the treatment of udder health problems (e.g. intrammammary infusion form). The pastoralists were dependent on intramuscular injectable antibiotics for the treatment of general animal health problems and the commonly stated drugs were intramuscular injections of oxytetracycline and penicillin (often combined formulation with streptomycin). Some studies showed better efficacy and increased bioavailability in udder when intramammary infusion is used compared with injectables; which further can minimize the potential occurrence of antimicrobial resistance [37]. Moreover, previous studies elsewhere showed that oxytetracycline is ineffective for mastitis treatment [38–40]. In this regard, further strategies should be designed by local government to avail efficacious antimicrobials in correct formulation for the treatment of mastitis in the area. But considering the low hygiene standards around milking and udder health in the Borana community, well planned training should be given to the animal health professionals as well as the pastoralists before implementing such treatment regime. In intramammary treatment, strict hygiene must be followed by careful cleaning and sanitation of the teats before infusing to avoid the introduction of pathogens [6]. During intrammary infusion the delicate tissues lining the teat duct can be traumatized with further complications and due to this incorrect procedure is one of the factors responsible for mastitis treatment failures [41].

Therefore, improvement of milking hygiene is the aspect which necessitates due consideration is preventive measures for mastitis. Poor milking hygiene in terms of not washing and drying udder, and milkers' hand before milking had significantly increased the prevalence of mastitis in dairy cattle [8]. The high prevalence of mastitis causing contagious pathogens (e.g. *Staphylococcus aureus*) likely is a direct result of poor hygienic practices. Previous studies in Ethiopia showed that among the different bacterial pathogens isolated from cases of mastitis, *Staphylococcus aureus* (a contagious pathogen) represents the commonest one. For example, 24.1% [8] and 41.5% [7] of the bacteria isolated from cows suffering from mastitis were *Staphylococcus aureus*. Similarly, *Staphylococcus aureus* has been isolated from 26.3% [10] and 12.8% [12] of camels in pastoral areas of Ethiopia; in both reports showing the highest proportion among the different pathogens. During our present fieldwork, it was observed that most pastoralists do not follow hygienic precautions such as milking of mastitis affected animals at the end and washing hands between milking different animals, thus enabling cross-contamination from diseased to healthy animals. Since pastoralists are not aware of pathogens and how they are transmitted during milking, the role of hygiene during milking is not obvious. Therefore, awareness creation on good hygienic and milking practices is necessary to reduce the high incidence and impacts of udder health problems in dairy animals in the area.

Interestingly, the pastoralists associated udder health problem with tick infestation which is compatible with existing scientific evidences. They correctly described biology of ticks and the damage ticks cause on udder or teat with subsequent occurrence of mastitis. Ticks can cause direct physical damage on the skin of the udder or teat favoring bacterial infection and occurrence of mastitis. Different epidemiological studies identified high prevalence of subclinical mastitis associated with the presence of lesions on the skin of udder or teat of cattle [7, 8] and camels [12] in Ethiopia. Abera et al. [10] also reported higher prevalence of mastitis in animals having udder or teat lesion and infested with ticks in camels in eastern Ethiopia. Thus, optimal control of tick infestation through acaricides application or careful hand removal in dairy animals can reduce the prevalence of mastitis and the resulting economic losses.

The negative consequences of mastitis on milk quality, animal productive longevity and the difficulty in milk processing were well described by the pastoralists, and in line with the results of other scientific studies [20, 42]. Mastitis directly affects milk composition often reducing lactose, non-fat solids and total solids [43] which can compromise milk processing and causes low yield of milk products. The change in milk composition can have direct negative consequences for milk quality and nutritional security of the people.

Udder health problem is an important disorder of dairy animals reducing milk production and subsequently causing food safety and food security problems in livestock keeping communities. The effect of mastitis on food and nutritional security of pastoral community is more serious due to the facts that milk and milk products largely contribute to the diet of the communities and is essential for child nutrition.

Conclusions

The good understanding of the pastoralists about the association of tick infestation with udder health problems can be better utilized in the preparation and implementation of educational messages targeting skin problems and udder health. On the other hand, the misperception of evil eye as a cause for acute udder inflammation and administration of traditional treatments through nostrils should be focus of future attempts to improve the awareness level of the pastoralists. Therefore, there is clearly a need to design effective and culturally sensitive communication and capacity development strategies to increase knowledge on causes of udder health problems, on proper treatment, management and prevention, and in general good milk production practices. Our study also revealed that specific herbs, only known by the herbalists, were used for traditional treatment of mastitis and this aspect should also be further investigated to determine their effectiveness.

Abbreviations
FGD: Focus Group Discussion; IDI: In-depth Interviews; Informal: Informal discussion

Acknowledgments
This study was funded through the CGIAR Research Program on Livestock and Fish. Kebede Amenu was supported through DAAD (German Academic Exchange Service) Fellowship. Mrs. Ayantu Chali is acknowledged for her wonderful assistance during the fieldwork activities. We would like to thank the pastoralists for their time and sharing their experiences. Dr. Samson Leta is thanked for preparing the map of the location of the study area.

Authors' contributions
KA: conceived the research idea, wrote the protocol; drafted data collection tools, collected data, performed data analysis and drafted initial manuscript, revised the manuscript; BS: involved in proposal write-up, preparation of study tools and manuscript preparation; BW, DG: supervised the research and made substantial intellectual contribution in the designing of the study, preparation of the study tools; result interpretation and manuscript preparation. All authors read and approved the final manuscript.

Competing interests
The authors declare that they do not have any competing interests.

Consent for publication
The authors declare that the manuscript does not contain any personally identifiable information and all the personal data were anonymized.

Author details
[1]School of Veterinary Medicine, Hawassa University, P.O. Box 5, Hawassa, Ethiopia. [2]International Livestock Research Institute, P. O. Box 5689, Addis Ababa, Ethiopia. [3]International Livestock Research Institute, P. O. Box 30709, Nairobi, Kenya. [4]Present address: Department of Microbiology, Immunology and Veterinary Public Health, College of Veterinary Medicine and Agriculture, Addis Ababa University, P.O.Box 34, Bishoftu, Ethiopia.

References
1. FAO. World livestock 2011: livestock in food security. Food and Agriculture Organization (FAO); 2011. http://www.fao.org/docrep/014/i2373e/i2373e.pdf. Accessed 21 Dec 2016.

2. Gustafson CR, VanWormer E, Kazwala R, Makweta A, Paul G, Smith W, Mazet JA. Educating pastoralists and extension officers on diverse livestock diseases in a changing environment in Tanzania. Pastoralism. 2015;5(1). doi: 10.1186/s13570-014-0022-5.

3. Sadler K, Kerven C, Calo M, Manske M, Catley A. Milk matters: a literature review of pastoralist nutrition and programming responses. Addis Ababa: Feinstein International Center, Tufts University and Save the Children; 2009. http://www.alnap.org/pool/files/ua197-018-018-00001-archival-(1).pdf. Accessed 4 Dec 2016.

4. Tolera A, Abebe A. Livestock production in pastoral and agro-pastoral production systems of southern Ethiopia. Livestock Research for Rural Development. 2007. 19(12). http://www.lrrd.cipav.org.co/lrrd19/12/tole19177.htm. Accessed 21 Dec 2016.

5. Seegers H, Fourichon C, Beaudeau F. Production effects related to mastitis and mastitis economics in dairy cattle herds. Vet Res. 2003;34(5):475–91.

6. Radostits OM, Gay CC, Hinchcliff KW, Constable PD. Veterinary medicine: a textbook of the diseases of cattle, horses, sheep, pigs and goats. 10th ed. London: Elsevier Health Sciences; 2007.

7. Getahun K, Kelay B, BekanaM Lobago F. Bovine mastitis and antibiotic resistance patterns in Selalle smallholder dairy farms, central Ethiopia. Trop Anim Health Prod. 2008;40(4):261–8.

8. Lakew M, Tolosa T, Tigre W. Prevalence and major bacterial causes of bovine mastitis in Asella, south eastern Ethiopia. Trop Anim Health Prod. 2009;41(7):1525–30.

9. Bedane A, Kasim G, Yohannis T, Habtamu T, Asseged B, Demelash B. Study on prevalence and risk factors of bovine mastitis in Borana pastoral and agro-pastoral settings of Yabello District, Borana zone, Southern Ethiopia. J Agric Environ Sci. 2012;12(10):1274–81.

10. Abera M, Abdi O, Abunna F, Megersa B. Udder health problems and major bacterial causes of camel mastitis in Jijiga, eastern Ethiopia: implication for impacting food security. Trop Anim Health Prod. 2010;42:341–7.

11. Regassa A, Golicha G, Tesfaye D, Abunna F, Megersa B. Prevalence, risk factors, and major bacterial causes of camel mastitis in Borana zone, Oromia regional state, Ethiopia. Trop Anim Health Prod. 2013;45(7):1589–95.

12. Megersa B, Tadesse C, Abunna F, Regassa A, Mekibib B, Debela E. Occurrence of mastitis and associated risk factors in lactating goats under pastoral management in Borana. Southern Ethiopia Trop Anim Health Prod. 2010;42(6):1249–55.

13. Tesfaye GY, Regassa FG, Kelay B. Milk yield and associated economic losses in quarters with subclinical mastitis due to *Staphylococcus aureus* in Ethiopian crossbred dairy cows. Trop Anim Health Prod. 2010;42(5):925–31.

14. Tangka F, Jabbar M, Shapiro BI. Gender roles and child nutrition in livestock production systems in developing countries: a critical review. International Livestock Research Institute (ILRI) Socio-Economics and Policy Research Working Paper. no. 27. 64p. Nairobi (Kenya): ILRI. 2000.

15. Coppock DL. The Borana plateau of southern Ethiopia: synthesis of pastoral research development and changes, 1980–90. Addis Ababa, Ethiopia: ILCA (International Livestock Centre for Africa; 1994.

16. Central Statistical Authority (CSA). Summary and statistical report of the 2007 population and housing census. Population size by age and sex. Addis Ababa: Population Census Commission, Federal Democratic Republic of Ethiopia; 2008.

17. Berhanu W, Colman D, Fayissa B. Diversification and livelihood sustainability in a semi-arid environment: a case study from southern Ethiopia. J Dev Stud. 2007;43(5):871–89.

18. Megersa B, Markemann A, Angassa A, Zárate AV. The role of livestock diversification in ensuring household food security under a changing climate in Borana, Ethiopia. Food Secur. 2014;6(1):15-28.

19. Péladeau N. QDA miner qualitative data analysis software, user's guide. Montreal: Provalis Research; 2004.

20. Phillips CJ. Principles of cattle production. UK: CABI. Oxfordshire; 2010.

21. Chilonda P, Van Huylenbroeck G. A conceptual framework for the economic analysis of factors influencing decision-making of small-scale farmers in animal health management. Rev Sci Tech. 2001;20(3):687–95.

22. Sollod AE, Stem C. Appropriate animal health information systems for nomadic and transhumant livestock populations in Africa. Rev Sci Tech. 1991;10(1):89–101.

23. Catley A. Use of participatory epidemiology to compare the clinical veterinary knowledge of pastoralists and veterinarians in East Africa. Trop Anim Health Prod. 2006;38(3):171–84.

24. Gakuya F, Ombui J, Heukelbach J, Maingi N, Muchemi G, Ogara W, Mijele D, Alasaad S. Knowledge of mange among Masai pastoralists in Kenya. PLoS One, 2012;17;7(8). doi:10.1371/journal.pone.0043342.

25. Ahorlu CK, Dunyo SK, Afari EA, Koram KA, Nkrumah FK. Malaria-related beliefs and behaviour in southern Ghana: implications for treatment, prevention and control. Tropical Med Int Health. 1997;2(5):488–99.

26. Beiersmann C, Sanou A, Wladarsch E, De Allegri M, Kouyaté B, Müller O. Malaria in rural Burkina Faso: local illness concepts, patterns of traditional treatment and influence on health-seeking behaviour. Malaria J, 2007;6(1). doi:10.1186/1475-2875-6-106.

27. Dell'Arciprete A, Braunstein J, Touris C, Dinardi G, Llovet I, Sosa-Estani S. Cultural barriers to effective communication between indigenous communities and health care providers in northern Argentina: an anthropological contribution to Chagas disease prevention and control. Int J Equity Health, 2014:13(1). doi:10.1186/1475-9276-13-6.

28. Shayo EH, Rumisha SF, Mlozi MR, Bwana VM, Mayala BK, Malima RC, Mlacha T, Mboera LE. Social determinants of malaria and health care seeking patterns among rice farming and pastoral communities in Kilosa District in central Tanzania. Acta Trop. 2015;144:41–9.

29. Masika PJ, Sonandi A, Van Averbeke W. Tick control by small-scale cattle farmers in the central eastern Cape Province, South Africa. J S Afr Vet Assoc. 1997;68(2):45–8.

30. Hillerton JE, Berry EA. Treating mastitis in the cow–a tradition or an archaism. J App Micro. 2005;98(6):1250–5.

31. Hodes R. Cross-cultural medicine and diverse health beliefs: Ethiopians abroad. West J Med. 1997;166(1):29–36.

32. Abebe G, Deribew A, Apers L, Woldemichael K, Shiffa J, Tesfaye M, Abdissa A, Deribie F, Jira C, Bezabih M, Aseffa A. Knowledge, health seeking behavior and perceived stigma towards tuberculosis among tuberculosis suspects in a rural community in southwest Ethiopia. PLoS One. 2010: 5(10). doi:10.1371/journal.pone.0013339.

33. Mesfin T, Shiferaw S. Indigenous veterinary practices of south Omo Agropastoral communities. Addis Ababa: Shama Books; 2009.

34. Teferra S, Shibre T. Perceived causes of severe mental disturbance and preferred interventions by the Borana semi-nomadic population in southern Ethiopia: a qualitative study. BMC Psychiatry. 2012;12(1). doi:10.1186/1471-244X-12-79.

35. Comoro C, Nsimba SE, Warsame M, Tomson G. Local understanding, perceptions and reported practices of mothers/guardians and health workers on childhood malaria in a Tanzanian district—implications for malaria control. Acta Trop. 2003;87(3):305–13.

36. Barkema HW, Schukken YH, Zadoks RN. Invited review: the role of cow, pathogen, and treatment regimen in the therapeutic success of bovine Staphylococcus Aureus mastitis. J Dairy Sci. 2006;89(6):1877–95.

37. Hillerton JE, Kliem KE. Effective treatment of streptococcus uberis clinical mastitis to minimize the use of antibiotics. J Dairy Sci. 2002;85(4):1009–14.

38. Erskine RJ, Bartlett PC, Crawshaw PC, Gombas DM. Efficacy of intramuscular oxytetracycline as a dry cow treatment for Staphylococcus Aureus mastitis. J Dairy Sci. 1994;77(11):3347–53.

39. Dueñas MI, Paape MJ, Wettemann RP, Douglass LW. Incidence of mastitis in beef cows after intramuscular administration of oxytetracycline. J Anim Sci. 2001;79(8):1996–2005.

40. Lents CA, Wettemann RP, Paape MJ, Vizcarra JA, Looper ML, Buchanan DS, Lusby KS. Efficacy of intramuscular treatment of beef cows with oxytetracycline to reduce mastitis and to increase calf growth. J Anim Sci. 2002;80(6):1405–12.

41. du Preez JH. Bovine mastitis therapy and why it fails. J S Afr Vet Ass. 2000; 71(3):201–8.

42. Neerhof HJ, Madsen P, Ducrocq VP, Vollema AR, Jensen J, Korsgaard IR. Relationships between mastitis and functional longevity in Danish black and white dairy cattle estimated using survival analysis. J Dairy Sci. 2000;83(5):1064–71.

43. dos Reis CB, Barreiro JR, Mestieri L, de Felício Porcionato MA, dos Santos MV. Effect of somatic cell count and mastitis pathogens on milk composition in Gyr cows. BMC Vet Res, 2013;9(1). https://bmcvetres.biomedcentral.com/articles/10.1186/1746-6148-9-67.

9

Aberrant use and poor quality of trypanocides: a risk for drug resistance in south western Ethiopia

T. Tekle[1], G. Terefe[3]*[ID], T. Cherenet[2], H. Ashenafi[3], K. G. Akoda[4], A. Teko-Agbo[4], J. Van Den Abbeele[5], G. Gari[1], P.-H. Clausen[6], A. Hoppenheit[6], R. C. Mattioli[7], R. Peter[8], T. Marcotty[9], G. Cecchi[10] and V. Delespaux[11]

Abstract

Background: Trypanocidal drugs have been used to control African animal trypanosomosis for several decades. In Ethiopia, these drugs are available from both authorized (legal) and unauthorized (illegal) sources but documentation on utilization practices and quality of circulating products is scanty. This study looked at the practices of trypanocidal drug utilization by farmers and the integrity of active ingredient in trypanocides sold in Gurage zone, south western Ethiopia. The surveys were based on a structured questionnaire and drug quality determination of commonly used brands originating from European and Asian companies and sold at both authorized and unauthorized markets. One hundred farmers were interviewed and 50 drug samples were collected in 2013 (Diminazene aceturate = 33 and Isometamidium chloride = 17; 25 from authorized and 25 from unauthorized sources). Samples were tested at the OIE-certified Veterinary Drug Control Laboratory (LACOMEV) in Dakar, Senegal, by using galenic standards and high performance liquid chromatography.

Results: Trypanosomosis was found to be a major threat according to all interviewed livestock keepers in the study area. Diminazene aceturate and isometamidium chloride were preferred by 79% and 21% of the respondents respectively, and 85% of them indicated that an animal receives more than six treatments per year. About 60% of these treatments were reported to be administered by untrained farmers. Trypanocidal drug sources included both unauthorized outlets (56%) and authorized government and private sources (44%). A wide availability and usage of substandard quality drugs was revealed. Twenty eight percent of trypanocidal drugs tested failed to comply with quality requirements. There was no significant difference in the frequency of non-compliance between diminazene-based and isometamidium chloride products ($P = 0.87$) irrespective of the marketing channel (official and unofficial). However, higher rates of non-compliant trypanocides were detected for drugs originating from Asia than from Europe ($P = 0.029$).

Conclusion: The findings revealed the presence of risk factors for the development of drug resistance, i.e. wide distribution of poor quality drugs as well as substandard administration practices. Therefore, it is strongly recommended to enforce regulatory measures for quality control of veterinary drugs, to expand and strengthen veterinary services and to undertake trypanocidal drug efficacy studies of wider coverage.

Keywords: Diminazene, Isometamidium, Trypanocide, Drug quality assessment, Drug utilization practice, Ethiopia

* Correspondence: getachew_terefe@yahoo.com;
getachew.terefe@aau.edu.et
[3]Department of Pathology & Parasitology, Addis Ababa University College of Veterinary Medicine and Agriculture, P.O.Box 34, Bishoftu, Ethiopia
Full list of author information is available at the end of the article

Background

Ethiopia has the largest livestock population in Africa with 53.9 million cattle, 24.6 million goats, 25.5 million sheep 6.8 million donkeys and 1.9 million horses [1]. Hence, livestock is a significant contributor to economic and social development of the country. Livestock accounts for 15 to 17% and 35 to 49% of the total and agricultural GDP respectively [1]. Unfortunately, the development and intensification of livestock production is hampered by transboundary epizootic diseases such as African animal trypanosomosis (AAT). It is estimated that, should AAT be eliminated in Ethiopia, direct benefits would exceed 800 million USD over a 20-year period [2].

Various efforts to control the disease and the associated economic losses have been directed mainly against the parasite through trypanocidal drugs and against the vector through odour-baited and insecticide- impregnated targets/traps and insecticide-treated cattle [3–5]. The main drugs used for treating the disease, i.e. diminazene aceturate (DA) and isometamidium chloride (ISM) have been on the market for more than half a century and the parasites' resistance to both trypanocidal drugs is increasing [6].

In Ethiopia, several authors reported prevalence of drug resistance against one or both of the drugs (DA and ISM) in several AAT-affected areas [7–12]. It is believed that the emergence of drug resistance is linked to bad handling and utilization practices as well as to poor drug quality [3, 13, 14], which severely reduces the effectiveness of chemotherapy. Drug resistance to ISM is more widespread than to DA [15], but multiple drug resistance is being increasingly reported from different parts of Africa [16, 17].

None of the previous studies has assessed the quality of trypanocidal drugs in Ethiopia. The present study aimed to fill this knowledge gap. The study was initiated as part of a joint action for trypanocidal drug quality control in Africa whose partners include the Global Alliance for Livestock Veterinary Medicine (GALVmed), the Food and Agriculture Organization of the United Nations (FAO), the International Federation of Animal Health (IFAH), the International Atomic Energy Agency (IAEA) and the Trypanosomosis Rational Chemotherapy (TRYRAC) project, funded by the European Commission. The specific objectives of this study were to assess trypanocidal drug utilization practices in south-western Ethiopia and evaluate the quality of diminazene aceturate- and isometamidium chloride-based brands.

Methods

Study sites

The study was conducted in south-western Ethiopia, at the boundary between the Southern Nations, Nationalities and Peoples' Region (Gurage zone) and Oromia Region (Jimma zone). In these areas, three species of tsetse flies are present (*G. pallidipes, G. fuscipes fuscipes, G. morsitans submorsitans*) [18]. According to information obtained from local authorities, there were 124 tsetse- and trypanosome-affected villages in the study areas. After randomly assessing 40 villages among the 124, five villages with trypanosome prevalence ≥10% were identified for the questionnaire survey: Borer 4, Borer 5, Misreta, Wolaita and Wuhalimat. Trypanocidal drugs were collected from authorized (veterinary clinics and drug stores) and unauthorized markets of the selected areas (Fig. 1) and wholesalers supplying the areas. The drugs were analysed at the Veterinary Drug Control Laboratory (LACOMEV) in Dakar, Senegal.

Questionnaire survey

A structured questionnaire was used to collect data on the practice of trypanocidal drug usage amongst farmers in the five selected study villages. A total of 100 farmers selected by systematic random sampling technique (20 farmers in each village) were interviewed. The respondents were selected from farmers who were voluntarily presenting their cattle for trypanosomosis screening during the study period.

Purchase of trypanocidal products for quality assessment

Fifty samples of trypanocidal drugs were collected from different vendors (authorized and unauthorized) and wholesalers in October 2013. The sampling points were purposely selected for consistency with the drug supply facilities mentioned in the questionnaire survey. Sampling points included authorized markets (Veterinary Clinics, pharmacies and wholesalers) and unauthorized sources (open markets) (Table 1). The drugs were sealed in plastic bags identified by a unique number. Information such as trade name, manufacturer, origin, date of manufacture, expiry date, and place of purchase were recorded. To ensure sufficient quantity, each sample contained at least 5–10 sachets depending on the content per sachet (Five for the 23.6 g DA and 1 g Ism, 10 for the 2.36 g DA and 125 mg ISM) to ensure sufficient sample for analysis.

Assessment of trypanocidal drug quality

Drug quality was assessed by (i) galenic tests, (ii) identification and (iii) measurement of concentration of the respective active ingredient according to standard operating procedures prepared by GALVmed, FAO and IFAH in collaboration with Manchester Metropolitan University and IAEA [19]. The galenic testing included pH measurement, solubility/limpidity of solutions prepared from DA granules or ISM powder according to the manufacturers' recommendations. The pH was measured using a Metler MP 230 pH meter with a pH between 4 and 7 considered

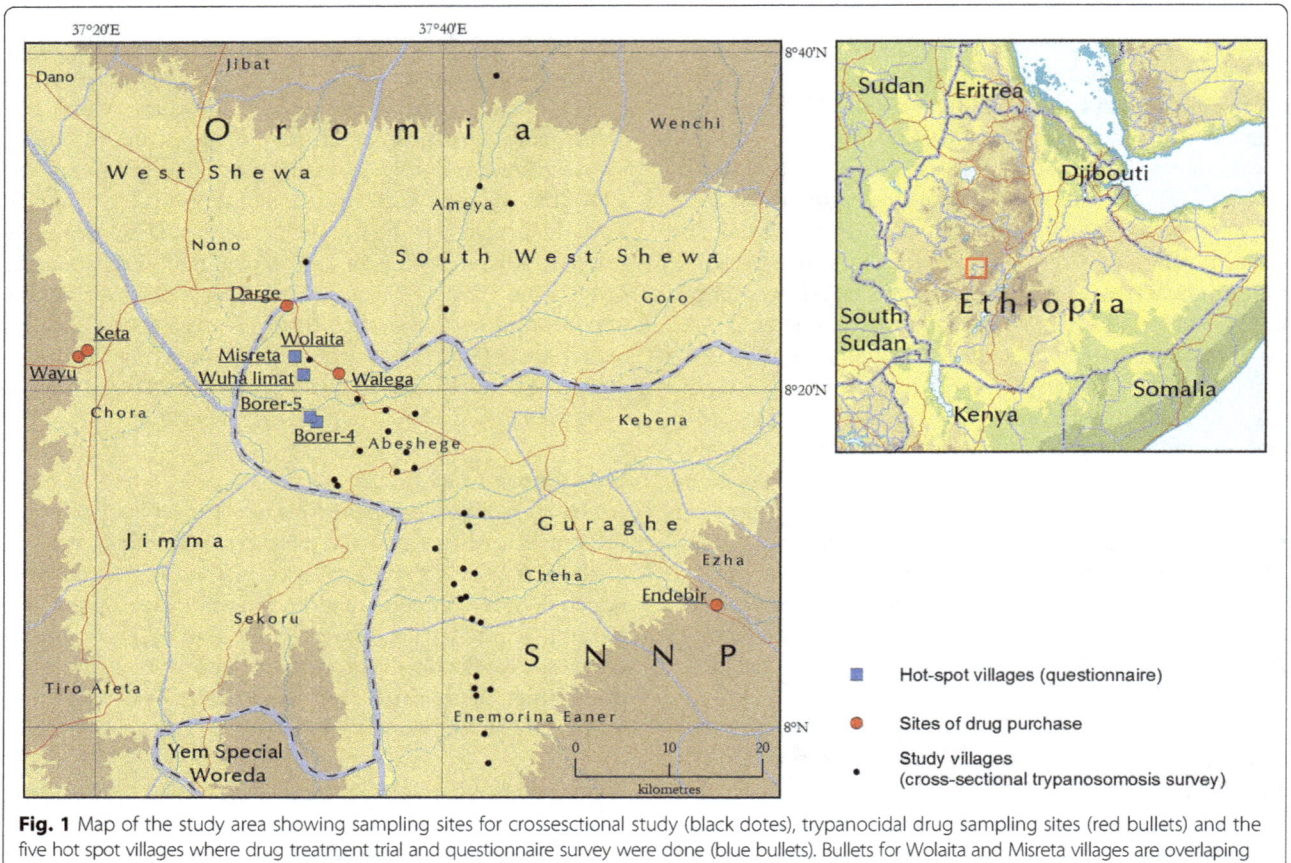

Fig. 1 Map of the study area showing sampling sites for crossesctional study (black dotes), trypanocidal drug sampling sites (red bullets) and the five hot spot villages where drug treatment trial and questionnaire survey were done (blue bullets). Bullets for Wolaita and Misreta villages are overlaping

as compliant. The limpidity of solutions was assessed visually with the naked eye for the absence of visible solid particles Identification and concentration of the active ingredient were assessed using high performance liquid chromatography (HPLC) [19]. Each sample was simultaneously measured with a reference standard. The standards for diminazene and isometamidium were manufactured by VETOQUINOL (Paris, France) and CEVA (Libourne, France), respectively [19] and were provided to LACO-MEV by the consortium GALVmed/FAO/IAEA/IFAH. They were stored in a refrigerator at 4° +/− 1 °C until use.

Trypanocide samples were dissolved in ultrapure water for DA and 25% acetonitrile in ultrapure water for ISM to obtain a solution of 0.1 mg/ml of active ingredient. The solution was poured into vials and introduced into the HPLC machine that was programmed to automatically conduct the process of identification and concentration measurement. The mobile phase for the analysis of DA used a mix of 10% methanol, 10% acetonitrile and 80% ammonium formate buffer (20 mM, pH 4.0). Similarly, the mobile phase for ISM used 25% acetonitrile and 75% ammonium formate buffer (50 mM, pH 2.8). A Water Kromasil C18® HPLC column (150 × 4.6 mm, 5 μm - AkzoNobel, Separation Products, Bohus, Sweden) was used to run the test. The procedure was performed twice with a maximum acceptable divergence between analyses of 2%. In case of higher divergence, the procedure was repeated until it fell within the 2% threshold. For DA, a measured concentration within ±10% variation from the manufacturers' label claim was considered as compliant [20]. For ISM, the following criteria were used to declare compliance: (i) presence of

Table 1 Sources of trypanocidal drugs used for quality analysis (drug samples were collected from nearby authorized vendors and suppliers as well as open markets/illegal sources)

	Wholesaler	Governmental clinic[a]	Private veterinary pharmacy[a]	Open market[b]	Total
ISM	5	3	2	7	17
DA	5	2	8	18	33
Total	**10**	**5**	**10**	**25**	**50**

[a]authorized, [b]unauthorized

ISM Isometamidium chloride, *DA* Diminazene aceturate

the four isomers (I, II, III, IV), (ii) a proportion of isomer I (principal component) equal to or greater than 55%, (iii) a proportion of isomers II, III and IV equal to or less than 40% and (iv) a proportion of the four isomers falling between 95 and 102%. All along the quality assessment process, names of pharmaceutical companies producing the drugs were kept anonymous.

Statistical analysis

Confidence intervals of proportions were calculated assuming binomial distributions of the variables. For drug quality compliance study, a logistic regression was employed. The response was drug compliance whereas binary explanatory variables were the drugs (DA/ISM), the marketing channel (official/unofficial) and the origin of the drugs (Asia/Europe). Non-significant explanatory variables ($p > 0.05$) were removed from the model. In the event that a category contained no observation, the exact test was used to evaluate the significance of the difference and the exact method was used to calculate the confidence interval.

Ethical considerations

The objectives of this study were well explained to all selected farmers and those who expressed their consent to participate in the questionnaire survey were recruited. The identity of study participants and data on their livestock population were kept confidential.

Results

Trypanocidal drug utilization practices

For all respondent farmers communal or free grazing is the predominant livestock management practice in the study villages and trypanosomosis was ranked as biggest animal health constraint.

For the control of trypanosomosis, all respondents reported to depend mainly on the two trypanocidal drugs diminazene aceturate and isometamidium chloride rather than other control methods (Additional file 1: Annex I). The majority of respondents in all villages (56%; 95% CI = 45.7–65.9%) get trypanocidal drugs from unauthorized markets (Fig. 2). Diminazene aceturate was the preferred drug (79%) over ISM (21%). Respondents explained that diminazene was cheaper and available from all sources in single dose/sachet. On the other hand, significant numbers of farmers in the study villages (59%; 95%; CI = 48.7–68.7%) administer trypanocidal drugs by themselves or through family members (Fig. 3) and about 85% of them (95% CI = 76.5–91.4%) indicated that they treat their animals seven or more times per year (Fig. 4). All respondents perceived treatment failures as common. In this respect, 58% of the interviewed farmers believe that treatments were more likely to be successful when the drugs are sourced from

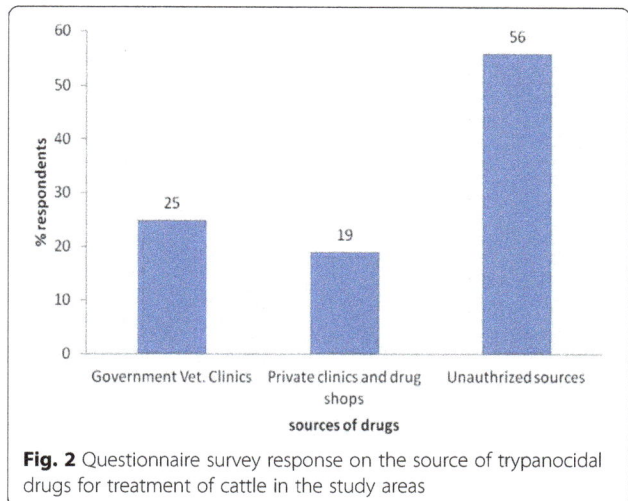

Fig. 2 Questionnaire survey response on the source of trypanocidal drugs for treatment of cattle in the study areas

government veterinary clinics and authorized private sources than from unauthorized open markets.

Trypanocidal drug quality

Of the 50 trypanocidal drug samples collected, 26 were from European companies and 24 from Asian companies. The samples represent 19 different trade names (9 for ISM and 10 for DA). Overall, result showed that 28% (14/50) of the drugs were non-compliant due to insufficient active ingredient detected by HPLC. The difference in non-compliance between the two drugs was not significant ($P = 0.87$ in univariate model), being 27.3% for DA and 29.4% for ISM. Also, clients of authorized and non-authorized markets are at similar risk of purchasing non-compliant trypanocides (Table 2; $P = 0.53$ in univariate model). On the other hand, logistic regression analysis shows that the proportion of non-compliant drugs was significantly more important amongst Asian drugs

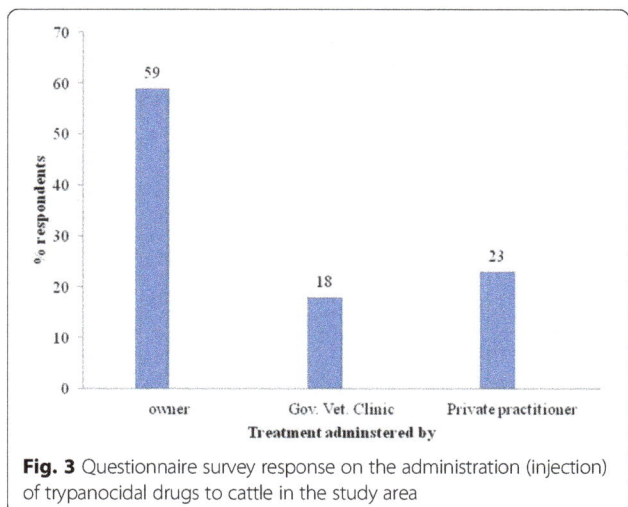

Fig. 3 Questionnaire survey response on the administration (injection) of trypanocidal drugs to cattle in the study area

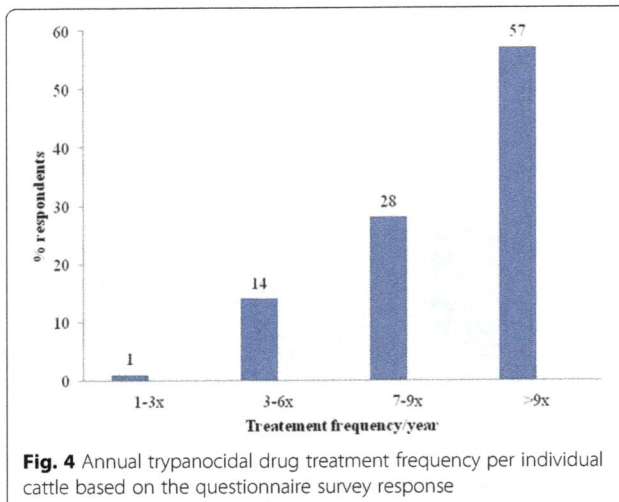

Fig. 4 Annual trypanocidal drug treatment frequency per individual cattle based on the questionnaire survey response

($P = 0.01$ in multivariate model) especially for ISM (interaction not tested since all European ISM samples were found compliant) (Fig. 5). Among the 14 non-compliant samples, nine were from Indian origin (four brands) and two were from UK (one brand). The correlation between marketing channel (official and unofficial) and trypanocidal origin (Asia and Europe) is not significant ($P = 0.26$ in a Chi Square test).

Discussion

Several control approaches are available to overcome the impacts of AAT [21]. Treatment by the use of trypanocidal drugs such as diminazene aceturate and isometamidium chloride is oftentimes the only method available to farmers for containing the disease in many parts of Ethiopia. In this respect, there is a growing risk of drug resistant trypanosomes as already reported by previous studies [7–12]. In this study, it was observed that risk factors for drug resistance such as the presence of unofficial drug sources and frequent treatments are widespread in the study areas. Diminazene aceturate was reported to be the most frequently used compared to isometamidium chloride. This is believed to expose the drug to more risk of trypanocidal resistance. Inadequate veterinary services and the higher price of drugs from legal sources might have contributed for the greater frequency of respondents opting for unauthorized/illegal sources to easily access trypanocidal drugs. A similar

report was documented by Zewdu et al. [22] where annual treatment frequencies range between one and 12 injections/animal and where a significant number of farmers directly gave injections without counseling from a veterinarian. According to Uilenberg [14] and Holmes et al. [3], a high number of annual trypanocidal treatments is suggestive of drug resistance in a given area. Therefore, the high frequency of trypanocidal treatments (more than six times per year), access to drugs from unauthorized sources and the practice of treating animals by untrained personnel is likely to increase the risk of trypanocidal resistance in the study areas and neighbouring localities.

Quality assessment of trypanocides is a prerequisite to ensure better management of trypanosomosis and the prevention of drug resistance [23]. Therefore, trypanocidal drugs were bought anonymously in order to avoid attracting the attention of vendors, who otherwise might have denied access to the products. The 28% of non-compliance observed in this study was below the 71.4% reported in Ivory Coast [24], 100% in Cameroon [25], 70% in Senegal [26], 40% in Togo [27] and 42.3% observed in Burkina Faso [28]. However, since the drugs were supplied by big companies who are national distributers, non-compliant products pose a serious threat to successful AAT treatments countrywide.

The non-compliant veterinary products identified by this study were found in both authorized and unauthorized markets which is different from reports from Togo and Mali [27–29] where non-conform drugs were significantly more often found on unauthorized markets. The non-compliance may be attributed to poor storage conditions and the doubtful sources of supply in this market channel. The 2012 Proclamation for Veterinary Drugs and Feed Administration and Control of Ethiopia capitalizes on regulating "the production, distribution and use of veterinary drugs to ensure safety, efficacy and quality of the products". It also focuses on "prevention and control of the illegal production, distribution and use of veterinary drugs" [30]. Therefore, the observation of such a significant frequency of non-compliant trypanocides undermines the basic objective of the existing law. Although there are encouraging signs, the existing situation signals the urgent need to enforce the legislative framework at all levels to reduce and prevent illegal marketing and use of veterinary

Table 2 Proportions of non-compliant samples according to the marketing channel and drug type

Molecule	Illegal(unauthorized)		Legal(authorized)	
	Complies	Not complies	Complies	Not complies
Diminazene Aceturate	13	5	11	4
Isometamidium chloride	4	3	8	2

Number of observations

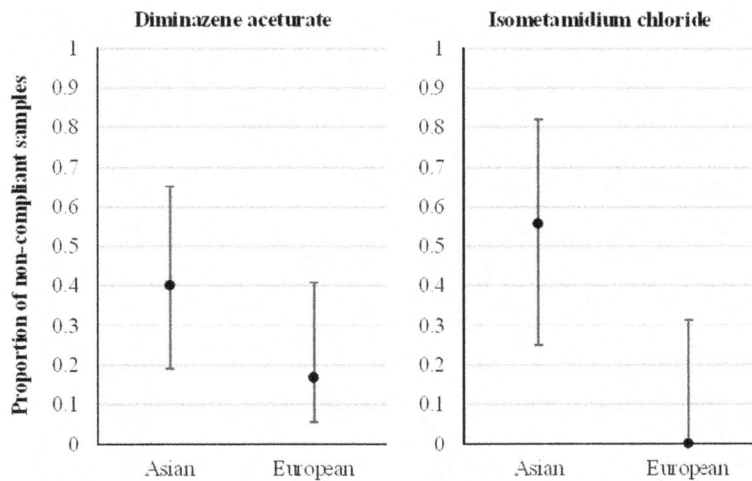

Fig. 5 Proportions of non-compliant samples according to continental origin (Asian and European) of the drugs with 95% confidence intervals (based on a logistic regression using drug, origin and the interaction between them as explanatory variables; the exact method was used for categories with no observations)

drugs. A similar failure to adequately regulate and control medicinal products for human use has already been reported in the country [31].

Although this study did not demonstrate significant variation in non-compliance between the two drugs, coupled with the high pressure on diminazene due to its short biological half-life and hence more frequent administration [32, 33], this loss of quality would mean a greater risk for the development of resistance. Various brands of trypanocides originating from several countries in Europe and Asia have shown different levels of compliance to drug quality casting doubt on the reputability of some of the manufacturers of veterinary products [27]. In this respect, the detection of non-compliant drugs within samples sourced from wholesalers strongly suggests that some of these drugs might be defective right from their origin and in such case unauthorized vendors in rural areas might have obtained them from legal importers/suppliers. On the other hand, since some of the brand names are identical between both authorized and unauthorised sources, it is also possible that counterfeit drugs are being sold carrying brand names of genuine manufacturers.

The impact of such quality defects in veterinary trypanocidal drug products can be very significant. A lack of efficacy will cause financial harm due to under dosage and residues in the muscle tissue damaging consumers when the concentration is above the prescribed limit [34, 35]. Some specialists [20] quite rightly believe that the circulation of counterfeit drugs in most sub-Saharan African countries inevitably leads to the persistence of animal diseases. This is proof of the urgent need to introduce systematic controls on the supply, distribution and utilization channels for veterinary trypanocidal products in Ethiopia.

Conclusion

Although the results obtained by this study may not represent trypanocidal drug use of the entire country, they clearly show the effects of poor quality veterinary trypanocidal drugs which are a result of the lack of a proper import control and country administration of veterinary controls in the study area and probably beyond. Unsafe utilization practices and poor handling of the drugs by farmers is also believed to aggravate the situation. Non-compliant trypanocidal drugs are known to compromise the efficacy of treatment and escalate the development of drug resistance, which has already been reported in different parts of Ethiopia. Corrective measures must therefore be taken towards effective implementation of regulatory and legislative frameworks, quality control of veterinary trypanocidal drugs and improving the coverage and quality of veterinary services and regular monitoring of trypanocidal drug efficacy.

Abbreviations
AAT: African animal trypanosomosis; DA: Diminazene aceturate; FAO: Food and Agriculture Organization; GALVmed: Global Alliance for Livestock Veterinary Medicine; GDP: Gross domestic product; HPLC: High performance liquid chromatography; IAEA: International Atomic Energy Agency; IFAH: International Federation for Animal Health; ISM: Isometamidium; LACOMEV: Laboratoire de Contrôle des Médicaments Vétérinaires; TRYRAC: Trypanosomosis Rational Chemotherapy

Acknowledgements
Special thanks go to the National Animal Health Diagnostic and Investigation Center of Ethiopia including Mr. Tesfaye Mulatu and Ms. Genet Bogale for their great assistance during data collection, LACOMEV in Senegal for drug quality analysis, and the CVMA of Addis Ababa University and Mr. Ayalew

Zelelew of the District Bureau of Agriculture in Gurage Zone for their all rounded assistance.

Funding
This research was funded by the European Union through the European Commission-funded Trypanosomosis Rational Chemotherapy (TRYRAC) project, DCI-FOOD/2011/279–754. FAO's assistance to this study was provided in the framework of the Programme against African Trypanosomosis (PAAT), and supported by the Government of Italy (Project 'Improving food security in sub-Saharan Africa by supporting the progressive reduction of tsetse-transmitted trypanosomosis in the framework of the NEPAD', codes GTFS/RAF/474/ITA and GCP/RAF/502/ITA).The European Commission and the FAO did not participate in the study process.

Authors' contributions
TT collected the data during field work analyzed and interpreted the results and drafted the manuscript. TG Developed a questionnaire, supervised the project and edited the manuscript. CT: Developed and supervised the project. HA Supervised the project and edited the manuscript. AKG and TAA analyzed drug samples in OIE lab in Dakar. CP-H and HAN developed the project and edited the manuscript. GG supported field data collection. MRC and PR facilitated laboratory analysis and revised the manuscript. MT analyzed the data and edited the manuscript. CG proof read the manuscript and constructed map of the study sites. VDAJ and DV developed the project and the study protocol, proof read the manuscript. All authors read and approved the final manuscript.

Competing interest
The authors declare that they have no competing interests.

Ethics approval and consent to participate
The objectives of this study were well explained to all selected farmers and those who expressed their consent to participate in the questionnaire survey were taken. The identity of study participants was kept confidential. Similarly, the identities of trypanocidal drug sources and company names remain confidential. The Research Ethics Review Committee (RERC) of the College of Veterinary Medicine and Agriculture of the Addis Ababa University (Ref. No. VM/ERC/05/07/06/2013) reviewed the study protocol. The committee waived the need for ethics approval and stated that the questionnaire survey part does not involve collecting information about the respondents and their wealth and hence does not require proper clearance.

Consent for publication
Not applicable

Author details
[1]National Animal Health Diagnostic and Investigation Center-Protozoology unit, P.O. Box 8615, Addis Ababa, Ethiopia. [2]Minstry of Livestock and Fisheries, Addis Ababa, Ethiopia. [3]Department of Pathology & Parasitology, Addis Ababa University College of Veterinary Medicine and Agriculture, P.O.Box 34, Bishoftu, Ethiopia. [4]Ecole Inter- Etats des Sciences et Médecine vétérinaires de Dakar, P.O.Box 5077, Dakar, Fann, Senegal. [5]Department of Biomedical Sciences Veterinary Protozoology, Institute of Tropical Medicine, Unit 155 Nationalestraat, B-2000 Antwerp, Belgium. [6]Institute for Parasitology and Tropical Veterinary Medicine, Freie Universitaet Berlin, Robert-von-Ostertag Str. 7-13, 14163 Berlin, Germany. [7]Food and Agriculture Organization of the United Nations, Viale delle Terme di Caracalla, 00153 Rome, Italy. [8]Global Alliance for Livestock Veterinary Medicines (GALVmed), Doherty Building, Pentlands Park, Bush Loan, Edinburgh EH26 0PZ, UK.

[9]Veterinary Epidemiology and Risk Analysis - Research and Development (VERDI-R&D), Rue du Gravier 7, 4141 Sprimont, Belgium. [10]Food and Agriculture Organization of the United Nations, Sub-Regional Office for Eastern Africa, Addis Ababa, Ethiopia. [11]Faculty of Sciences and Bio-engineering Sciences, Vrije Universiteit Brussel, Pleinlaan 2, B-1050 Brussels, Belgium.

References
1. Central Statistical Agency. Agricultural sample survey. Report on livestock and livestock characteristics. 2013; Volume II and IV, Addis Ababa, Ethiopia.
2. Shaw APM, Cecchi G, Wint GRW, Mattioli RC, Robinson TP. Mapping the economic benefits to livestock keepers from intervening against bovine trypanosomosis in eastern Africa. Prev Vet Med. 2014;113:197–210.
3. Holmes PH, Eisler MC, Geerts S. Current chemotherapy of animal trypanosomiasis. In: Maudlin I, Holmes PH, Miles MA, editors. The trypanosomiasis. Wallingford, United Kingdom, CABI Publishing, 2004. p. 431.
4. Meyer A, Holt HR, Selby R, Guitian J. Past and ongoing tsetse and animal trypanosomiasis control operations in five African countries: a systematic review. PLoS Negl Trop Dis. 2016;10:e0005247. https://doi.org/10.1371/journal.pntd.0005247.
5. Shaw AP, Wint GR, Cecchi G, Torr SJ, Mattioli RC, Robinson TP. Mapping the benefit-cost ratios of interventions against bovine trypanosomosis in eastern Africa. Prev Vet Med. 2015;122:406–16.
6. Giordani F, Morrison LJ, Rowan TG, De Koning HP, Barrett MP. The animal trypanosomiases and their chemotherapy: a review. Parasitology. 2016;143:1862–89.
7. Shimelis D, Getachew T, Getachew A, Dave Barry D, McCulloch R, Goddeeris B. In vivo experimental drug resistance study in Trypanosoma vivax isolates from tsetse infested and non-tsetse infested areas of Northwest Ethiopia. Acta Trop. 2015;146:95–100.
8. Moti Y, Fikru R, Van Den Abbeele J, Buscher P, Duchateau L, Delespaux V. Gibe river basin in Ethiopia: present situation of trypanocidal drug resistance in T.congolense using tests in mice and PCR-RFLP. Vet Parasitol. 2012;189:197–203.
9. Desalegn WY, Etsay K, Getachew A. Study on the assessment of drug resistance on Trypanosoma vivax in Selemti wereda, Tigray, Ethiopia. Ethiop Vet J. 2010;14:15–30.
10. Shimelis D, Sangwan AK, Getachew A. Assessment of trypanocidal drug resistance in cattle of the Abay (Blue Nile) basin areas of north West Ethiopia. Ethiop Vet J. 2008;12:45–59.
11. Tewelde N, Abebe G, Eisler MC, Mcdermott J, Grainer M, Afework Y, et al. Application of field methods to assess isometamidium resistance of trypanosomes in cattle in western Ethiopia. Acta Trop. 2004;90:163–70.
12. Afework Y, Clausen PH, Abebe G, Tilahun G, Mehlitz D. Multiple drug resistant Trypanosoma congolense population in village cattle of Metekel district, north-West Ethiopia. Acta Trop. 2000;76:231–8.
13. Van Den Bossche P, Doran M, Connor RJ. An analysis of trypanocidal drug use in Eastern Province of Zambia. Acta Trop. 2000;75:247–58.
14. Uilenberg G. A field guide for the diagnosis, treatment and prevention of African animal trypanosomosis. Rome: Food and Agriculture Organization of the United Nations; 1998. p. 158.
15. Geerts S, Holmes PH, Eisler MC, Diall O. African bovine trypanosomosis: the problem of drug resistance. Trends Parasitol. 2001;17:25–8.
16. Mungube EO, Vitouley HS, Allegye-Cudjoe E, Diall O, Boucoum Z, Diarra B, et al. Detection of multiple drug resistant Trypanosoma congolense populations in village cattle of south east Mali. Parasit Vectors. 2012;5:155.
17. Geerts S, Holmes PH. Drug management and parasite resistance in bovine trypanosomiasis in Africa. PAAT Technical Science Series, No. 1, FAO, Rome, 1998. p.31.
18. Cecchi G, Paone M, Argiles Herrero R, Vreysen MJ, Mattioli RC. Developing a continental atlas of the distribution and trypanosomal infection of tsetse flies (Glossina species). Parasit Vectors. 2015;8:284.
19. Sutcliffe OB, Skellern GG, Araya F, Cannavan A, Sasanya JJ, Dungu B, et al. Animal trypanosomosis: making quality control of trypanocidal drugs possible. RevSci Tech Int des Epizoot. 2014;33:1–42.
20. Tettey J, Chizyuka G, Atsriku C, Slingenbergh J. Non-conformance of diminazene preparations to manufacture's label claims: an extra factor in the development of parasite resistance? ICPTV Newsl. 2002;5:24–5.

21. Diall O, Cecchi G, Wanda G, Argilés-Herrero R, MJB V, Cattoli G, et al. Opinion: developing a progressive control pathway for African animal trypanosomosis. Trends Parasitol. 2017;33:499–509.
22. Zewdu S, Getachew T, Hagos A. Farmers' perception of impacts of bovine trypanosomosis and tsetse fly in selected districts in Baro-Akobo and Gojeb river basins. South-western Ethiopia BMC Vet Res. 2013;9:214.
23. Clausen P-H, Bauer B, Zessin KH, Diall O, Bocoum Z, Sidibe I, et al. Preventing and containing trypanocide resistance in the cotton zone of West Africa. Transbound Emerg Dis. 2010;57:28–32.
24. Assoumy AM, Teko-Agbo A, Akoda K, Niang EMM, Oulai J. Qualité pharmaceutique des médicaments vétérinaires en Côte d'Ivoire: cas du district d'Abidjan. Rev Africaine Santé Prod Anim. 2010;8:149–53.
25. Teko-Agbo A, Akoda K, Assoumy AM, Kadja MC, Niang EMM, Messomo Ndjana F, et al. Qualité des médicaments vétérinaires en circulation au Cameroun et au Sénégal. Dakar Med. 2009;54:226–34.
26. Akoda K, Walbadet L, Niang EMM, Teko-Agbo A. Qualité des médicaments vétérinaires en circulation au Sénégal. Rev Africaine Santé Prod Anim. 2008;6:29–33.
27. Tchamdja E, Kulo AE, Akoda K, Teko Agbo A, Assoumi AM, Niang E, et al. Drug quality analysis through high performance liquid chromatography of isometamidium chloride hydrochloride and diminazene diaceturate purchased from official and unofficial sources in northern Togo. Prev Vet Med. 2016;126:151–8.
28. Teko-Agbo A, Assoumy AM, Akoda K, EMM N, PLJ BH. Qualité pharmaceutique des médicaments antiparasitaires vétérinaires au Burkina-Faso. Rev Africaine Santé Prod Anim. 2011;9:3–7.
29. Abiola FA. Qualité des anthelminthiques et des trypanocides au Mali (étude préliminaire par sondage limité). Rapport d'expertise. Dakar: EISMV; 2002. 13 pp.
30. Federal Democratic Republic of Ethiopia. A proclamation to provide for veterinary drug and feed administration and control, Proclamation No. 728/2011. Federal Negarit Gazeta, Addis Ababa. 2012. 18th year, No. 14.
31. Suleman S, Woliyin A, Woldemichael K, Tushune K, Duchateau L, Degroote A, et al. Pharmaceutical regulatory framework in Ethiopia: a critical evaluation of its legal basis and implementation. Ethiop J Health Sci. 2016;26:259–76.
32. Kaur G, Chaudhary RK, Srivastava AK. Pharmacokinetics, urinary excretion and dosage regimen of diminazene in cross bred calves. Acta Vet Hung. 2000;48:187–92.
33. Eisler MC, Maruta J, Nquindi J, Connor RJ, Ushewokunze-Obatolu U, Holmes PH, et al. Isometamidium concentrations in the sera of cattle maintained under a chemo-prophylactic regime in a tsetse-infested area of Zimbabwe. Tropical Med Int Health. 1996;1:535–41.
34. Monteiro AM. Pharmaceutical quality of veterinary anthelminthics sold in Kenya. Vet Rec. 1998;142:392–8.
35. Boisseau J. Contrôle de la qualité pharmaceutique des médicaments vétérinaires. In: Séminaire sur le contrôle de la qualité des médicaments vétérinaires Niamey, 8–12 December 1997. Paris: OIE. p. 1997, 93.

10

Prevalence and clonal relationship of ESBL-producing *Salmonella* strains from humans and poultry in northeastern Algeria

Samia Djeffal[1,2,3], Sofiane Bakour[3], Bakir Mamache[4], Rachid Elgroud[1], Amir Agabou[5], Selma Chabou[3], Sana Hireche[1], Omar Bouaziz[1], Kheira Rahal[6] and Jean-Marc Rolain[3*]

Abstract

Background: The aims of this study were to investigate *Salmonella* contamination in broiler chicken farms and slaughterhouses, to assess the antibiotic resistance profile in avian and human *Salmonella* isolates, and to evaluate the relationship between avian and human Extended Spectrum β-Lactamase (ESBL)-producing isolates. *Salmonella* was screened in different sample matrices collected at thirty-two chicken farms and five slaughterhouses. The human isolates were recovered from clinical specimens at the University Teaching Hospital of Constantine (UTH). All suspected colonies were confirmed by MALDI-TOF (Matrix Assisted Laser Desorption Ionization Time OF light) and serotyped. Susceptibility testing against 13 antibiotics including, amoxicillin/clavulanic acid, ticarcillin, cefoxitin, cefotaxime, aztreonam, imipenem, ertapenem, gentamicin, amikacin, ciprofloxacin, colistin, trimethoprim/sulfamethoxazole and fosfomycin, was performed using the disk diffusion method on Mueller-Hinton agar. ESBL-production was screened by the double-disk synergy test and confirmed by molecular characterization using PCR (polymerase chain reaction) amplification and sequencing of ESBL encoding genes. Clonality of the avian and human strains was performed using the Multi Locus Sequencing Typing method (MLST).

Results: Forty-five isolated avian *Salmonella* strains and 37 human collected ones were studied. Five *S. enterica* serotypes were found in avian isolates (mainly Kentucky) and 9 from human ones (essentially Infantis). 51.11% and 26.6% of the avian isolates were resistant to ciprofloxacin and cefotaxime, respectively, whereas human isolates were less resistant to these antibiotics (13.5% to ciprofloxacin and 16.2% to cefotaxime). Eighteen (12 avian and 6 human) strains were found to produce ESBLs, which were identified as $bla_{CTX-M-1}$ ($n = 12$), $bla_{CTX-M-15}$ ($n = 5$) and bla_{TEM} group ($n = 8$). Interestingly, seven of the ESBL-producing strains (5 avian and 2 human) were of the same ST (ST15) and clustered together, suggesting a common origin.

Conclusion: The results of the combined phenotypic and genotypic analysis found in this study suggest a close relationship between human and avian strains and support the hypothesis that poultry production may play a role in the spread of multidrug-resistant *Salmonella* in the human community within the study region.

Keywords: *Salmonella*, poultry, human, serotype, antimicrobial resistance, resistance genes, clonality

* Correspondence: jean-marc.rolain@univ-amu.fr
[3]Unité de recherche sur les maladies infectieuses et tropicales émergentes (URMITE), UM 63, CNRS 7278, IRD 198, INSERM 1095, IHU Méditerranée Infection, Faculté de Médecine et de Pharmacie, Aix-Marseille-Université, Marseille, France
Full list of author information is available at the end of the article

Background

Salmonella infections are a major public health problem with a significant social and economic impact. Many animal species are potential reservoirs for this bacterium, especially chickens, pigeons and reptiles [1]. Humans can commonly acquire the infection through the food chain [2]. Young and immunocompromised patients are the most exposed to dangerous complications which are generally treated with fluoroquinolones and extended-spectrum cephalosporins that are largely used in veterinary medicine [3].

In Algeria, the poultry industry has grown remarkably since 1980. However, as a result of the deleterious hygienic conditions, many infectious diseases, such as Salmonellosis, were detected in broilers and constitute a risk to the human health [4]. Due to a lack of surveillance programs, information on the prevalence of *Salmonella* and other food pathogens is incomplete. For instance, a study by the Pasteur Institute of Algeria revealed that 11% of food poisoning cases were caused by *Salmonella* spp. in 2011 [5]. 47% of these cases were mainly related to the consumption of chicken meat, as reported by Mouffok et al., [5]. Eggs and ovoproducts are also among the major sources of human infections, which require a thorough assessment and control measures for *Salmonella spp.* in the poultry industry [6]. These measures are well detailed in the Official Journal of the Algerian Republic (No. 36 of June 8, 2003); but were only applied to the poultry industry facilities in the public sector as well as to private poultry production units and facilities [7].

Selective pressure due to the misuse of antibiotics in humans and domestic livestock is one of the many factors that has led to the emergence of antibiotic resistance in commensal and pathogenic bacteria; thus, multidrug-resistant (MDR) *Salmonella* have increasingly been isolated from various food products worldwide [8].

Drug resistance is growing and has affected critically important classes of antibiotics, such as the β-lactams, which are among the most significant bactericidal antibiotics used to treat bacterial infections in humans [9]. Extended spectrum β-lactamases (ESBLs) were identified following the introduction of extended-spectrum oxyimino-cephalosporins in the 1980s for the treatment of severe human infections [10]. In veterinary medicine, a variety of these drugs are currently authorized for use, resulting in the emergence of ESBL-producing Gram-negative bacteria [11, 12].

TEM, SHV and CTX-M are the most prevalent ESBL types. Over the last decade, rates of CTX-M producing bacteria have increased worldwide in comparison with TEM and SHV [13]. This situation is rendered more complicated as these enzymes confer co-resistance to other drug classes [13, 14].

In Algeria, the first ESBLs were detected in non-typhoidal *Salmonella* in 1994 from humans [15], but until now there has been no information on the magnitude of this problem in animals and humans. This situation drove us to identify the contamination status of *Salmonella* serotypes in chicken farms and slaughterhouses, to assess their sensitivity to antimicrobials and ESBL-production and, finally, to evaluate their clonality with human pathogenic strains.

Methods

Study area

The present study was carried out between December 2011 and May 2013. Thirty-two chicken farms and five chicken slaughterhouses located in the province of Skikda, Algeria, took part in this survey. The choice of this study area was motivated by the size of the poultry industry and the frequent occurrence of infectious gastrointestinal pathologies as reported by local veterinary practitioners [16].

Twenty-seven poultry houses had concrete walls and floors and corrugated metal sheet roofs while the remaining five houses had earth floors with walls and roofs made of straw and reeds covered with plastic foil. None of the farming sites was fenced, allowing free access for domestic and wild animals. Their rearing capacities vary from 3500 to 20,000 birds per house.

All slaughterhouses had concrete floors with earthenware walls. The slaughtering capacity ranges from 2000 to 7000 chickens per day and the broilers are brought from different poultry farms located in several neighboring provinces. The number of chickens and slaughterhouses is statistically representative of the study region and governed by the capacity of the laboratory processing the samples.

The human *Salmonella* strains were kindly provided by the University Teaching Hospital (UTH) of Constantine and included in the study.

Study design

For technical reasons, including access to the sampling sites, a total of 1194 samples were collected during two sampling periods (between December 2011 and September 2012 and between December 2012 and May 2013). The poultry houses were visited at two periods (when the birds were aged 15–30 days and 45–60 days).

From the chicken farms, a total of 320 samples of water were taken from drinking vessels, 160 samples of feed were taken from the feeding vessels, 330 cloacal swabs were taken and 320 droppings were collected and placed in sterile containers. 64 surface wipes (25 cm × 25 cm, AES Chemunex, Combourg, France) were also obtained from a height of 30 cm from the ground over a 400 cm^2 area of the four walls of the poultry house and placed into sterile Stomacher bags.

The five poultry slaughterhouses were visited once for sampling. Due to limited financial resources, in each poultry slaughterhouse, we pooled individual samples to minimize study costs. The samples were randomly taken from three organs from five chickens (5 g of 5 caeca, 5 g of 5 livers, 5 g of 5 neck skins), and from the environment (one sample of carcass rinsing water, one swab from a sticking knife and one wipe from the walls). The slaughtered animals were brought from several neighboring provinces.

All samples were transported to the laboratory, on ice packs within a period not exceeding two hours, to be treated on the same day or kept in the refrigerator overnight.

Salmonella isolation and identification

Bacteriological analyses were performed according to the EN/ISO 6579−2002/Amd1:2007 protocol for *Salmonella* detection in food and animal feedstuffs. 25 g of samples (droppings, feed, liver, caeca, neck skin) were individually pre-enriched with 225 mL of buffered peptone water broth (PWB) (Fluka, Sigma Aldrich, France). The swabs were individually placed in 10 mL PWB, while 100 mL of drinking and carcass rinsing water was individually mixed with 100 mL of double strength PWB for pre-enrichment according to NF U 47−101 Standard (2005) [17]. All samples were incubated at 37 °C for 18−20 h. From each pre-enrichment solution, 1 mL and 0.1 mL were respectively transferred into 10 mL of enrichment Muller-Kauffmann tetrathionate/novobiocin broth (AES Chemunex Combourg, France) and 10 mL of Rappaport Vassiliadis broth (Merck Darmstadt, Germany), incubated respectively at 37 °C and 42 °C for 24 h. Both enriched samples were then streaked on XLD (Fluka analytical Steinheim, Switzerland) and Hektoen agars (Pasteur Institute of Algeria) and incubated at 37 °C for 24 h [18]. Suspected colonies were first identified with the API 20E System (bioMérieux, France), then with MALDI-TOF (Bruker Daltonics GmbH, Germany) [19]. Confirmed *Salmonella* isolates were serotyped according to Kauffmann-White-Le Minor's scheme [20].

Human clinical strains

We selected 37 (non-repetitive) strains recovered from clinical specimens at the UTH of Constantine over a decade (2005−2015). These were isolated from stool samples collected from different wards and among which 26 strains derived from the diarrheic stools of infants admitted to the neonatology ward. The Main characteristics of patients are shown in Additional file 1.

The strains were confirmed using MALDI-TOF mass spectrometry for prescreening *Salmonella* species and sub-species and to identify epidemiologically important serovars which were further tested with conventional serotyping method.

Antimicrobial susceptibility testing and ESBL detection

All human and avian strains were submitted to susceptibility testing against antibiotics using the disk diffusion method on Mueller-Hinton (MH) agar, and the results were interpreted according to the European Comitee on Antimicrobial Suceptibility Testing (EUCAST) [21] (Additional file 2). Thirteen antibiotics (Bio-Rad, France) were tested: amoxicillin/clavulanic acid AMC (20/10 µg), ticarcillin TIC (75 µg), cefoxitin FOX (30 µg), cefotaxime CTX (5 µg), aztreonam ATM (30 µg), imipenem IPM (10 µg), ertapenem ETP (10 µg), gentamicin GEN (10 µg), amikacin AK (30 µg), ciprofloxacin CIP (5 µg), colistin CT (50 µg), trimethoprim/sulfamethoxazole SXT (1.25 µg/23.75 µg) and fosfomycin FF (50 µg).

Extended spectrum β-lactamase production was screened by the double-disc synergy test (DDST) [22].

PCR detection of ESBL genes

Total nucleic acids were extracted using a BioRobot EZ1 Advanced XL instrument (QIAGEN, Hilden, Germany) according to the manufacturer's instructions.

Detection of β -lactamase genes (including bla_{TEM}, bla_{SHV}, and bla_{CTX-M}) was carried out by polymerase chain reaction (PCR) using specific primers: $bla_{CTX-M-1}$ group [23], $bla_{CTX-M-9}$ group [24], bla_{TEM} group [25] and bla_{SHV} [26] and the master mix QuantiTect Probe PCR Kit (QIAGEN, Hilden, Germany).

Amplification products were detected by electrophoresis using agarose gels containing SYBR safe (Invitrogen, Leek, the Netherlands), along with a DNA molecular weight marker (BenchTop pGEM®DNA Marker, Promega, Madison, Wisconsin, USA). Visualization of gels was carried out using the BenchTop pGEM®DNA Marker (Promega, Madison, Wis- consin, USA) under ultraviolet illumination.

ESBL genes sequencing and multilocus sequence typing

ESBL-positive PCR products were purified using the NucleoFast 96 PCR plate (Machery-Nagel EURL, France) and sequenced using the BigDye terminator chemistry on an ABI3730 automated sequencer (Applied Biosystems, Foster City, California, USA). The obtained sequences were analyzed with the ARG-ANNOT database [27].

Multilocus sequence typing (MLST) is useful in assessing the role of specific STs in human and animal disease and assessing overlap between these hosts. It was carried out by PCR amplification and sequencing of seven housekeeping genes: *thrA* (aspartokinase + homoserine dehydrogenase), *purE* (phosphoribosylaminoimidazole carboxylase), *sucA* (alpha ketoglutarate dehydrogenase), *hisD* (histidinol dehydrogenase), *aroC* (chorismate synthase), *hemD* (uroporphyrinogen III cosynthase), and *dnaN* (DNA polymerase III beta subunit), as described by Kidgell et al. 2002 [28].

Briefly, template DNA prepared from bacterial isolates was amplified by PCR with the use of an oligonucleotide sequence for seven housekeeping genes (available in the MLST database: http://mlst.warwick.ac.uk/mlst/dbs/Senterica). Sequencing with the same automated sequencer of the PCR product was carried out by the dideoxynucleotide chain termination method using the master mix QuantiTect Probe PCR Kit (QIAGEN, Hilden, Germany). Forward and reverse DNA sequences were assembled, trimmed, edited, and analyzed for each gene fragment using the ARG-ANNOT database [27]. Allelic profile and sequence type determinations were assigned according to the *Salmonella* MLST database: http://mlst.warwick.ac.uk/mlst/dbs/Senterica.

Clonality analysis

Protein mass profiles were obtained using a Microflex LT MALDI-TOF mass spectrometer (Bruker Daltonics, Germany), with Flex Control software (Bruker Daltonics). The spectrum profiles obtained were visualized with Flex analysis v.3.3 software and exported to ClinProTools software v.2.2 and MALDI-Biotyper v.3.0 (Bruker Daltonics, Germany) for data processing (smoothing, baseline subtraction and spectra selection) and evaluation with cluster analysis.

The phyloproteomic analysis of ESBL-positive *Salmonella* strains from human and poultry origins was assessed through construction and comparison of their characteristic reference spectra (main spectra) with the MALDI-Biotyper v.3.0 software (Bruker Daltonics, Germany). Cluster analysis was performed based on pairwise comparisons of specific main spectra (MSP: mean spectra projection dendrogram) of the different strains to generate a dendrogram of similarities among spectra profiles using the software default correlation function. A distance level of 560 was selected for clustering evaluation of the isolates.

Statistical analysis

Differences in contamination levels of poultry houses at two sampling periods (15–30 days versus 45–60 days), and the antimicrobial resistance patterns between avian and human *Salmonella* strains, were assessed by the Chi square test (at 95% CI and $p < 0.05$) or Fisher's exact test if N is less than 20 and one expected cell is less than or equal 5. All statistical analyses were performed using IBM SPSS Statistics version 24 software (2016).

Results

Frequency of isolation of *Salmonella* serotypes in broilers, slaughterhouses and human samples

Forty-five *Salmonella enterica* from slaughterhouses and poultry farms and 37 of human clinical origin were studied. 34.37% of the poultry farms and all slaughterhouses were contaminated with *Salmonella* and the isolation rate varied depending on the sampling matrix. The samples taken at the age of 15–30 days were more contaminated than those collected at 45–60 days; however, the difference was not significant *(p > 0.05)*.

The isolated *Salmonella* strains belonged mainly to two serotypes: Kentucky and Heidelberg, and the remaining strains were Enteritidis, Virginia and Newport. There was an evident heterogeneous distribution of serotypes in poultry farms and slaughterhouses (Table 1).

Among the human *Salmonella* strains, Infantis was the most frequent serotype, followed by Senftenberg, Enteritidis, Kedougou, Tyhimurium, Heidelberg, Kentucky, Ohio and Arizona. Most of these strains were from infants and the others were from adult diarrheic stools (especially *Salmonella enterica* serotypes Enteritidis and Typhimurium).

Drug resistance patterns of the isolated *Salmonella* strains

A high frequency of resistance to ciprofloxacin (51.1%) was noted in *Salmonella* isolates from both chicken farms and slaughterhouses. These strains were resistant to cephalosporins (26.6% to cefotaxime), aztreonam (26.6%), ticarcillin (46.6%) and gentamicin (22.2%). ESBL-production was found in 26.6% of these avian isolates (11 *Salmonella* ser. Heidelberg and one *Salmonella* ser. Newport).

The human *Salmonella* isolates were highly resistant to ticarcillin (56.75%), amoxicillin (45.94%) and a less extent to trimethoprim/sulfamethoxazole (24.3%), gentamicin (24.3%), aztreonam (18.9%), cefotaxime (16.2%), ciprofloxacin (13.5%) and amikacin (10.8%). Six ESBL-positive strains were detected (Fig. 1).

The difference in resistance to antibiotics, between poultry strains and human ones was significant for: amoxicillin/clavulanic acid $(P < 0.05)$, gentamicin $(P < 0.05)$, ciprofloxacin $(P < 0.05)$ and fosfomycin $(P < 0.05)$. However, it was not significant for cefotaxime, ticarcillin, aztreonam and amikacin $(P > 0.05)$.

The characteristics of the 18 ESBL-producing strains isolated from chicken farms/slaughterhouses and patients are shown in Table 2. Eleven of the *Salmonella* ser. Heidelberg strains harbored $bla_{CTX-M-1}$ genes. The bla_{TEM} group was identified in one avian *Salmonella* ser. Newport isolate as well as in S. Heidelberg, in which it was coupled with the $bla_{CTX-M-1}$ gene. Human strains harbored the $bla_{CTX-M-15}$ gene in association with bla_{TEM}, except one strain which carried instead the $bla_{CTX-M-1}$. None of the strains were positive for $bla_{CTX-M-9}$ and bla_{SHV} genes.

The MLST analysis showed that the *salmonella* isolates belonged to six different sequence types (ST) including, ST14, ST15, ST16, ST32, ST38 and ST198 (Fig. 2). The results demonstrate that ST15 represents the predominant clone. Indeed, this ST was found in 13 *Salmonella* ser. Heidelberg strains (two human and 11 avian) (Fig. 2). In

Table 1 Frequency of isolation of *Salmonella enterica* subsp. *enterica* serotypes in different sample matrices from poultry farms and slaughterhouses

Serotype	Poultry farms and slaughterhouses		Samples	
	N° of positive (ID n°)	(%)	N° of positive samples	(%)
Kentucky	7 Poultry farms (F1, F6, F13, F14, F18, F20, F32)	21.87	5 cloacal swabs 6 droppings 2 wipes 3 water samples	1.51 1.87 3.12 0.93
	4 Slaughterhouses (S1, S2, S3, S4)	80	1 sticking knife 2 caeca 1 liver 1 wipe	20 8.0 4.0 20
Heidelberg	4 Poultry farms (F12, F13, F15, F27)	12.5	6 cloacal swabs 4 droppings 2 water samples 1 wipe	1.81 1.25 0.62 1.56
	(0) Slaughterhouse	0	-	-
Virginia	1 Poultry farm (F31)	3.12	1 water sample	0.31
	2 Slaughterhouses (S1,S4)	40	1 wipe 2 neck skins 1caeca	20 40 4.0
Enteritidis	3 Poultry farms (F18, F20, F27)	9.37	2 cloacal swabs 1 wipe	0.61 1.56
	1 Slaughterhouse (S3)	20	1 rinse water sample	
Newport	1 Poultry farm (F15)	3.12	1 wipe	1.56
	(0) Slaughterhouse	0	-	-
Total	11 Poultry farm (F1, F6, F12, F13, F14, F15, F18, F20, F27, F31, F32)	34.7	13 cloacal swabs 10 droppings 5 wipes 6 water samples	3.93 3.12 7.81 1.87
	5 Slaughterhouse (S1, S2, S3, S4, S5)	100	1 sticking knife 2 neck skins 1 rinse water sample 3 caeca 1 liver 2 wipe	20 8.0 4.0 12 4.0 40

F Farm, *S* Slaughterhouse

addition, the phylogenetic tree shows that seven *Salmonella* ser. Heidelberg strains (two human and five avian) clustered together and belonged to the same sequence type ST15 (Fig. 2), which suggests a possible crossing of this serotype, and particularly this ST between the poultry and the human community in northeastern Algeria.

Discussion

This study involved thirty-two poultry farms and five slaughter houses. It covered most of the districts in the Skikda province. The selection of the sites was based on the managers' willingness to cooperate with the study and to spend a significant amount of time and effort to perform and collect the various samples. This work provides epidemiological data on *Salmonella* serotype contamination in poultry farms and slaughterhouses in the region of Skikda, in order to investigate the molecular mechanisms of ß-lactam resistance in avian and human ESBL-producing *S. enterica* and to analyze the genetic relatedness of avian and human isolates.

The recorded prevalence rates (34.37% for poultry farms and 100% for slaughterhouses) are in accordance with those reported in Constantine and Batna provinces (northeastern Algeria): 36.6 and 60%, respectively [29, 30], but were higher than those reported in several European countries (Italy (9.2%), France (3.4%), Germany (2.7%), Spain (1.02%)) and Morocco (24%) [31–33]. This high prevalence can be attributed to the absence of a *Salmonella* infection-control plan (especially in healthy chicken flocks) [31] and to the poor hygienic state in poultry farms, where *Salmonella* can persist during several grow-outs [34]. Furthermore, our findings are in accordance with those of Gardel et al. (2003), who found that samples taken at the third week of the grow-out to be more contaminated by *Salmonella spp.* than those collected at advanced ages [35]. According to many surveys, contamination of poultry products with *Salmonella* may take place at different stages

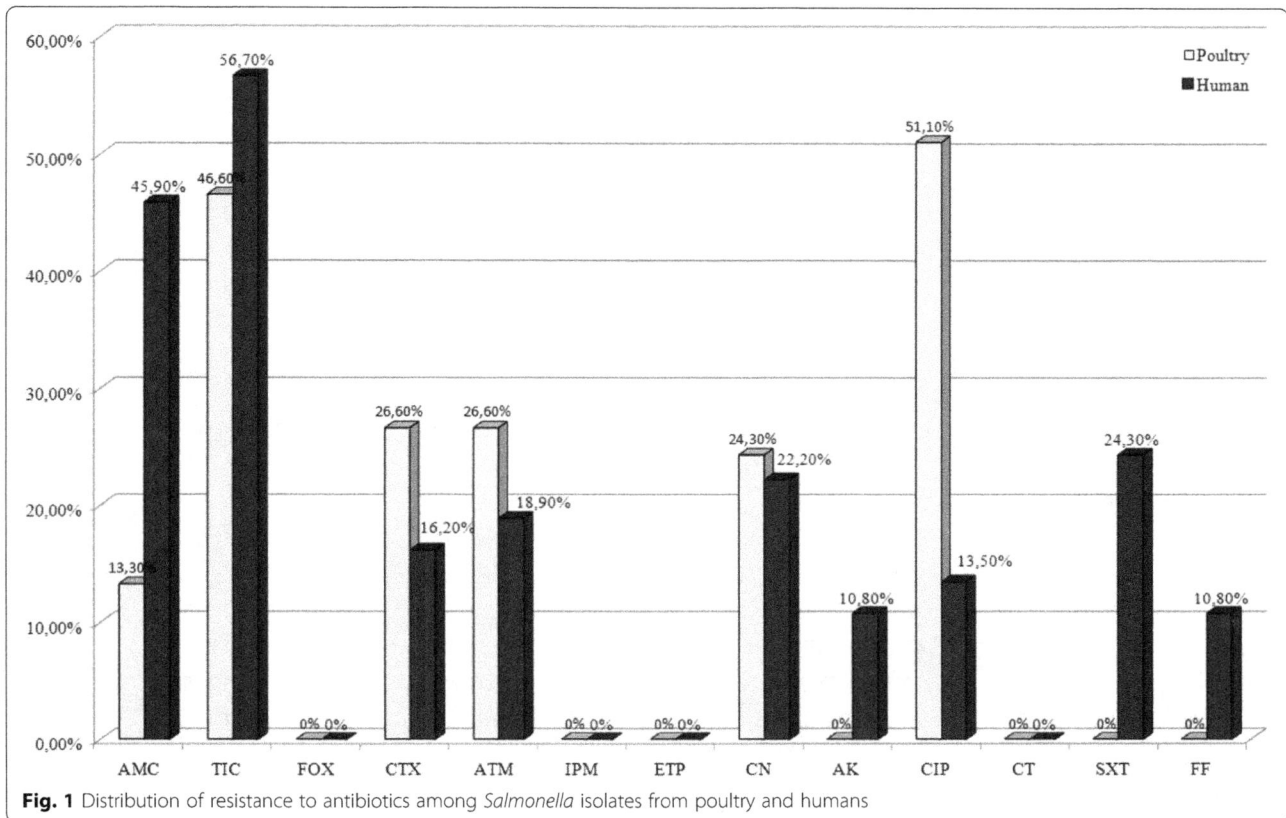

Fig. 1 Distribution of resistance to antibiotics among *Salmonella* isolates from poultry and humans

of the production process [36]. After contamination of the birds at the farm, bacteria colonize their intestines and can infect their carcasses at slaughter [37].

Elgroud et al. [29] found that 73.3% of the poultry slaughterhouses in Constantine were *Salmonella*-positive [29]. Among the avian serotypes isolated in our study, Kentucky was the most predominant. Currently, this serotype is distributed worldwide, especially its ST198 [3]. *Salmonella* ser. Heidelberg has been isolated from broiler carcasses collected at four provinces in the center of Algeria [38]. Despite the fact that we have isolated *Salmonella* ser. Enteritidis in only 9% of the samples, it is worth noting that this serotype is the most common in animal products, especially poultry [31]. Our *Salmonella* strains exhibited a high resistance rate to fluoroquinolones, and interestingly, a Kentucky ciprofloxacin-resistant serotype was isolated in France from a patient who had previously stayed in Algeria [39]. In the present study, we report the presence of ESBLs in *Salmonella* ser. Heidelberg, Senftenberg, Infantis and Newport of avian and human origins, with the CTX-M groups as the most prevalent. To our knowledge, this is the first report of $bla_{\text{CTX-M-1}}$ genes in avian *Salmonella* strains in Algeria. In fact, this group is the principal ESBL type in human *Salmonella* encountered in Europe [9]. CTX-M-15 was identified in all human *Salmonella* strains. This

finding corroborates well with several studies performed in Algeria showing that different *Salmonella* serotypes (Heidelberg, Kedougou, Infantis and Enteritidis) isolated from humans harbored this gene [40–43].

The present study had demonstrated the presence of bla_{TEM} genes in one avian *Salmonella* ser. Newport, one avian *Salmonella* ser. Heidelberg strain and all human ESBL-positive strains. Olesen et al. (2004) reported that in Denmark that the major recorded ESBL was the TEM group [44].

In Algeria, cephalosporin use is uncommon in poultry production [22], and the fact that ESBL-positive *Salmonella* strains of avian origin were isolated suggest that these resistant strains may have been introduced into the poultry production chain from other sources, or resulted from the acquisition by avian *Salmonella* strains of ESBL resistance determinants that are generally carried on mobile genetic elements (such as plasmids) [45].

The fact that seven avian and human *Salmonella* ser. Heidelberg strains were of the same ST and clustered together suggests that this clone is circulating in the poultry production chain as well as in the human community. Chicken-to-human transmission of *Salmonella* during farming has been widely demonstrated. It may take place through the food chain [46], and also through occupational exposure from direct contact with live

Table 2 Antimicrobial resistance and resistant genes profiles of ESBLs producing *Salmonella enterica* strains isolated from poultry and humans

Strain ID N°	Origin	Antimicrobial Resistance Pattern	Serotype	B-Lactamase	ST
162	Poultry	TIC, CTX, ATM,	*Heidelberg*	CTX-M-1	15
163	Poultry	TIC, CTX, ATM,	*Heidelberg*	CTX-M-1	15
164	Poultry	TIC, CTX, ATM,	*Heidelberg*	CTX-M-1	15
165	Poultry	TIC, CTX, ATM,	*Heidelberg*	CTX-M-1	15
167	Poultry	TIC, CTX, ATM,	*Heidelberg*	CTX-M-1	15
169	Poultry	TIC, CTX, ATM,	*Newport*	TEM	198
170	Poultry	TIC, CTX, ATM,	*Heidelberg*	CTX-M-1	15
171	Poultry	TIC, CTX, ATM,	*Heidelberg*	CTX-M-1	15
172	Poultry	TIC, CTX, ATM,	*Heidelberg*	CTX-M-1, TEM	15
174	Poultry	TIC, CTX, ATM,	*Heidelberg*	CTX-M-1	15
177	Poultry	TIC, CTX, ATM,	*Heidelberg*	CTX-M-1	15
178	Poultry	TIC, CTX, ATM,	*Heidelberg*	CTX-M-1	15
305	Human	AMC,TIC,CTX, ATM, GEN,FF	*Senftenberg*	CTX-M-15	14
476	Human	AMC, TIC, CTX, ATM,GEN,AK,SXT	*Infantis*	CTX-M-15, TEM	38
883	Human	CTX, ATM, GEN, AK, SXT	*Heidelberg*	CTX-M-15, TEM	15
884	Human	AMC,TIC,CTX, ATM, GEN, AK, SXT	*Heidelberg*	CTX-M-15, TEM	15
YFA	Human	AMC, TIC, CTX,GEN	*Infantis*	CTX-M-1	32
1577	Human	AMC,TIC,CTX, GEN, AK, SXT	*Infantis*	CTX-M-15, TEM	16

AMC Amoxicillin/Clavulanic Acid, *TIC* Ticarcillin, *FOX* Cefoxitin, *CTX* Cefotaxime, *ATM* Aztreonam, *IPM* Imipenem, *ETP* Ertapenem, *GEN* Gentamicin, *AK* Amikacin, *CIP* Ciprofloxacin, *CT* Colistin, *SXT* Trimethoprim/Sulfamethoxazole, *FF* Fosfomycin

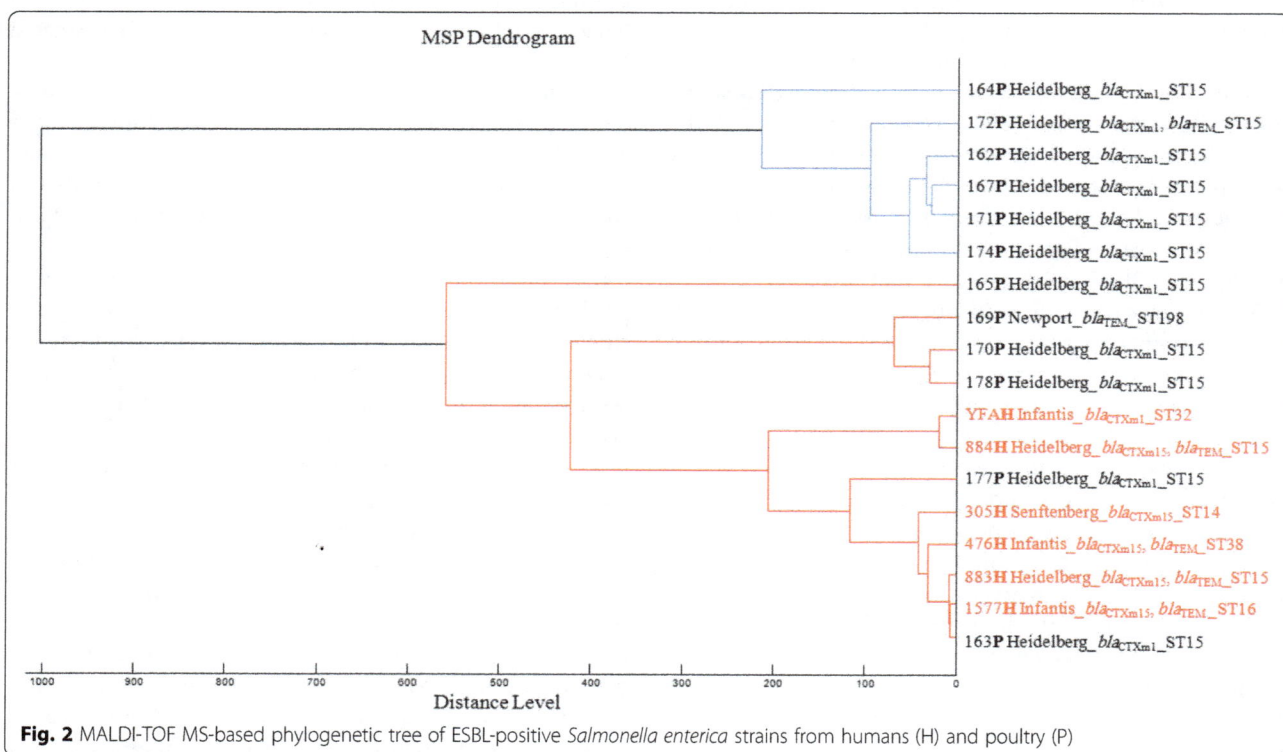

Fig. 2 MALDI-TOF MS-based phylogenetic tree of ESBL-positive *Salmonella enterica* strains from humans (H) and poultry (P)

animals and their environment in the broiler chicken industry [47].

Conclusion

The results of this study demonstrate that *Salmonella* contamination is highly prevalent in broiler poultry farms and slaughterhouses in the region of Skikda (northeastern Algeria), with an increasing resistance to medically important antibiotics. To the best of our knowledge, our results present for the first time, the emergence of ESBL-producing *S. enterica* isolates in poultry in this region. In addition, despite the different sampling times of avian and human *Salmonella* strains, their relatedness has been clearly demonstrated. The clonal relationship between human and avian strains indicates that the poultry industry may act as an important reservoir for ESBL-producing *Salmonella* that are transmitted to humans by direct contact or essentially through the food chain, but more discriminatory typing methods may be able to add more information as to the epidemiology of ESBL-producing *Salmonella* strains in Algeria. Infections caused by multidrug-resistant *Salmonella* species and therapeutic failures increase the risk of death. This is why surveillance programs, rational use of antibiotics and strict biosecurity measures have to be implemented in order to identify the sources, the exact routes of bacterial transmission and to limit the spread of these health-threatening bacteria in the local and national poultry industries.

Abbreviations

DDST: Double-disc synergy test; ESBLs: Extended spectrum β-lactamases; MALDI-TOF: Matrix-assisted laser desorption/ionization time-of-flight.; MLST: Multilocus sequence typing; PCR: Polymerase chain reaction; PWB: Peptone water broth; ST: Sequence type; UTH: University Teaching Hospital

Acknowledgments

The authors would like to thank the poultry farmers and slaughterhouses managers for their precious cooperation in collecting samples at both sites and C. Belkader and the entire staff of the bacteriology laboratory at the Pasteur Institute of Algeria for their technical assistance. Prof. C. Bentchouala (Microbiology Laboratory- University Teaching Hospital of Constantine) is highly respectful for providing the human *Salmonella* strains used in this study. The authors also thank TradOnline for corrections to the English language manuscript.

Funding

This work was partly funded by the CNRS and IHU Méditerranée Infection.

Authors' contributions

SD has actively worked on the isolation of avian *Salmonella* strains, identification of strains by mass spectrometry and their characterization (antibiotic-susceptibility testing, molecular typing of ESBL genes, sequencing, clonality assessment and Multilocus sequence typing MLST), data interpretation, drafting the paper and revising it. SB performed the data analysis, including sequence checking, Dendrogram of ESBL-positive *Salmonella* strains and revision of the article. BM, RE, and OB conceived and designed the study. AA participated actively in data interpretation, drafting the paper and critically revising it. SH contributed actively to the draft writing and revision of the article. SC participated in strains analysis (Identification by mass spectrometry and PCR characterization of human ESBL-positive *Salmonella* strains). KR contributed in part to the study design and data analysis (Serotyping and antibiotic- susceptibility testing of avian strains). JMR conceived and designed the study of ESBL-producing isolates, the clonality testing, antibiotic-susceptibility testing of the strains and contributed to the draft writing and the revision of the article. All the authors read and approved the article.

Competing of interest

The authors declare that they have no conflicts of interest.

Consent for publication

"Not applicable" in this section.

Author details

[1]GSPA research Laboratory (Management of Animal Health and Productions), Institute of Veterinary Sciences, University Frères Mentouri Constantine 1, Constantine, Algeria. [2]Institute of Veterinary and Agronomic Sciences, University Chadli Bendjedid, Eltarf, Algeria. [3]Unité de recherche sur les maladies infectieuses et tropicales émergentes (URMITE), UM 63, CNRS 7278, IRD 198, INSERM 1095, IHU Méditerranée Infection, Faculté de Médecine et de Pharmacie, Aix-Marseille-Université, Marseille, France. [4]Institute of Veterinary and Agronomic Sciences, University Hadj Lakhdar, Batna, Algeria. [5]PADESCA Research Laboratory, Institute of Veterinary Sciences, University Frères Mentouri, Constantine, Algeria. [6]Pasteur Institute, Medical Bacteriology Service, Algiers, Algeria.

References

1. Sanchez S, Hofacre CL, Lee MD, Maurer JJ, Doyle MP. Animal sources of salmonellosis in humans. J Am Vet Med Assoc. 2002;4:492–7.
2. Majowicz SE, Musto J, Scallan E, Angulo FJ, Kirk M, O'Brien SJ, Jones TF, Fazil A, Hoekstra RM. The global burden of nontyphoidal *Salmonella* gastroenteritis. Clin Infect Dis. 2010;50:882–9.
3. Le Hello S, Hendriksen RS, Doublet B, Fisher I, Nilsen EM, Whichard JM. International spread of an epidemic population of *Salmonella enterica* serotype Kentucky ST198 resistant to ciprofloxacin. J Infect Dis. 2011;204: 675–84.
4. Mekademi K, Saidani K. Contamination of broilers by *Salmonella* non Typhi in Mitidja. J Curr Res Sci. 2013;4:213–7.
5. Mouffok F. Situation en matière TIA en Algérie de 2010 à 2011. In: Dèuxième congrés magrébin sur les TIA Tunis; 2011. http://mail.pasteur.dz/pasteur-old-2048/presentation.php?page=72. Accessed 26 June 2015.
6. Refregier-Petton J, Kemp GK, Nebout JM, Allo JC, Salvat G. Post treatment effects of a SANOVA immersion treatment on turkey carcases and subsequent influence on recontamination and cross contamination of breast fillet meat during turkey processing. Br Poult Sci. 2003;44:790–1.
7. Journal officiel de la république Algérienne N°36. Arrêté interministériel de 17 Dhou El Kaada 1423 correspondant au 20 Janvier 2003 définissant les mesures de prévention et de luttes spécifiques au salmonellose aviaires à *Salmonella* Entéritidis, Typhimurium, Typhi, arizona, dublin, paratyphi et pullorum gallinarum. 2003. http://www.joradp.dz/FTP/jo-francais/2003/F2003036.pdf.Accessed 20 November 2016.
8. Murgia M, Bouchrif B, Timinouni M, Al-Qahtani A, Al-Ahdal MN, Cappuccinelli P, Rubino S, Paglietti B. Antibiotic resistance determinants and genetic analysis of *Salmonella enterica* isolated from food in Morocco. Int J Food Microbiol. 2015; 215:31–9.
9. Livermore DM, Canton R, Gniadkowski M, Nordmann P, Rossolini GM, Arlet G, Ayala J, Coque TM, Kern-Zdanowicz I, Luzzaro F, Poirel L, Woodford N. CTX-M: changing the face of ESBLs in Europe. J Antimicrob Chemother. 2007;59:165–74.
10. Bradford PA. Extended-spectrum beta-lactamases in the 21st century: characterization, epidemiology, and detection of this important resistance threat. Clin Microbiol Rev. 2001;14:933–51.
11. Bush K, Jacoby GA, Medeiros AA. A functional classification scheme for beta-lactamases and its correlation with molecular structure. Antimicrob Agents Chemother. 1995;39:1211–33.
12. Livermore DM. beta-Lactamases in laboratory and clinical resistance. Clin Microbiol Rev. 1995;8:557–84.

13. Cantón R, Coque TM. The CTX-M beta-lactamase pandemic. Curr Opin Microbiol. 2006;9:466–75.

14. Paterson DL, Bonomo RA. Extended-spectrum beta-lactamases: a clinical update. Clin Microbiol Rev. 2005;18:657–86. https://www.ncbi.nlm.nih.gov/pmc/articles/PMC1265908/pdf/0016-05.pdf. Accessed 3 June 2015.

15. Rahal K, Reghal A. A nosocomial epidemic of Salmonella Mbandaka which produces various broad spectrum beta-lactamases: preliminary results. Med Trop. 1994;54:227–30.

16. Bulletin Sanitaire vétérinaire, Ministère de l'agriculture et du développement rural, Algérie. Avril 2011. http://www.fao.org/fileadmin/user_upload/remesa/library/Bulletin%20Sanitaire%20Vétérinaire%20Avril%202011.pdf. Accessed 20 November 2016.

17. Norme française NF U47-101. Isolement et identification de tout sérovar ou de sérovar(s) spécifié(s) de salmonelles chez les oiseaux. France: Editions AFNOR; 2005. p. 34.

18. ISO.(Comité international de normalisation AW/9 ISO6579) Microbiology General Guidance on methods for the detection of Salmonella –2002/Amendment 1, Annex D: Detection of Salmonella spp. in animal feces and in environmental samples from the primary production stage. 2007.

19. Seng P, Drancourt M, Gouriet F, La SB, Fournier PE, Rolain JM, Raoult D. Ongoing revolution in bacteriology: routine identification of bacteria by matrix-assisted laser desorption ionization time-of-flight mass spectrometry. Clin Infect Dis. 2009;49:543–51.

20. Guibourdenche M, Roggentin P, Mikoleit M, Fields PI, Bockemuhl J, Grimont PA, Weill FX. Suppl 47 2003-2007 to the White-Kauffmann-Le Minor scheme. Res Microbiol. 2010;161:26–9.

21. EUCAST. (European Comitee on Antimicrobial Suceptibility Testing) Breakpoint tables for interpretation of MICs and zone diameters. Version 5.0, 2015. http://www.eucast.org. Accessed 10 Mar 2016.

22. SAEN (ED). Standardisation de l'antibiogramme à l'échelle nationale (médecine vétérinaire et humaine), Algérie. 2011.

23. Roschanski N, Fischer J, Guerra B, Roesler U. Development of a Multiplex Real-Time PCR for the Rapid Detection of the Predominant Beta-Lactamase Genes CTX-M, SHV,TEM and CIT-Type AmpCs in Enterobacteriaceae. PLoS One. 2014;9:e100956.

24. Edelstein M, Pimkin M, Palagin I, Edelstein I, Stratchounski L. Prevalence and molecular epidemiology of CTX-M extended-spectrum beta-lactamase producing Escherichia coli and Klebsiella pneumoniae in Russian hospitals. Antimicrob Agents Chemother. 2003;47:3724–32.

25. Kruger T, Szabo D, Keddy KH, Deeley K, Marsh JW, Hujer AM, Bonomo RA, Paterson DL. Infections with non typhoidal Salmonella species producing TEM-63 or a novel TEM enzyme, TEM-131, in South Africa. Antimicrob Agents Chemother. 2004;48:4263–42670.

26. Yagi T, Kurokawa H, Shibata N, Shibayama K, Arakawa Y. A preliminary survey of extended-spectrum-lactamase (ESBLs) in clinical isolates of Klebsiella pneumoniae and Escherichia coli in Japan. FEMS Microbiol Lett. 2015;184:53–6.

27. Gupta SK, Padmanabhan BR, Diene SM, Lopez-Rojas R, Kempf M, Landraud L, Rolain JM. ARG-ANNOT, a new bioinformatic tool to discover antibiotic resistance genes in bacterial genomes. Antimicrob Agents Chemother. 2014;58:212–20.

28. Kidgell C, Reichard U, Wain J, Linz B, Torpdahl M, Dougan G, Achman M. Salmonella Typhi, the causative agent of typhoid fever, is approximately 50,000 years old. Infect Genet Evol. 2002;2:39–45.

29. Elgroud R, Zerdoumi F, Benazzouz M, Bouzitouna-Bentchouala C, Granier SA, Fremy S, Brisabois A, Dufour B, Millemann Y. Characteristics of Salmonella contamination of broilers and slaughterhouses in the region of Constantine (Algeria). Zoonoses Public Health. 2009;56:84–93.

30. Ayachi A, Alloui N, Bennoune O, Kassah-Laouar A. Survey of Salmonella serovars in broilers and laying breeding reproducers in East of Algeria. J Infect Dev Ctries. 2010;4:103–6.

31. European Food Safety Authority (EFSA). Scientific Report of EFSA and ECDC the European Union summary report on trends and sources of zoonoses, zoonotic agents and food-borne outbreaks in 2001. EFSA J. 2013;11:3129.

32. Lamas A, Fernandez-No IC, Miranda JM, Vazquez B, Cepeda A, Franco CM. Prevalence, molecular characterization and antimicrobial resistance of Salmonella serovars isolated from northwestern Spanish broiler flocks (2011-2015). Poult Sci 2016;0:1-9.

33. Chaiba A, Rhazi Filali F. Prévalence de la contamination par Salmonella des élevage. de poulet de chair au Maroc. Cah Agric. 2016;doi:10.1051/cagri/2016017. https://www.cahiersagricultures.fr/articles/cagri/pdf/2016/03/cagri160054.pdf. Accessed 13 July 2016.

34. Lahellec C, Colin P. Relationship between serotypes of Salmonellae from hatcheries and rearing farms and those from processed poultry carcases. Br Poult Sci. 1985;26:179–86.

35. Gradel KO, Rattenborg E. A questionnaire-based, retrospective field study of persistence of Salmonella Enteritidis and Salmonella Typhimurium in Danish broiler houses. Prev Vet Med. 2003;56:267–84.

36. Salvat G. Prévention des problèmes de santé publique liés aux produits issus de la filière avicole. Acad Vet France. 1997;70:43–8.

37. Van Scothorst M, Notermans S. Food-borne diseases associated with poultry in: MEAD, G.C and Freeman, B.M (Editors) Meat quality in poultry and game birds. Edinburgh Brit Poultry Sci Ltd. 1980;90:79–90.

38. Bounar-Kechih S, Hamdi TM, Mezali L, Assaous F, Rahal K. Antimicrobial resistance of 100 Salmonella strains isolated from Gallus gallus in 4 wilayas of Algeria. Poult Sci. 2012;91:1179–85.

39. Weill FX and LE Hello S. Rapport d'activité annuel Centre National de Référence de Salmonella CNR. 2011.

40. Kermas R, Touati A, Brasme L, Le Magrex-Debar E, Mehrane S, Weill FX. De champs C. Characterization of extended-spectrum beta-lactamase-producing Salmonella enterica serotype Brunei and Heidelberg at the Hussein Dey hospital in Algiers (Algeria). Foodborne Pathog Dis. 2012;9:803–8.

41. Touati A, Benallaoua S, Gharout A, Amar AA, Le Magrex DE, Brasme L, Madoux J. De champs C, Weill FX. First report of CTX-M-15 in Salmonella enterica serotype Kedougou recovered from an Algerian hospital. Pediatr Infect Dis J. 2008;27:479–80.

42. Naas T, Bentchouala C, Cuzon G, Yaou S, Lezzar A, Smati F, Nordmann P. Outbreak of Salmonella enterica serotype Infantis producing ArmA 16S RNA methylase and CTX-M-15 extended-spectrum beta-lactamase in a neonatology ward in Constantine. Algeria Int J Antimicrob Agents. 2011;38:135–9.

43. Bouzidi N, Aoun L, Dekhil M, Granier SA, Poirel L, Brisabois A, Nordmann P, Millemann Y. Co-occurrence of aminoglycoside resistance gene armA in non-Typhi Salmonella isolates producing CTX-M-15 in Algeria. J Antimicrob Chemother. 2011;66:2180–1.

44. Olesen I, Hasman H, Aarestrup FM. Prevalence of beta-lactamases among ampicillin-resistant Escherichia coli and Salmonella isolated from food animals in Denmark. Microb Drug Resist. 2004;10:334–40.

45. Bae D, Cheng CM, Khan AA. Characterization of extended-spectrum beta-lactamase (ESBL) producing non-typhoidal Salmonella (NTS) from imported food products. Int J Food Microbiol. 2015;214:12–7.

46. Currie A, MacDougall L, Aramini J, Gaulin C, Ahmed R, Isaacs S. Frozen chicken nuggets and strips and eggs are leading risk factors for Salmonella Heidelberg infections in Canada. Epidemiol Infect. 2005;133:809–16.

47. Kim A, Lee YJ, Kang MS, Kwag SI, Cho JK. Dissemination and tracking of Salmonella spp. in integrated broiler operation. J Vet Sci. 2007;8:155–61.

11

Effect of Ketoprofen on acute phase protein concentrations in goats undergoing castration

Umit Karademir[1*], Ibrahim Akin[2], Hasan Erdogan[3], Kerem Ural[3] and Gamze Sevri Ekren Asici[4]

Abstract

Background: The objective of this study was to determine the effect of ketoprofen on acute phase protein (APPs) concentrations in goats undergoing castration. A total of 16 clinically healthy, male and 12 months old goats were enrolled and each case received ketoprofen (group I) or control (group II) in a randomized fashion. Goats were sedated with Xylazine-HCl, afterwards ketoprofen (3 mg/kg) was injected via jugular vein in group I, whereas physiological saline solution was administered to group II. Goats were castrated by the Burdizzo method. Hematological parameters were determined with a blood cell counter and plasma fibrinogen (Fb), serum haptoglobin (Hp), serum amyloid A (SAA) and ceruloplasmin (Cp) concentrations were measured Millars technique, ELISA kit or p-phenylenediamine oxidase activity prior to castration and throughout the study on 0 to 96 h.

Results: There were no differences in pre-treatment serum Cp, SAA and Fb concentrations among the groups. Contrarily, there were significant differences in plasma Hp concentrations on 0 to 96 h onwards post-castration. There were no differences in WBC and PCV between groups. Cp, Fb, and SAA were almost constant or showed slight changes at various stages of the study with no significant differences between groups.

Conclusions: The results revealed that, levels of Cp, Fb and SAA may not be affected by castration such as the confounding parameters similarly to stress. More investigations possessing different surgical or non-surgical castration techniques with larger number of goats and focusing on specific markers for stress are suggested for precise analysis.

Keywords: Acute Phase Protein, Castration, Goat, Ketoprofen

Background

Goats are one of the most important food-producing animal species in developing countries. The castration of male goats is a routine practice in many countries aimed at reducing management problems with aggressive and sexual behaviour, as well as improving meat quality [28]. The main techniques used to castrate goats include surgical or nonsurgical/ischemic (elastrator, burdizzo or emasculatome) methods [6, 8]. Castration has been shown to elicit inflammatory reactions, physiological stress, suppression of immune function, pain-associated behaviour, and a reduction in performance [15, 16, 28].

The acute phase response (APR) refers to nonspecific and complex reaction of an animal that includes changes in concentration of numerous liver derived plasma proteins, called acute phase proteins (APPs) [20]. APPs are a group of blood proteins that change in concentration in animals subjected to external or internal challenges, such as infection, inflammation, surgical trauma or stress and they are classified as positive (major, moderate and minor) or negative depending on the increase or decrease in the serum concentration, respectively, during the APR [3]. Positive APPs, such as haptoglobin, C-reactive protein, serum amyloid A, ceruloplasmin, fibrinogen, and alpha 1-acid glycoprotein, increase in concentration in response to inflammation. The "negative" APPs decrease in concentration in response to inflammation and include proteins like albumin and transferrin [30]. Quantification of APP concentration in plasma or serum can provide valuable

* Correspondence: umitkarademir@yahoo.com
[1]Department of Pharmacology and Toxicology, Faculty of Veterinary Medicine, University of Adnan Menderes, Isikli, Aydin, Turkey
Full list of author information is available at the end of the article

diagnostic information in the detection, prognosis, and monitoring of disease in several animal species [12]. In addition, the use of APPs for screening in ante- or post-mortem inspection to identify animals that should be subjected to a more thorough inspection or to ensure the health of animals prior to entry to the human food chain has been suggested [38].

Nonsteroidal anti-inflammatory drugs (NSAID) including ketoprofen (KTP) are among the most widely drugs in veterinary medicine. They block the activity of cyclooxygenase (COX) enzymes and reduce prostaglandin concentrations through the body. As a consequence, inflammation, pain and fever are reduced [17, 27]. These drugs make them ideal for the clinical management of inflammation and postoperative pain in animals. However, patients condition (eg, respiratory, renal or hepatic insufficiency, dehydration, ascites, coagulopathies, pregnancy or gastric ulcer) and drugs selection must be considered prior to NSAID use due to their potential of adverse effects (eg, antithrombotic activity, gastro-duodenal erosion and ulceration, nephropathy, delayed healing or nonunion of the wound and fracture) [2, 5, 27]. KTP, a propionic acid derivate, is a NSAID which used for the treatment and management symptoms associated with musculoskeletal inflammation and pain in animals [5]. Ruminants have been investigated on the effects of APPs in experimental inflammation models and stress by many authors which the administration of a different dose of lipopolysaccharide [43], turpentine [20], some *Pasteurella spp.* (e.g., *P. haemolytica, P. multocida*) [7, 25] and virus spp. (e.g. respiratory syncytial, viral diarrhoea) [18, 22], vaccination [11] or restricted feeding [26] and transportation [29]. Although the effects of NSAIDs on APPs in castrated ruminants have been well recognized in some of the prior research articles [9, 40, 41], the effects of NSAIDs on castration induced increases in APPs in goats have not been yet reported and limited data is a currently available. Furthermore there is scarce information in goats on APPs, clearly indicating that there is a need to evaluate goat model. Hence it was hypothesized that ketoprofen has probably effects on some of the positive APPs (Hp, SAA, Fb and Cp) in relation to castration, which is a frequent procedure of goats with Ketoprofen administration.

However, to the best of our knowledge, there is no published data for alterations of APPs in castration of goats with NSAID administration. The aim of this study was to investigate the alterations of some positive APPs (Hp, SAA, Fb and Cp) in castration of goats with KTP administration.

Results

All goats remained healthy through the study. Through available evidence suggested that castration lead to an increase in APPs in group II. Changes in mean values for serum concentrations of Hp, SAA and Cp, plasma concentrations of Fib, WBC and PCV counts were determined over the sample collection period for the two groups (Table 1). There were no differences in pre-treatment serum Cp, SAA and Fib concentrations among the groups (Fig. 1).

At the beginning of the study, there was no significant difference regarding Hp on 24 h of study. Afterwards through 48 to 96 h of the completion of the study there was a statistically significant difference between group I and II regarding Hp. The present authors also reported that there was significant group by time interaction regarding Cp (P= 0,001), Fb (P= 0,003) and WBC (P= 0,001) values. For other parameters (Hp, SAA and PCV) there was no group by time interaction.

Regarding SAA there was no significance between group I and II throughout the study whereas there was a time interaction in group I among 0.hr and 48 to 96 h. Besides this difference exists at 24th hour in group II. Taking into account Cp values there was a time interaction through 0. and 48 to 96 h in group I. Fibrinogen values possessed significant alterations on 0 and 6th hours. WBC values presented time interaction and group by time interaction (P<0,001).

Discussion

Regarding veterinary literature in farm animals APPs are important diagnostic indicators of inflammatory disorders also in goats [1, 4, 10]. In the analytical methods of prior study for measuring Hp, SAA, acid soluble glycoprotein (ASG), Fb, and albumin concentrations in goats were validated, in an attempt to assess their response to an inflammatory stimulus in goats [20]. In a recent article establishing reference intervals for acute phase proteins in healthy goats; Hp was be interpreted with caution in unknown pregnancy status, besides it was also suggested that APPs were recommended as useful biomarkers in goat diseases [4]. Hp increases were reported in several diseases of goats; i.e. helminth infestations [42], ruminal acidosis [19], sarcoptic mange [36], besnoidosis [31], coccidiosis [21] and gangrenous mastitis [13]. On the other hand there was no statistically significant differences were found regarding Hp concentrations in Caprine arthritis encephalitis positive and negative goats [24].

The present authors interest to this subject was aroused following receipt of goats referred for castration. At that time a through literature search revealed studies in cattle subjected to castration in relation to APPs. Contrarily the present authors were unaware of finding documented reports regarding APPs and castration procedure in goats.

As aforementioned above cattle studies largely took a part in the literature. In a prior trial determining the effect of repeated KTP administration to surgically castrated bulls on APPs revealed increased plasma Hp and

Table 1 Mean ± SD concentrations of Hp, SAA, Cp and Fb as well as WBC and PCV counts undergoing castration in male goats ($n = 8$ goats/group)

	Group	0.hr	6.hr	12.hr	24.hr	48.hr	72.hr	96.hr	Interactions	p value
Hp (mg/dl)	Ketoprofen	0,15 ± 0,10	0,63 ± 0,24	0,41 ± 0,09	0,63 ± 0,33	1,06 ± 0,86	1,39 ± 0,92	1,39 ± 1,15	Group	0,020
	Control	0,10 ± 0,11	0,49 ± 0,08	0,37 ± 0,12	0,47 ± 0,33	0,50 ± 0,17	0,41 ± 0,22	0,40 ± 0,23	Time	0,020
									Group by Time	0,212
SAA (mg/dl)	Ketoprofen	21,10 ± 8,01	30,05 ± 18,86	34,55 ± 38,34	51,07 ± 69,60	149,18 ± 147,04	114,43 ± 117,21	69,46 ± 89,74	Group	0,392
	Control	14,52 ± 4,42	29,37 ± 22,90	75,82 ± 67,67	140,48 ± 97,44	173,33 ± 136,34	120,72 ± 110,72	71,14 ± 54,55	Time	0,021
									Group by Time	0,433
Cp (mg/dl)	Ketoprofen	21,60 ± 7,88	21,86 ± 7,17	24,23 ± 8,62	22,78 ± 7,34	26,63 ± 8,67	27,23 ± 9,55	29,12 ± 11,16	Group	0,753
	Control	22,93 ± 7,58	25,66 ± 11,52	24,65 ± 11,27	21,92 ± 10,82	21,75 ± 9,19	23,29 ± 9,73	23,52 ± 7,22	Time	0,012
									Group by Time	0,001
Fib (mg/dl)	Ketoprofen	159,75 ± 29,64	181,78 ± 44,50	156,78 ± 35,78	165,96 ± 28,20	159,70 ± 25,48	160,29 ± 25,48	161,56 ± 25,34	Group	0,096
	Control	149,03 ± 26,18	160,58 ± 6,97	153,38 ± 7,51	139,93 ± 3,83	143,68 ± 7,59	140,48 ± 3,56	139,00 ± 6,14	Time	0,038
									Group by Time	0,003
WBC (×10⁹ cells/l)	Ketoprofen	17,04 ± 6,47	16,60 ± 7,49	16,58 ± 4,44	18,53 ± 6,50	16,08 ± 5,52	15,67 ± 4,77	18,46 ± 7,25	Group	0,571
	Control	14,96 ± 1,20	16,74 ± 9,99	23,64 ± 2,15	22,14 ± 3,35	15,91 ± 2,99	14,07 ± 2,35	17,35 ± 2,09	Time	0,000
									Group by Time	0,000
PCV (%)	Ketoprofen	14,20 ± 2,99	12,01 ± 3,19	16,15 ± 2,40	16,52 ± 2,53	14,54 ± 2,18	14,72 ± 2,53	14,09 ± 1,75	Group	0,112
	Control	14,48 ± 0,58	13,29 ± 3,89	14,00 ± 0,97	14,96 ± 1,39	13,22 ± 1,50	13,36 ± 1,54	13,18 ± 0,67	Time	0,036
									Group by Time	0,630

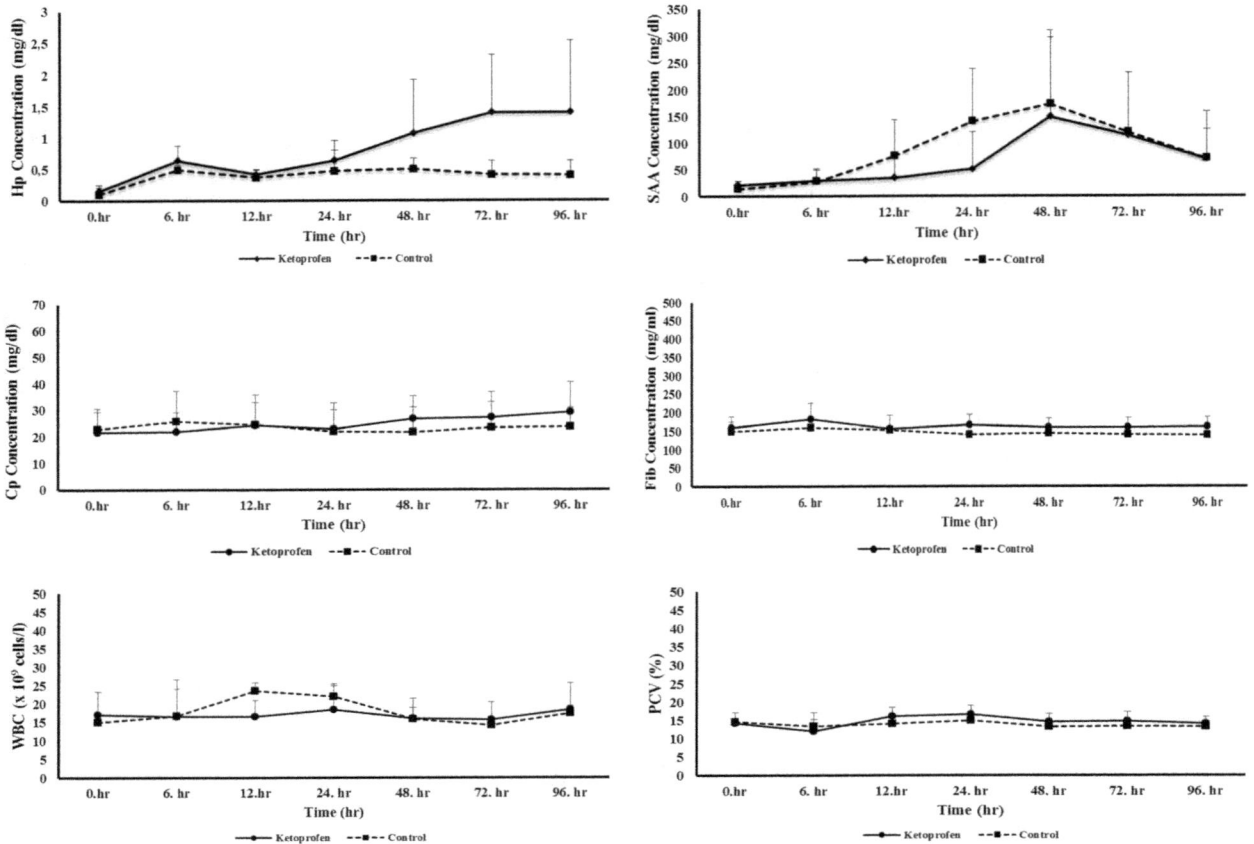

Fig. 1 Mean ± SD concentrations of Hp, SAA, Cp and Fb as well as WBC and PCV counts in male goats (*n* = 8 goats/group) receiving iv administered control treatment and ketoprofen (3 mg/kg) before and 96 h after undergoing castration

Fb concentrations were increased (*P*<0.05) on day 3 in the castration groups in comparison to the controls, in which were attributed to tissue trauma induced by castration. In the latter study surgical castration increased plasma cortisol and acute-phase proteins. On the other hand repeated KTP dose 24 h after treatment did not have influence on alteration in APPs [41]. Similarly, the effects of carprofen administration before banding or burdizzo castration of bulls on APPs were investigated. In that study Hp concentrations presented similarity (*P*= 0.58) among treatments before the time of castration. Afterwards on day 1, no differences in Hp concentration was detected and castrated and control groups. On day 3 band group showed elevated (*P*<0.05) Hp in comparison to control. On the other hand no differences in Hp concentrations were detected among treatments on d 7, 14, 21, and 28. finally on day 35 banded group showed greater (*P*<0.05) Hp concentrations compared with Band+C and control groups [34].

In the present study, Hp was the solely affected APPs deemed statistically significant in KTP administered goats (*P*<0.05) in comparison to controls. This may be briefly explained. Although recognized of the concentrations of the Hp, Cp and Fb may be useful in the diagnosis of tissue injury [35], according to the results of the present study unlike Hp, levels of Cp, Fb, and SAA may not be affected by tissue injury through Burdizzo castration. Increased Hp levels observed in this study might be related to the immediate tissue trauma, inflammation, and probably psychological (pain) stress in response to castration [32, 34]. Pang et al. [33] reported banding or burdizzo castration did not effect plasma Hp and Fb levels. Previous reports presented an increase in Hp and Fb levels on days 1, 3, and 7 post-castration in younger animals [14, 15, 34, 40]. Horadagoda et al. [23] reported that APPs, such as SAA and Hp are excellent markers for indicating acute inflammatory conditions in cattle. Pang et al. [33] stated that unchanged Hp and Fb levels in castrates, might be related to the dynamics (increased followed by a return to normal) of APPs during injury. In addition WBC and PCV values were deemed statistically unaffected between groups, revealed that tissue damage or injury was not significant, nor stress leukogram appeared in castrated animals participated in the present study.

Conclusion

Cp, Fb and SAA were almost constant or showed slight changes at various stages of the study with no significant difference between groups. Levels of Cp, Fb and SAA is not affected by castration such as the confounding parameters similarly to stress. More investigations possessing different surgical or non-surgical castration techniques with larger number of goats and focusing on specific markers for stress are suggested for precise analysis.

Methods

The present authors ensured that their manuscript reported adheres to the arrive guidelines for the reporting of animal experiments. This statement address to their manuscript that these guidelines were followed.

Animals and housing

The study was approved by the Animal Ethics Committee of Adnan Menderes University (with no: 64583101/2015/030). A total of 16 clinically healthy, male, 12 months old and weighing 25–30 kg Alpine Goats were used in the study. The animals were obtained from the faculty farm, belonging to the Adnan Menderes University, Faculty of Veterinary Medicine. Written owner consent was available through farm manager. All goats were considered clinically healthy after a thorough clinical check together with blood and serum chemistry profile and urinalysis. The goats were given a quarantine anthelmintic drench (ricabendazole – Rizal Enjectabl, Sanovel, Istanbul, Turkey; ivermectine - Vilmectin® Enjektabl, Vilsan Veteriner Ilaclari, Ankara, Turkey) at the manufacturers recommended doses in an animal house for a 2-week period before the commencement of the study. The animals were fed twice daily at 8:00 and 16:00 with a ration of commercial goat pellets and alfa alfa hay. Water was supplied ad libitum and mineral licks were provided for free access.

Study design, castration and treatment

Goats were enrolled, and each case received KTP (group I, $n = 8$) or control (group II, $n = 8$) in a randomized fashion, similarly to what have been described elsewhere [39]. Each group of goats was kept in suitable single boxes, which were then marked by ear tags. Goats were sedated with 0.3 mg/kg dose Xylazine-HCl intravenously [8]. Afterwards KTP (3 mg/kg) was injected via jugular vein in group I, indeed physiological saline solution (1 ml) was administered to group II. Goats were castrated by the Burdizzo (emasculatome) method. All castrations were performed by the same surgeon, who was experienced with the technique.

Collection of blood samples and laboratory analyses

Blood samples (4 ml/sample) from all goats were collected from the jugular vein via 20-gauge 25 mm needles into 2 evacuated tubes (one that contained EDTA-K, and another that contained a coagulation activator). Blood samples were obtained 30 min before injection of Xylasine-HCl (baseline: time 0) and 6, 12, 24, 48, 72, 96 h after the end of castration.

Blood samples contained EDTA-K were used to determine hematologic variables and Fb concentration. Hematologic parameters were performed with a blood cell counter (Abacus Junior Vet 5, Diatron Messtechnik GmbH, Vienna, Austria) calibrated for goat blood; WBC and PCV counts were used for statistical analysis. Plasma Fb concentration was measured via the Millars technique [1]. Plasma Fb concentrations and hematologic variables were determined within 6 h of the same day.

Other blood samples contained a coagulation activator were used to determine other APPs. Each blood sample was centrifuged at 3000 g for 10 min and the resulting serum was transferred to plastic tubes and stored at –20 °C for analysis. All serum samples were analysed on the same day after the sample collection period.

Serum Hp and SAA concentrations were measured with a commercially available ELISA kit (Cat no: TP-801 and TP-802 for Hp and SAA, respectively, Tridelta Development Ltd., Kildare, Ireland) at the manufacturers' recommended assay procedure. Hp and SAA concentrations were evaluated reference value versus at 630 and 450 nm, respectively, in a microplate reader (ELX-808, BioTek Instruments Inc., Vermont, USA) as mentioned in the method. Free hemoglobin possesses peroxidase activity that might be inhibited at low pH. Hp present in the blood sample reacts with hemoglobin, with a low pH demonstrates peroxidase activity by bounding to hemoglobin. SAA kit, a solid sandwich Enzyme Linked Immuno Sorbent Assay (ELİSA) performed in automated format. By the manufacturer a monoclonal antibody specific for SAA has been coated onto the wells of the microtitre strips. Obtained specimens [involving calibrators of known SAA content], were incubated into micro-wels at 37 °C together with a HRP labeled anti-SAA antibody. The presence SAA was captured between the the labeled antibody and coated microplate. The plate was washed following sampling and antibody-HRP incubation were removed within unbound material. Afterwards adding TMB, a blue product generating the colour, to those of direct proportion to the amount of SAA present in the original sample/calibrator. The reaction was finalized within the addition of stop reagent. The serum concentration of Cp was determined by measuring p-phenylenediamine oxidase activity as described by Ravin [37] with a spectrophotometer (UV-1601 UV-VIS Spectrophotometer, Shimadzu Corporation Tokyo, Japan).

Statistical analysis

Statistical analysis was performed with a statistical software program (SPSS-Version 21.0, SPSS Inc., Chicago, USA). A Kolmogorov-Smirnov test was used to assess all variables for normality. For data that were not distributed normally, transformations were applied to normalize the distribution. The effects of time, group (i.e., treatment), and group-by-time interaction were assessed via and ANOVA for repeated measures. When a significant group-by-time interaction was detected, Tukey multiple comparison tests were used to compare treatments within each time period. Within each group, the baseline value was compared with the values at various time points after castration and isotonic-NaCl/ketoprofen by use of the Bonferroni correction method. Results were considered significant at values of $P < 0.05$. Comparisons within and between groups were based on the final statistical model.

Abbreviations
APPs, Acute phase proteins; APR, The acute phase response; ASG, Acid soluble glycoprotein; Cp, Ceruloplasmin; Fb, Fibrinogen; Hp, Serum haptoglobin; KTP, Ketoprofen; NSAID, Nonsteroidal anti-inflammatory drugs; SAA, Serum amyloid A

Acknowledgements
The authors would like to thank Prof. Pinar Alkim Ulutas for her assistance.

Funding
This study was self-funded, no external resources were used.

Authors' contributions
All authors have made substantial contributions to all of the following: the conception and design of the study (UK, IA), the animal phase of the experiments (IA, UK, HE), the analytical phase of the study (UK, KU, GSEA), drafting the article (UK, KU). All authors have read and approved the final manuscript.

Competing interests
The authors declare that they have no competing interests.

Consent for publication
Not applicable.

Author details
[1]Department of Pharmacology and Toxicology, Faculty of Veterinary Medicine, University of Adnan Menderes, Isikli, Aydin, Turkey. [2]Department of Surgery, Faculty of Veterinary Medicine, University of Adnan Menderes, Isikli Koyu, Aydin, Turkey. [3]Department of Internal Medicine, Faculty of Veterinary Medicine, University of Adnan Menderes, Isikli Koyu, Aydin, Turkey. [4]Department of Biochemistry, Faculty of Veterinary Medicine, University of Adnan Menderes, Isikli Koyu, Aydin, Turkey.

References
1. Benjamin MM. Fibrinogen, third ed. Outline of Veterinary Clinical Pathology. USA: The Iowa State University Press; 1978. p. 117.

2. Bergh MS, Budsberg SC. The coxib NSAIDs: potential clinical and pharmacologic importance in veterinary medicine. J Vet Intern Med. 2005; 19(5):633.

3. Ceciliani F, Ceron JJ, Eckersall PD, Sauerwein H. Acute phase proteins in ruminants. J Proteom. 2012;75:4207–31.

4. Cray C, Zaias J, Altman NH. Acute phase response in animals: a review. Comp Med. 2009;59(6):517.

5. Curry SL. Nonsteroidal anti-inflammatory drugs: a review. J Am An Hos Assoc. 2005;41:298–309.

6. Dawson LJ. Preferred Management Practices. In: Solaiman SG, editor. Goat Science and Production. Iowa: Blackwell Publishing; 2010.

7. Dowling A, Hodgson JC, Schock A, Donachie W, Eckersall PD, McKendrick IJ. Experimental induction of pneumonic pasteurellosis in calves by intratracheal infection with Pasteurella multocida biotype A:3. Res Vet Sci. 2002;73:37–44.

8. Duncanson GR. Veterinary Treatment of Sheep and Goats. Cambridge: CABI; 2012.

9. Earley B, Crowe MA. Effects of ketoprofen alone or in combination with local anesthesia during the castration of bull calves on plasma cortisol, immunological, and inflammatory responses. J Anim Sci. 2002;80:1044–52.

10. Eckersall PD, Bell R. Acute phase proteins: biomarkers of infection and inflammation in veterinary medicine. Vet J. 2010;185:23–7.

11. Eckersall PD, Lawson FP, Kyle CE, Waterston M, Bence L, Stear MJ, Rhind SM. Maternal undernutrition and the ovine acute phase response to vaccination. BMC Vet Res. 2008;4:1–10.

12. Eckersall PD. Recent advances and future prospects for the use of acute phase proteins as markers of disease in animals. Rev Med Vet. 2000;151:577–84.

13. El-Deeb WM. Clinicobiochemical investigations of gangrenous mastitis in does: immunological responses and oxidative stress biomarkers. J Zhejiang Univ Sci B. 2013;14(1):33–9.

14. Faulkner DB, Eurell T, Tranquilli WJ, Ott RS, Ohl MW, Cmarik GF, Zinn G. Performance and health of weanling bulls after butorphanol and xylazine administration at castration. J Anim Sci. 1992;70:2970–4.

15. Fisher AD, Crowe MA, Alonso de la Varga ME, Enright WJ. Effect of castration method and the provision of local anaesthesia on plasma cortisol, scrotal circumference, growth and feed intake of bull calves. J Anim Sci. 1996;74:2336–43.

16. Fisher AD, Crowe MA, O'Nuallain EM, Monaghan ML, Larkin JA, Kiely PO, Enright WJ. Effects of cortisol on in vitro interferon-γ production, acute phase proteins, growth, and feed intake in a calf castration model. J Anim Sci. 1997;75:1041–7.

17. Friton GM, Cajal C, Ramirez-Romero R. Long-term effects of meloxicam in the treatment of respiratory disease in fattening cattle. Vet Rec. 2005;156:809–11.

18. Ganheim C, Hulten C, Carlsson U, Kindahl H, Niskanen R, Waller KP. The acute phase response in calves experimentally infected with Bovine Viral Diarrhoea Virus and/or Mannheimia Haemolytica. J Vet Med B. 2003;50(4):183–90.

19. González FHD, Ruipérez FH, Sánchez JM, Souza JC, Martínez-Subiela S, Cerón JJ. Haptoglobin and serum amyloid a in subacute ruminal acidosis in goats. Rev Med Vet Zoot. 2010;57:168–77.

20. Gonzalez FHD, Tecles F, Martinez-Subiela S, Tvarijonaviciute A, Soler L, Ceron JJ. Acute phase protein response in goats. J Vet Diagn Invest. 2008;20:580–4.

21. Hashemnia M, Khodakaram-Tafti A, Razavi SM, Nazifi S. Alternative formats. Korean J Parasitol. 2011;49(3):213–9.

22. Heegaard PMH, Godson DL, Toussaintc MJM, Tjùrnehùj K, Larsen LE, Viuff B, Rùnsholt L. The acute phase response of haptoglobin and serum amyloid A (SAA) in cattle undergoing experimental infection with bovine respiratory syncytial virus. Vet Immun Immunopat. 2000;77:151–9.

23. Horadagodo NU, Knox KM, Gibbs HA, Reid SW, Horadagoda A, Edwards SE, Eckersall PD. Acute phase proteins in cattle: discrimination between acute and chronic inflammation. Vet Rec. 1999;144:437–41.

24. Kaba J, Stefaniak T, Bagnicka E, Czopowicz M. Haptoglobin in goats with caprine arthritis-encephalitis. Centr Eur J Immunol. 2011;36:76–8.

25. Katoh N, Nakagawa H. Detection of haptoglobin in the high-density lipoprotein and the very high-density lipoprotein fractions from sera of calves with experimental pneumonia and cows with naturally occurring fatty liver. J Vet Med Sci. 1999;6:119–24.

26. Katoh N, Oikawa S, Oohashi T, Takahashi Y, Itoh F. Decreases of apolipoprotein B-100 and A-I concentrations and induction of haptoglobin and serum amyloid A in nonfed calves. J Vet Med Sci. 2002;64:51–5.

27. Mathews KA. Nonsteroidal anti-inflamatory analgesics: a review of current practice. J Vet Emerg Crit Care. 2002;12:89–97.

28. Molony V, Kent JE, Robertson IS. Assessment of acute and chronic pain after different methods of castration of calves. Appl Anim Behav Sci. 1995;46:33–48.

29. Murata H, Miyamoto T. Bovine haptoglobin as a possible immunomodulator in the sera of transported calves. Br Vet J. 1993;149:277–83.

30. Murata H, Shimada N, Yoshioka M. Current research on acute phase proteins in veterinary diagnosis: an overview. Vet J. 2004;168:24–40.

31. Nazifi S, Oryan A, Namazi F. Hematological and serum biochemical analyses in experimental caprine besnoitiosis. Korean J Parasitol. 2011;49(2):133–8.

32. Obled C. Amino acid requirements in inflammatory states. Can J Anim Sci. 2003;83:365–73.

33. Pang WY, Earley B, Gath V, Crowe MA. Effect of banding or burdizzo castration on plasma testosterone, acute-phase proteins, scrotal circumference, growth, and health of bulls. Livest Sci. 2008;117:79–87.

34. Pang WY, Earley B, Sweeney T, Crowe MA. Effect of carprofen administration during banding or burdizzo castration of bulls on plasma cortisol, in vitro interferon-gamma production, acute-phase proteins, feed intake, and growth. J Anim Sci. 2006;84:351–9.

35. Pfeffer A, Rogers KM. Acute phase response of sheep: changes in the concentrations of ceruloplasmin, fibrinogen, haptoglobin and the major blood cell types associated with pulmonary damage. Res Vet Sci. 1989;46:118–24.

36. Rahman MM, Lecchi C, Fraquelli C, Sartorelli P, Ceciliani F. Acute phase protein response in Alpine ibex with sarcoptic mange. Vet Parasitol. 2010; 168(3):293–8.

37. Ravin HA. An improved colorimetric enzymatic assay of ceruloplasmin. J Lab Clin Med. 1961;58:161–8.

38. Saini PK, Webert DW. Application of acute phase reactants during antemortem and post-mortem meat inspection. J Am Vet Med Assoc. 1991; 198:1898–901.

39. Thomas J, Doherty WA Will, Barton W, Rohrbach DR, Dennis Geiser R. Effect of morphine and flunixin meglumine on isoflurane minimum alveolar concentration in goats. Vet Anaesth Analg. 2004;31:97-101.

40. Ting ST, Earley B, Crowe MA. Effect of repeated ketoprofen administration during surgical castration of bulls on cortisol, immunological function, feed intake, growth, and behavior. J Anim Sci. 2003;81:1253–64.

41. Ting STL, Earley B, Hughes JML, Crowe MA. Effect of ketoprofen, lidocaine local anesthesia, and combined xylazine and lidocaine caudal epidural anesthesia during castration of beef cattle on stress responses, immunity, growth, and behavior. J Anim Sci. 2003;81(5):1281–93.

42. Ulutas PA, Voyvoda H, Ulutas B, Aypak S. Haptoglobin, serum amyloid-a and ceruloplasmin concentrations in goats with mixed helminth infection. Acta Paras Tur. 2008;32:229–33.

43. Werling D, Sutter F, Arnold M, Kun G, Tooten PCJ, Gruys E, Kreuzer M. Characterisation of the acute phase response of heifers to a prolonged low dose infusion of lipopolysaccharide. Res Vet Sci. 1996;61:252–7.

12

Association of Circulating Transfer RNA fragments with antibody response to *Mycoplasma bovis* in beef cattle

Eduardo Casas[1]* (ID), Guohong Cai[1], Larry A. Kuehn[2], Karen B. Register[1], Tara G. McDaneld[2] and John D. Neill[1]

Abstract

Background: High throughput sequencing allows identification of small non-coding RNAs. Transfer RNA Fragments are a class of small non-coding RNAs, and have been identified as being involved in inhibition of gene expression. Given their role, it is possible they may be involved in mediating the infection-induced defense response in the host. Therefore, the objective of this study was to identify 5′ transfer RNA fragments (tRF5s) associated with a serum antibody response to *M. bovis* in beef cattle.

Results: The tRF5s encoding alanine, glutamic acid, glycine, lysine, proline, selenocysteine, threonine, and valine were associated ($P < 0.05$) with antibody response against *M. bovis*. tRF5s encoding alanine, glutamine, glutamic acid, glycine, histidine, lysine, proline, selenocysteine, threonine, and valine were associated ($P < 0.05$) with season, which could be attributed to calf growth. There were interactions ($P < 0.05$) between antibody response to *M. bovis* and season for tRF5 encoding selenocysteine (anticodon UGA), proline (anticodon CGG), and glutamine (anticodon TTG). Selenocysteine is a rarely used amino acid that is incorporated into proteins by the opal stop codon (UGA), and its function is not well understood.

Conclusions: Differential expression of tRF5s was identified between ELISA-positive and negative animals. Production of tRF5s may be associated with a host defense mechanism triggered by bacterial infection, or it may provide some advantage to a pathogen during infection of a host. Further studies are needed to establish if tRF5s could be used as a diagnostic marker of chronic exposure.

Keywords: Cattle, RNA-seq, Selenocysteine, Small non-coding RNA, tRF

Background

Bovine respiratory disease complex is the most expensive condition in cattle, costing up to $1 billion annually in the United States [1]. Despite the development of vaccines and antibiotics, the condition is responsible for significant morbidity and mortality losses to the cattle industry [2]. Miles [3], suggests that perhaps it is time to look for ways to reduce losses by focusing on the animal's response to related pathogens, instead of continuing to focus on the pathogens themselves.

Mycoplasma bovis (M. bovis) has been identified as a prime pathogen causing respiratory disease of cattle, along with *Pasteurella multocida, Mannheimia haemolytica,* and *Histophilus somni* [2, 4, 5]. Common problems with cattle infected with *M. bovis* are chronic sickness, insensitivity to treatment, and inability to reach target weights. *M. bovis* is one of the most common pathogens recovered from lung samples in the abattoir [6].

High throughput sequencing allows identification of small non-coding RNAs [7, 8]. Transfer RNA Fragments (tRFs) are a class of small interfering RNA that were originally considered a degradation product of the translation process, but their role in regulation of gene translation in the cell has now been recognized [9, 10]. Their classification is based on the processing site of the transfer RNA (tRNA): tRFs processed from the 5′ end of the mature tRNA are denoted tRF5; tRFs cleaved at the 3′ end of the mature tRNA are referred to as tRF3; those produced from the beginning of the 3′ end, cleaved from the immature tRNA are designated tRF1

* Correspondence: Eduardo.casas@ARS.USDA.GOV
[1]USDA, ARS, National Animal Disease Center, Ames, IA 50010, USA
Full list of author information is available at the end of the article

[8, 11, 12]. These tRFs are the second most abundant in tissues, after microRNAs [8]; however, tRFs are the most abundant sncRNAs in serum in cattle, with tRF5s being the predominant group among the three types of tRFs [13].

EN Nolte-'t Hoen et al. [14], proposed that tRFs are produced in bone marrow and immune cells; however, it has been suggested that other cells may also have the ability to produce them [15, 16]. tRFs have been identified as being involved in inhibition of gene expression in stressed cells and in virus replication [15, 17]. Given their production site and their role in inhibiting gene expression, it is possible they may be involved in mediating the infection-induced defense response [18]. Therefore, our objective was to identify tRF5s associated with serum antibody response to *M. bovis* in beef cattle.

Methods
Animals
Bleeding of animals was done according to the management protocol approved by the Animal Care and Use Committee of the Institution. Sera from sixteen beef steers born during spring, 2013, were obtained from the US Meat Animal Research Center, Clay Center Nebraska. Animals were bled at three time points: during summer, 2013, while in the pasture with the dam, at weaning in the fall of the same year, and during summer, 2014. Blood was obtained by jugular venipuncture using a syringe. The samples were centrifuged at 1300 X g for 25 min at 4 °C and serum was aspirated and frozen at − 20 °C until used. Samples were shipped to the National Animal Disease Center, Ames, Iowa.

Antibody response against *M. bovis*
Cattle sera were tested for antibodies reactive with *M. bovis* using a direct ELISA, as previously reported [19], except that 0.5 µg of antigen was used per well, anti-bovine IgG-peroxidase conjugate (KPL, Inc.), diluted 1:3000 in wash buffer, was used to detect cattle IgG and color development was halted after 45 min. The *M. bovis* isolate M23 was used as the source of antigen [20]. The presence or absence of serum antibody to *M. bovis* was confirmed in each animal using a commercially available ELISA (Biovet, Inc.) prior to selection for inclusion in the appropriate pool. Sera included in the positive pool were 3+ or 4+ positive, on a scale of 1+ to 4+, as described by the ELISA manufacturer. The pool itself tests at 4+ with the Biovet ELISA and has a level of IgG higher than that of the positive control serum provided with the kit. A positive result in our in-house ELISA was defined as an average absorbance at 405 nm greater than the average plus 3 standard deviations of the negative control, calculated independently for each plate analyzed. Sera from the sixteen beef calves collected in summer were ELISA negative for IgG reactive with

M. bovis. By the fall, eight animals were seropositive (positive group), while eight remained negative (negative group). By spring, all animals in both groups were seropositive.

tRF isolation
The tRFs were isolated from 200 µl of each serum sample using the miRNeasy Serum/Plasma kit (QIAGEN, Germantown, MD). The tRFs were extracted according to the manufacturer's direction and the samples were eluted in 14 µl of RNase free water. After extraction 1 µl of each sample was run using the Small RNA chip on an Agilent 2100 Bioanalyzer (Agilent Technologies, Santa Clara, CA), to quantify the tRFs extracted from the samples. The tRFs concentration was determined by using a 10–40 nucleotide gate.

Library preparation
A sequence library was prepared for each extracted sample. The libraries were prepared using the NEBNext Multiplex Small RNA Library Prep Set for Illumina Set 1 and 2 (New England BioLabs, Ipswich, MA). Each set comprises of 24 unique sequences or barcodes, therefore, 48 unique barcodes were used to identify each sample. Six microliters of each animal's isolated small RNA fraction was used in library preparation according to manufacturer's instructions. After the library preparation, libraries were cleaned up and concentrated using the QIAquick PCR purification kit (QIAGEN, Germantown, MD) from 100 µl to a final volume of 27.5 µl. The quality and quantity of the libraries was determined by running 1 µl of each library on a DNA 1000 chip on an Agilent 2100 Bioanalyzer (Agilent Technologies, Santa Clara, CA). The concentration of each indexed library was determined by using a 135–170 nucleotide gate. All the indexed libraries were then pooled and size selected. Five nanograms of each indexed library were used to make the pool, with the total volume of the pool being 246.5 µl. The pool was concentrated using the QIAquick PCR purification kit (QIAGEN, Germantown, MD) to 35 µl of RNase free water. The pool was then size selected using the Pippin Prep on a 3% Agarose gel without added ethidium bromide (SAGE Sciences, Beverly, MA) with a size selection of 142–170 nucleotides according to the manufacturer's instructions. After the gel was run the pools were concentrated using the QIAquick PCR purification kit (QIAGEN, Germantown, MD) by eluting in 32 µl of RNase free water. One microliter of the size selected library pool was run using a High sensitivity DNA chip Agilent 2100 Bioanalyzer (Agilent Technologies, Santa Clara, CA). The concentration was determined by using a 135–170 nucleotide gate. The final concentration of the size selected pool library was 1.5 nM and the pool was stored at − 20 °C.

Sequencing the library pool

The pooled size selected library was sequenced using the Hi-Seq Sequencing Kit v2 50 Cycles (Illumina, San Diego, CA) in the Sequencing Core Facility at the National Animal Disease Center (NADC).

Data analysis

The quality of Illumina sequences was inspected using FastQCv0.11.22 program in the fastx toolkit3. The Illumina adapter was removed using fastx_clipper. Multiple occurrences of unique reads were merged using a custom script. Reads 18–40 bp in size were used in downstream analysis. These reads were first mapped to *Bos taurus* genome (ENSEMBL UMD3.1.75) using Novoalign software (Novocraft Technologies) allowing two mismatches. Those aligned to the genome were then mapped to a database containing different annotated genome features to determine their origin: genomic tRNA sequences were downloaded from http://gtrnadb.ucsc.edu/; mitochondrial tRNA, cDNA, and other non-coding RNA sequences were downloaded from ENSEMBL version 75. The Illumina reads that aligned to tRNA genes or their flanking sequences were further characterized. They were initially aligned to a *Bos taurus* tRNA database using BLASTN and the results were processed using a custom script. Those perfectly aligned to the beginning of mature tRNAs were classified as tRF5. After tRF5 sequences had been determined, their occurrences in Illumina sequences from individual animals were obtained using a custom script. Sequences have been submitted to NCBI Short Read Archive4, under BioProject accession PRJNA319677. Read counts were normalized to library size to reads per million prior to statistical analysis.

Statistical analysis

Analysis was done using the Mixed procedure of SAS (SAS Inst. Inc., Cary, NC). The model included the effects of ELISA status (positive or negative), season (summer, fall, or spring), and the interaction between ELISA status and season. The present study accounted for the minimum number of samples that could be run and that could provide significant statistical differences. Eight biological replicates (animals) per group suffice the requirements of the study. The power to establish differences at the $P < 0.05$ level, between both groups with a sample size of $n = 16$ is: 1-beta = 0.94.

Probability values shown are nominal and uncorrected for multiple testing. Sixteen animals were used to ascertain the association of tRFs with ELISA status in the present study. Additional experimental units would be needed if significance was adjusted for multiple comparisons. Although next generation sequencing allows profiling tRFs in each experimental unit, the cost associated with embarking on a large scale study is still a limiting factor. The present study was designed to ascertain nominal significant differences with the minimal number of samples. For this reason it was deemed relevant to present un-corrected significances in the present study. Significances should be taken in consideration when interpreting results.

Results

There were 452,264,204 sequences obtained in the present study. From these, 416,296,523 sequences mapped to tRNA. Only sequences that matched 100% with the tRNA genes and their flanking sequences were included in the study, therefore there were a total of 263,556,821 sequences that matched these criteria. Of these, 261,502,003 sequences were identified as tRF5s. Table 1 shows the number of tRF5s corresponding to the anticodons of each amino acid. The tRF5s with the most number of sequences were tRF5-Glu ($n = 989,044 + 67,677,835 = 68,666,879$), tRF5-Gly ($n = 156,913,852$), tRF5-His ($n = 23,770,874$), tRF5-Lys ($n = 3,223,746$), and tRF5-Val ($n = 8,103,320$). The number of sequences for these five tRF5s comprise 99.7% of the total number of tRF5 sequences in the study. The tRF5s with fewer than 1000 sequences were tRF5-Arg ($n = 283 + 1 + 42 = 326$), tRF5-Ile ($n = 20$), tRF5-Leu ($n = 165$), tRF5-Phe ($n = 3$), tRF5-Ser ($n = 9$), tRF5-Trp ($n = 11$), and tRF5-Tyr ($n = 4$). The remaining tRF5s had more than 1000, but less than 1000,000 sequences.

For identification purposes, the nomenclature of tRF5s will be by amino acid and anticodon. As an example, for the tRF5 for alanine, anticodon AGC, the abbreviation tRF5-AlaAGC will be used.

There were nine tRF5s with different anticodons associated with the outcome of ELISA status (Table 2). The positive group had a greater count of tRF5s when compared to the negative group for all anticodons, with the exception of tRF5-AlaCGC and tRF5-GluCTC. The most significant associations were observed with tRF5-LysCTT ($P = 0.0002$), tRF5-LysTTT ($P = 0.0057$), and with tRF5-SelCys ($P = 0.002$).

Table 3 shows differences in tRF5 counts by season. There were five tRF5s with different anticodon for the same amino acid (tRF5-Ala, tRF5-Gln, tRF5-Gly, tRF5-Pro, and tRF5-Val), associated ($P = 0.05$) with season. For tRF5-AlaAGC, tRF5-AlaTGC, tRF5-ValAAC, tRF5-VAlTAC, tRF5-LysTTT, and tRF5-ProCGG, the greatest number of sequences was observed in summer and fall, 2013, declining by the spring, 2014. For tRF5-GlnTTG, tRF5-GlnTTC, tRF5-GlyTCC, tRF5-SelCysUGA, and tRF5-ThrTGT, the number of copies was the lowest during summer, 2013, consistently increasing throughout the fall, 2013, and spring, 2014. For tRF5-ProAGG, and tRF5-ProTGG, the number of sequences during summer, 2013, and spring, 2014 were the lowest, having the greatest numbers in fall, 2013.

The interaction for tRF5-SeCys, between ELISA status and season is shown in Fig. 1. During summer, there was

Table 1 Number of 5′-transfer RNA fragments (tRF5) sequences by amino acid and anticodon

Transfer RNA Fragment	Anticodon	Number of sequences
tRF5-Ala	AGC	142,824
tRF5-Ala	CGC	279,970
tRF5-Ala	TGC	2189
tRF5-Arg	CCT	283
tRF5-Arg	TCG	1
tRF5-Arg	TCT	42
tRF5-Asp	GTC	1248
tRF5-Cys	GCA	29,406
tRF5-Gln	CTG	67,818
tRF5-Gln	TTG	25,446
tRF5-Glu	CTC	989,044
tRF5-Glu	TTC	67,677,835
tRF5-Gly	ACC	1
tRF5-Gly	CCC	127,016,832
tRF5-Gly	GCC	29,840,312
tRF5-Gly	TCC	56,707
tRF5-His	GTG	23,770,874
tRF5-Ile	AAT	20
tRF5-Leu	AAG	112
tRF5-Leu	CAG	27
tRF5-Leu	TAG	26
tRF5-Lys	CTT	2,740,190
tRF5-Lys	TTT	483,556
tRF5-Met	CAT	5063
tRF5-Phe	GAA	3
tRF5-Pro	AGG	211,844
tRF5-Pro	CGG	1667
tRF5-Pro	TGG	47,038
tRF5-SelCys	UGA	4579
tRF5-Ser	CGA	4
tRF5-Ser	GCT	1
tRF5-Ser	TGA	4
tRF5-Thr	CGT	16
tRF5-Thr	TGT	3685
tRF5-Trp	CCA	11
tRF5-Tyr	GTA	4
tRF5-Val	AAC	31,138
tRF5-Val	CAC	6,639,535
tRF5-Val	TAC	1,432,647
Total		261,502,003

no difference in counts for tRF5-SelCysUGA between groups. However, in fall, the positive group had an increased number of sequences, compared to the negative group. The difference in number of sequences of tRF5-SelCysUGA between the positive and negative groups increased further during spring, 2014. In Mycoplasmas the codon UGA, typically a stop codon, is instead translated as tryptophan [21]. It was determined that tRF5-SelCys identified in the present study were of bovine origin.

Figure 2 shows the interaction for tRF5-ProCGG, between ELISA status and season. During summer, 2013, both groups have similar number of sequences. A steady decline in the number of sequences was observed for the positive group throughout fall, 2013, and spring, 2014. For the negative group, the number of sequences increased from summer to fall. In spring, the negative group has a decline of tRF5-ProCGG was also detected.

A different pattern was observed for tRF5-GlnTTG (Fig. 3). There was no difference in the number of sequences between groups in summer, 2013. In fall, the positive group had a greater number of sequences when compared to the negative group. In spring, 2013, there was no difference in the number of sequences between groups. The slope of the increase from summer to fall in the positive group, and from fall to spring in the negative group was similar.

Discussion

It has been indicated that tRFs, tRF5s specifically, may be a significant regulator of gene silencing when animals are faced with pathogens [17]. There is a limited number of studies that have evaluated tRFs in species other than cattle [7, 11, 22], and a single study that has evaluated tRF response to a virus in cell culture [17]. Production of tRFs has been observed when cells are under stress, regardless of it being physical, chemical, or by a viral infection [15, 17, 18]. It has been established that under normal conditions, tRF5s are the predominant molecules in serum, compared to tRF3s and tRF1s [13]. Therefore, the present study characterizes the tRF5 response in cattle to exposure to the respiratory pathogen *M. bovis*.

In the present study, tRF5s for selenocysteine were identified as being differentially expressed between groups positive or negative with a *M. bovis* ELISA, throughout the growth period of the animals (Fig. 1). Selenocysteine has been recognized as the 21st amino acid [23]. It has been established that selenoproteins, which comprise selenocysteine, are produced by prokaryotic and eukaryotic organisms [24, 25]. It has been shown that pectoral muscles of poultry express selenoproteins, which makes them a primary target of selenium deficiency diseases [26]. It has also been established that selenoproteins in macrophages protect mice from dextran sodium sulfate colitis

Table 2 5'-transfer RNA fragment (tRF5), anticodon, normalized count of tRF5s in serum by group, standard error (SE) and their association (P-value) with ELISA status

tRF5	Anticodon	ELISA status		SE	P-value
		Negative (RPM)[a]	Positive (RPM)[a]		
tRF5-Ala	CGC	616	494	31	0.0081
tRF5-Glu	CTC	2584	1492	306	0.0155
tRF5-Gly	TCC	65	148	21	0.0082
tRF5-Lys	CTT	3800	6933	534	0.0002
tRF5-Lys	TTT	728	1193	113	0.0057
tRF5-Pro	TGG	78	118	10	0.0093
tRF5-SelCys	UGA	4.9	13.7	1.9	0.0020
tRF5-Thr	TGT	6.4	8.2	0.6	0.0499
tRF5-Val	CAC	11,409	15,104	1275	0.0468

[a]RPM Reads per million

[27]. However, the biological implication of structural differences between transfer RNA and selenocysteine transfer RNA still remains to be fully understood [28]. Figure 1 shows a greater number of tRF5-SelCysUGA in the group that became exposed to the pathogen between summer and fall, 2013, with the number increasing from fall to spring in the same group. However, although the negative group was negative in the fall, 2013, and became positive by the spring, 2014, no increase in the counts of tRF5-SelCysUGA was observed in the latter group. This could indicate that chronically exposed animals are prone to increase counts of tRF5-SelCysUGA. Selenocysteine is a rarely used amino acid that is incorporated into proteins by the opal codon (UGA). Production of tRF5-SelCysUGA may be associated with a host defense mechanism triggered by the bacterial infection; however, it is also possible that it may provide some advantage to the pathogen during the infection of the host. Further studies would be needed to ascertain if the tRF5-SelCysUGA could be used as a diagnostic indicator of chronic exposure.

The number of sequences for tRF5-ProCGG, declined as animals became ELISA positive (Fig. 2). A proline-

Table 3 5'-transfer RNA fragment (tRF5), anticodon, normalized count of tRF5s in serum by season (summer and fall, 2013, and spring, 2014), standard error (SE) and their association (P-value)

tRF5	Anticodon	Season			SE	P-value
		Summer, 2013 (RPM)[c]	Fall, 2013 (RPM)[c]	Spring, 2014 (RPM)[c]		
tRF5-Ala	AGC	288[a,b]	386[a]	190[b]	36	0.0017
tRF5-Ala	TGC	4.6[a]	5.9[a]	2.1[b]	0.7	0.0012
tRF5-Gln	CTG	41.6[a]	163.8[b]	196.6[b]	29.9	0.0017
tRF5-Gln	TTG	28.6[a]	53.8[b]	65.9[b]	6.0	0.0003
tRF5-Glu	TTC	161,604[a]	126,622[b]	128,748[b]	8341	0.0074
tRF5-Gly	GCC	47,848[a]	54,304[a]	74,013[b]	5349	0.0035
tRF5-Gly	TCC	65[a]	93[a,b]	161[b]	26	0.0330
tRF5-His	GTG	41,999[a,b]	70,356[a]	35,304[b]	10,090	0.0428
tRF5-Lys	TTT	948[a,b]	1219[a]	714[b]	138	0.0446
tRF5-Pro	AGG	406[a]	562[b]	335[a]	53	0.0124
tRF5-Pro	CGG	4.13[a]	4.14[a]	1.95[b]	0.60	0.0185
tRF5-Pro	TGG	92[a]	138[b]	63[a]	13	0.0006
tRF5-SelCys	UGA	3.9[a]	9.1[a,b]	14.9[b]	2.3	0.0068
tRF5-Thr	TGT	5.7[a]	8.8[b]	7.4[a,b]	0.8	0.0285
tRF5-Val	AAC	68[a]	71[a]	46[b]	7	0.0425
tRF5-Val	TAC	2,666[a,b]	3621[a]	2344[b]	336	0.0277

[a, b] Means without a common superscript within row are statistically different (P < 0.05)
[c]RPM Reads per million

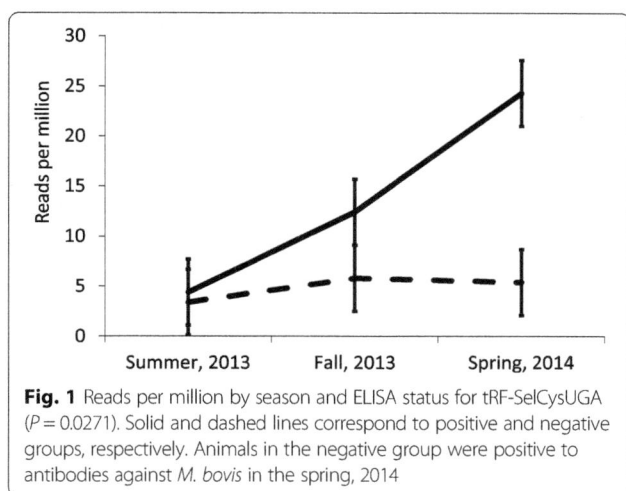

Fig. 1 Reads per million by season and ELISA status for tRF-SelCysUGA ($P = 0.0271$). Solid and dashed lines correspond to positive and negative groups, respectively. Animals in the negative group were positive to antibodies against *M. bovis* in the spring, 2014

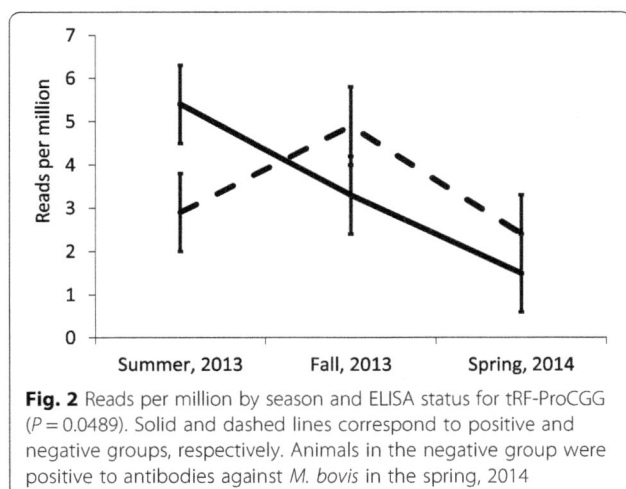

Fig. 3 Reads per million by season and ELISA status for tRF5-GlnTTG ($P = 0.0406$). Solid and dashed lines correspond to positive and negative groups, respectively. Animals in the negative group were positive to antibodies against *M. bovis* in the spring, 2014

rich peptide (PRP), isolated from the bovine neurohypophysis, has been shown to regulate immune activity, preventing death of mice infected with Gram-negative bacteria [29]. The PRP has a regulatory role in the oxidative burst induction of normal and relapsing inflammatory diseases cells such as neutrophils and monocytes [30, 31]. *Mycoplasma bovis* is a Gram-negative bacterium that could initiate or disrupt the regulation of PRP in infected cattle. Diminishing counts of tRF5-ProCGG may be a sign the host is producing additional mature transfer RNA for proline to generate additional PRP molecules to defend against *M. bovis*, thus, depleting the circulating amounts of this tRF5. However, it is also possible that diminishing counts of tRF5-ProCGG in ELISA positive animals to *M. bovis* could be a defense mechanism from the bacterium trying to inhibit the production of PRP in the host. Additional studies need to be developed to fully understand the role of this tRF5 in the defense mechanism of the host.

The counts of tRF5-GlnTTG increased as animals became ELISA positive (Fig. 3). There was a similar response in the positive and negative groups. The positive group increased the number of sequences of tRF5-GlnTTG between summer and fall, 2013; whereas the negative group increased it between fall, 2013, and spring, 2014. A review of the effect of administration of glutamine in acute respiratory disease syndrome in humans suggests that it reduces lung inflammation and mortality, while increasing alveolar barriers and oxygenation [32]. GP Oliveira, MG de Abreu, P Pelosi and PR Rocco [32], indicate that administration of exogenous glutamine may be beneficial in respiratory disease, representing a potential therapeutic tool for the condition. The association of number of sequences of tRF5-GluTTG could be associated with the use of glutamine by the host in response to a *M. bovis* infection.

When Tables 2 and 3 are compared, it is recognized that tRF5-Ala and tRF5-Val have a distinctive pattern. For tRF5s of both amino acids, two of the three anticodons are associated with season, while the third anticodon (tRF5-AlaCGC and tRF5-ValCAC) is associated with antibody response to *M. bovis*. This pattern can also be observed between tRF5-GluCTC (associated with antibody response to *M. bovis*), and tRF5-GluTTC (associated with season). This comparison shows that one of the tRF5s for each amino acid is not modified by season. These tRF5s could be likely targets as diagnostic indicators of exposure to *M. bovis*.

The ELISA used here to identify cattle as positive or negative for *M. bovis* is based on the reactivity of serum IgG with an extract enriched in membrane proteins of the bacterium. ELISAs of similar design are commonly used for detection of *M. bovis*-infected animals in both research and diagnostic settings. However, it is not known whether cross-reactive antibodies that may be elicited by infection with related commensal species,

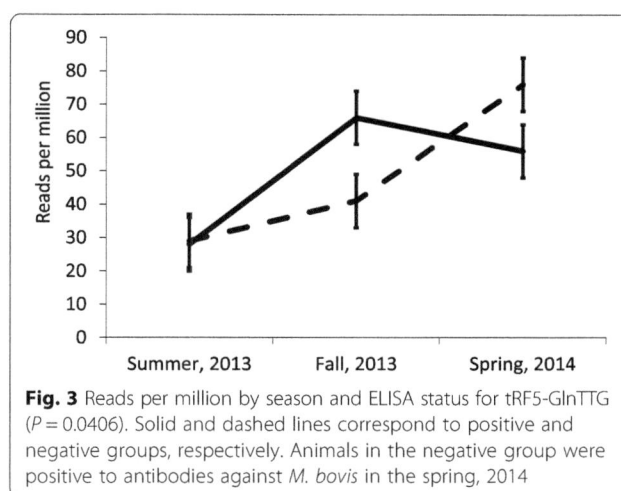

Fig. 2 Reads per million by season and ELISA status for tRF-ProCGG ($P = 0.0489$). Solid and dashed lines correspond to positive and negative groups, respectively. Animals in the negative group were positive to antibodies against *M. bovis* in the spring, 2014

such as *M. bovirhinis*, could falsely contribute to estimates of *M. bovis*-specific antibody in individual animals, including those in this study. Nonetheless, seroconversion on a group level is predictive of *M. bovis* infection, especially when antibody titers are high [33], as we found here for sera from animals in the positive group. At the fall, 2013 sampling, when positive and negative groups were defined, the level of serum antibody in all 8 positive animals exceeded the level in the pool of sera used as a positive control, in most cases by at least two-fold. Therefore, we expect any cross-reactive antibodies detected by the ELISA to have had a negligible effect on the accurate identification of cattle infected with *M. bovis*.

Growth of the animal could be responsible for expression of tRF5s in different seasons. There have been no studies involving the expression of tRF5s in growing cattle. Expression of other small non-coding RNAs such as microRNAs has been compared between muscle tissue from fetal and adult cattle [34, 35], but there are no studies evaluating tRF5s. There were seven tRF5s (tRF5-AlaAGC, tRF5-AlaTGC, tRF5-HisGTG, tRF5-LysTTT, tRF5-ProCGG, tRF5-ValAAC, and tRF5-ValTAC), that were down-regulated in spring, 2014, when compared to summer, and fall, 2013. These tRF5s could be essential for development of the calves at initial stages of growth, and their relevance diminishes as the animal reaches later physiological stages (i.e. puberty). There were other group of tRF5s that had the opposite profile. These tRF5s (tRF5-GlnCTG, tRF5-GlnTTG, tRF5-GlyTCC, tRF5-SelCysUGA, and tRF5-ThrTGT), were down-regulated in summer, 2013, when compared to fall, 2013, and spring, 2014. This group of tRF5s could be relevant as the animal reaches latter stages, while being unimportant during initial phases of growth. Two tRF5s (tRF5-ProAGG and tRF5-ProTGG) were upregulated only during fall, 2013. It is possible these tRF5s are related to weaning because their up-regulation coincided with the age at which calves are naturally weaned from the mother. tRF5-GluTTC and tRF5-GlyGCC were upregulated in summer, 2013, and in spring, 2014, respectively. It is unknown if their importance is due to their upregulation during these seasons or the growth stage at which the animals are in. Additional studies should be developed to address this question. Animals in this study were born in spring, 2013, and raised under semicommercial conditions until reaching slaughter weight, at approximately 1.25 years of age. Given there is no prior information of tRF5 production in growing beef cattle, it can only be assumed that growth is responsible for differences in tRF5 production by season.

The tRF5s with the most sequences were tRF5-Glu, tRF5-Gly, tRF5-His, tRF5-Lys, and tRF5-Val, while the tRF5s with the fewest counts were tRF5-Arg, tRF5-Ile, tRF5-Leu, tRF5-Phe, tRF5-Ser, tRF5-Trp, and tRF5-Tyr.

This is a similar pattern previously observed [13]. Casas et al. [13], used mature dairy cows to establish the proportion of tRF5s for each amino acid, while the present study comprises growing crossbred beef cattle. Given that proportions of each tRF5 are similar between both studies, it is likely the proportions of tRF5s for each amino acid observed are characteristic of the species.

Diagnosis of exposure to pathogens is important in cattle production. There is a need to identify biomarkers for early diagnosis of disease. Using small non-coding RNAs as diagnostic markers has been proposed as an alternative to traditional methods [36, 37]. Serum is a readily available biological sample, and it has been established that serum small non-coding RNAs such as microRNAs are stable even after being exposed to conditions that degrade RNA [38]. Transfer RNA Fragments are a novel group of small non-coding RNAs that circulate in serum. These molecules may even be a better candidate for use as a biomarker given their abundance in serum of cattle [13].

Conclusions

We identified nine tRF5s for which expression levels were associated with ELISA status. There were tRF5s associated with differences in expression due to season, which could be attributed to calf growth. A selenocysteine tRF5 was also identified as differentially expressed between animals positive and negative to a *M. bovis* ELISA. It is unclear the role of this tRF5, but it is known that selenocysteine plays an important role in the defense mechanism of the host when colonized by a pathogen. These data suggest that the selenocysteine tRF5 could be a potential biomarker to identify cattle exposed to *M. bovis*. 5′ transfer RNA fragments that were differentially expressed by ELISA status could also be considered as potential biomarkers. Further studies should be conducted to establish if these transfer RNA fragments could be used as a diagnostic indicator of exposure to *Mycoplasma bovis*.

Abbreviations

Ala: Alanine; Arg: Arginine; Asp: Aspartic acid; Cys: Cysteine; ELISA: Enzyme-linked Immunosorbent Assay; Gln: Glutamine; Glu: Glutamic acid; Gly: Glycine; His: Histidine; IgG: Immunoglobulin G; Ile: Isoleucine; Leu: Leucine; Lys: Lysine; *M. bovis*: *Mycoplasma bovis*; Met: Methionine; Phe: Phenylalanine; Pro: Proline; RNA: Ribonucleic acid; SelCys: Selenocysteine; Ser: Serine; Thr: Threonine; tRF5s: 5′ transfer RNA fragments; tRFs: Transfer RNA Fragments; tRNA: Transfer RNA; Trp: Tryptophan; Tyr: Tyrosine; Val: Valine

Acknowledgements

The authors thank Sandra Nejezchleb from the USMARC, and Randy Atchison and William Boatwright from the NADC for remarkable technical assistance. Mention of trade name, proprietary product, or specified equipment does not constitute a guarantee or warranty by the USDA and does not imply approval to the exclusion of other products that may be suitable. USDA is an Equal Opportunity Employer.

Funding
This study was part of an intramural research project of the USDA, Agricultural Research Service. The USDA had no role in the study design, data collection, analysis, interpretation of results, or preparation of manuscript.

Authors' contributions
EC, GC, KBR, and JDN proposed the experiment. LAK and TGM provided the experimental units for the experiment. KBR assessed the response to antibodies against *M. bovis*. GC and EC analyzed the information. EC wrote, and GC, LAK, KBR, JDN, and TGM reviewed the manuscript. All authors read and approved the final manuscript.

Consent for publication
Not applicable.

Competing interests
EC, GC, LAK, KBR, TGM, and JDN, are USDA employees and declare no competing interest. EC is an editorial board member of BMC Veterinary Research.

Author details
[1]USDA, ARS, National Animal Disease Center, Ames, IA 50010, USA. [2]USDA, ARS, U.S. Meat Animal Research Center, Clay Center, NE 68933, USA.

References
1. Griffin D. Economic impact associated with respiratory disease in beef cattle. Vet Clin North Am Food Anim Pract. 1997;13(3):367–77.
2. Taylor JD, Fulton RW, Lehenbauer TW, Step DL, Confer AW. The epidemiology of bovine respiratory disease: what is the evidence for preventive measures? Can Vet J. 2010;51(12):1351–9.
3. Miles DG. Overview of the north American beef cattle industry and the incidence of bovine respiratory disease (BRD). Anim Health Res Rev. 2009;10(2):101–3.
4. Caswell JL, Bateman KG, Cai HY, Castillo-Alcala F. Mycoplasma bovis in respiratory disease of feedlot cattle. Vet Clin North Am Food Anim Pract. 2010;26(2):365–79.
5. Maunsell FP, Woolums AR, Francoz D, Rosenbusch RF, Step DL, Wilson DJ, Janzen ED. Mycoplasma bovis infections in cattle. J Vet Intern Med. 2011;25(4):772–83.
6. Shahriar FM, Clark EG, Janzen E, West K, Wobeser G. Coinfection with bovine viral diarrhea virus and mycoplasma bovis in feedlot cattle with chronic pneumonia. Can Vet J. 2002;43(11):863–8.
7. Cole C, Sobala A, Lu C, Thatcher SR, Bowman A, Brown JW, Green PJ, Barton GJ, Hutvagner G. Filtering of deep sequencing data reveals the existence of abundant dicer-dependent small RNAs derived from tRNAs. RNA. 2009; 15(12):2147–60.
8. Lee YS, Shibata Y, Malhotra A, Dutta A. A novel class of small RNAs: tRNA-derived RNA fragments (tRFs). Genes Dev. 2009;23(22):2639–49.
9. Garcia-Silva MR, Cabrera-Cabrera F, Guida MC, Cayota A. Hints of tRNA-derived small RNAs role in RNA silencing mechanisms. Genes. 2012;3(4):603–14.
10. Sobala A, Hutvagner G. Small RNAs derived from the 5' end of tRNA can inhibit protein translation in human cells. RNA Biol. 2013;10(4):553–63.
11. Kumar P, Anaya J, Mudunuri SB, Dutta A. Meta-analysis of tRNA derived RNA fragments reveals that they are evolutionarily conserved and associate with AGO proteins to recognize specific RNA targets. BMC Biol. 2014;12:78.
12. Haussecker D, Huang Y, Lau A, Parameswaran P, Fire AZ, Kay MA. Human tRNA-derived small RNAs in the global regulation of RNA silencing. RNA. 2010;16(4):673–95.
13. Casas E, Cai G, Neill JD. Characterization of circulating transfer RNA-derived RNA fragments in cattle. Front Genet. 2015;6:271.
14. Nolte-'t Hoen EN, Buermans HP, Waasdorp M, Stoorvogel W, Wauben MH, 't Hoen PA. Deep sequencing of RNA from immune cell-derived vesicles uncovers the selective incorporation of small non-coding RNA biotypes with potential regulatory functions. Nucleic Acids Res. 2012;40(18):9272–85.
15. Thompson DM, Parker R. Stressing out over tRNA cleavage. Cell. 2009;138(2):215 9.
16. Ivanov P, Emara MM, Villen J, Gygi SP, Anderson P. Angiogenin-induced tRNA fragments inhibit translation initiation. Mol Cell. 2011;43(4):613–23.
17. Wang Q, Lee I, Ren J, Ajay SS, Lee YS, Bao X. Identification and functional characterization of tRNA-derived RNA fragments (tRFs) in respiratory syncytial virus infection. Mol Ther. 2013;21(2):368–79.
18. Zhang Y, Zhang Y, Shi J, Zhang H, Cao Z, Gao X, Ren W, Ning Y, Ning L, Cao Y, et al. Identification and characterization of an ancient class of small RNAs enriched in serum associating with active infection. J Mol Cell Biol. 2014;6(2):172–4.
19. Register KB, Sacco RE, Olsen SC. Evaluation of enzyme-linked immunosorbent assays for detection of mycoplasma bovis-specific antibody in bison sera. Clin Vaccine Immunol. 2013;20(9):1405–9.
20. Vanden Bush TJ, Rosenbusch RF. Characterization of the immune response to mycoplasma bovis lung infection. Vet Immunol Immunopathol. 2003;94(1–2):23–33.
21. Yamao F, Muto A, Kawauchi Y, Iwami M, Iwagami S, Azumi Y, Osawa S. UGA is read as tryptophan in mycoplasma capricolum. Proc Natl Acad Sci U S A. 1985;82(8):2306–9.
22. Loss-Morais G, Waterhouse PM, Margis R. Description of plant tRNA-derived RNA fragments (tRFs) associated with argonaute and identification of their putative targets. Biol Direct. 2013;8:6.
23. Stadtman TC. Selenium biochemistry. Science. 1974;183(4128):915–22.
24. Taylor EW, Nadimpalli RG, Ramanathan CS. Genomic structures of viral agents in relation to the biosynthesis of selenoproteins. Biol Trace Elem Res. 1997;56(1):63–91.
25. Cravedi P, Mori G, Fischer F, Percudani R. Evolution of the Selenoproteome in helicobacter pylori and Epsilonproteobacteria. Genome Biol Evol. 2015;7(9):2692–704.
26. Yao HD, Wu Q, Zhang ZW, Zhang JL, Li S, Huang JQ, Ren FZ, Xu SW, Wang XL, Lei XG. Gene expression of endoplasmic reticulum resident selenoproteins correlates with apoptosis in various muscles of se-deficient chicks. J Nutr. 2013;143(5):613–9.
27. Kaushal N, Kudva AK, Patterson AD, Chiaro C, Kennett MJ, Desai D, Amin S, Carlson BA, Cantorna MT, Prabhu KS. Crucial role of macrophage selenoproteins in experimental colitis. J Immunol. 2014;193(7):3683–92.
28. Commans S, Bock A. Selenocysteine inserting tRNAs: an overview. FEMS Microbiol Rev. 1999;23(3):335–51.
29. Galoyan A. Neurochemistry of brain neuroendocrine immune system: signal molecules. Neurochem Res. 2000;25(9–10):1343–55.
30. Davtyan TK, Manukyan HM, Hakopyan GS, Mkrtchyan NR, Avetisyan SA, Galoyan AA. Hypothalamic proline-rich polypeptide is an oxidative burst regulator. Neurochem Res. 2005;30(3):297–309.
31. Davtyan TK, Manukyan HA, Mkrtchyan NR, Avetisyan SA, Galoyan AA. Hypothalamic proline-rich polypeptide is a regulator of oxidative burst in human neutrophils and monocytes. Neuroimmunomodulation. 2005;12(5): 270–84.
32. Oliverira GP, de Abreu MG, Pelosi P, Rocco PR. Exogenous glutamine in respiratory diseases: Myth or reality? Nutrients. 2016;8(2):76. https://doi.org/10.3390/nu8020076
33. Martin SW, Bateman KG, Shewen PE, Rosendal S, Bohac JG, Thorburn M. A group level analysis of the associations between antibodies to seven putative pathogens and respiratory disease and weight gain in Ontario feedlot calves. Can J Vet Res. 1990;54(3):337–42.
34. Sun J, Sonstegard TS, Li C, Huang Y, Li Z, Lan X, Zhang C, Lei C, Zhao X, Chen H. Altered microRNA expression in bovine skeletal muscle with age. Anim Genet. 2015;46(3):227–38.
35. Huang YZ, Sun JJ, Zhang LZ, Li CJ, Womack JE, Li ZJ, Lan XY, Lei CZ, Zhang CL, Zhao X, et al. Genome-wide DNA methylation profiles and their relationships with mRNA and the microRNA transcriptome in bovine muscle tissue (Bos taurine). Sci Rep. 2014;4:6546.
36. Oksuz Z, Serin MS, Kaplan E, Dogen A, Tezcan S, Aslan G, Emekdas G, Sezgin O, Altintas E, Tiftik EN. Serum microRNAs; miR-30c-5p, miR-223-3p, miR-302c-3p and miR-17-5p could be used as novel non-invasive biomarkers for HCV-positive cirrhosis and hepatocellular carcinoma. Mol Biol Rep. 2015;42(3):713–20.
37. Cortez MA, Bueso-Ramos C, Ferdin J, Lopez-Berestein G, Sood AK, Calin GA. MicroRNAs in body fluids-the mix of hormones and biomarkers. Nat Rev Clin Oncol. 2011;8(8):467–77.
38. Chen X, Ba Y, Ma L, Cai X, Yin Y, Wang K, Guo J, Zhang Y, Chen J, Guo X, et al. Characterization of microRNAs in serum: a novel class of biomarkers for diagnosis of cancer and other diseases. Cell Res. 2008;18(10):997–1006.

13

Whole genome sequencing of an ExPEC that caused fatal pneumonia at a pig farm in Changchun, China

Ling-Cong Kong, Xia Guo, Zi Wang, Yun-Hang Gao, Bo-Yan Jia, Shu-Ming Liu and Hong-Xia Ma*

Abstract

Background: In recent years, highly frequent swine respiratory diseases have been caused by extraintestinal pathogenic *Escherichia coli* (ExPEC) in China. Due to this increase in ExPECs, this bacterial pathogen has become a threat to the development of the Chinese swine industry. To investigate ExPEC pathogenesis, we isolated a strain (named SLPE) from lesioned porcine lungs from Changchun in China, reported the draft genome and performed comparative genomic analyses.

Results: Based on the gross post-mortem examination, bacterial isolation, animal regression test and 16S rRNA gene sequence analysis, the pathogenic bacteria was identified as an ExPEC. The SLPE draft genome was 4.9 Mb with a G + C content of 51.7%. The phylogenomic comparison indicated that the SLPE strain belongs to the B1 monophyletic phylogroups and that its closest relative is *Avian Pathogenic Escherichia coli* (APEC) O78. However, the distribution diagram of the pan-genome virulence genes demonstrated significant differences between SLPE and APEC078. We also identified a capsular polysaccharide synthesis gene cluster (CPS) in the SLPE strain genomes using blastp.

Conclusions: We isolated the ExPEC (SLPE) from swine lungs in China, performed the whole genome sequencing and compared the sequence with other *Escherichia coli* (*E. coli*). The comparative genomic analysis revealed several genes including several virulence factors that are ExPEC strain-specific, such as fimbrial adhesins (*pap*G II), *ire*A, *pgt*P, *hly*F, the *pix* gene cluster and *fec*R for their further study. We found a CPS in the SLPE strain genomes for the first time, and this CPS is closely related to the CPS from *Klebsiella pneumoniae*.

Keywords: ExPEC, Pneumonia, Pigs, Genome

Background

According to genetic and clinical criteria, *E. coli* has been classified into commensal, intestinal pathogenic and ExPEC groups [1]. ExPEC has become a bacterial pathogen that threatens the health of humans and animals. These strains can cause a wide range of extraintestinal infections, including in the urinary tract, central nervous system, circulatory system and respiratory system [2]. In recent years, some reports have demonstrated that ExPEC has been frequently isolated from clinical samples in the swine industry in China [3, 4], leading to significant economic losses.

To date, several *E. coli* isolates have had their whole genomes sequenced, including intestinal pathogenic *E.* coli and extraintestinal pathogenic *E. coli*. Based on the whole sequenced genomes, *E. coli* can be classified into four major phylogroups (A, B1, D1, D2 and B2). ExPEC are mainly distributed into phylogroups B2, D1, D2 and F, but occasionally into other phylogroups such as A and B1 depending on the virulence gene repertoire [5]. However, to date, little work has been conducted to characterize the ExPEC that caused fatal pneumonia at a pig farm in Changchun, China. In this paper, we isolated the ExPEC from swine lungs from China and describe the genome characteristics of the isolate.

Results

Gross post mortem examination and histological examination

The pig carcass was in fair nutritional condition. There was no evidence of any viruses or external parasites.

* Correspondence: hongxia0731001@163.com
College of Animal Science and Technology, Jilin Agricultural University, Changchun, China

Only the lungs had some congestion and bleeding. The stomach was devoid of contents. There was nothing remarkable about any of the other visceral systems. The histological preparations of the lungs showed the presence of inflammatory cells in the bronchioles and the surrounding alveoli (red arrow), and presence of some bleeding and edema (Fig. 1). Histological examinations of pneumonia-like presentation may be classified as bronchopneumonia.

Bacterial isolates and serotyping
Some colonies were obtained from the pig lung, in colonies, only one colony was selected, which contained two of the ExPEC virulence markers, iutA and papC, named SLPE. The SLPE isolate was positive for serogroup O149.

Experimental challenge studies
The SLPE isolate killed all of the mice at the 5×10^7 CFU, 5×10^6 CFU and 5×10^5 CFU challenge doses, but the surviving time of each group is different, the 5×10^7 CFU dose group only survived 6 to 8 h and the 5×10^6 CFU and 5×10^5 CFU dose group survived 12 to 18 h. All of the mice in the negative control group survived without any clinical signs. All tested were dissected and examined for lesion, the lungs had some bleeding and swelling, but there was nothing remarkable lesions in other tissues and organs. The SLPE presented as a highly virulent ExPEC strain in the mouse model. Chen et al., has isolated 315 ExPEC in china, only 2 isolated of these killed 5/5 mice at a challenge dose of 10^5 CFU [3].

Antimicrobial resistance
The MIC values for the 8 antimicrobial agents obtained from the examinations of the SLPE isolates are shown. Cefotaxime was found to be the most active compound in vitro (MIC < 0.03 μg/mL). The isolates was already resistant to the rest of the antimicrobials and displayed the following MIC values: (1) florfenicol, MIC = 128 μg/mL; (2) doxycycline, MIC = 128 μg/mL; (3) ciprofloxacin, MIC = 64 μg/mL; (4) tilmicosin, MIC = 512 μg/mL; (5) enrofloxacin, MIC = 128 μg/mL; (6) sulfamethoxazole, MIC>512 μg/mL and (7) amikacin, MIC = 128 μg/mL.

Genomic features
A total of 336 contigs from the genome were assembled onto 152 scaffolds. The predicted genome size was 4.9 Mb, with an N50 value of 106,661 for the assembly. Ombining the glimmer 3.02, Genemark and Z-Curve programs annotations gave 4806 genes with approximately 51.7% GC content, which is similar to other reported E. coli genomes. This Whole Genome Shotgun project has been deposited at DDBJ/EMBL/GenBank under the accession LJCG00000000.1. The SLPE genome harbour some resistance genes, including MarC, SoxR, MATE, ABC genes, tet, FloR and also carrie class Iintegrons. but the gene cassettes of Iintegrons genes, that conferred resistance to tetracycline, fluoroquinolone, aminoglycosides and so on.

Phylogenomic genome analysis of SLPE with other E. coli pathotypes
The phylogenomic tree was constructed to examine the SLPE strain in the context of the E. coli population structure. Figure 2 shows that the 43 E. coli strains were divided into five monophyletic phylogroups (A, B1, B2, D and E). The SLPE strain belongs to the B1 monophyletic phylogroup, and its closest relative is APEC 078.

Distribution diagram of the pan-genome virulence genes
The distribution of the ExPEC virulence factors were analysed among the 43 sequenced E. coli strains. Figure 2 shows that ExPEC-specific virulence factors were classified into the following six categories: (1) adhesins, (2) invasins, (3) toxins, (4) iron acquisition/transport

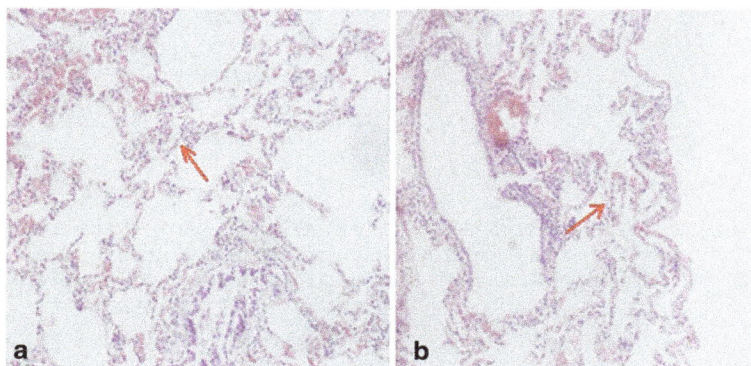

Fig. 1 Histological preparation of swine lungs (haematoxylin and eosin). Note the presence of inflammatory cells in the bronchioles (**a**) and the surrounding alveoli (**b**)

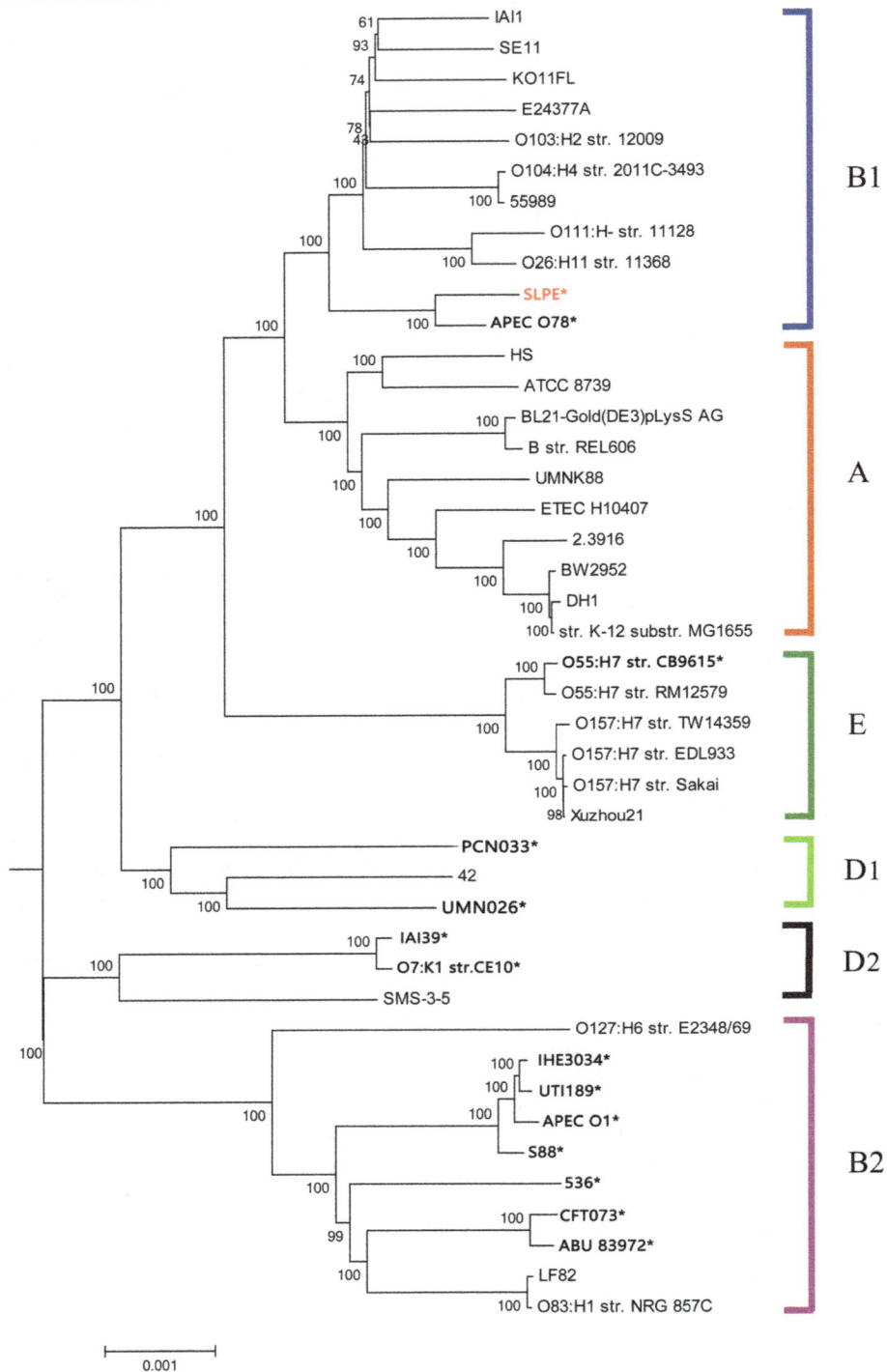

Fig. 2 Phylogenomic tree of the 43 *E. coli* strains. Blastp, perl, mcl, clustalW and MEGA were used to reconstruct the phylogenomic tree. The phylogenomic tree of different *E. coli* based on the comparison of 2043 core genes. The 43 *E. coli* strains are divided into five monophyletic phylogroups (A, B1, B2, D and E). The SLPE strains are highlighted in blackbody. SLPE is the closest relative to APEC 078

systems, (5) polysialic acid synthesis and (6) other virulence genes. The 51 virulence genes from the 43 sequenced strains are shown in Fig. 3. The distribution diagram shows that even though there were significant differences between the SLPE and APEC078 strains, that these two strains are most closely related to one another.

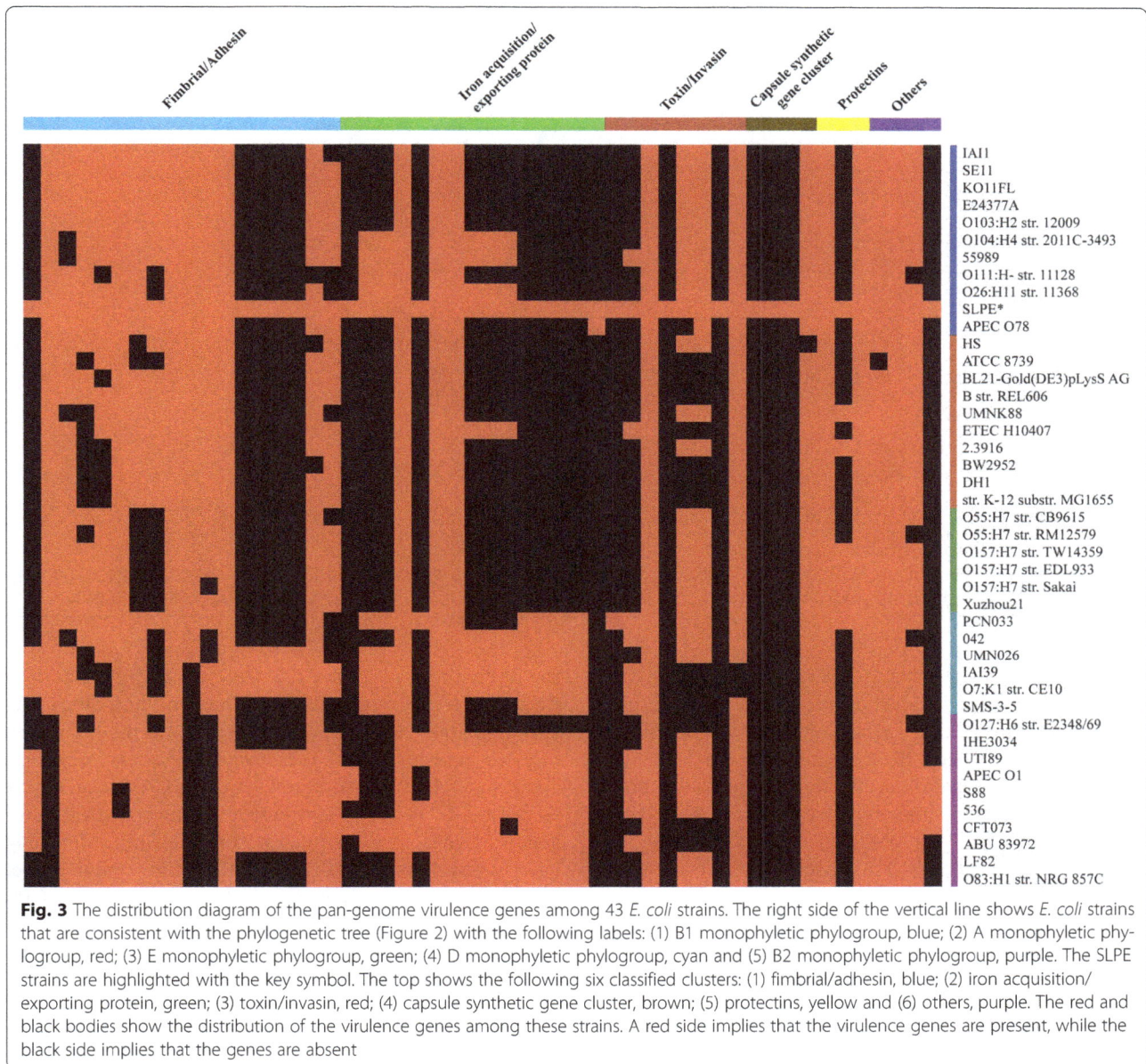

Fig. 3 The distribution diagram of the pan-genome virulence genes among 43 *E. coli* strains. The right side of the vertical line shows *E. coli* strains that are consistent with the phylogenetic tree (Figure 2) with the following labels: (1) B1 monophyletic phylogroup, blue; (2) A monophyletic phylogroup, red; (3) E monophyletic phylogroup, green; (4) D monophyletic phylogroup, cyan and (5) B2 monophyletic phylogroup, purple. The SLPE strains are highlighted with the key symbol. The top shows the following six classified clusters: (1) fimbrial/adhesin, blue; (2) iron acquisition/exporting protein, green; (3) toxin/invasin, red; (4) capsule synthetic gene cluster, brown; (5) protectins, yellow and (6) others, purple. The red and black bodies show the distribution of the virulence genes among these strains. A red side implies that the virulence genes are present, while the black side implies that the genes are absent

We also found that SLPE has more ExPEC-specific virulence factors, including several virulence factors that are ExPEC strain-specific, such as fimbrial adhesins (*papG* II), *ire*A, *pgt*P, *hly*F, the *pix* gene cluster and *fec*R (Fig. 3).

Analysis of the capsular polysaccharide synthesis gene cluster

We found a capsular polysaccharide synthesis gene cluster (CPS) in the SLPE genomes using blastp. This CPS was closely related to the *Klebsiella pneumonia* CPS. The *Klebsiella pneumonia* CPS is closely related to *Klebsiella pneumonia* pathogenesis. The similarity and GC content of the 5 CPSs from *Klebsiella pneumoniae*

genomes and 1 CPS from an *E. coli* strain were analysed. The results are shown in Fig. 4.

Discussion

In this study, we isolated ExPEC isolates (ALPE) from pig farms and demonstrated that the SLPE isolates are pneumonia pathogens. Then, we performed genome sequencing and comparative genomic analysis with the SLPE isolates. Our analysis identified differences between SLPE and other ExPEC strains including various differences in the toxin and capsular polysaccharide synthesis gene clusters.

The phylogenomic comparison indicated that the SLPE strain belongs to the B1 monophyletic phylogroup,

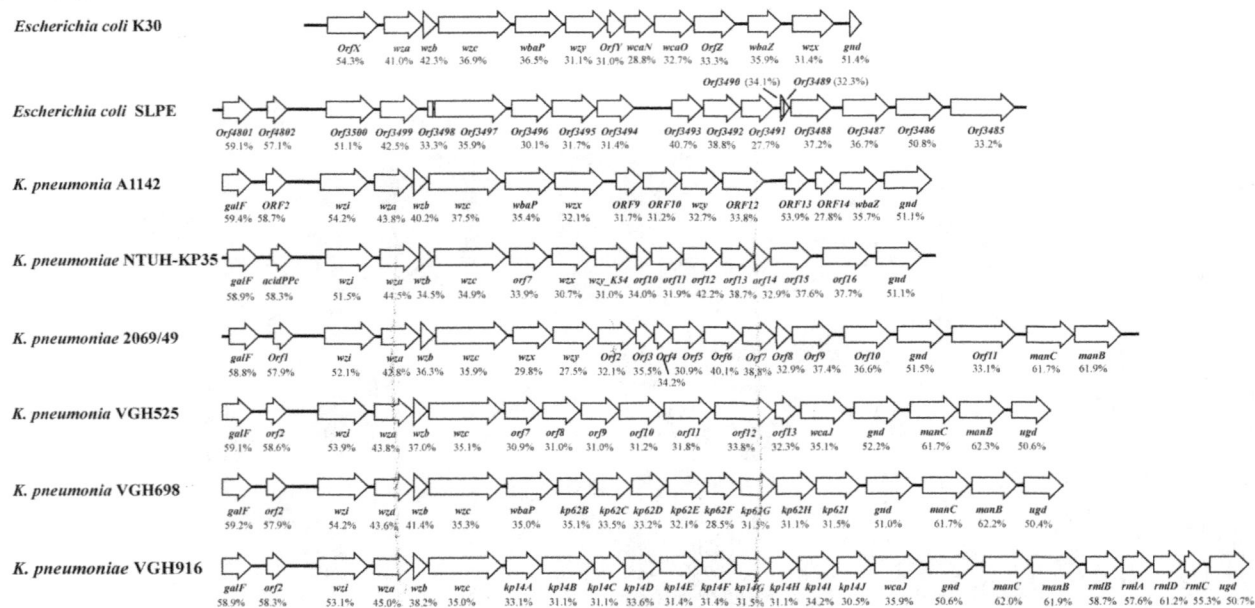

Fig. 4 The genetic context for seven different types of capsular polysaccharide synthesis gene clusters. Shaded areas show conservative genes. Arrows represent the coding sequences and indicate the direction of transcription. The percentage under arrow represents the G/C content of the gene

with its closest relative being APEC O78. APEC O78, an O78 strain, is an avian pathogenic *E. coli* isolated from the lungs of turkey that caused extensive animal and financial losses globally [6, 7]. Although the SLPE and APEC O78 hosts are different, the strains have similar evolutionary relationships and both contain 4033 homologous genes. Now the question remains whether they have a common ancestor that is worthy of our attention.

ExPEC strains contain certain specific virulence traits that enable them to invade and colonize extraintestinal sites and cause a wide range of infections [8]. In this study, the distribution diagram of the pan-genome virulence genes showed that there were significant differences between the SLPE and APEC078, although they are most closely related to one another. We also found that SLPE has more ExPEC-specific virulence factors, including several virulence factors that are ExPEC strain-specific, such as fimbrial adhesins (*pap*G II), *ire*A, *pgt*P, *hly*F, the *pix* gene cluster and *fec*R. Of these genes, only *pap*G II has been proven to have contact with urethral and avian pathogenicity [9]. These genes' functions in pig ExPECs have yet to be explored.

In China, Liu et al., has reported the full genome characterization of ExPEC PCN033 strain isolated from pigs. The ExPEC virulence markers are different between PCN033 and SLPE, and the comparative genomic analysis results are different. SLPE was assigned to group B1 and clustered with APEC 078, while PCN033 was assigned to group D1 [10]. Furthermore, we found a capsular polysaccharide synthesis gene cluster (CPS) in

the SLPE strain genomes, this CPS was closely related to the Klebsiella pneumonia CPS. The gene homology was approximately 96% (*wzi*, *wza* and *wzc*) and the homology of genes responsible for the serotype (*wzx* and *wzy*) was approximately 99%. Additionally, the GC content of the CPS was 51.7%, which is significantly lower than the full genome. These results suggest that the SLPE CPS may have been transferred from *Klebsiella pneumoniae*. Based on further analysis, the *Klebsiella pneumonia* CPS is the best characterized virulence factor in pneumonia and urinary tract infections [11, 12]. Functional analysis of the SLPE CPS will be forthcoming.

Conclusions

The SLPE isolated from swine lungs in China, was assigned to group B1 and clustered with APEC 078. The comparative genomic analysis revealed several genes for their further study. We found a CPS in the SLPE strain genomes, and this CPS is closely related to the CPS from *Klebsiella pneumoniae*. In conclusion, the mechanisms of ExPEC virulence still remain largely unknown. Studies on ExPEC pathogenesis will be enhanced by public access to high-quality genomic sequences.

Methods

Gross post-mortem examination

The swine industry was affected by pneumonia in the Changchun district of China in 2014. The pigs presented with pneumonia for 4–6 days prior to death. The autopsy performed after death included gross examination and observations of the lesions according to standard operating procedures. Lung tissue samples were quickly removed and fixed in formalin 10%. The diagonal section of samples was obtained and sectioned at 4 μm thickness, then the samples were stained using hematoxylin and eosin(H&E) stains, assessed using a light microscopy.

Bacterial isolation, culture conditions and serogroup
Bacteria were isolated from the surface of heat sterilized tissues, including lung, lymph, brain and heart blood samples.. All isolates were cultured on Mueller Hintom (MH), MacConkey, chocolate and 5% sheep's blood agar plates. The isolates were identified by 16S rRNA gene sequence analysis and all isolates were identified in PCR for virulence marker genes of ExPEC (papA/papC, sfa/foc, afa/dra, kpsMTII and iutA) and ExPECs were defined as the isolates containing two or more virulence markers. Then, the serogroup was determined using antisera from the China Institute of Veterinary Drug Control (Beijing, China).

Experimental challenge studies
To test the pathogenicity of the porcine ExPEC strains, Experimental challenge studies was performed. The protocol was approved by the Committee on the Ethics of Animal Experiments of the Jilin Agricultural University. A total of 20 BALB/c mice weighing 20(±2) g were randomly divided into 4 groups, every group contained 5 mice (purchased from the Animal Centre of Jilin University, Changchun, China) according to the institutional guidelines for the use of experimental animals (Laboratory animal—requirements of environment and housing facilities number GB 14925–2010). Approximately 5×10^7 CFU, 5×10^6 CFU and 5×10^5 CFU SLPE suspensions were prepared, and the bacterium were centrifuged and cleaned by physiological saline. Then, the strains were injected intraperitoneally into BALB/c mic, physiological saline was used as a negative control. Mice for health status were observed twice daily during 3 days. Deaths were recorded, and histological examination of dead mice lung was performed.

Antimicrobial resistance
Antimicrobial susceptibility testing was performed using the microplate dilution method [13]. The antimicrobial agents included cefotaxime, florfenicol, doxycycline, ciprofloxacin, tilmicosin, enrofloxacin, sulfamethoxazole and amikacin. The reference strain E. coli ATCC 25922 served as a standard for quality control. Each experiment was repeated four times.

DNA extraction
Total DNA was extracted using the Rapid Bacterial Genomic DNA Isolation Kit (Sangon Biotech, Shanghai, China) and stored at −80 °C until use. The DNA preparation was used for genome sequencing.

Genome sequencing and genome assembly
A whole genome shotgun library was produced with 1 μg of genomic DNA according to the instructions of the TruSeq™ DNA Sample Prep Kit-Set A (Illumina, USA) and the TruSeq PE Cluster Kit (Illumina, USA). The libraries were sequenced on an Illumina MiSeq platform at the Laboratory of Chinese National Human Genome Center (Shanghai, China). Afterward, the sequences were assembled using the SOAPdenovo software package [14, 15]. A total of 9,569,782 paired-end reads were assembled into 336 contigs.

Gene prediction and functional annotation
Coding sequences were predicted by combining the results obtained with the glimmer 3.02, Genemark and Z-Curve programs [16, 17]. Translational products of the coding sequences were annotated with the GenBank database. Putative functions of translation products were confirmed using the KEGG and Clusters of Orthologous Groups (COGs) databases [18, 19]. Putative genes of interest were identified with NCBI ORF finder and iterative BLASTn and BLASTp searches.

Phylogenomic comparison of SLPE with other E. coli pathotypes
Forty-three E. coli strain genomes were downloaded from the NCBI GenBank, fourteen strains was ExPEC, belong to different groups. The common genes were considered by the BLAST-like alignment tool (e-value = $1\text{-}e^{-10}$) and mcl [20]. All of the common genes were aligned by MUSCLE and concatenated together [21]. Then, clustalW and MEGA were used to reconstruct the phylogenomic tree. Genomes of Escherichia coli used for phylogenetic construction and core genome content used to construct phylogenetic tree were listed in Additional file 1: Table S1 and Additional file 2: Table S2.

The distribution diagram of pan-genome virulence genes
To examine whether SLPE strains harboured unique ExPEC-specific virulence factors, the pan-genome analysis of 51 virulence factors were performed, which are all positive in SLPE isolate [22]. ExPEC-specific virulence factors were classified into the following six categories: (1) adhesins, (2) invasins, (3) toxins, (4) iron acquisition/transport systems, (5) polysialic acid synthesis and (6) other virulence genes. All of the virulence factors were conducted with blastp and the virulence genes were clustered with PGAP. Then, a distribution diagram was created with the R program (http://www.r-project.org/). Six groups of genes that were used for heat map

construction were listed in Additional file 1: Table S1 and Additional file 3: Table S3.

Analysis of capsular polysaccharide synthesis gene cluster

We found a CPS in the genome of the SLPE strain using blastp. The SLPE CPS was a close relative to the CPS from *Klebsiella pneumoniae* in the NCBI GenBank. The *Klebsiella pneumoniae* CPS is considered to be closely related to pathogenesis. Therefore, analysis of the CPS gene cluster between SLPE and other *Klebsiella pneumoniae* was important. Six CPSs from *Klebsiella pneumoniae* genomes and 1 CPS from an *E. coli* strain were downloaded from the NCBI GenBank. The GC content was calculated by GC content.pl and analysed with clc sequence viewer. Then, the similarity was analysed by mauve. CPS gene clusters were listed in Additional file 4: Table S4.

Additional files

Additional file 1: Table S1. Genomes of *Escherichia coli* used for phylogenetic construction.

Additional file 2: Table S2. Core genome content used to construct phylogenetic tree.

Additional file 3: Table S3. Six groups of genes that were used for heat map construction.

Additional file 4: Table S4. Comparison of CPS gene cluster of SLPE with closely related CPS gene clusters from other genomes.

Abbreviations

APEC: Avian pathogenic *Escherichia coli*; CPS: Capsular polysaccharide synthesis; *E. coli*: *Escherichia coli*; ExPEC: extraintestinal pathogenic *Escherichia coli*

Acknowledgements

We thank Dr. LUAN Wei-min (Jilin Agricultural University, Changchun) for critical review of this manuscript.

Funding

This research was funded by National Key Research and Development Plan (Grant number: 2016YFD0501301), the Science and Technology Development Plan of Jilin Province (Grant number 20170520074JH), and the Science and Technology Development Plan of Jilin Province (Grant number 20150101109JC).

Authors' contributions

LCK and XG contributed equally, there were involved in devising the experimental. HXM contributed to the analysis of the data. ZW, YHG and SML contributed to the experimental challenge studies and antimicrobial resistance. All authors read and approved the final manuscript and accepted the final version of this manuscript.

Competing interests

The authors declare that they have no competing interests.

Consent for publication

Not applicable.

References

1. Smith JL, Fratamico PM, Gunther NW. Extraintestinal pathogenic *Escherichia coli*. Foodborne Pathog Dis. 2007;4(2):134–63.
2. Orskov F, Orskov I. *Escherichia coli* serotyping and disease in man and animals. Can J Microbiol. 1992;38(7):699–704.
3. Tan C, Tang X, Zhang X, Ding Y, Zhao Z, Wu B, et al. Serotypes and virulence genes of extraintestinal pathogenic *Escherichia coli* isolates from diseased pigs in China. Vet J. 2012;192(3):483–8.
4. Liu C, Chen Z, Tan C, Liu W, Xu Z, Zhou R, et al. Immunogenic characterization of outer membrane porins OmpC and OmpF of porcine extraintestinal pathogenic *Escherichia coli*. FEMS Microbiol Lett. 2012;337(2):104–11.
5. Ge XZ, Jiang J, Pan Z, Hu L, Wang SH, Wang HJ, et al. Comparative Genomic Analysis Shows That Avian Pathogenic *Escherichia coli* Isolate IMT5155 (O2: K1: H5; ST Complex 95, ST140) Shares Close Relationship with ST95 *APEC* O1: K1 and Human ExPEC O18: K1 Strains. PLoS One. 2014;9(11):e112048.
6. Mangiamele P, Nicholson B, Wannemuehler Y, Seemann T, Logue CM, Li G, et al. Complete genome sequence of the avian pathogenic *Escherichia coli* strain *APEC* O78. Genome Announc. 2013;1(2):e0002613.
7. Lamarche MG, Dozois CM, Daigle F, Caza M, Curtiss R 3rd, Dubreuil JD, et al. Inactivation of the pst system reduces the virulence of an avian pathogenic *Escherichia coli* O78 strain. Infect Immun. 2005;73(7):4138–45.
8. Orskov I, Orskov F. *Escherichia coli* in extra-intestinal infections. J Hyg (Lond). 1985;95(03):551–75.
9. Norinder BS, Luthje P, Yadav M, Kadas L, Fang H, Nord CE, et al. Cellulose and PapG are important for *Escherichia coli* causing recurrent urinary tract infection in women. Infection. 2011;39(6):571–4.
10. Liu C, Zheng H, Yang M, et al. Genome analysis and in vivo virulence of porcine extraintestinal pathogenic *Escherichia coli* strain PCN033. BMC Genomics. 2015;16(1):1–18.
11. Tomas A, Lery L, Regueiro V, Perez-Gutierrez C, Martinez V, Moranta D, et al. Tournebized Re, Bengoecheaf JA. Functional Genomic Screen Identifies *Klebsiella pneumoniae* Factors Implicated in Blocking Nuclear Factor kappaB (NF-kappaB) Signaling. J Biol Chem. 2015;290(27):16678–97.
12. Arakawa Y, Wacharotayankun R, Nagatsuka T, Ito H, Kato N, Ohta M. Genomic organization of the *Klebsiella pneumoniae* cps region responsible for serotype K2 capsular polysaccharide synthesis in the virulent strain Chedid. J Bacteriol. 1995;177(7):1788–96.
13. Clinical and Laboratory Standards Institute. Performance Standards for Antimicrobial Disk and Dilution Susceptibility Tests for Bacteria Isolated From Animals. Approved Standard, 4th ed. In: CISI document. Wayne: CLSI; 2013. p. VET01–A4.
14. Li R, Zhu H, Ruan J, Qian W, Fang X, Shi Z, et al. De novo assembly of human genomes with massively parallel short read sequencing. Genome Res. 2010;20(2):265–72.
15. Li R, Li Y, Kristiansen K, Wang J. SOAP: short oligonucleotide alignment program. Bioinformatics. 2008;24(5):713–4.
16. Delcher AL, Harmon D, Kasif S, White O, Salzberg SL. Improved microbial gene identification with GLIMMER. Nucleic Acids Res. 1999;27(23):4636–41.
17. Lukashin AV, Borodovsky M. GeneMark.hmm: new solutions for gene finding. Nucleic Acids Res. 1998;26(4):1107–15.
18. Tatusov RL, Fedorova ND, Jackson JD, Jacobs AR, Kiryutin B, Koonin EV, et al. The COG database: an updated version includes eukaryotes. BMC Bioinformatics. 2003;4(1):41.
19. Kanehisa M, Goto S, Kawashima S, Okuno Y, Hattori M. The KEGG resource for deciphering the genome. Nucleic Acids Res. 2004;32:D277–80.
20. Kent WJ. BLAT–the BLAST-like alignment tool. Genome Res. 2002;12(4):656–64.
21. Edgar RC. MUSCLE: multiple sequence alignment with high accuracy and high throughput. Nucleic Acids Res. 2004;32(5):1792–7.
22. Logue CM, Doetkott C, Mangiamele P, Wannemuehler YM, Johnson TJ, Tivendale KA, et al. Genotypic and phenotypic traits that distinguish neonatal meningitis-associated *Escherichia coli* from fecal *E. coli* isolates of healthy human hosts. Appl Environ Microbiol. 2012;78(16):5824–30.

Seroepidemiology and associated risk factors of *Toxoplasma gondii* in sheep and goats in Southwestern Ethiopia

Dechassa Tegegne*, Amin kelifa, Mukarim Abdurahaman and Moti Yohannes

Abstract

Background: *T.gondii* is a global zoonotic disease and is considered as the most neglected tropical disease in sub-Saharan countries. The exact seroepidemiological distribution and risk factors for the infection of food animals and humans in Ethiopia was less studied although, such studies are important. The objective of the current study was to determine the seroprevalence and potential risk factors of *T. gondii* infection in sheep and goats in Southwestern Ethiopia.

Methods: Cross sectional study was conducted from November 2014 to March 2015 in South west Ethiopia in four selected districts of Jimma zone ($n = 368$). Slide agglutination test (Toxo-latex) was used to detect anti-*T.gondii* antibodies. Logistic regression was used to determine potential risk factors.

Results: An overall seroprevalence of 57.60% (212/368; 95% CI: 52.55–62.6) was detected. 58.18% (148/252; 95% CI: 52.75–64.88) and 55.18% (64/116; 95% CI: 46.13–64.23) sero prevalence was found in sheep and goats respectively. Multivariable logistic regression analysis showed that the risk of *T. gondii* infection was significantly higher in adult sheep and goats [(sheep: Odds Ratio (OR) = 2.5, confidence interval (CI): 1.19–5.23; $p = 0.015$), (goats: OR = 3.9, confidence interval (CI):1.64–9.41: $p = 0.002$)] than in young sheep and goats, in female [(sheep: OR = 1.93, CI: 1.11–3.36, $p = 0.018$, (goats: OR = 2.9, CI: 121–6.93, $p = 0.002$)] than in males sheep and goats, in Highland [(sheep: OR = 4.57, CI: 1.75–12.66, $P = 0.000$, (goats: OR = 4.4, CI: 1.75–13.66, $p = 0.004$)] than sheep and goats from lowland.

Conclusion: This study indicates that seroprevalence of latent toxoplasmosis in small ruminants is high, therefore, it is decidedly indispensable to minimize risk factors exposing to the infection like consumption of raw meat as source of infection for humans.

Keywords: Goat, Sheep, Toxo-latex, *T. gondii*, Seroprevalence, Southwestern Ethiopia

Background

Toxoplasmosis is zoonotic disease caused by an obligate intracellular parasite known as *T. gondii* [1]. It is the most prevalent parasitic infections in human and veterinary medicine and has negative impacts on public health and animal production. *T. gondii* is believed to be the most triumphant parasitic pathogen in large scale [1]. Despite having adverse health effects analogous to those of salmonellosis and campylobacteriosis, toxoplasmosis is still a neglected and underreported parasitic infection. Human vaccines are not available and the results of the usage of the current anti-parasitic therapies are quite disappointing [2].

Toxoplasmosis is found globally; almost one third of the human population [1, 3]. The occurrence of toxoplasmosis has been significantly increasing as a result of the opportunistic infection of immune compromised patients, for instance, acquired immune deficiency syndrome (AIDS). In these people deaths usually result from rupture of cysts that lead to continued multiplication of tachyzoites [4]. Hence, encephalitis was presented as the main clinical manifestation of toxoplasmosis in AIDS patients as a result of reactivation of latent infection [5]. Majority of ocular cases at the present are associated with acquired

* Correspondence: dachassat@yahoo.com
Department of Veterinary Microbiology and Public Health, College of Agriculture and Veterinary Medicine, School of Veterinary medicine, Jimma University, PO. Box 307, Jimma, Ethiopia

toxoplasmosis, thus preventive strategies should be focused not only on pregnant women but also in the general population [2].

From wide range of farm animals, sheep and goats are more commonly infected with *T.gondii* than cattle and chicken. This parasite causes abortion and neonatal death in major monetary losses to sheep, goat and pig farming [3, 5]. This is more serious especially when primary infection occurs during pregnancy [6].

For evaluating the comparative significance of wide causes of toxoplasmosis in human's epidemiological survey still remains the main important approach. There have been a wide range of serological surveys conducted in different countries to determine the prevalence of toxoplasmosis in farm animals and humans; from North and South America [7–13], Europe [14–16], Africa [17–23] Asia [24, 25]. According to Australian Centre for International Agricultural Research (ACIAR) [26] *T.gondii* is extensively spread among farm animals and human with variable seroprevalence rates of 11–61% in goats, less than 10% in cows, 35–73% in cats, 75% in dogs, 11–36% in pigs, and 35–73% in humans.

In Africa different reports indicate widespread occurrence of toxoplasmosis. Thirty percent infection rate of toxoplasmosis was reported in goats in Botswana [20]. Limited studies have been carried out to investigate the magnitude of toxoplasmosis in animals and humans in Ethiopia so far. A preliminary serological study made in sheep and goat population around Nazareth showed an overall seroprevalence of 54.7% in sheep and 26.7% in goats using Enzyme linked Immunosorbent Assay (ELISA) and Modified Agglutination Test (MDAT) [21]. In another seroprevalence study of human toxoplasmosis of workers at Addis Ababa abattoir [22], reported a prevalence of 96.8% using an indirect haemagglutination assay [16], 80.7% in HIV patients in Agaro Health Centre in Jimma Zone [27] on top of this 6.6, 22.9 and 11.6% prevalence was reported in cattle, sheep and goats in central Ethiopia respectively [17].

In general, there is a scarcity of data on sero- epidemiology of toxoplasmosis in animals and humans in Ethiopia though numerous literatures associate human toxoplasmosis with utilization of raw/undercooked meat products of animal origin. The exact seroepidemiological distribution and risk factors for the infection of food animals in Jimma are unknown but, such studies are indispensable because consumption of raw meat is a popular tradition in Jimma. Thus, human toxoplasmosis in Jimma might have strong linkage with seroprevalence of the infection in food animals on top of this little is known about seroprevalence of *T.gondii* infection in sheep and goat Jimma zone. Therefore, the present study was intended with objective of estimating the seroprevalence and risk factors of *T.gondii* infection in small ruminants in Jimma zone.

Methods

Study areas

The study was carried out in four districts of Jimma zone namely Seka Chokorsa 20 km in Southeast, Mena 22 km in North east, Kersa 18 km in Northwest and Goma 50 km west of the Jimma. Jimma is a capital city of of Jimma zone which is located in Oromia regional state found 352 km away from the capital city (Addis Ababa) in the southwest Ethiopia. It is located at latitude of 7°13′– 8°56′ N and longitude of 35°52′–37°37′ E, and at an altitude ranging from 880 m to 3360 m above sea level (masl). The study areas were purposively selected to represent three agro-ecological zones such as highland (≥2300, masl), midland (1500–2300 masl) and lowlands (≤1500 masl). The area receives about 1530 mm rainfall that comes from the long and short rainy seasons. The mean temperature yearly ranges from 25 to 30 °C and 7 to 12 °C [28] (Additional file 1: Figure S1).

As it was reported by Central Statistical Agency census [29], the total population of Jimma zone is 2,642,114, from these Jimma town populations accounts 177,900 (Ethiopian statistical agency 2015 projected from 2007 census), 49.7 and 49.3% females and males respectively). From the total population in the zone, 2,204,225 (88.66%) are rural community engaged in agricultural activities for their livelihood. Jimma zone is potential source of livestock which contribute to the country growth and domestic production; about 466,154 of sheep, 194,677 of goats, 1,718,284 of cattle, 40,555 of donkeys, 30,541 of mules and 74,774 of horses [30].

Study animals

The study subjects were sheep and goats. Small ruminant production in the study areas was mainly characterized by traditional and extensive type of management system. Mainly male's sheep and goats are known to be kept for mutton production in most parts of the country while females are for breeding. The study dealt with animals kept by peasants in four districts of Jimma zone.

Study design and sample size

Cross-sectional study design was used. Different age and sex groups of sheep and goats were included for this study. The study was conducted from November 2014 to March 2015. Serological investigation was used to detect anti-*T.gondii* antibodies from blood serum collected from sheep and goats in the districts under study. Since there was no previous expected prevalence in the area, sample size was calculated as it is stated by Thrusfield [31] using an expected prevalence of 50.0% a desired precision of 5% and with 95% level of confidence. Hence, the sample size was 384. Basically due to the fact that goat's population at study area was very limited in number, therefore, only 116 goats and 252 of sheep sera were subjected to analyses, totally 368 samples were analyzed. Simple random sampling technique was carried out to collect sera from small ruminants.

Blood collection

Sheep and goats was aseptically bled (approximately 5 ml) from the jugular vein by using 10 ml vacutainer tubes which contained no anti-coagulants or preservatives and vein puncture needle and needle holder was used and properly labelled with water proof marker with the necessary information. Blood sample was transported to Jimma University, microbiology and veterinary public health laboratory and was kept overnight at room temperature and then centrifuged at 3000 rpm for 10 min to get serum. The serum was collected in 1.5 ml Eppendorf tubes and kept at −20 °C until serologically tested for the presence of anti *T. gondii* antibodies.

Serological examination

T. gondii antibodies were detected by the Toxo Latex slide Agglutination test following the procedure described by manufacturer SPINREACTGirona/Spain. Briefly, 50 μL sera samples was placed on toxo-latex agglutination slide and one drop of each positive and negative control into separate circles on the slide test and mixed thoroughly. 25 μL of toxo-latex reagent was added to 50 μL sera samples into separates circles and mixed thoroughly with stirrer then the mixture was spread over entire circle. The slide was placed on mechanical rotator at 96 rpm for 4 min. The presence or absence of visible precipitation was macroscopically examined immediately after removing the slide from the rotator. The formation of precipitation

was recorded as positive and it indicates an antibody concentration equal or greater than 4 IU/ml.

Data analysis

Data were recorded and coded using Microsoft Excel spreadsheet and analysed using SPSS version 20. Seroprevalence was calculated by dividing the number of animals possessing anti-*T.gondii* antibodies against the total number of animals tested. Relationship of risk factors with dependent variable was primarily assessed using cross tabulation. Univariable logistic regression analysis was performed and strength of association between risk factors and *T. gondii* infection were evaluated using odds ratios (OR). The 95% confidence interval (CI) and a significance level of $\alpha = 0.05$ were used.

Result

Overall seroprevalence

Out of 368 animals examined 212 (57.60%, 95% CI: 52.55–62.65) were seropositive for *T.gondii* antibody; 58.8% (148/252; 95% CI: 52.75–64.88) and 55.18% (64/116; 95% CI: 46.13–64.23) seroprevalence was found in sheep and goats respectively. Both serum samples from sheep and goats showed positive reactions of similar proportions (Table 1). Their variation is statistically insignificant $(P > 0.05)$.

Risk factors

Results of logistic regression analysis indicates that the potential risk factors related to sex, age and altitude revealed that the likelihood of *T. gondii* infection was higher in adult sheep (OR = 2.5), female sheep (OR = 1.95) and highland (OR = 4.57) when compared with male, young

Table 1 Seroprevalence of *T. gondii* antibody and logistic regression analysis of risk factors for sheep

Risk factors	Total	No of positive	Univariable		Multivariable	
Species[a]			COR (95% CI)	P-Value	AOR (95% CI)	P-Value
Sheep	252	148 (58.73%)	-		-	
Goat	116	64 (55.18%)	-		-	
Total	368	212 (57.60%)				
Age						
Young (<1 year)	40	15 (37.5%)	1		1	
Adult (>1 year)	212	133 (62.7%)	2.8 (1.39,5.63)	0.004	2.5 (1.19,5.23)	0.015
Sex						
Male	99	46 (46.5%)	1			
Female	153	102 (66.7%)	2.3 (1.37, 3.87)	0.002	1.93 (1.11,3.36)	0.018
Altitude						
Highland	113	79 (69.9.9%)	5.07 (2.23,1163)	0.000	4.57 (1.75,12.66)	0.000
Midland	104	58 (55.8%)	2.75 (1.22,6.19)	0.015	2.8 (1.23,6.54)	0.014
Lowland	35	11 (31.4%)	1			-

COR crude odd ratio, *AOR* adjusted odd ratio, *CI* Confidence Interval
[a]Species, $P = 0.213$

and lowland sheep respectively (Table 1). On top of this, the likelihood of *T. gondii* infection was higher in adult goats (OR = 2.5), female goats (OR = 1.95) and highland (OR = 4.4) when compared with male, young and lowland goats, respectively (Table 2).

Discussion

Out of 368 animals examined the seroprevalence of anti-*T.gondii* antibody was found 212 (57.60%). This is comparable to the previous report from Ethiopia [32]. Different studies revealed that the prevalence from 0 to 100% was recorded in different areas of the world [33]. This difference in prevalence is depending up on cat density, climate condition, age of the animals, species, sex, altitude and management of animal production [3, 5, 34].

Lower prevalence values of 3.8, 4.3, 11.2, 12.1 and 16.9% were recorded by Sharma [35] in India, Samra [36] in South Africa, Ramzan [37] in Pakistan, Dubey and Foreyt [38] in the North America and Márcia de Figueiredo [12] in Brazil respectively. The current study estimated sero-prevalence of *T.gondii* antibody in sheep was 58.73% and this is in agreement with the finding of 56.00% [32] in Central Ethiopia and higher seroprevalence in sheep has been reported when compared to the present finding [3, 12, 39, 40].

The seroprevalence of *T.gondii* antibody in goats was 55.18% is closely related to the finding of 59.4% from Egypt [41], but it is higher than the finding of 19.70 and 37.20% reported by Zewdu *et al.* [39, 42] and Yibeltal, [43] in Central Ethiopia and in South Wollo respectively. It is also higher than the 39, 31.7, 28.9, 27. 9 and 25.4%, reported from Thailand [44], Pakistan [37] and Brazil [12, 13, 23] respectively. In contrast, the prevalence of the present study is lower than the 67.9% reported from Zimbabwe [45]. The variations in the overall prevalence observed in the current study and the above studies could be due to differences in the access of small ruminants to

contaminated feed and water, the climatic variation and the diagnostic techniques used [34, 46].

The present study indicated that sheep from the highland (OR = 4.57, CI: 1.75–12.66) and midland (OR = 2.8, CI: 1.23–6.54) areas of Southwest Ethiopia have significantly (*P* = 0.000) high risk of infection of *T.gondii* than those of from the lowland. Similarly, goats from the highland (OR = 4.4, CI: 1.75–15.5) and midland (OR = 3.9, CI: 1.64–9.41) areas of Southwest Ethiopia have significantly higher risk of *T. gondii* infection than those from the lowland (*P* = 0.004). This study was in agreement with the finding of 46.91% in highland and 46.24% in midland but lower prevalence in lowland 13.36% [39, 42] in Central Ethiopia. This variation among risk factors can be described by the variation in temperature and moisture in these areas. It is well known that the epidemiology of toxoplasmosis is influenced by the environment [3, 34]. Humidity increases, the chance of oocyst survival in the environment, thereby contributing to the higher sero-prevalence. A dry climate has an impact on the survival and epidemiological distribution of the parasite [1, 47].

Correspondingly, seroprevalence of *T.gondii* antibody was high in adults (62.7.4% sheep and 65.8% goats) than in young animals (37.5.09% sheep and 35% goats). Statistically significant variation was observed among them. This finding is relatively similar to [48, 49] that reported prevalence (10.00%) in young and (26.76%) in adult. Multivariable logistic regression analysis showed that the likelihood of acquiring infection was higher in adult [(sheep: OR = 2.5, CI: 1.19–5.23; *P* = 0.015), (goats: OR = 3.9, CI: 1.64–9.41: *p* = 0.002)] than in young sheep and goats. Seroprevalence of *T.gondii* antibody increases with age in both species. This could be attributed to the reality that increment of the disease prevalence in older animals is due to exposure of animals to the risk factors for longer period of time than the younger ones [1, 12, 13, 39, 42]. With regard to sex risk factor, the

Table 2 Seroprevalence of *T.gondii* and logistic regression analysis of risk factors for goats

Risk factors	Total	No of positive	Univariable		Multivariable	
			COR (95% CI)	*P*-value	AOR (95% CI)	*P*-Value
Age						
Young (<1 year)	40	14 (35%)				
Adult (>1 year)	76	50 (65.8%)	3.57 (1.59,7.98)	0.002	3.9 (1.64,9.41)	0.002
Sex						
Male	42	18 (42.9%)	1			
Female	74	46 (62.2%)	2.19 (1.01, 4.73)	0.035	2.9 (1.21,6.93)	0.002
Altitude						
Highland	59	38 (67.9%	4.52 (57,13.5)	0.005	4.4 (1.75,15.5)	0.004
Midland	38	19 (50%)	2.14 (0.714,6.43)	0.174	3.9 (1.64,9.41)	0.10
Lowland	22	7 (31.8%)	1			-

COR crude odd ratio, *AOR* adjusted odd ratio, *CI* Confidence Interval

study showed that the seroprevalence of anti-*T.gondii* antibody is higher in females (60.9%) than in males (54.64%) in both species. The result was similar to the finding of Zewdu *et al.* [39, 42] in female (34.39%) and males (19.43%).

Multivariable logistic regression analysis revealed that the likelihood of acquiring infection was higher in females [(sheep: OR = 1.93, CI: 1.11–3.36, p = 0.018), (goats: OR = 2.9, CI: 121–6.93, p = 0.002)] than in males sheep and goats. A small number of male sheep and goats are kept for breeding whilst others are culled and sold. In addition, the stress of lactation and pregnancy due to hormonal difference lead to immune-suppression that may expose the female sheep and goats to toxoplasmosis [50].

Conclusions

This study indicated that seroprevalence of latent toxoplasmosis in small ruminants is high. Therefore, it is particularly requisite providing health education about its public health importance and risk factors that expose to humans the infection such as consumption of raw meat of small ruminants.

Acknowledgement
The author highly acknowledges Jimma University College of Agriculture and Veterinary Medicine, laboratory of Microbiology for the facility provision as well as those laboratory attendants who technically assisted this study. The authors again highly acknowledge the institute which financially supports this study.

Funding
This work was supported financially by Jimma University College of Agriculture and Veterinary medicine (JUCAVM).

Authors' contributions
DT conceived and designed the study protocol. DT, AK, MA and MY carried out sample collection and serological examination. DT, AK and MY interpreted the results of data analysis and drafted the manuscript. DT and MY prepared study area map using GIS and performed statistical analysis. DT, MA, and MY compiled the results, improved and corrected the manuscript. All authors read, commented and approved the final manuscript. All authors read and approved the final manuscript.

Competing interest
The authors declare that they have no competing interests.

Consent for publication
Not applicable.

References
1. Dubey JP. Toxoplasmosis of animals and humans. 2nd ed. Beltsville: CRC Press; 2010. p. 1–338.

2. Kijlstra A, Jongert E. Toxoplasma-safe meat: close to reality? Trends Parasitol. 2008;25:18–22.

3. Tenter AM, Heckerotha AR, Weiss LM. *Toxoplasma gondii*: from animals to humans. Int J Parasitol. 2000;30:1217–58.

4. Roberts LS, Janovy J, Schimidt GD. Foundation of Parasitology. 7 thth ed. USA: McGraw-Hill Companies, Inc; 2000. p. 135–8.

5. Jithendran KP. Seroprevalence of *Toxoplasma* antibodies in domestic animals- an indicator of *Toxoplasma gondii* in the environment and human. ENVIS Bulletin Himalayan Ecol. 2004;12:22–8.

6. Radostits OM, Gay CC, Hinchcliff KW, Constable PD. Veterinary medicine: a textbook of the diseases of cattle, horses, sheep, pigs and goats. 10 thth ed. London: Sounders; 2006. p. 1518–22.

7. Dubey JP, Hill DE, Jones JL, Hightower AW, Kirkland E, Roberts JM, Marcet PL, Lehmann T, Vianna MCB, Miska K, Sreekumar C, Kwok OCH, Shen SK, Gamble HR. Prevalence of viable *Toxoplasma gondii* in beef, chicken, and pork from retail meat stores in the United States: Risk assessment to consumers. J Parasitol. 2005;91:1082–93.

8. Ragozo AMA, Yai LEO, Oliveira LN, Dias RA, Dubey JP, Gennari SM. Seroprevalence and isolation of *Toxoplasma gondii* from sheep from Sao Paulo State, Brazil. J Parasitol. 2008;96:1259–63.

9. Carneiro AC, Carneiro M, Gouveia AM, Guimaraes AS, Marques AP, Vilas-Boas LS, Vitor RW. Seroprevalence and risk factors of caprine toxoplasmosis in Minas Gerais, Brazil. Vet Parasitol. 2009;160:225–9.

10. Alvarado-Esquivel C, Torres-Castorena A, Liesenfeld O, Garcia-Lopez CR, Estrada-Martinez S, Sifuentes-Alvarez A, Marsal-Hernandez JF, Esquivel-Cruz R, Sandoval-Herrera F, Castaneda JA, Dubey JP. Sero-epidemiology of *Toxoplasma gondii* infection in pregnant women in rural Durango, Mexico. J Parasitol. 2009;95:271–4.

11. Lopes WDZ, Dos Santos TR, Da Silva RDS, Rossanese WM, De Souza FA, Rodrigues JDF, De Mendonca RP, Soares VE, Da Costa AJ. Seroprevalence and risk factors for *Toxoplasma gondii* in sheep raised in the Jaboticabal microregion, Sao Paulo State, Brazil. Res Vet Sci. 2010;88:104–6.

12. Pereira MF, Peixoto RM, Langoni H, Greca H, Azevedo SS, Porto WJN, Medeiros ES, Mota RA. Risk factors for *Toxoplasma gondii* infection in sheep and goats in Pernambuco, Brazil. Pesq Vet Bras. 2012;32(2):140–6.

13. Anderlini GA, Mota RA, Faria EB, Cavalcanti EF, Valença RM, Pinheiro Júnior JW, de Albuquerque PP, de Souza Neto O. Occurrence and risk factors associated with infection by *Toxoplasma gondii* in goats in the State of Alagoas, Brazil. Rev Soc Bras Med Trop. 2011;44(2):157–62.

14. Acici M, Babur C, Killie S, Hokelek M. Prevalence of antibodies to *Toxoplasma gondii* infection in humans and domestic animals in Samsun province, Turkey. Trop Anim Health Prod. 2008;40:311–5.

15. Pereira Bueno J, Quintanilla Gozalo A, Perez Perez V, Alvarez Gracia G, Collantes Fernandez E, Ortega Mora LM. Evaluation of ovine abortion associated with Toxoplasma gondii in Spain by different diagnostic techniques. Vet Parasitol. 2004;121:33–43.

16. Gilot-Fromont E, Aubert D, Belkilani S, Hermitte P, Gibout O, Geers R, Villena I. Landscape, herd management and within-herd seroprevalence of *Toxoplasma gondii* in beef cattle herds from Champagne-Ardenne, France. Vet Parasitol. 2009;161:36–40.

17. Bekele T, Kasali OB. Toxoplasmosis in sheep, goats and cattle in Central Ethiopia. Vet Res Commun. 1989;3:371–5.

18. Guebre-xabier M, Nurilign A, Gebre-Hiwot A, Hailu A, Sissay Y, Getachew E, Frommel D. Sero-epidemiological survey of *Toxoplasma gondii* infection in Ethiopia. Ethiop Med J. 1993;31:201–8.

19. Achu-Kwi MD, Ekue NF. Prevalence of *Toxoplasma gondii* antibodies in Djallonke sheep flocks in the Vina division, Cameroon. Bull Anim Health Prod Afr. 1994;42:89–92.

20. Sharma SP, Baipoledi EK, Nyange JF, Tlagae L. Isolation of *Toxoplasma gondii* from goats with a history of reproductive disorders and the prevalence of *Toxoplasma* and *chlamydial* antibodies. Onderstepoort J Vet Res. 2003;90:65–8.

21. Negash T, Tilahun G, Medhin G. Seroprevalence of *Toxoplasma gondii* in Nazaret Town, Ethiopia. East Afr J Public Health. 2008;5:211–4.

22. Yimer E, Abebe P, Kassahun J, Woldemichael T, Bekele A, Zewdie B, Beyene M. Seroprevalence of human toxoplasmosis in Addis Ababa, Ethiopia. Ethiopian Vet J. 2005;9:123–8.

23. Bisson AS, Maley CM, Rubaire-Akiiki J, Watling M. The seroprevalence of antibodies to *Toxoplasma gondii* in domestic goats in Uganda. Acta Trop. 2000;76:33–8.

24. Yang H, Jin KN, Park YK, Hong SC, Bae JM, Lee SH, Choi HS, Hwang HS, Chung JB, Lee NS, Nam HW. Seroprevalence of toxoplasmosis in the residents of Cheju island, Korea. Korean J Parasitol. 2000;38:91–3.

25. Huang CQ, Lin YY, Dai AL, Li XH, Yang XY, Yuan ZG, Zhu XQ. Seroprevalence of *Toxoplasma gondii* infection in breeding sows in Western Fujian Province, China. Trop Anim Health Prod. 2010;42:115–8.

26. ACIAR (Australian Centre for International Agricultural Research). Assessment of zoonotic diseases in Indonesia. Canberra: ACIAR; 2007. p. 34–5.

27. Tegegne D, Abdurahaman M, Mosissa T, Yohannes M. Anti-Toxoplasma antibodies prevalence and associated risk factors among HIV patients. Asian Pac J Trop Med. 2016;9(5):460–4.

28. Tesema E. Zonal diagnosis and intervention plan report for Jimma zone. Ethiopia: LIVES. 2013. p. 13-20.

29. Central Statistical Authority (CSA). Summary and statistical report of the 2007 population and housing census results. Addis Ababa: EFDRE, Population census commission; 2008. p. 25–7.

30. Central Statistical Authority (CSA). Federal Democratic Republic of Ethiopia Central Statistical Agency Agricultural Sample Survey 2007/08. Volume II, report on livestock and livestock characteristics. Addis Ababa. Stat Bull. 2008;2:23–9.

31. Thrusfield M. Veterinary epidemiology. 3rd ed. London: Blackwell Science Ltd; 2005. p. 227–47.

32. Negash T, Tilahun G, Patton S, Prevot F, Dorchies PH. Serological survey of toxoplasmosis in sheep and goats in Nazareth, Ethiopia. Rev Med Vet. 2004;55:486–7.

33. Olivier A, Herbert B, Sava BC, Pierre D, John C, Aline DK. Surveillance and monitoring of Toxoplasma in humans, food and animals: a scientific opinion of the panel on biological hazards. European Food Safety Assoc J. 2007;583:1–64.

34. Dubey JP. Toxoplasmosis - a waterborne Zoonosis. Vet Parasitol. 2004;26:57–72.

35. Sharma S, Sandhu KS, Bal MS, Kumar H, Verma S, Dubey JP. Serological Survey of Antibodies to *Toxoplasma gondii* in Sheep, Cattle, and Buffaloes in Punjab, India. J Parasitol. 2008;94:1174–5.

36. Samra NA, McCrindle CM, Penzhorn BL, Cenci-Goga B. Seroprevalence of toxoplasmosis in sheep in South Africa. J South African Vet Assoc. 2007;78:116–20.

37. Ramzan M, Akhtar M, Muhammad F, Hussain FI, Hiszczynska-Sawicka E, Haq AU, Mahmood MS, Hafeez MA. Seroprevalence of *Toxoplasma gondii* in sheep and goats in Rahim Yar Khan (Punjab), Pakistan. Trop Anim Health Prod. 2009;41:1225–9.

38. Dubey JP, Foreyt WJ. Seroprevalence of *Toxoplasma gondii* in Rocky Mountain big horn sheep (Ovis canadiansis). J Parasitol. 2000;86:622–3.

39. Gebremedhin EZ, Agonafir A, Tessema TS, Tilahun G, Medhin G, Vitale M, et al. Seroepidemiological study of ovine toxoplasmosis in East and West Shewa Zones of Oromia regional state, Central Ethiopia. BMC Vet Res. 2013;9:117.

40. Mason S, Quinnell JR, Smith JE. Detection of *Toxoplasma gondii* in lambs via PCR screening and serological follow –up. UK. Vet Parasitol. 2010;4:192–5.

41. Barakat AM, Abdelaziz MM, Fadaly M. Comparative diagnosis of toxoplasmosis in Egyptian small ruminants by indirect hemagglutination assay and EIISA. Global Veterinaria. 2009;3:9–14.

42. Zewdu E, Agonafir A, Tessema TS, Tilahun G, Medhin G, Vitale M, et al. Seroepidemiological study of caprine toxoplasmosis in East and West Shewa Zones, Oromia Regional State, Central Ethiopia. Res Vet Sci. 2013;94:43–8.

43. Yibeltal MM. Seroprevalence study of toxoplasmosis in small ruminants and humans (HIV/AIDS patient) in selected district of South Wollo, Ethiopia, MSc. Thesis. Debre-Zeit: Addis Ababa University, Faculty of Veterinary Medicine; 2008. p. 25–40.

44. Jittapalapong S, Sangvaranond A, Pinyopanuwat N, Chimnoi W, Khachaeram W, Koizumi S, Maruyama S. Seroprevalence of *Toxoplasma gondii* infection in domestic goats in Satun province, Thailand. Vet Parasitol. 2005;127:17–22.

45. Hove T, Lind P, Mukaratirwa S. Seroprevalence of *Toxoplasma gondii* infection in goats and sheep in Zimbabwe. Onderstepoort J Vet Res. 2005;72:267–72.

46. Innes EA, Bartley PM, Buxton D, Katzer F. Ovine toxoplasmosis. Vet Parasitol. 2009;136:1884–7.

47. Jones JL, Moran KD, Wilson M, McQuillan G, Navin T, James B, McAuley JB. *Toxoplasma gondii* infection in the United States: Seroprevalence and risk Factors. Am J Epidemiol. 2001;154:357–65.

48. O'Donoghue PJ, Riley MJ, Clarke JF. Serological survey for *Toxoplasma* infections in sheep. Aust Vet J. 1987;64:40–5.

49. Van der Puije WNA, Bosompem KM, Canacoo EA, Wastling JM, Akanmori BD. The prevalence of anti-*Toxoplasma Gondii* antibodies In Ghanaian sheep and goats. Acta Trop. 2000;76:21–6.

50. Dubey JP, Lappin MR. Toxoplasmosis and Neosporosis. In: Greene CE, editor. Infectious Diseases of the Dog and Cat. Philadelphia: WB Saunders; 1998. p. 493–509.

15

African animal trypanosomiasis as a constraint to livestock health and production in Karamoja region: a detailed qualitative and quantitative assessment

Dennis Muhanguzi[1]* [iD], Albert Mugenyi[2], Godfrey Bigirwa[1], Maureen Kamusiime[3], Ann Kitibwa[1], Grace Gloria Akurut[1], Sylvester Ochwo[1], Wilson Amanyire[1], Samuel George Okech[1], Jan Hattendorf[4,5] and Robert Tweyongyere[1]

Abstract

Background: Nagana (African Animal Trypanosomiasis-AAT) and tick-borne diseases (TBDs) constrain livestock production in most parts of sub-Saharan Africa. To this realisation, Uganda government set up an African trypanosomiasis (AT) control unit, which among other activities generates national tsetse control priority maps using apparent tsetse density data. Such maps underestimate mechanically transmitted AAT and thus ought to be refined using actual AT prevalence data. We therefore set out to generate up-to-date cattle and donkey trypanosomiasis prevalence data as well as find out the constraints to livestock production in Karamoja region in a bid to re-define AT control priority in this region.

Results: Livestock keepers and animal health workers indicated that TBDs and AAT were the most important livestock diseases in Karamoja region. The prevalence of *Trypanosoma* spp. in cattle and donkeys was 16.3% (95% CI: 12.4–21.1%) and 32.4% (95% CI; 20.2–47.6%) respectively. *Trypanosoma vivax* (12.1%) and *Trypanosoma congolense* savannah (29.6%) were the most prevalent *Trypanosoma* spp. in cattle and donkeys respectively. Majority of the cattle (85.7%) and more than half of the donkey (57.1%) herds were positive for *Trypanosoma* spp.

Conclusions: African animal trypanosomiasis and TBDs are the most important constraints to livestock production in Karamoja region. In order to improve livestock production and hence Karamajong livelihoods, government of Uganda and her development partners will need to invest in livestock health programs particularly targeting tsetse and TBD control.

Keywords: African animal trypanosomiasis, Tick-borne diseases, Control, ITS1-PCR, Prevalence, Karamoja region

Background

African trypanosomiasis (AT) constrains livestock production and human health in 37 sub-Saharan African countries [1–3]. In these areas, about 60 million people are at moderate to high risk of acquiring sleeping sickness (human African trypanosomiasis HAT). Sleeping sickness is caused by *Trypanosoma brucei gambiense* (chronic form) and *Trypanosoma brucei rhodesiense* (acute form) [4, 5]. Both nagana and sleeping sickness are transmitted by 30 species of tsetse flies (Diptera: Glossinidae) which inhabit about 10 million km^2 of land in the humid regions of Africa [6]. This reduces the livestock production potential for such an expanse of land [7].

Tsetse and AT are considered a major livestock production [8–10] and public health constraint in Uganda [11]. For this reason, the Coordinating Office for Control of Trypanosomiasis in Uganda (COCTU), was set up to coordinate tsetse and trypanosomiasis control [12]. As part of her mandate, COCTU produces and avails tsetse and trypanosomiasis control priority maps (e.g. Fig. 1) so as to heighten vector and disease control advocacy [13]. These African trypanosomiasis (AT) control priority maps are generated from apparent tsetse density, socio-

* Correspondence: luckydenno@covab.mak.ac.ug
[1]College of Veterinary Medicine Animal Resources and Biosecurity, Makerere University, P.O. Box 7062, Kampala, Uganda

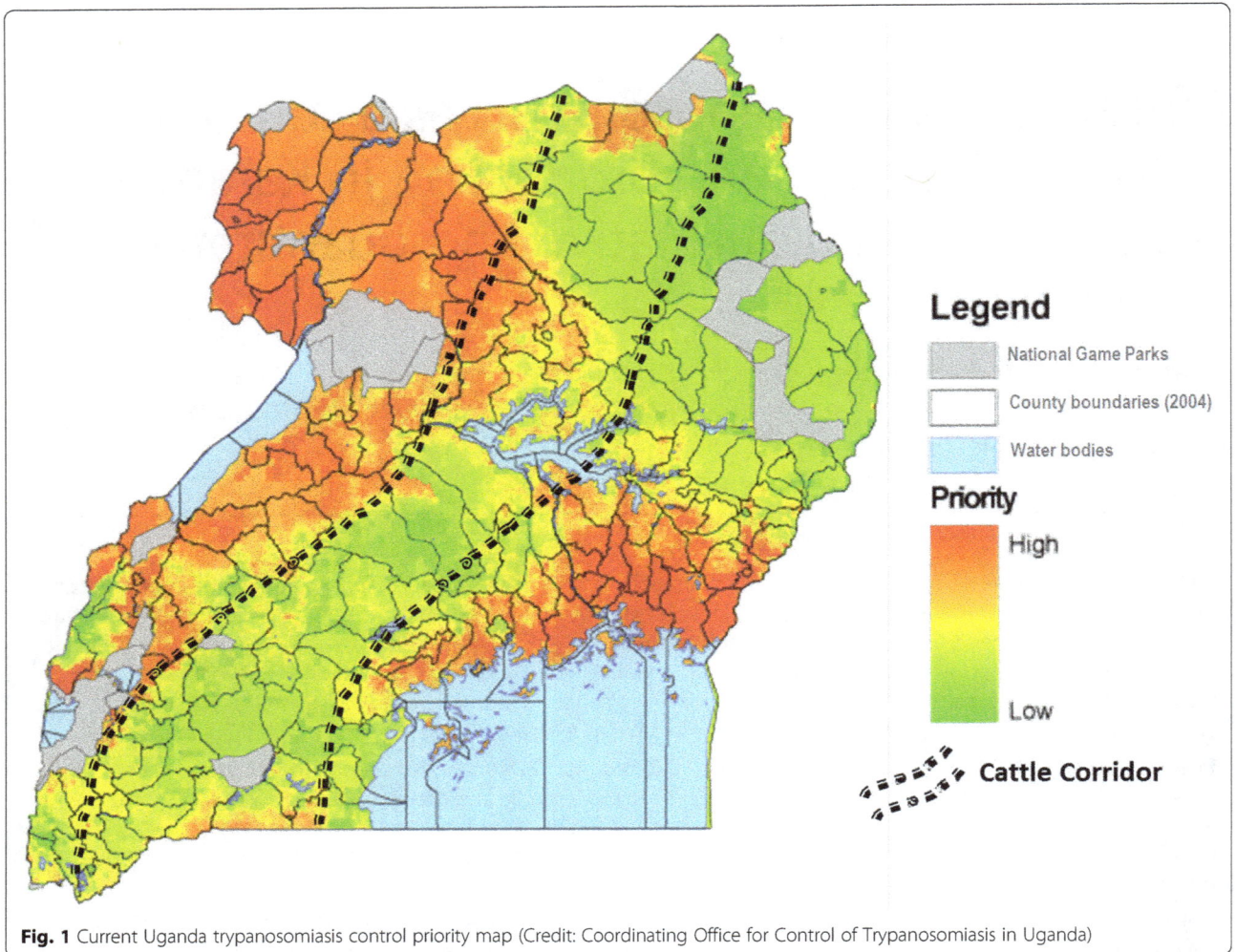

Fig. 1 Current Uganda trypanosomiasis control priority map (Credit: Coordinating Office for Control of Trypanosomiasis in Uganda)

economic and environmental variables and not from actual AT prevalence data. This approach underestimates the extent of AT. This partly explains why the cattle corridor is endemic for AAT yet it is currently given low to medium priority for tsetse and AT control (Fig. 1). As such, a recent aggregate of AAT prevalence data for the East African region including Uganda presents Karamoja region with no AAT prevalence data and presumably no risk of the disease [14].

Several biting flies including tabanids can mechanically transmit *T. vivax* [15, 16]. As a result, *T. vivax* distribution often extends beyond tsetse belts [17, 18]. This indicates that tsetse distribution patterns cannot accurately predict the distribution of *T. vivax* AAT [15]. In addition, AT spatial distribution cannot be sufficiently predicted by apparent tsetse density (tsetse flies caught per trap per day) alone [19, 20]. The level of challenge (product of apparent tsetse density, mean tsetse fly infection rate and the proportion of feeds taken by these tsetse from livestock) is a better

predictor of AT distribution [19, 20]. However, the level of challenge is often not used to predict AT distribution and to determine the level (no priority, low, moderate or high priority, etcetera) of control priority for different parts of the country. As a result, farmers in low to medium AT control priority areas including the Karamoja region have indicated that AAT is one of the major constraints to animal health and production [21, 22]. There is therefore urgent need to generate and use up-to-date AT prevalence data to refine Uganda AT control priority maps. This is particularly important for Karamoja region where such studies were not previously possible due to insecurity associated with cattle rustling. We therefore carried out this study to generate up-to-date cattle and donkey *Trypanosoma* spp. prevalence data and used it to suggest how AT control priority map for Karamoja region can be refined. In addition, we interviewed animal health providers and key farmers so as to explore the main constraints to livestock health and production in this region.

Methods

Study area description

Karamoja region is a largely remote area in north-eastern Uganda covering about 25,000 km^2 of land extending over 7 districts between 33 and 35 East and latitude 1–4 North. It is composed of seven districts namely; Kaabong, Abim, Kotido, Moroto, Napak, Nakapiripirit and Amudat. The road network is at its development stage leaving the commonest local means of transport to be footpath trekking by the Karamojong pastoralists and their livestock. Camels and donkeys provide not only animal proteins to the Karamojong but also the most needed means of transport for their agricultural products and livestock production inputs. For this reason, Karamoja region is the only home of the 32,000 camels in Uganda and about 90% (0.134 /0.15 millions) of all the donkey population [23, 24]. This study was carried out in 8 of the 31 sub-counties of Kaabong, Kotido, Nakapiripirit and Amudat districts (Fig. 2). The 8 study sub-counties contained an estimated 75,250 head of cattle distributed within 149 cattle-containing villages (4 target districts): an average of 250 cattle per village. In addition, there were 111,000 head of donkeys; an average of 2 donkeys per village [23].

The climate of the region is semi-arid with a characteristic uni-modal rainfall pattern. The region receives average rainfall of 745 mm (varies widely from 600 mm in the north to 1000 mm in the southern and western parts). The rainy season spans from April to September with scanty rains in June; a main peak in July/August and a minor peak in May. The rest of the months present with an extended dry season (6.5 months) with characteristic high temperatures ranging from 28 to 32.5 °C and an average minimum of 18 °C [25, 26].

Karamoja region is divided into pastoral, agro-pastoral and agricultural livelihood zones depending on the degree of aridity and dependence on livestock. The pastoral region is largely semi-arid and entirely dependent on livestock. It extends from the Kenyan boarder and runs southwards through large parts of Kaabong, Moroto and Amudat districts. This zone experiences prolonged dry seasons and very erratic rainfall. As a result, pastoralists have to move livestock from this zone to other zones for most parts of the prolonged dry season. The agro-pastoral zone extends from the border with South Sudan and covers central parts of Kotido, Kaabong, Moroto, Napak, Amudat and Nakapiripirit districts. This zone receives an average annual rainfall of 500 –800 mm and is largely dependent on livestock and rain-fed crop production. The agricultural zone, the the smallest of the three, covers the western side of Karamoja. The zone supports most tropical food crops because of fertile soils and an average annual rainfall of 700–1000 mm [25, 26].

Study design and sampling methods

Three stage cluster sampling was used to randomly select study districts, parishes and villages. The initial step involved selection of 4 districts from 7. The second and third stages of sampling included selection of 16 (from 164) and 35 (from 149) parishes and villages respectively. The sampling frame (list of all villages) was obtained from COCTU. With the communal livestock obtaining in Karamoja region, or elsewhere for that matter, AAT tends to cluster at primary cattle grazing units which are villages. A village was therefore taken as the epidemiological unit.

Sample size estimation

Cluster sampling methodology [27] was implemented in *C Survey* version 2.0 [28] to determine the minimum number of villages (clusters) needed for this survey. In so doing, the following parameters were used; – anticipated prevalence of 5.2% [21], the precision of the sample estimate (one half-length of the 95% confidence interval) of 3 percentage points and an intracluster correlation coefficient-ICC (degree of homogeneity among cattle within the study villages) of 0.15 [10, 20]. Thirty-five (35) clusters (55 cattle each) were required to estimate the apparent prevalence with the set precision. All the Seventy-one (71) donkeys that were presented in the 35 sampled villages were bled.

Key informant' and kraal leaders' interviews

Semi-structured interviews were administered to 21 key informants in order to understand the major constraints to livestock production and health. In addition, 20 focus group discussions were completed with kraal leaders. We selected key informants from key animal health workers in the region. These included community based animal health workers-CBAHWs ($n = 7$), District veterinary officers, heads of production and marketing department ($n = 5$), animal production officers ($n = 2$) and Veterinary officers ($n = 7$). Key informants were employed either by local government departments, non-governmental organisations and intergovernmental organisations (e.g. Food and Agriculture Organization of the United Nations-FAO) operating in Karamoja region. All the interviews were conducted during May 2016.

Livestock are kept in communal herds during the day and in large protected kraals at night. The farmer hierarchy and roles are well structured with the young Karamojong males (5–25 years) in charge of cattle grazing during the day. They return the cattle to communal kraals at night for protection from cattle rustlers as well as for milking. We interviewed kraal leaders (experienced and communally respected farmers) in focus groups ($n = 10–15$) so as to gain an insight of what this category of farmers finds to be the major constraints to

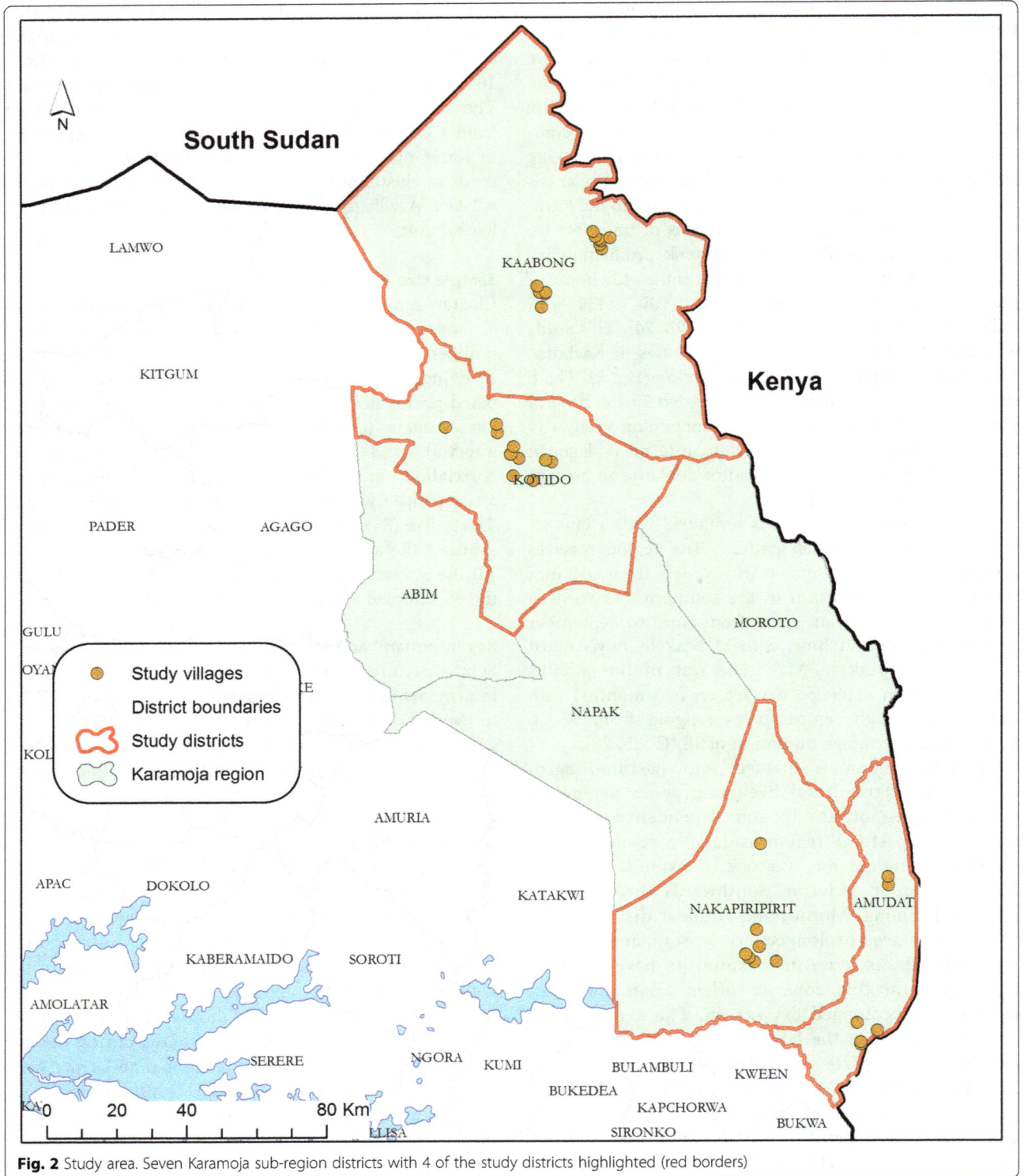

Fig. 2 Study area. Seven Karamoja sub-region districts with 4 of the study districts highlighted (red borders)

livestock production in the region. We also probed them to get an insight of what they find to be the cost of dealing with different animal health problems.

Key informant and kraal leader interview guides

We asked key informants to enumerate the commonest livestock diseases they often encountered during veterinary practice in their order of importance. They were also asked to explain the methods and cost of dealing with the stated diseases. In addition, we asked them to indicate who often treated (prophylactic and therapeutic) livestock and the cost of their professional fees for each of the diseases. In order to get the key informants' views about the topical issues under the study, they were given opportunity to ask questions and to freely express such views at the end of each interview. These similar questions were paraphrased in Ngakarimajong (Karamoja local language) in a way understandable by kraal leaders and administered to them by a local Ngakarimajong speaking veterinarian. Kraal leaders' responses were triangulated with those of key informants. All interview guides were pre-tested and improved to fit their intended purpose before use.

Individual animal sampling procedure

Given the required sample size ($n = 55$ cattle per village × 35 villages =1925 head of cattle) and an average of 250 animals per village, every 4th presented animal was sampled. On the other hand, all presented donkeys ($n = 71$) were sampled. A temporary paint mark was applied to every sampled animal in order to avoid accidental double sampling. The mean number of cattle and donkeys per village was 252 and 2 respectively. Blood samples were taken from randomly selected cattle ($n = 2030$) and donkeys (n = 71) from 35 villages in the four districts of Karamoja region. Almost all the cattle (99.9%) were of the short horn East African zebu breed (Table 1).

Cattle and donkey blood sample collection

Middle ear venepuncture was done on each of the sampled animals and 125 µl of blood collected into capillary tubes. The samples were then transferred onto Flinders Technology Associates (FTA) MiniCards (GE Healthcare, Chalfont Station road, UK) using capillary tubes as previously described [29, 30]. The samples were air-dried and sequentially labelled. A summary data collection form (district, county, sub county, parish, village name, coordinates, animal species, age, sex and breed) was completed at each sampling site. Samples were then packed in foil pouches (GE Healthcare) with a silica gel desiccant (Sigma Aldrich, Co., Life sciences, USA) and transported to Makerere university for analysis.

Table 1 Description of the cattle and donkey populations

Population attributes	Cattle [N (%)]	Donkeys [N (%)]
A) Age		
0–12 months	276 (13.6)	5 (7.0)
13-24 months	341 (16.8)	10 (14.1)
25-36 months	316 (15.6)	7 (9.9)
>36 months	1097 (54.0)	49 (69.0)
B) Sex		
Female	1273 (62.7)	49 (69.0)
Male	621 (30.6)	22 (31.0)
Neutered	136 (6.7)	0 (0)
C) Breed		
Short horn East African Zebu	2028 (99.9)	N/A
Friesian	2 (0.1)	N/A
Equus africanus asinus	N/A	71 (100)

DNA extraction and *Trypanosoma* spp. detection

DNA was extracted and eluted in Chelex®100 resin (Sigma Aldrich) from five 3 mm FTA sample discs as previously described [29, 31]. Briefly, five discs were punched out from each of the individual samples using a Harris 3.0-mm MicroPunch (GE Healthcare) and discharged into 1.5 ml Eppendorf tubes. The MicroPunch cutting edge was decontaminated after each sample by punching out twice the number of discs from unused filter paper. The sample or negative control discs were incubated twice, each time for 15 min in 1.0 ml FTA Purification Reagent (GE Healthcare). This was followed by two rinses of 15 min in 1.0 ml of 1×TE buffer (10 mM Tris–HC$_1$, 0.1 mM EDTA, pH 8.0). Incubation was then done with agitation at room temperature. Thereafter, the discs were dried at 37 °C for 30 min in an incubator (Heldoph, Schwabach, Germany) or were left to dry at room temperature overnight. DNA was eluted from the discs in 100 µl of 5% w/v Chelex/RNA and DNase-free water at 90 °C in a thermocycler (My cycler, Bio-Rad, USA). Eluted DNA samples were kept at −20 °C for long term PCR analyses or 4 °C if they were to be analysed within a week of extraction.

Trypanosome DNA detection

Eluted DNA samples were screened for different trypanosome species using a single pair of internal transcribed spacer 1 (ITS1) CF/BR primers and thermo-cycling conditions as previously described [32]. The ITS1- PCR was carried out in 25 µl reactions. Each reaction contained 5 µl of the test sample, negative or positive control, 1×–reaction buffer (670 mM Tris–HC$_1$ pH 8.8, 166 µM (NH4)$_2$SO$_4$, 4.5% Triton X-100, 2 mg/ml gelatin) (Bioline, Humber Road, London, UK), 2.5 mM MgCl$_2$, 200 µM of each dNTP, 5 µM each of the CF and BR primers, 0.5 U of

*BioTaq*DNA polymerase (Bioline, Humber Road, London, UK), and 15.2 μl of RNA and DNase-free water.

The multiplex serum-resistance associated gene (SRA)-PCR was carried out on each of the samples from which a 480 bp fragment was detected by ITS1-PCR. The multiplex SRA-based PCR simultaneously detects 324, 669 and 800 bp fragments of the glycosylphosphatidylinositol-phospholipase C (GPI-PLC), SRA and variable surface glycoprotein (VSG) genes respectively. Serum-resistance associated gene is specific to *T. b. rhodesiense* while GPI-PLC is a Trypanozoon specific marker. Glycosylphosphatidylinositol-phospholipase C is included as a sample DNA quality control. A sample would be considered positive for *T. b. rhodesiense* only if SRA gene band size were detected; otherwise its absence, but with presence of the GPI-PLC or VSG bands would imply presence of other Trypanozoon trypanosomes [*T. B. brucei* or *T. evansi*]. The samples from which VSG and GPI-PLC bands were amplified were subjected to *T. evansi* specific PCR [33, 34] so as to determine whether they were positive for *T. B. brucei* or *T. evansi*. Samples were not checked for presence of *T. b. gambiense* because *T. b. gambiense* is known to be limited to north-western Uganda districts. Multiplex PCR was carried out in 25 μl reactions using primers and conditions as previously described [35].

All samples from which a ≥ 600 bp fragment was amplified on ITS1-PCR were screened for *T. congolense* savannah / kilifi and forest using TC1/2, TK1/2 and TF1/2 primer sets. This was done in order to determine the commonest *T. congolense* genotype(s) circulating in Karamoja region. The congolense strain specific PCRs were completed in 25 μl reactions as previously described [36]. Each reaction contained 5 μl of the test sample, negative or positive control, 1×−reaction buffer (670 mM Tris−HCl pH 8.8, 166 μM (NH4)$_2$SO$_4$, 4.5% Triton X-100, 2 mg/ml gelatin) (Bioline, Humber Road, London, UK), 0.75 mM MgCl$_2$, 200 μM of each dNTP, 12.5 μM each of the TC1/2, TK1/2, TF1/2primers, 1 U of BioTaqDNA polymerase and 13.05 μl of RNase-free water.

Donkey and cattle samples from which VSG and GPI-PLC bands were amplified by multiplex SRA-PCR were subjected to to *T. evansi* specific PCR using TeRo-Tat920F/TeRo-Tat1070R primers that amplify a 151 bp fragment of the *T. evansi* RoTat 1.2 VSG gene. The *T. evansi* specific PCR was completed in 25 μl reactions using thermal cycling conditions as previously described [33, 34].

PCR products for the six sets of PCRs were separated in 1.5% agarose (Bio Tolls Inc. Japan), stained in ethidium bromide (Sigma Aldrich, Co., Life sciences, USA) and visualised on an ultraviolet transilluminator (Wagtech International, Thatcham, UK) for fragment size determination.

Statistical analyses

Key informants' and kraal leaders' responses were coded and analysed in Microsoft Office Excel 2016 by coding and memo writing methods [37, 38]. Trypanosome prevalence and their corresponding confidence intervals (CIs) were estimated using generalised estimating equation models with binary outcome and logit link function. Robust sandwich standard errors accounted for correlation within villages. The analysis was performed in R statistical software (geepack-package) version 3.3.1. ArcMap v10.3 (spatial analyst extension) was used to map prevalence estimates in different villages.

Results

Livestock health constraints in Karamoja region

Tick-borne diseases namely; − East Coast Fever (ECF), Anaplasmosis, Cowdriosis and Babesiosis were ranked by 75% (14/21) of the key informants as the most commonly encountered diseases during their veterinary practice. A quarter (5/21) of this group also ranked AAT as the second most commonly encountered livestock disease. African animal trypanosomiasis was reported to be mostly encountered in Kotido and Amudat districts. Key informants reported Foot and Mouth Disease (FMD), Contagious Bovine Pleuropneumonia (CBPP), Caprine Pleuropneumonia (CCPP), Peste des Petits Ruminants (PPR), Helminthiasis, Lumpy Skin Disease (LSD), Foot Rot and Brucellosis as the other important diseases in descending order of importance.

On the other hand, livestock keepers in 8 of the 20 focus group discussions intimated that TBDs were the most important causes of losses in their livestock. The rest of the farmers split in equal halves (of 6/20) indicated that either AAT or CBPP were the most important constraints to livestock production. These responses were fairly uniform in the farmers' groups within the four study districts. Just like the animal health providers, farmers reported LSD, FMD, Helminthiasis, foot rot, black quarter and Brucellosis as some of the other livestock diseases that constrain livestock production in the region.

Livestock disease control methods

Key informants indicated that farmers controlled livestock diseases using three main strategies. These included vaccination (for example against FMD, LSD and CBPP), insecticide / acaricide application (TBDs and AAT) and non-characterised herbal extracts (ticks and tsetse). Livestock production inputs including vaccines, acaricides, insecticides were largely provided by government (district veterinary offices), non-governmental organisations or FAO and least by farmers themselves. All key informants indicated that chemotherapeutic and chemoprophylactic management of the most common livestock diseases was largely (95% of the time) done by CBAHWs and farmers. The majority (98.9%) of the kraal leaders' responses on ways of dealing with major livestock diseases were similar to those of key informants.

Livestock diseases associated with highest financial losses in Karamoja region

Whereas this study was not designed to provide an extensive financial or economic analysis of different causes of losses to the livestock sector in Karamoja region, we sought kraal leaders' and key informants' experiences of the major causes of financial / economic losses to the sector. TBDs, AAT, CBBP, CCPP, PPR and FMD were reported to be the most important livestock diseases in descending order. The main causes of losses to the livestock industry (albeit not quantified) were reported to be due to mortalities, morbidities, quarantine (FMD, CBPP, CCPP & PPR) and costs of prevention and treatment of these diseases. Tick-borne disease treatment was reported to be associated with the highest financial losses (UGX 2,5000–75,000 [USD 7–21] depending on the individual TBD. On the other hand, AAT case management cost UGX 3000–4000 [USD 0.8–1.1] if diminazene or Isometamidium was administered respectively (Table 2). Professional fees were not included in these estimates since majority of them were administered by farmers or CBAHWs. The latter were only paid the cost of drugs if they helped farmers administer treatments.

Key informants reported the average cost of treating a single ECF case as UGX 75,000 [USD 21] and that of other TBDs to be UGX 25,000 [USD 7]. In addition, they reported the cost of treating AAT case to be UGX 3000–4000 [USD 0.8–1.1] depending on whether curative (diminazene) or prophylactic (Isometamidium) trypanocidal treatments were administered. Since farmers treated sick animals in almost all cases, it was hard to quantify the cost of professional fees for managing most important livestock diseases in Karamoja region. In addition, CBAHWs charged non-monetary items including milk and other food items that were had to express in monetary terms since their quantities varied widely.

Trypanosome species prevalence in cattle and donkeys in Karamoja region

Sixteen percent (95% CI; 12.4–21.1%) of the cattle were infected with at least one of the three detected trypanosome species namely *T. vivax*, *T. congolense savannah* (*T. congolense*) and *T. brucei brucei* (*T. brucei*). Twenty-two cattle had mixed trypanosome infections. The most common co-infections observed were *T. vivax* and *T. congolense* (17/2030), *T. vivax* and *T. brucei* (3/2030). Only one cow was co-infected with *T. congolense* and *T. brucei*. *T. brucei* was the least prevalent (0.9%, 95% CI: 0.5–1.7%) trypanosome species. *T. congolense savannah* was the only circulating *T. congolense* strain in the cattle sampled (Table 3).

Forty percent (14/35) of the villages sampled kept donkeys. The donkey to cattle proportion was about 1:100. Four of the seventy-one donkeys (5.6%, 95% CI: 2.2–13.9%) were positive for *T. vivax* while 21 (29.6%, 95% CI: 18.6–43.6%) were positive for *T. congolense*. Only 2 donkeys were co-infected with *T. vivax* and *T. congolense*. The overall prevalence of different trypanosome species in donkeys tested was 32.4% (95% CI: 20.2–47.6%) (Table 3). Donkey herd as well as the district level distribution of trypanosome infections were highly clustered in Kotido district (Table 4).

Spatial distribution of different bovine trypanosome species

The spatial distribution of *Trypanosoma* spp. in cattle and donkeys tested was highly clustered at both village

Table 2 Relative cost of managing main livestock diseases in Karamoja region

Disease	Karamoja region cattle	Number affected annually[c]	Direct Cost (UGX) of managing a case	Total Direct costs (billion UGX)	Total Indirect costs[a] (billion UGX)	Total costs	
						Total Costs (billion UGX)	Total (million USD)
Anaplasmosis, Babesiosis, Cowdriosis	2,600,000	468,000.0	25,000.0	11.7	9.1	20.8	6.1
ECF		468,000.0	75,000.0	35.1	27.4	62.5	18.3
Nagana		452,400.0[d]	3-4000[b]	5.4	4.2	9.6	2.8
Total	2,600,000	1,388,400.0		52.2	40.7	92.9	27.2

Uganda shillings (*UGX*), United States Dollars (*USD*)

Costs were triangulated from responses of 20 kraal leaders' focus groups as well as 21 individual key informants. Abroad brush analysis was undertaken using already published literature to generate indicative costs of dealing with AAT and TBDs

[a] Direct cost of managing vector-borne diseases like TBDs and AAT have previously been noted to be ~77.6% of the direct costs [9, 41]

[b] 4 Curative (diminazene) treatments per year (total UGX 12,000) or 3 prophylactic (isometamidium) treatments per year (UGX 12, 000)

[c] Tick-borne diseases (ECF, Anaplasmosis, Babesiosis) have been reported recently to have an incidence rate of 18% in the Karamoja region [22]

[d] Represents 17.4% of all cattle in Karamoja region; the proportion that this study found to be infected with at least a single economically important trypanosome

Table 3 Prevalence of *Trypanosoma* spp. in cattle and donkeys in Karamoja region (May 2016)

Trypanosoma spp.	Cattle (n = 2030)			Donkeys (n = 71)		
	Positive	% prevalence (95%CI)	% Herd prevalence[a]	Positive	% prevalence (95%CI)	% Herd prevalence[a]
Overall	331	16.3 (12.4–21.1)	85.7	23	32.4 (20.2–47.6)	57.1
T. vivax	246	12.1 (9.0–16.1)	80.0	4	5.6 (2.2–13.6)	21.4
T. congolense	90	4.4 (3.0–6.5)	68.6	21	29.6 (18.6–43.6)	57.1
T. brucei s.l.	17	0.9 (0.5–1.7)	25.7	0	0.0	0.0

[a]Due to the local farming systems, all animals within a village are considered as a herd

and district levels. Six out of the 35 villages recorded no trypanosomes in cattle sampled. The highest prevalence of trypanosomes in cattle was recorded in Kotido district while the highest cattle herd prevalence of 65% was recorded in Kotarukot village in Kaabong district (Table 5). The ICC of *Trypanosoma* spp. was 0.14. *T. vivax* was the most common (12.1%, 95% CI; 9.0–16.1%) trypanosome detected in all cattle sampled. In addition, 80% of all the 35 sampled cattle herds were positive for *T. vivax*. *T. congolense* and *T. brucei* were detected in 69% and 26% of all the 35 cattle herds sampled respectively. None of the *T. brucei* s.l. DNA positive samples were positive for the SRA gene indicating that they were all *T. b. brucei* positive.

Discussion

Livestock health workers reported in this study, just like in previous studies [22, 39], that TBDs (ECF, Anaplasmosis, Cowdriosis and Babesiosis), AAT and CBPP are

the top three important livestock health constraints in Karamoja region. East coast fever has recently been reported to be non-endemically stable within Karamoja region explaining why the disease was reported to be associated with highest incidence and mortalities in the region [22, 39]. This study indeed confirmed that AAT is a major cause of losses to the livestock sector since 17.4% of the cattle sampled were positive for economically important trypanosome DNA.

Byaruhanga and colleagues reported that livestock which suffer from either TBDs or AAT are likely to die (mean case fatality rates; 67–90%) [22]. On the other hand, this study just like previous studies [22, 39, 40], found out that animal health workers and farmers' have high level of understanding of TBDs and AAT clinical signs, right control and treatment methods. Failure to effectively control and treat TBDs and AAT much as the different actors have sufficient knowledge to do so is most likely to be as a result of

Table 4 Donkey *Trypanosoma* spp. herd level prevalence in Karamoja region

Village	District	Sampled (n)	Percentage prevalence			
			T. spp.	*T. vivax*	*T. brucei*	*T. congolense*
Nasinyon	Kotido	16.0	56.3	6.3	0.0	50.0
Nadomeo	Kotido	6.0	50.0	0.0	0.0	50.0
Nakumoit	Kotido	4.0	25.0	25.0	0.0	25.0
Lokiding	Kotido	6.0	16.7	0.0	0.0	16.7
Lotanyat	Kotido	3.0	66.7	0.0	0.0	66.7
Kalogwel	Kotido	5.0	40.0	20.0	0.0	40.0
Kanameriongor	Kotido	8.0	37.5	25.0	0.0	25.0
Locheger East	Kaabong	2.0	0.0	0.0	0.0	0.0
Morudikae	Kaabong	4.0	0.0	0.0	0.0	0.0
Nariwore	Kaabong	2.0	0.0	0.0	0.0	0.0
Lochoto	Kaabong	2.0	0.0	0.0	0.0	0.0
Lolelia centre	Kaabong	4.0	50.0	0.0	0.0	50.0
Lokadangan	Nakapiripirit	4.0	0.0	0.0	0.0	0.0
Moruarengan	Amudat	5.0	0.0	0.0	0.0	0.0
Totals	14	71.0	32.4	5.6	0.0	29.6

T. spp. Trypanosoma spp. namely; –*T congolense, T vivax* and *T brucei s.l*

Table 5 Bovine *Trypanosoma* spp. herd level prevalence in Karamoja region

Village	District	Sampled (n)	Percentage trypanosome prevalence			
			T. spp	T. vivax	T. brucei	T. congolense
Loputuk	Kotido	56	30.4	19.6	0.0	10.7
Nasinyon	Kotido	56	23.2	14.3	0.0	8.9
Nadomeo	Kotido	62	33.9	19.4	1.6	12.9
Nakumoit	Kotido	56	12.5	8.9	1.8	1.8
Namukur	Kotido	56	33.9	16.1	7.1	10.7
Lokiding	Kotido	56	19.6	7.1	5.4	7.1
Lotanyat	Kotido	60	35.0	16.7	0.0	18.3
Poet	Kotido	56	10.7	10.7	0.0	0.0
Kalogwel	Kotido	60	28.3	21.7	0.0	6.7
Kanameriongor	Kotido	62	17.7	11.3	0.0	6.5
Kanakuruk	Kaabong	56	1.8	0.0	0.0	1.8
Locheger East	Kaabong	50	2.0	2.0	0.0	0.0
Locheger West	Kaabong	50	2.0	0.0	0.0	2.0
Loputuk	Kaabong	56	1.8	1.8	0.0	0.0
Morudikae	Kaabong	56	0.0	0.0	0.0	0.0
Napeichokei	Kaabong	50	2.0	2.0	0.0	0.0
Nariwore	Kaabong	56	10.7	8.9	0.0	1.8
Kotarukot	Kaabong	60	65.0	53.3	1.7	10.0
Lobalangit	Kaabong	56	16.1	14.3	0.0	1.8
Lochoto	Kaabong	56	25.0	17.9	1.8	5.4
Lolelia Centre	Kaabong	60	13.3	10.0	0.0	3.3
Nakwakwa	Kaabong	56	16.1	16.1	0.0	0.0
Aoyalira	Nakapiripirit	60	0.0	0.0	0.0	0.0
Apeicorait	Nakapiripirit	60	0.0	0.0	0.0	0.0
Lokitela	Nakapiripirit	60	0.0	0.0	0.0	0.0
Lokadangan	Nakapiripirit	60	10.0	8.3	0.0	1.7
Nakiloro	Nakapiripirit	60	5.0	5.0	0.0	0.0
Naabore-B	Nakapiripirit	46	30.4	23.9	4.3	2.2
Arengesiep	Nakapiripirit	60	0.0	0.0	0.0	0.0
Abongai	Amudat	82	31.7	18.3	3.7	9.8
Korenyang	Amudat	60	20.0	11.7	0.0	8.3
Moruarengan	Amudat	60	15.0	13.3	0.0	1.7
Morumodo	Amudat	60	35.0	15.0	3.3	16.7
Angarab	Amudat	60	30.0	30.0	0.0	0.0
Tingas	Amudat	60	18.3	18.3	0.0	0.0
Totals		2030	16.3	12.1	0.9	4.4

T. spp. Trypanosoma spp. namely; −*T. congolense, T. vivax* and *T. brucei s.l*

farmers and CBAHWs treating over 95% of the sick animals. As such, farmers and key informants indicated that veterinarians only treat 5% of the sick animals. Low farmers' willingness to pay veterinarians to treat their cattle is a likely reason why most animals are treated by farmers and CBAHWs resulting into infective treatment/control of TBDs and AAT.

Non-governmental and inter-governmental organisations (mainly FAO) working in Karamoja region train CBAHWs (who are largely farmers with not much formal training) to equip them with basic knowledge about livestock disease clinical signs and their control methods. This level of training does not seem to be sufficient to help CBAHWs be able to administer right

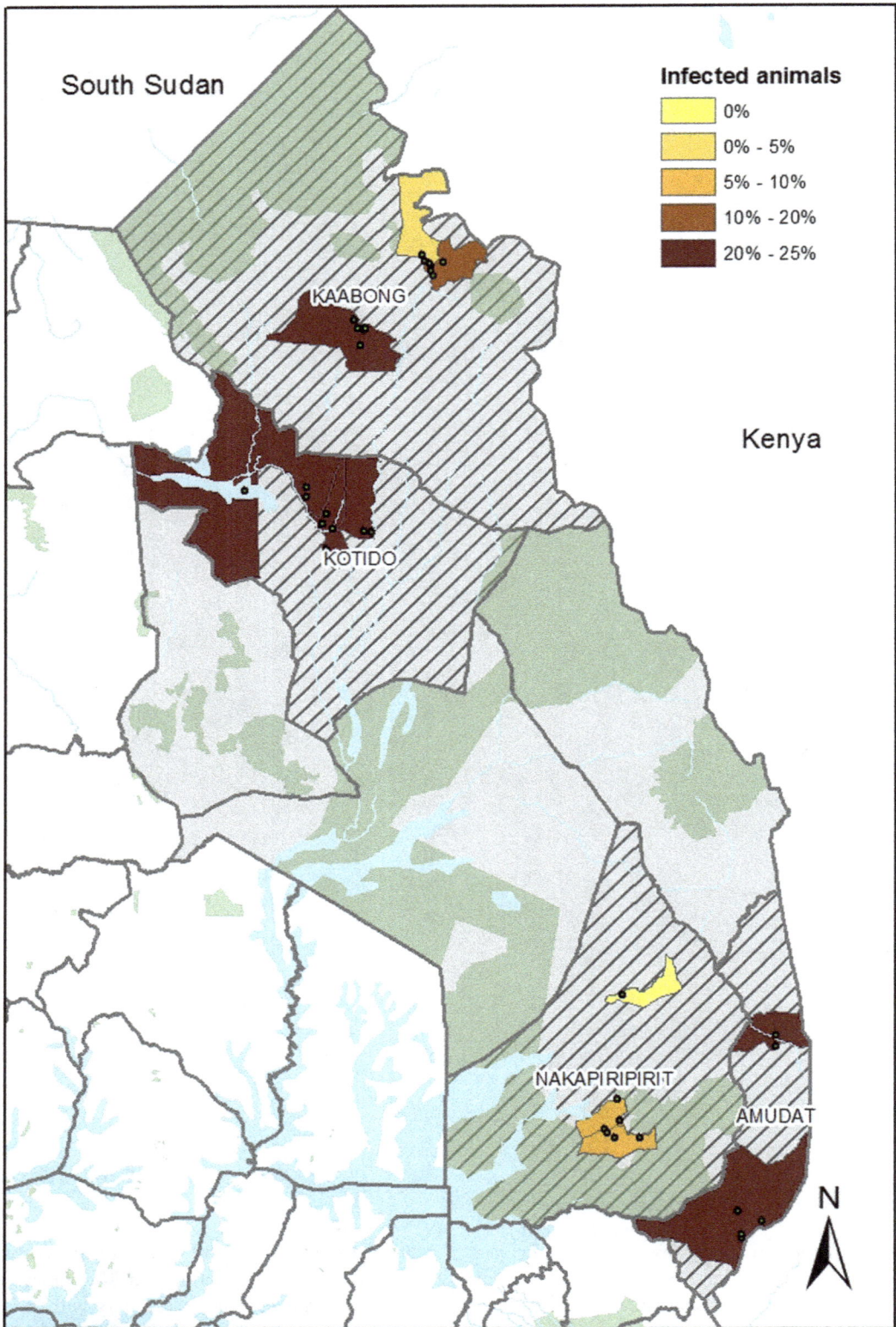

Fig. 3 Proportion of animals infected by any *Trypanosoma* spp. Data were aggregated at parish level because selected villages were within the interpolation distance and could not otherwise give enough spatial resolution

treatment regimens for different livestock diseases. There is therefore need to retrain them about effective treatment regimens of TBDs and AAT as well as establish and maintain a regulatory framework for the involvement of CBAHWs in animal health.

Given that recent studies have indicated that the indirect costs associated with tsetse and tick-borne disease management are up to 77.6% of direct costs [9, 41], the indirect costs associated with AAT and TBD management by farmers in Karamoja region is about UGX 40.7 billion (US$ 11.9 million). These costs would be higher if adjusted to include costs due to loss in production, mortalities and vector control. Even at indicative costs (before they are refined in more exhaustive financial and economic studies) level, the total costs of dealing with AAT and TBDs (UGX 93 billion, US$ 27.2 million) are about 7 times the annual government support to the department of production and marketing (UGX 14.4 billion; ~US$ 4.32 million) in the whole Karamoja region. This explains why farmers and livestock health professionals in this region unmistakably ranked TBDs and AAT as the two most important constraints to livestock health.

A previous small trypanosomiasis survey ($n = 196$) in Kotido district [21] reported an overall prevalence of 5.2% much lower than is reported in this study. However, the trypanosomes species detected during the previous survey are largely similar to those detected in this study. The large difference (12.5%) in the overall prevalence of bovine trypanosome species previously reported in Kotido district [21] and this study is likely to be due to the differences in the sizes (number of cattle sampled, methods of sampling and the districts included) of both surveys as well as the time differences. To our knowledge, this is the first large (covering cattle and donkeys in 4/7 districts of Karamoja region) AAT survey in Karamoja region. This survey indicates that AAT is a major constraint to livestock production in Karamoja region (Fig. 3). This implies that there is need to re-define Karamoja region as a high priority area for trypanosomiasis control.

T. vivax undergoes a very short life cycle in the proboscis of tsetse flies [42] and is associated with rapid build-up of parasitaemia in its mammalian hosts [10, 16]. These two features make it very easy to transmit cyclically. *T. vivax* is also known to be mechanically transmitted by a range of hematophagous flies making it appear in areas beyond known tsetse belts [16]. This partly explains why *T. vivax* was the most prevalent trypanosome species detected in this study just like in recent studies in other AAT endemic regions of Uganda [10, 20].

T. congolense savannah and *T. b. brucei* were detected in low overall and herd prevalence in the four districts of Karamoja region. *T. brucei s.l.* and *T. congolense savannah* low herd and overall prevalence have been reported elsewhere in AAT endemic regions of Uganda [10, 20]. *T. congolense savannah* is highly pathogenic to cattle which partly explains its low prevalence in most routine molecular epidemiological studies involving apparently healthy cattle [10]. None of the *T. brucei s.l. DNA* positive samples were positive for the serum resistance antigen (SRA) gene indicating that none of them was human infective (*T. b. rhodesiense*). *T. b. rhodesiense* infections can be very focal in the cattle and wild life reservoir especially in non-epidemic states [20]. The current indication of no risk of sleeping sickness in this region needs to be confirmed by large-scale surveys involving all districts in the Karamoja region.

Donkeys and camels provide animal proteins as well as transport of agricultural products and inputs [Fig. 4]. There is need to reflect this key role of donkeys and camels in providing for their health, in this case looking

Fig. 4 Donkeys used as the main form of transport for agricultural products and inputs in Kotido district, northern Karamoja (Credit: Dennis Muhanguzi)

at the most important trypanosomes that constrain donkey health, production and management. *T. congolense savannah* and *T. vivax* were detected in high proportions in donkeys at herd and district levels. Both *Trypanosoma* spp. are reported to cause chronic AAT in donkeys [43, 44]. Given that the two trypanosome species were the most prevalence in cattle as well, there is need to study the role of donkeys in the epidemiology of other livestock AAT.

Bovine and equine trypanosome species were highly clustered at both village and district levels (Fig. 4). This kind of clustering pattern has recently been reported in the AAT endemic regions of south-eastern Uganda [10, 20]. Clustered distribution of trypanosome infections can be attributed to different factors. These include but are likely not limited to; –level of challenge and differences in livestock management practices practiced by farmers in different villages and districts [10, 19, 20].

Conclusion

Livestock production remains the mainstay of the Karamajong livelihoods. On the other hand, livestock production potential in this region is limited by endemic livestock diseases including AAT and TBDs. We report here that 16.3% and 32.4% of all cattle and donkeys sampled respectively were infected with different *Trypanosoma* spp. This was in strong agreement with farmers' and key informants' observations that AAT is only second to TBDs in constraining livestock production in Karamoja region. In order to improve livestock health and production, it is therefore apparent that government of Uganda needs to invest in livestock health and production programs particularly AAT and TTBD control. Karamoja region AT priority control map should be refined using the new AT prevalence data (arising from this survey and those in future) so as to highlight the region as a high priority region for trypanosomiasis control.

Abbreviations
AAT: Animal African Trypanosomiasis; AT: African Trypanosomiasis; CBAHWs: Community based animal health workers; CBPP: Contagious Bovine Pleuropneumonia; CCPP: Caprine Pleuropneumonia; CI: Confidence interval; ECF: East Cost Fever; FAO: Food and Agriculture Organization of the United Nations; FMD: Foot and Mouth Disease; FTA : Flinders Technology Associates; GPI-PLC: Glycosylphosphatidylinositol-phospholipase C; HAT: Human African Trypanosomiasis; ICC: Intracluster Correlation Coefficient; ITS1-PCR: Internal transcribed spacer 1- based polymerase chain reaction; LSD: Lumpy Skin Disease; PPR: Peste des petits ruminants; SRA: Serum-resistance associated gene; TBDs: Tick-borne diseases; TTBDs: Ticks and tick-borne diseases; UGX: Uganda shillings; USAID: United States Agency for International Development; USD: United States Dollars; VSG: Variable surface glycoprotein

Acknowledgements
Authors wish to acknowledge study area District Veterinary officers for their help during sample collection and completion of key informants' semi-structured interviews. We would also like to acknowledge the administrative support of all the Mercy Corps Uganda staff and that of Prof. Mugisha Anthony, Associate Prof. Acai Okwe and Dr. Peninah Nsamba of School of Veterinary Medicine, Makerere University. In addition, we would also like to acknowledge

Dr. Ward Bryssinckx of Avia-GIS, Risschotlei 33, B-2980 Zoersel, Belgium for helping with some of the spatial analyses. We also acknowledge farmers in the four study districts for offering their cattle and donkeys to be sampled and spending time to provide answers to the semi-structured interviews.

Funding
The research leading to this manuscript received funding from United States Agency for International Development (USAID) through Mercy Corps Uganda to MD.USAID had no role in the design and execution of the study as well as the decision to publish this manuscript.

Authors' contributions
DM, SGO, MK, JH, WA and RT conceived and designed this study. DM, WA, SGO collected blood samples and administered all interviews. DM, GB, AK, GGA, SO did all the molecular analyses. DM, WA and AM completed spatial analyses. DM and WA designed, administered, coded and analysed the semi-structured key informants' and focus group discussion interviews. All authors read and approved the final version of the manuscript.

Ethics approval and consent to participate
Verbal informed consent was obtained from farmers before they were interviewed or samples taken from their cattle or donkeys. Verbal informed consent was chosen over written informed consent because more than 90% of all farmers in Karamoja region can neither write nor read and have no legally appointed representatives. Verbal informed consent was obtained through an Ngarakimajong speaker [Karamoja region local language] who explained to all study participants and participating communities the benefits and risks involved with participating in this study. The verbal consent documentation as well as the main protocols leading to results presented in this manuscript were reviewed and approved (SBLS/REC/16/137) by the Makerere University School of Biosecurity, Biotechnical and Laboratory Sciences (SBLS) Research and Ethics committee. In addition, they were reviewed, approved and the study registered (A 514) by the Uganda National Council of Science and Technology.

Consent for publication
All authors consented to publication.

Competing interests
The authors declare that they have no competing interests.

Author details
[1]College of Veterinary Medicine Animal Resources and Biosecurity, Makerere University, P.O. Box 7062, Kampala, Uganda. [2]Coordinating Office for Control of Trypanosomiasis in Uganda, Ministry of Agriculture, Animal Industry and Fisheries, Plot 78, Buganda Road, P. O. Box: 16345 Wandegeya, Kampala, Uganda. [3]Mercy Corps Uganda, PO Box 32021, Clock Tower, Kampala, Uganda. [4]Swiss Tropical Institute, Socinstrasse 57, -4002 Basel, CH, Switzerland. [5]University of Basel, Petersplatz 1, 4003 Basel, Switzerland.

References
1. Meyer A, Holt HR, Selby R, Guitian J. Past and ongoing tsetse and animal Trypanosomiasis control operations in five African countries: a systematic review. PLoS Negl Trop Dis. 2016; https://doi.org/10.1371/journal.pntd.0005247.
2. Franco JR, Simarro PP, Diarra A, Jannin JG. Epidemiology of human African trypanosomiasis. Clin Epidemiol. 2014;6:257–75.
3. Simarro PP, Cecchi G, Franco JR, Paone M, Diarra A, Ruiz-Postigo JA, Fèvre EM, Mattioli RC, Jannin JG. Estimating and mapping the population at risk of sleeping sickness. PLoS Negl Trop Dis. 2012;6:e1859.

African animal trypanosomiasis as a constraint to livestock health and production in Karamoja...

139

4. Fèvre EM, Wissmann BV, Welburn SC, Lutumba P. The burden of human African Trypanosomiasis. PLoS Negl Trop Dis. 2008; https://doi.org/10.1371/journal.pntd.0000333.

5. Fèvre EM, Odiit M, Coleman PG, Woolhouse MEJ, Welburn SC. Estimating the burden of rhodesiense sleeping sickness during an outbreak in Serere, eastern Uganda. BMC Public Health. 2008;8:96.

6. Brun R, Blum J, Chappuis F, Burri C. Human African trypanosomiasis. Lancet. 2010;375:148–59.

7. Swallow B. Impacts of trypanosomiasis on African agriculture. Int Livest Res Institute, Nairobi, Kenya. 1999:1–46.

8. Okello WO, Muhanguzi D, MacLeod ET, Welburn SC, Waiswa C, Shaw AP. Contribution of draft cattle to rural livelihoods in a district of southeastern Uganda endemic for bovine parasitic diseases: an economic evaluation. Parasit Vectors. 2015;8:571.

9. Muhanguzi D, Okello WO, Kabasa JD, Waiswa C, Welburn SC, Shaw APM. Cost analysis of options for management of African animal Trypanosomiasis using interventions targeted at cattle in Tororo District; south-eastern Uganda. Parasit Vectors. 2015;8:387.

10. Muhanguzi D, Picozzi K, Hattendorf J, Thrusfield M, Kabasa JD, Waiswa C, Welburn SC. The burden and spatial distribution of bovine African trypanosomes in small holder crop-livestock production systems in Tororo District, south-eastern Uganda. Parasit Vectors. 2014;7:603.

11. Welburn SC, Bardosh KL, Coleman PG. Novel financing model for neglected tropical diseases: development impact bonds applied to sleeping sickness and rabies control. PLoS Negl Trop Dis. 2016;10:e0005000.

12. Uganda Trypanosomiasis Control Council Act 1992 | ULII [Internet]. [cited 2017 Nov 22]. Available from: http://www.ulii.org/ug/legislation/consolidated-act/211.

13. Albert M, Wardrop NA, Atkinson PM, Torr SJ, Welburn SC. Tsetse fly (G.F. Fuscipes) distribution in the Lake Victoria Basin of Uganda. PLoS Negl Trop Dis. 2015;9:e0003705.

14. Cecchi G, Paone M, Feldmann U, Vreysen MJB, Diall O, Mattioli RC. Assembling a geospatial database of tsetse-transmitted animal trypanosomosis for Africa. Parasit Vectors. 2014;7:39.

15. Wells EA. The importance of mechanical transmission in the epidemiology of nagana: a review. Trop Anim Health Prod. 1972;4:74–89.

16. Desquesnes M, Dia ML. Trypanosoma vivax: mechanical transmission in cattle by one of the most common African tabanids, Atylotus agrestis. Exp Parasitol. 2003;103:35–43.

17. Jones TW, Dávila AMR. Trypanosoma vivax - out of Africa. Trends Parasitol. 2001;17:99–101.

18. Gardiner PR, Wilson AJ. Trypanosoma (Duttonella) vivax. Parasitol Today. 1987;3:49–52.

19. Rogers DJ. Trypanosomiasis "risk" or "challenge": a review. Acta Trop. 1985;42:5–23.

20. Muhanguzi D, Picozzi K, Hatendorf J, Thrusfield M, Welburn SC, Kabasa JD, Waiswa C. Improvements on restricted insecticide application protocol for control of human and animal African Trypanosomiasis in eastern Uganda. PLoS Negl Trop Dis. 2014;8:e3284.

21. Asaku ST, MacLeod E, Mwiine FN. A report of PCR typed trypanosome species and subspecies in Kotido district, nothern Karamoja. Germany: Lambert academic publishing, 66111 Saarbrücken; 2007.

22. Byaruhanga C, Oosthuizen MC, Collins NE, Knobel D. Using participatory epidemiology to investigate management options and relative importance of tick-borne diseases amongst transhumant zebu cattle in Karamoja region, Uganda. Prev Vet Med. 2015;122:287–97.

23. Uganda Bureau of Statistics (UBOS) (2008) The National Livestock Census Report 2008. doi: www.ubos.org

24. UBOS (2013) Uganda Bureau of Statistics; 2013 statistical abstract.

25. Egeru A, Wasonga O, Kyagulanyi J, Majaliwa G, MacOpiyo L, Mburu J. Spatio-temporal dynamics of forage and land cover changes in Karamoja sub-region, Uganda. Pastor Res Policy Pract. 2014;4:6.

26. Mubiru D. Climate change and adaptation options in Karamoja. Organ: Food Agric; 2010.

27. Bennett S, Woods T, Liyanage WM, Smith DL. A simplified general method for cluster-sample surveys of health in developing countries. World Health Stat Q. 1991;44:98–106.

28. Farid M, Frerichs RR (2007) Csurvey Software. http://www.ph.ucla.edu/epi/csurvey.html. Accessed 13 Apr 2016.

29. Ahmed HA, MacLeod ET, Hide G, Welburn SC, Picozzi K. The best practice for preparation of samples from FTA cards for diagnosis of blood borne infections using African trypanosomes as a model system. Parasit Vectors. 2011;4:68.

30. Picozzi K, Tilley A, Fèvre EM, Coleman PG, Magona JW, Odiit M, Eisler MC, Welburn SC. The diagnosis of trypanosome infections: applications of novel technology for reducing disease risk. Afr J Biotechnol. 2002;1:39–45.

31. Becker S, Franco JR, Simarro PP, Stich A, Abel PM, Steverding D. Real-time PCR for detection of Trypanosoma brucei in human blood samples. Diagn Microbiol Infect Dis. 2004;50:193–9.

32. Njiru ZK, Constantine CC, Guya S, Crowther J, Kiragu JM, Thompson RCA, Dávila AMR. The use of ITS1 rDNA PCR in detecting pathogenic African trypanosomes. Parasitol Res. 2005;95:186–92.

33. Salim B, Bakheit MA, Kamau J, Nakamura I, Sugimoto C. Molecular epidemiology of camel trypanosomiasis based on ITS1 rDNA and RoTat 1.2 VSG gene in the Sudan. Parasit Vectors. 2011;4:31.

34. Ngaira JM, Njagi ENM, Ngeranwa JJN, Olembo NK. PCR amplification of RoTat 1.2 VSG gene in Trypanosoma evansi isolates in Kenya. Vet Parasitol. 2004;120:23–33.

35. Picozzi K, Carrington M, Welburn SC. A multiplex PCR that discriminates between Trypanosoma brucei brucei and zoonotic T. b. rhodesiense. Exp Parasitol. 2008;118:41–6.

36. Masiga DK, Smyth AJ, Hayes P, Bromidge TJ, Gibson WC. Sensitive detection of trypanosomes in tsetse flies by DNA amplification. Int J Parasitol. 1992;22:909–18.

37. Padgett DK. Qualitative and mixed methods in public health. SAGE Publ. 2014; https://doi.org/10.4135/9781483384511.

38. Bourgeault I (2010) The SAGE handbook of qualitative methods in Health Research [electronic resource].

39. Byaruhanga C, Collins NE, Knobel D, Kabasa W, Oosthuizen MC. Endemic status of tick-borne infections and tick species diversity among transhumant zebu cattle in Karamoja region, Uganda: support for control approaches. Vet Parasitol Reg Stud Reports. 2016;1:1–10.

40. Gradé JTT, Tabuti JRSS, Van Damme P. Ethnoveterinary knowledge in pastoral Karamoja, Uganda. J Ethnopharmacol. 2009;122:273–93.

41. Shaw APM, Torr SJ, Waiswa C, Cecchi G, Wint GRW, Mattioli RC, Robinson TP. Estimating the costs of tsetse control options: an example for Uganda. Prev Vet Med. 2013;110:290–303.

42. Jefferies D, Helfrich MP, Molyneux DH. Cibarial infections of Trypanosoma vivax and T. congolense in Glossina. Parasitol Res. 1987;73:289–92.

43. Murray M, Gray AR. The current situation on animal trypanosomiasis in Africa. Prev Vet Med. 1984;2:23–30.

44. Faye D, Pereira de Almeida PJL, Goossens B, Osaer S, Ndao M, Berkvens D, Speybroeck N, Nieberding F, Geerts S. Prevalence and incidence of trypanosomosis in horses and donkeys in the Gambia. Vet Parasitol. 2001;101:101–14.

16

Seroprevalence of Rift Valley fever virus in livestock during inter-epidemic period in Egypt, 2014/15

Claudia Mroz[1], Mayada Gwida[2], Maged El-Ashker[3], Mohamed El-Diasty[4], Mohamed El-Beskawy[5], Ute Ziegler[1], Martin Eiden[1] and Martin H. Groschup[1*]

Abstract

Background: Rift Valley fever virus (RVFV) caused several outbreaks throughout the African continent and the Arabian Peninsula posing significant threat to human and animal health. In Egypt the first and most important Rift Valley fever epidemic occurred during 1977/78 with a multitude of infected humans and huge economic losses in livestock. After this major outbreak, RVF epidemics re-occurred in irregular intervals between 1993 and 2003. Seroprevalence of anti-RVFV antibodies in livestock during inter-epidemic periods can be used for supporting the evaluation of the present risk exposure for animal and public health. A serosurvey was conducted during 2014/2015 in non-vaccinated livestock including camels, sheep, goats and buffalos in different areas of the Nile River Delta as well as the furthermost southeast of Egypt to investigate the presence of anti-RVFV antibodies for further evaluating of the risk exposure for animal and human health. All animals integrated in this study were born after the last Egyptian RVF epidemic in 2003 and sampled buffalos and small ruminants were not imported from other endemic countries.

Results: A total of 873 serum samples from apparently healthy animals from different host species (camels: $n = 221$; sheep: $n = 438$; goats: $n = 26$; buffalo: $n = 188$) were tested serologically using RVFV competition ELISA, virus neutralization test and/or an indirect immunofluorescence assay, depending on available serum volume. Sera were assessed positive when virus neutralization test alone or least two assays produced consistent positive results. The overall seroprevalence was 2.29% (95%CI: 1.51–3.07) ranging from 0% in goats, 0.46% in sheep (95%CI: 0.41–0.5), and 3.17% in camels (95%CI: 0.86–5.48) up to 5.85% in buffalos (95%CI: 2.75–8.95).

Conclusion: Our findings assume currently low level of circulating virus in the investigated areas and suggest minor indication for a new RVF epidemic. Further the results may indicate that during long inter-epidemic periods, maintenance of the virus occur in vectors and also most probably in buffaloes within cryptic cycle where sporadic, small and local epidemics may occur. Therefore, comprehensive and well-designed surveillance activities are urgently needed to detect first evidence for transition from endemic to epidemic cycle.

Keywords: Rift Valley fever virus, Livestock, Inter-epidemic period, Surveillance, Egypt

Background

Rift Valley fever is a mosquito-borne zoonotic disease in ruminants, camels and humans caused by Rift Valley fever virus (RVFV), a Phlebovirus within the family *Bunyaviridae* [1, 2]. The viral disease was identified for the first time in 1930 in Kenya and is characterized by high fever and abortion in livestock and high neonatal mortality mainly in sheep [3–6]. Infected humans show a mild febrile illness, however in 1–2% of cases the patients develop severe complications such as ocular disease, hemorrhagic fever syndrome or encephalitis [7]. Typically the general case fatality is low (1–3%). But patients with hemorrhagic fever syndrome show fatality rates up to 50% [8].

It has been reported that more than 30 mosquito species from 6 genera can transmit the virus to susceptible hosts [7]. Bites of infected mosquitos play the most important

* Correspondence: martin.groschup@fli.bund.de
[1]Institute of Novel and Emerging Infectious Diseases, Friedrich-Loeffler-Institut, Südufer 10, 17493 Greifswald - Isle of Riems, Germany

role for ruminant infection [7, 9]. The direct contact with infectious materials when handling with sick or dead infected animals, abortion material or other fresh tissues represents the main transmission route in humans. Due to climatic changes and high level livestock trade, the virus is widespread in Africa and spread also in 2000 to Saudi Arabia and Yemen [5, 9, 10]. Climatic and environmental conditions like heavy rainfalls with increasing mosquito population redound consistently to new RVF outbreaks. Severe outbreaks occurred for instance in Mauritania and in South Africa in 2010, in Kenya, Tanzania and Somalia in 2007 as well as in Sudan in 2008 and 2010 [11–14].

The RVFV was introduced to Egypt in 1977 and caused an extensive epidemic with thousands of infected humans, more than 600 deaths and high economic losses in livestock affecting five governorates in the Nile Delta (Sharqia, Aswan, Qalyubia, Giza and Assiut [5, 15–18]. Up to now, it has been considered the major outbreak for Egypt and one of the largest epidemics in the RVF history of Africa. After a long inter-epidemic period, the RVF re-occurred in the Nile Delta of Egypt in 1993 in Aswan and Damietta governorates [19–21]. Further outbreaks recurred in 1994 (Beheira and Kafr el Sheikh governorates) as well as in 1997 (Assuit and Aswan governorates) and most recently in 2003 (Kafr el Sheikh governorate) [19, 21–23]. The sources of the diverse outbreaks are broadly discussed but the maintenance of the virus during inter-epidemic periods is still poorly understood [21, 24]. It has been reported that the presence of unvaccinated susceptible livestock in combination with favorable conditions for mosquito breeding and spread are facilitating conditions for the persistence of the RVFV in Egypt [21]. Detection of RVFV specific antibodies in non-immunized animals a long time after the last RVF epidemic indicates endemic maintenance of the virus in inter-epidemic periods and seroconversion often occurs without any clinical signs in the livestock population [25, 26].

Evidence of circulating virus in the current inter-epidemic phase has been found by Ramadan [27] in 2009 who proves the presence of anti-RVFV-antibodies in Dakahlia governorate in different livestock species (sheep = 20%, goats = 17%, cattle = 5% and buffalos = 11% respectively). An additional survey from Marawan [28] in 2012 shows related prevalence rates in non-immunized sheep, goats, camels, cattle and buffalos (17, 7, 0, 19 and 10%, respectively) in four governorates in the Nile delta of Egypt (Qalyubia, Dakahlia, Sharkia, Kafr El Sheikh). A compilation of outbreak sites and sites of previous seroepidemiological studies in egypt are indictaed in Additional file 1.

Seroepidemiological studies could merely give a brief insight into the infection status for a short period in which the study was carried out. Therefore, the need for continuous inspections of the antibody prevalence in susceptible

species is highly recommended in endemic areas. This paper presents an overview about the antibody presence in non-vaccinated susceptible hosts including sheep, goats and buffalos from Dakahlia governorate, a vulnerable part for new RVF epidemic in Egypt. Furthermore camel sera from an abattoir near Cairo and from southeast of Egypt, near the border to Sudan, were investigated. Aim of this study was to determine the current status of anti-RVFV antibodies in non-immunized hosts for further evaluating the exposure risk for animal and human health.

Methods
Animal population, study areas and sample collection
The present seroepidemiological study included a total of 873 non-immunized, apparently healthy animals including small ruminants, buffalos and camels, which were sampled during 2014 to 2015 in different areas of Egypt. Due to the high susceptibility of sheep and their suitability for the use as sentinel animals [9, 29–31], 438 sheep samples from 8 different small holding herds were collected during 2015 (Table 1). All animals were settled in Dakahlia governorate, a central part of the Nile River Delta, in open yard holdings, with movement restrictions in the vicinity (Fig. 1). The number of animals in the flocks varied from 20 to 87 animals (mean (M) = 54.75; standard deviation (SD) =23.24). Additionally sheep herd 1 and 2 included a total of 26 goats (17 in herd 1 and 9 in herd 2). Small holders were characterized by keeping less than 50 animals, often in family farming with poor resources, no health management and prevalently with more than one species.

Further ruminant samples were collected from 188 Asian water buffalos (Table 1, Fig. 1). 88 serum samples derived from four small holdings located in Dakahlia governorate (61 samples collected in 2014 and 27 samples in 2015). Small holder flocks included 9 to 31 buffalos (M = 22; SD = 8.3). Additional 100 samples were collected from buffalos from a farm in Ismailia governorate, situated in the south eastern of the Nile Delta at the Suez Canal. In general farms are characterized by herd sizes over 50 animals and a good health management and restricted movements of the animals (only to local markets in the country). In this study the sampled farm kept 1200 animals. All buffalos and small ruminants included in this study were born after the last Egyptian RVF epidemic in 2003 and were not imported from other endemic countries.

To investigate the role of camels in Egypt additional samples were collected from this species. Most of the Egyptian camels were imported from Sudan and were kept in quarantine near the border of Sudan in Red Sea governorate after importation (Fig. 1). We investigated 150 of those camel derived sera and additional 71 camel sera taken at an abattoir near Cairo (Table 1).

All blood samples were randomly collected from apparently healthy animals and an informed consent for

Table 1 Samples ordered by species, holding system, and region

Species	Holding	Herd number	Age	Region	Number of samples	Date of sample collection
Sheep	Open yard (small holder)	1	2–5	Dakahlia Governorate (Aga district)	87	April 2015
Goat			2–3		17	
Sheep	Open yard (small holder)	2	2–10	Dakahlia Governorate (Aga district)	82	May 2015
Goat					9	
Sheep	Open yard (small holder)	3	2–5	Dakahlia Governorate (Belkas district)	50	May 2015
Sheep	Open yard (small holder)	4	1–4	Dakahlia Governorate	78	June 2015
Sheep	Open yard (small holder)	5	1–5	Dakahlia Governorate	40	June 2015
Sheep	Open yard (small holder)	6	2–4	Dakahlia Governorate	49	June 2015
Sheep	Open yard (small holder)	7	3–7	Dakahlia Governorate	32	June 2015
Sheep	Open yard (small holder)	8	2–4	Dakahlia Governorate	20	February 2015
Buffalo	Small holder	1	3–7	Dakahlia Governorate (Belkas district)	31	2014
Buffalo	Small holder	2	3–7	Dakahlia Governorate	9	2014
Buffalo	Small holder	3	3–7	Dakahlia Governorate	21	2014
Buffalo	Small holder	2	3–5	Dakahlia Governorate (Belkas district)	27	2015
Buffalo	Farm	1	4–6	Ismailia Governorate	100	2015
Camel	Abattoir (El Basatine)		5–7	Cairo	71	2014
Camel	Imported from Sudan		2–7	Red Sea Governorate (Halayb and Shalatein)	150	2015

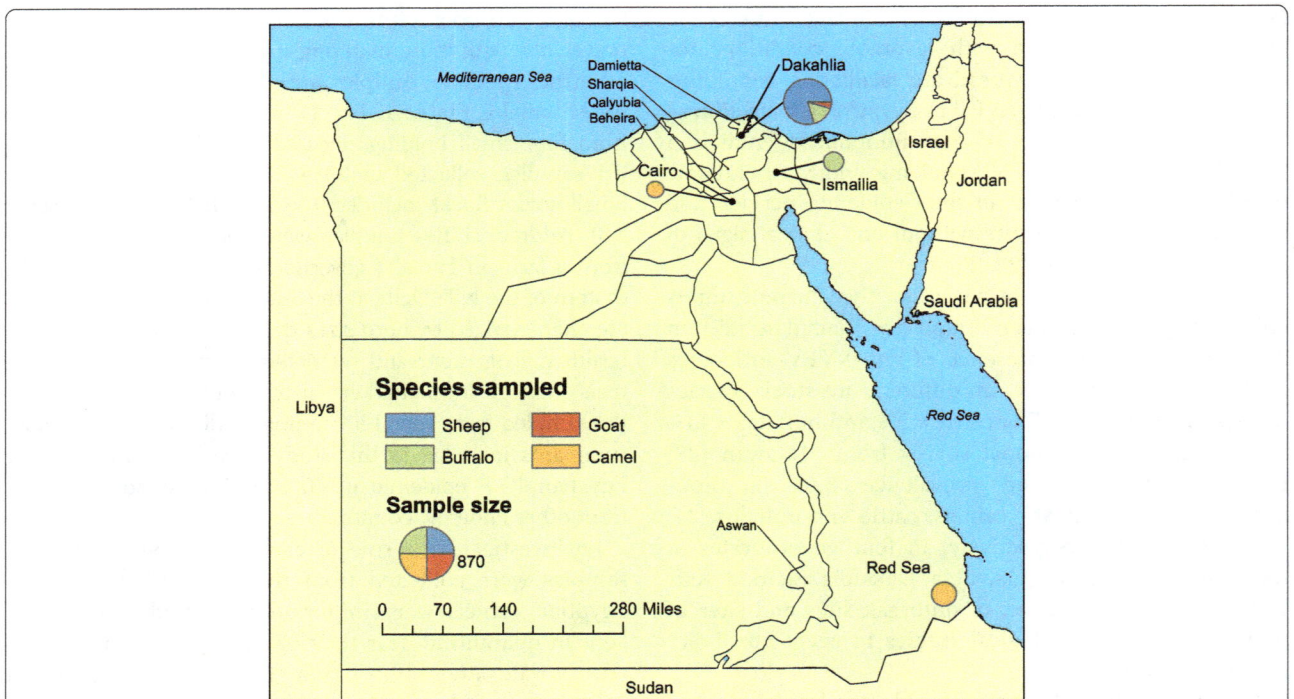

Fig. 1 Region of samples collection. Samples from small ruminants and buffalos from small holders were collected in Dakahlia governorate. Further buffalo samples, derived from farm animals were collected from Ismailia governorate. Camel derived sera were provided from an abattoir near Cairo and from Rea Sea governorate. Previous Egyptian RVFV outbreaks occurred in indicated governorates (Sharqia, Aswan, Qalyubia, Damietta and Beheira)

RVFV investigation was given by the owners. All procedures were performed in accordance with the principles and specific guidelines presented in the Guidelines for the Care and Use of Agricultural Animals in Research and Teaching, 3rd ed. [32].

Animal specimens were collected under the direction of the Mansoura University, Mansoura, Egypt, within the framework of the project 'Brucellosis, Q-fever and viral hemorrhagic fever infections in Egypt'. The collected blood samples were kept overnight at room temperature to allow blood clotting. On the next day, clear sera were collected and stored at −20 °C until shipping. In Germany, the sera were subjected to Gamma radiation (Synergy Health, Radeberg) and were irradiated for 24 h by 30 k gray. This treatment was carried out to ensure that the examinations can be done under BSL 2 conditions. After radiation, the sera were stored at −20 °C until serological examinations.

Serological testing's
Enzyme linked immunosorbent assay (ELISA)

Each sample was first analyzed with the commercial ID Screen® RVFV competition multispecies ELISA (ID VET, Montpellier, France) according to the manufacturers' instructions [30]. The cELISA is based on recombinant nucleoprotein and detects anti-RVFV antibodies (both IgM and IgG).

Virus neutralization test (VNT)

Investigation of neutralizing antibodies of the samples was carried out by VNT, the gold standard for serological RVF analysis. VNT was performed using the RVFV-MP12 vaccine strain according to the OIE terrestrial manual, 2015 [33]. Briefly, each serum samples were run in duplicate and titrated in three steps (1:5, 1:10, 1:20). The diluted virus (100 TCID$_{50}$ per well) was added to 1:10 and 1:20 serum dilution, 1:5 served as serum control. After 30 min incubation at 37 °C and 5% CO$_2$ 3×10^5 Vero 76 cells (Collection of Cell Lines in Veterinary Medicine, Friedrich-Loeffler-Institut, Germany) per ml were added to each well. Positive and negative control sera as well as cell and virus controls were included in each test cycle. After six days at 37 °C and 5% CO$_2$ cytopathic effects were revealed and the plates were fixed with formalin, colored with crystal violet and the VNT titer was calculated. According to the Behrens-Kaerber method the neutralizing antibody titer of the samples was defined as the 50% neutralization dose (ND$_{50}$). Serum samples with ND$_{50}$ values above 10 were determined as positive; samples with lower titer than 10 were determined as negative. While the endpoint titer was not in focus of interest all sera with a higher ND$_{50}$ value than 30 were not further diluted and were summarized as ≥30.

Indirect Immunofluorescence assay (IIFA)

Additional confirmation of positive and inconclusive results in previous tests was performed with an in-house immunofluorescence assay (IIFA) according to a previously published protocol [11]. In short diluted serum samples were added to commercial RVFV immunofluorescence slides from Euroimmun (Lübeck, Germany). After 30 min reaction time, the slides were washed and the conjugate was added. A donkey anti-sheep IgG Cy3 (Indocarbocyanine)-labelled antibody (Dianova, Hamburg, Germany) was used for small ruminants and a goat anti-bovine IgG Cy3-labelled antibody (Dianova) for buffalo samples, respectively. A polyclonal rabbit-anti-camel antiserum (Bethyl laboratories, Montgomery, TX, USA) was used as a second antibody for camel samples followed by a Cy3-conjugated goat anti-rabbit antibody (Dianova) according to the method described by Jäckel [11]. The slides were evaluated with a fluorescence microscope (Nikon, Tokyo, Japan).

IgM ELISA

To detect recent infection samples showing positive cELISA results (positive to IgG or IgM) in combination with negative IIFA results (detects only IgG) were further tested with the ID Screen® Rift Valley Fever IgM capture ELISA (ID VET, Montpellier, France) according to the manufacturer's instructions.

Interpretation of the results

Sera were classified as 'positive' when VNT alone or at least two assays produced consistent positive results. In a limited number of cases, due to low sample volume, application of combined analysis was not possible. Referred to the required serum volume for performing the different test systems, such sera were first tested by IIFA and then, if possible, confirmed by VNT and/or cELISA. If only one test could be carried out, the result of this test was used for classification. Combined test performance with result conclusion was considered to increase the total test accuracy. Positive results indicated seroconversion of the animals after contact to the antigen, independent of the point of exposure.

The true prevalence of each species/herds was calculated using the calculation tool of 'Working in Epidemiology' from the University of Zaragoza [20] (http://www.winepi.net/uk/index.htm).

Results

Antibodies against Rift Valley fever virus were detected with three different methods, an approach, which enabled the identification of antibodies against the nucleoprotein in case of competition ELISA, neutralizing antibodies (frequently against the glycoprotein Gn) in case of the VNT and of IIFA reactive antibodies.

Only two out of 438 sheep samples were tested positive displaying a seroprevalence of 0.46% (95% confidence interval [95%CI:] 0.414–0.5). In addition no seropositive goat was detected. Results from individual tests as well as the data interpretation are shown in Table 2. A detailed analysis of positive samples is shown in Table 3.

Four sera were subjected to IgM capture ELISA to detect recent infections. All sera gave negative results in this test.

Buffalo derived sera gained from four small holders in Dakahlia governorate revealed that 11 out of 88 sera were determined positive corresponding to a seroprevalence of 12.5% (95% CI: 11.25–13.75) (Tables 2 and 3). Individual small holding herds showed prevalence of 19.35 (95%CI: 15.93–22.78), 22.22% (95%CI: 22.22–22.22), 14.29% (95%CI: 11.09–17.48) and 0%, respectively. In contrast, no positive results were observed in 100 buffalo sera from the farm. The overall seropositivity for 188 buffalo samples was therefore 5.85% (95% CI: 2.75–8.95).

Finally also camel sera were subjected to serological analysis. Seven out of 71 sera of animals from the slaughterhouse showed positive results (five of them displayed ND_{50} values of 30 or higher [Table 3]); however none of the samples from imported camels were determined positive. The corresponding prevalence for camels from abattoir was 9.86% (95% CI: 2.92–16.79) and 3.17% (95% CI: 0.86–5.48) comprised the total camel number. The total seroprevalence of all investigated animals ($n = 873$) was 2.29% (95% CI: 1.30–3.28).

Discussion

The seroepidemiological study presented here provided insights into the current RVF antibody status in non-immunized livestock and enabled a preliminary evaluation of the exposure risk to the virus in susceptible host species.

Sheep are the most susceptible species to RVFV infections [9, 30] and often used as sentinel animals in endemic areas [29, 31]. Therefore the investigation of sheep samples in the Nile Delta, a vulnerable part of Egypt for new RVF epidemics was in the focus of this study. The presence of antibodies against RVFV in sheep samples was very low (0.46%). Studies conducted 7 years after the large epidemic/epizootic in 1977 encompassing 1714 sheep showed a prevalence of 1.2% in the governorate of Dakahlia [34]. These results are in line with the finding obtained in this study i.e. 12 years after the last RVF outbreak in Egypt which occurred in 2003.

Unlike to the results stated here, previous reports conducted in the present inter-epidemic period document a relatively higher seropositivity in small ruminants located in the Nile Delta of Egypt. Ramadan, [27] found in 2009 a prevalence of 19.9% ($n = 183$ sheep samples from

Dakahlia governorate) and also the results of Marawan, [28] in year 2012 indicated a considerably higher positivity of 12.3% ($n = 70$ sheep samples from the same governorate). Further studies in additional Nile Delta governorates showed likewise higher prevalence levels (a total of 17.6% in Qalyubia, Sharkia and Kafr el Sheikh in 2012; 21.4% in El Monofia, Beheira and Kafr el Sheikh in 2009/2010 [28, 35]). Deviating prevalence levels in small ruminants examined in 2009 and 2012 as compared to our data for sheep in 2015 indicate that the more recent virus circulation was negligible. Nevertheless, as a) none of these animals was imported and b) animals were transported only locally, a RVFV circulation can be assumed, especially also since all animals were born after the last reported RVFV outbreak.

The significant role of camels in transmission and spread of RVFV was already demonstrated during the Egyptian RVF outbreak in 1997. Due to often rare clinical manifestations in camels and the assumption that camels are less susceptible, camels, imported from other endemic countries in particular can carry the virus [7, 36]. Egypt imports camels for human consumption from endemic countries like Sudan or the Horn of Africa [21]. None of the imported camels included in the here presented study showed a positive result, whereas the collected specimens from abattoir-camels gave a prevalence of 9.86%. Elevated neutralization titers in these camels might suggest an antigen contact quite recently. Unfortunately there was no further information about the history of the origin of the abattoir-camels. Therefore it could not be ruled out, that those animals were previously imported. Previous studies in Egyptian camels during the present inter-epidemic period are conducted in 2009/2010 with 10 camel sera from an abattoir [35] and in 2012 encompassing 100 camels from Qalyubia governorate [28]. Likewise no further information about the origin of the animals is given in these reports. Both studies detected no antibodies in Egyptian camels. Only 10% of imported animals are tested against RVFV antibodies directly after the import as a routine disease control. Positive results in camels in this study were supposed to be imported positive animals from other endemic countries. The risk to import viremic animals should not be neglected; hence continuing investigations on the role of camels should be in focus of further investigations.

Buffalos presumably play a role as amplifying hosts during inter-epidemic periods. Antibodies against RVFV were detected in several studies in different African countries during inter-epidemic times (e.g. prevalence rates of 21, 15.6 and 12.5% in South Africa, Kenya and Botswana respectively [37–39]).

The analyzed buffalo sera in this study showed also a relatively slight prevalence of 5.85%. This finding corresponds to results obtained from Horton in 2009 [35],

Table 2 Serological analysis of Egyptian serum samples from small ruminants, buffalos and camels with cELISA (ID Vet Screen Rift Valley Fever competition ELISA (multispecies), indirect immunofluorescence assay (IIFA) and virus neutralization test (VNT) (A) Results grouped into species. Result conclusion (B). Detailed prevalence of herds is shown in part (C)

(A)

Species	ID screen® multispecies competition ELISA				Virus neutralization test			Immunofluorescence			
	Tested	Positive	Doubtful	Negative	Tested	Positive	Negative	Tested	Positive	Doubtful	Negative
Sheep	417	3	1	413	381	0	381	88	3	3	82
Goat	26	0	0	26	26	0	26	n.t.	n.t.	n.t.	n.t.
Buffalo (small holder)	78	8	0	70	56	2	54	32	8	0	24
Buffalo (farms)	95	1	0	94	88	0	88	16	0	0	16
Camels abattoir	1	1	0	0	59	6	53	71	2	0	69
Camels imported from Sudan	130	0	0	130	121	0	121	20	0	0	20
Total:	747	13	1	733	731	8	723	227	13	3	211

(B)

Result conclusion	animals in the herds	tested	positive	inconclusive	negative	prevalence (%)	95% CI
Sheep	440	438	2	1	435	0.46	0.41–0.5
Goat	17	26	0	0	26	0.00	0.00–0.00
Buffalo (small holder)	91	88	11	0	77	12.5	11.25–13.75
Buffalo (farm)	1200	100	0	0	100	0.00	0–2.83
Came l (abattoir)		71	7	0	64	9.86	2.92–16.79
Imported camel		150	0	0	150	0.00	0.00–0.00

(C)

Results in individual herds	animals in the herds	tested	positive	prevalence (%)	95% C I
sheep herd 4	78	78	1	1.28	1.28–1.28
sheep herd 5	42	40	1	2.5	1.44–3.56
buffalo small holder 1	33	31	6	19.35	15.93–22.78
buffalo small holder 2	9	9	2	22.22	22.22–22.22
buffalo small holder 3	22	21	3	14.29	11.09–17.48

which included 153 buffaloes from the Muneeb abattoir in central Egypt, who reported a prevalence of 3% and also to the results from Marawan, 2012, [28] who obtained a prevalence of 9.8% in 102 buffalo sera. Interestingly, there was a striking difference of the prevalence levels for different holding systems: it ranged from 0% in buffalo on the farm compared to animals owned by small holders (12.5%). Individual small buffalo holdings showed seroprevalences of up to 22%, which might be due to a lower health status of the animals in small holding flocks. Additionally, farmed buffalos were located in Ismailia governorate. Ismailia is located marginal to the

east lower Nile Delta, which could be a reason for lower presence of transmission of the virus to naïve animals.

Main vectors for the RVFV transmission in Egypt are mosquitos of genus Culex [40]. In 2009 Ramadan [27] collected these mosquitos from Dakahlia governorate and found 11.1% positive encompassing 806 mosquitos in 13 pools which indicates virus maintenance in the country. Due to more elaborate health management measures in larger farms, which might include also vector control programs, natural RVFV infections by Culex mosquitos in these farms are less likely as compared to small holder husbandry. We found a prevalence of 18%

Table 3 Individual positive results of Egyptian serum samples from small ruminants, buffalos and camels with cELISA (ID Vet Screen Rift Valley Fever competition ELISA (multispecies), indirect immunofluorescence assay (IIFA) and virus neutralization test (VNT)

Sample number			Herd	Age (years)	ID vet cELISA S/N%	IIFA	VNT ND50 value*	Result conclusion
Sheep								
EG	12	15/OV	1	2–7	*32,82*	negative	negative	negative
EG	247	15/OV	4	1–4	*17,79*	++	negative	*positive*
EG	256	15/OV	4	1–4	*16,24*	negative	negative	negative
EG	282	15/OV	4	1–4	49,48	negative	negative	negative
EG	299	15/OV	5	1–5	*not tested*	+/–	negative	negative
EG	304	15/OV	5	1–5	*not tested*	+/–	negative	negative
EG	321	15/OV	5	1–5	*not tested*	+/–	*not tested*	*inconclusive*
EG	328	15/OV	5	1–5	*not tested*	+	*not tested*	*positive*
EG	394	15/OV	7	1–5	*not tested*	++	negative	*inconclusive*
Buffalo - small holder								
EG	4	14/BF	1	3–7	88,81	negative	*10*	*positive*
EG	9	14/BF	1	3–7	*4,90*	not tested	*not tested*	*positive*
EG	14	14/BF	1	3–7	66,06	negative	*20*	*positive*
EG	15	14/BF	1	3–7	*4,03*	++	*not tested*	*positive*
EG	20	14/BF	1	3–7	*3,88*	++(+)	*not tested*	*positive*
EG	23	14/BF	1	3–7	*3,65*	++	*not tested*	*positive*
EG	34	14/BF	2	3–7	*4,01*	++(+)	*not tested*	*positive*
EG	38	14/BF	2	3–7	*18,46*	++(+)	*not tested*	*positive*
EG	41	14/BF	3	3–7	*37,71*	+	*not tested*	*positive*
EG	42	14/BF	3	3–7	*10,88*	++	*not tested*	*positive*
EG	53	14/BF	4	3–7	*not tested*	(+)	*not tested*	*positive*
Buffalo - farms								
EG	59	15/BF	1	4–6	8,80	negative	negative	Negative
Camel - abattoir								
EG	3	14/CM	abattoir	5–7	*not tested*	negative	*10*	*positive*
EG	4	14/CM	abattoir	5–7	*not tested*	negative	*>30*	*positive*
EG	18	14/CM	abattoir	5–7	*not tested*	negative	*>30*	*positive*
EG	32	14/CM	abattoir	5–7	*5,76*	++	*>30*	*positive*
EG	46	14/CM	abattoir	5–7	*not tested*	negative	*>30*	*positive*
EG	58	14/CM	abattoir	5–7	*not tested*	++	*not tested*	*positive*
EG	67	14/CM	abattoir	5–7	*not tested*	negative	*>30*	*positive*

cELISA results with a percentage of inhibition lower than 40 were defined as positive, between 40 and 50% as inconclusive and results higher than a percentage of inhibition of 50 were defined as negative. IIFA results were defined from low (+) to strong (+++) straining, or as inconclusive (+/–). ND_{50} values lower than 10 were defined as negative when performing VNT. Titers higher than ND_{50} values of 20 indicate a strong immune response
Positive and inconclusive findings and results are indiciated in italics

in buffalos sampled in 2014 ($n = 61$), whereas the small holder buffalos from 2015 ($n = 27$) showed only negative results. It would be of interest, to study in more detail, to which extent differences in the holding systems can influence the general infection risk for the animals. According to the farmer's documentation the buffalos were not imported from other endemic countries. Moreover buffalos were restricted in their movements and were only transported to local markets. Therefore also the data obtained for buffalos (similar to those found in small ruminants) suggests an active virus transmission in Dakahlia governorate during the present interepidemic period in Egypt.

Altogether, a comprehensive and well-designed surveillance program allows the early detection of first indications for the transition from endemic to epidemic cycle. Such a surveillance program should include investigations on the serological status of all susceptible animals in areas at risk, the monitoring of vectors and the intensification of import controls for animals coming from endemic countries.

Conclusions

The examination of 873 sera collected from sheep, goats, camels and buffalos gave an insight into the anti-RVFV antibody situation in Egypt during 2014/2015. All animals included in this study were born after the last Egyptian RVF epidemic in 2003 and, as to farmer's records, buffalos and sheep were not imported from other endemic countries. Therefore the antibody prevalence observed in buffalos and sheep were results of cryptic virus transmissions during the present inter-epidemic period. Based on a general low prevalence in all investigated animal species, a currently low level of circulating virus in the investigated areas can be assumed. Due to the general lack of detailed data about the role of camels for cryptic virus transmissions in Egypt further investigations are needed.

Assuming the high susceptibility of small ruminants to RVF, our data indicate, that small ruminants in Egypt are not the main source of inter-epidemic virus circulation, which means infections of alternative animal hosts that are less receptive to clinical manifestations.

Acknowledgement
The authors would like to thank Timo Homeier-Bachmann for preparation of the map.

Funding
We thank the German Office for Foreign Affairs for funding this study (German Partnership Program for Excellence in Biological and Health Security).

Authors' contributions
MG, MEA, MED and MEB realized the collection of well-documented ruminant sera. All authors read and approved the final manuscript.

Competing interests
The authors declare that they have no competing interests.

Consent for publication
Not applicable.

Ethics approval and consent to participate
The animals were sampled under the direction of the Mansoura University, Mansoura, Egypt, within the framework of the project 'Brucellosis, Q-fever and viral hemorrhagic fever infections in Egypt' [IIA6] during 2013 to 2015. Brucellosis, Q-fever and viral hemorrhagic fever infection like the Rift Valley fever virus are priority diseases for control and eradication by the Government of Egypt. Informed consents for RVFV investigations were given by all owners. All procedures were performed in accordance with the principles and specific guidelines presented in the Guidelines for the Care and Use of Agricultural Animals in Research and Teaching, 3rd ed.

Author details
[1]Institute of Novel and Emerging Infectious Diseases, Friedrich-Loeffler-Institut, Südufer 10, 17493 Greifswald - Isle of Riems, Germany. [2]Department of Hygiene and Zoonoses, Faculty of Veterinary Medicine, Mansoura University, Mansoura 35516, Egypt. [3]Department of Internal Medicine and Infectious Diseases, Faculty of Veterinary Medicine, Mansoura University, Mansoura 35516, Egypt. [4]Animal Health Research Institute-Mansoura Provincial Laboratory, Mansoura, Egypt. [5]Faculty of Veterinary Medicine, Mansoura University, Mansoura, Egypt.

References
1. Easterday BC. Rift Valley fever. Adv Vet Sci. 1965;10:65–127.
2. Pepin M, et al. Rift Valley fever virus (Bunyaviridae: Phlebovirus): an update on pathogenesis, molecular epidemiology, vectors, diagnostics and prevention. Vet Res. 2010;41(6):61.
3. Daubney R, Hudson JR, Garnham PC. Enzootic hepatitis or rift valley fever. An undescribed virus disease of sheep cattle and man from east africa. J Pathol. 1931;34:545–79.
4. Findlay GM, Daubney R. The virus of Rift Valley fever or enzootic hepatitis. Lancet. 1931;218(5651):1350–1.
5. Bird BH, et al. Rift Valley fever. J Am Vet Med Assoc. 2009;234(7):883–93.
6. Elliott RM. Emerging viruses: the Bunyaviridae. Mol Med. 1997;3(9):572–7.
7. Linthicum KJ, Britch SC, Anyamba A. Rift Valley Fever: an emerging mosquito-borne disease. Annu Rev Entomol. 2016;61:395–415.
8. Madani TA, et al. Rift Valley fever epidemic in Saudi Arabia: epidemiological, clinical, and laboratory characteristics. Clin Infect Dis. 2003;37(8):1084–92.
9. Chevalier V, et al. Rift Valley fever - a threat for Europe? Eurosurveillance. 2010;15(10):18–28.
10. Balkhy HH, Memish ZA. Rift Valley fever: an uninvited zoonosis in the Arabian peninsula. Int J Antimicrob Agents. 2003;21(2):153–7.
11. Jäckel S, et al. Molecular and serological studies on the Rift Valley fever outbreak in Mauritania in 2010. Transbound Emerg Dis. 2013;60(2):31–9.
12. Himeidan YE, et al. Recent outbreaks of Rift Valley fever in East Africa and the Middle East. Front Public Health. 2014;2:169.
13. Pienaar NJ, Thompson PN. Temporal and spatial history of Rift Valley fever in South Africa: 1950 to 2011. Onderstepoort J Vet Res. 2013;80(1):384.
14. Aradaib IE, et al. Rift Valley fever, Sudan, 2007 and 2010. Emerg Infect Dis. 2013;19(2):246–53.
15. Meegan JM. The Rift Valley fever epizootic in Egypt 1977–78. 1. Description of the epizzotic and virological studies. Trans R Soc Trop Med Hyg. 1979; 73(6):618–23.
16. Meegan JM, Hoogstraal H, Moussa MI. An epizootic of Rift Valley fever in Egypt in 1977. Vet Rec. 1979;105(6):124–5.
17. Darwish M, Hoogstraal H. Arboviruses infecting humans and lower animals in Egypt: a review of thirty years of research. J Egypt Publ Hlth Assoc. 1981;56:1–112.
18. Gerdes GH. Rift Valley fever. Rev Sci Tech. 2004;23(2):613–23.
19. Arthur RR, et al. Recurrence of Rift Valley fever in Egypt. Lancet. 1993;342:1149–50.
20. Abu-Elyazeed R, et al. Prevalence of anti-rift-valley-fever IgM antibody in abattoir workers in the Nile delta during the 1993 outbreak in Egypt. Bull World Health Organ. 1996;74(2):155–8.
21. Kamal AS. Observations on Rift Valley fever virus and vaccines in Egypt. Virol J. 2011;8(1):532.
22. Hanafi HA, et al. Virus isolations and high population density implicate Culex antennatus (Becker) (Diptera: Culicidae) as a vector of Rift Valley fever virus during an outbreak in the Nile Delta of Egypt. Acta Trop. 2011;119(2–3):119–24.
23. Abd el-Rahim IH, Abd el-Hakim U, Hussein M. An epizootic of Rift Valley fever in Egypt in 1997. Rev Sci Tech. 1999;18(3):741–8.
24. Drake JM, Hassan AN, Beier JC. A statistical model of Rift Valley fever activity in Egypt. J Vector Ecol. 2013;38(2):251–9.
25. Gerdes GH. Rift Valley fever. Vet Clin North Am Food Anim Pract. 2002; 18(3):549–55.
26. Roger M, et al. Evidence for circulation of the Rift Valley fever virus among livestock in the union of Comoros. PLoS Negl Trop Dis. 2014;8(7):e3045.
27. Ramadan HHE. Epidemiological study on Rift Valley fever as a zoonotic viral disease transmitted by arthropod and its public health importance (thesis).

In: Faculty of Veterinary Medicine, Department of Hygiene and Zoonoses. Mansoura: Mansoura University; 2009.

28. Marawan AM, et al. Epidemiological studies on Rift Valley fever disease in Egypt. Benha Vet Med J. 2012;23(1):171–84.

29. Al-Qabati AG, Al-Afaleq AI. Cross-sectional, longitudinal and prospective epidemiological studies of Rift Valley fever in Al-Hasa Oasis, Saudi Arabia. J Anim Vet Adv Vet Sci. 2010;9(2):258–65.

30. Kortekaas J, et al. European ring trial to evaluate ELISAs for the diagnosis of infection with Rift Valley fever virus. J Virol Methods. 2012;187(1):177–81.

31. Lichoti JK, et al. Detection of Rift Valley fever virus interepidemic activity in some hotspot areas of kenya by sentinel animal surveillance, 2009–2012. Vet Med Int. 2014;2014:379010.

32. FASS. Guide for the Care and Use of Agricultural Animals in Research and Teaching, Third Edition, January 2010. 2010 [cited 2016 1. April]; Available from: https://www.aaalac.org/about/Ag_Guide_3rd_ed.pdf.

33. OIE. Manual of Diagnostic Tests and Vaccines for Terrestrial Animals. 2015 [cited 2016 6. April]; Available from: http://www.oie.int/manual-of-diagnostic-tests-and-vaccines-for-terrestrial-animals/.

34. Botros BAM, et al. Rift Valley fever in Egypt 1986. Surveillance of sheep flocks grazing in the northeast Nile Delta. J Trop Med Hyg. 1988;91(4):183–8.

35. Horton KC, et al. Serosurvey for zoonotic viral and bacterial pathogens among slaughtered livestock in Egypt. Vector Borne Zoonotic Dis. 2014;14(9):633–9.

36. Davies FG, Koros J, Mbugua H. Rift Valley fever in Kenya: the presence of antibodies to the virus in camels (Camelus dromedarius). J Hyg (Lond). 1985;94:241–4.

37. LaBeaud AD, et al. Rift Valley fever virus infection in African buffalo (Syncerus caffer) herds in rural South Africa: evidence of interepidemic transmission. Am J Trop Med Hyg. 2011;84(4):641–6.

38. Evans A, et al. Prevalence of antibodies against Rift Valley fever virus in Kenyan wildlife. Epidemiol Infect. 2008;136(9):1261–9.

39. Jori F, et al. Serological evidence of Rift Valley fever circulation in domestic cattle and African buffalo in Northern Botswana (2010–2011). Fronties Vet Sci. 2015;2:63.

40. Zayed AB, et al. Mosquitoes and the environment in Nile delta villages with previous Rift Valley fever activity. J Am Mosq Control Assoc. 2015;31(2):139–48.

17

Seroprevalence of anti-*Toxoplasma gondii* antibodies in Egyptian sheep and goats

Yara M. Al-Kappany[1†], Ibrahim E. Abbas[1†], Brecht Devleesschauwer[2*] ⓘ, Pierre Dorny[3,4], Malgorzata Jennes[5†] and Eric Cox[5†]

Abstract

Background: Toxoplasmosis is a zoonotic disease that affects a wide range of animals, including small ruminants. Sheep and goats are considered as biological indicators for the contamination of the environment with *Toxoplasma gondii* oocysts. In addition, in countries such as Egypt, where sheep and goat meat is frequently consumed, *T. gondii* infection in small ruminants may also pose a public health risk. To establish baseline estimates of the prevalence of *T. gondii* infection in Egyptian small ruminants, we used an indirect immunofluorescence assay (IFA) and an enzyme-linked immunosorbent assay (ELISA) to assess the seroprevalence in 398 sheep from four Egyptian governorates (Cairo, Giza, Dakahlia and Sharkia) and in 100 goats from Dakahlia. The positive and negative agreements of both tests were calculated and the true prevalence was estimated using a Bayesian approach.

Results: The true prevalence of antibodies to *T. gondii* as determined by both tests was higher in Egyptian goats (62%) than in sheep for each province (between 4.1 and 26%). Sheep slaughtered at the Cairo abattoir had the lowest true prevalence (4.1%), while true prevalences in Dakahlia, Giza and Sharkia governorates (26%, 23% and 12%, respectively) were substantially higher.

Conclusions: The high prevalence of antibodies to *T. gondii* may indicate an important role of goat and sheep in the transmission of human toxoplasmosis in Egypt, given the habit of eating undercooked grilled mutton.

Keywords: Egypt, Filter paper, Small ruminants, *Toxoplasma gondii*, True prevalence

Background

Toxoplasmosis is a globally distributed zoonotic disease with important medical and economic implications for man and animals, respectively [1]. Infection of sheep and goats with *Toxoplasma gondii* may cause abortion, stillbirth and neonatal death [2]. *T. gondii* infection in small ruminants also poses a public health risk, since man can acquire *T. gondii* from infected sheep and goats through consumption of undercooked meat, drinking unpasteurized milk or handling raw meat [3].

Many studies have assessed the seroprevalence of *T. gondii* in sheep and goats in different parts of the world using different serological techniques [1]. In Egypt, around 10 million small ruminants are yearly produced with slightly more than 50% kept in small herds with less

than 10 animals, and 25% owned by people without agricultural land. The prevalence of *T. gondii* infection in sheep in Egypt has been shown to range between 34 and 100% [4, 5], and that in goats between 42 and 60% [5, 6]. However, these studies were limited to one of the Egyptian governorates (respectively, Cairo [4], Faiyum [5], and Giza [6]). Moreover, no previous report investigated the prevalence of *T. gondii* infection in the Dakahlia governorate, one of the Nile delta governorates (Fig. 1).

We aimed to assess the prevalence of anti-*T. gondii* antibodies in sera of Egyptian goats in the Dakahlia governorate and sheep in four Egyptian governorates (Cairo, Dakahlia, Giza and Sharkia).

Methods

Sample collection

Blood samples were obtained from 100 goats (reared at villagers' houses at Dakahlia governorate) and 398 sheep slaughtered at the main abattoir in four Egyptian governorates: Cairo (urban) ($n = 100$), Dakahlia ($n = 100$) and

---
* Correspondence: brechtdv@gmail.com; brecht.devleesschauwer@wiv-isp.be
[†]Equal contributors
[2]Department of Public Health and Surveillance, Scientific Institute of Public Health (WIV-ISP), Rue Juliette Wytsmanstraat 14, 1050 Brussels, Belgium
Full list of author information is available at the end of the article

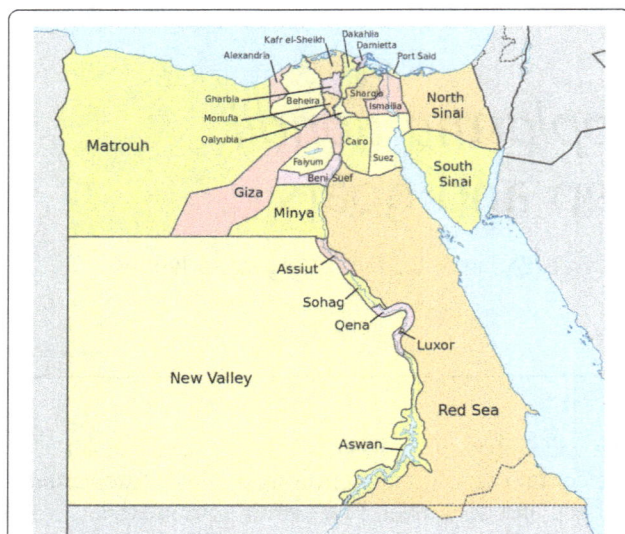

Fig. 1 Governorates of Egypt. CC BY-SA
3.0, https://commons.wikimedia.org/w/index.php?curid=32089528

Sharkia ($n = 99$) in the Nile delta and Giza ($n = 99$) (middle Egypt). The Cairo governorate is the most populated of the governorates and completely urbanized. Dakahlia and Sharkia governorates are both located in the Nile Delta, a highly populated agricultural region. The Giza governorate is one of the largest agricultural governorates in Egypt and represents middle Egypt. Sera were collected and stored at − 20 °C until preparation for transport to Belgium and further analysis. Long-term storage of serum normally needs refrigeration or freezing; furthermore, transport of sera from countries with transboundary animal diseases to other countries free of these diseases is not allowed. Therefore, sera were first filtered over a 0.2 μm filter (Pall Life Sciences, USA), thereby removing possible bacterial contaminants [7]. Then, 5 μl from each sample was spotted on Whatman filter paper No 4 (Whatman international Ltd., Maidstone, England), whereafter this paper was heat-treated in a household oven at 60 °C for 180 min to inactivate possible viral contaminants such as foot-and-mouth disease virus and Rift Valley fever virus [8, 9]. An import permit was obtained from the Federal Agency for the Safety of the Food Chain, Belgium (1069371) and the dried filter papers were transported in sealed bags to the Laboratory of Immunology, Faculty of Veterinary Medicine, Ghent University, Belgium. There, each serum spot was cut out the filter paper and dialyzed against 500 μl of phosphate buffered saline (PBS) supplemented with 0. 2% Tween®20 (PBS-Tw). Finally, this 1:100 serum dilution in PBS-Tw was used for antibody testing by an indirect immunofluorescence assay (IFA) and an enzyme-linked immunosorbent assay (ELISA).

Indirect immunofluorescence assay

Formalin-treated tachyzoites of an RH *T. gondii* strain coated on IFA slides (Toxo-Spot® IF, Bio-Mérieux, France) were incubated for 30 min at 37 °C with 50 μl of 1/100 in PBS diluted serum samples. After washing with PBS, the slides were incubated for 30 min at 37 °C with 30 μl of fluorescein isothiocyanate conjugated rabbit-anti-sheep IgG or rabbit-anti-goat IgG (Bethyl Laboratories. Inc., Montgomery USA), diluted in PBS-Evans Blue (counter dye). After washing and drying, the slides were interpreted using a fluorescence microscope. The cut-off read-out of the fluorescence test was established at a dilution of 1/50 with *Toxoplasma* negative and positive sheep reference sera (collected during the 2011 Maedi-Visna screening and validated with the MAT assay), according to the Toxo-Spot® IF guidelines.

Enzyme-linked immunosorbent assay

All collected sera were tested for the presence of IgG antibodies against *Toxoplasma* total lysate antigen (TLA) according to the ELISA described by Verhelst et al. [10]. Absorbance was read at 405 nm using an iMARK Microplate reader (Biorad, Nazareth, Belgium). The cut-off value, calculated as the mean optical density plus three times the standard deviation of three negative sheep and goat sera (collected during the 2011 Maedi-Visna screening and validated with the MAT assay) assayed at a 1/100 dilution, was 0.395 for sheep and 0. 159 for goat.

Data analysis

Based on the observed test results, we calculated prevalences and corresponding 95% exact confidence intervals (CI) using the prevalence package for R 3.4.0 [11, 12]. Agreement between the results of both serological tests was quantified as the positive agreement index, $PA = 2a/(2a + b + c)$, and the negative agreement index, $NA = 2d/(2d + b + c)$, with $\{a, b, c, d\}$ the cell values of the concerned two-by-two table [13]. Confidence intervals for PA and NA were obtained through bootstrapping using 1,000,000 Dirichlet random deviates, implemented via the mc2d package for R 3.4.0 [12, 14].

As serological assays may yield false positive or false negative results, the observed test results only represent an *apparent* prevalence estimate [15]. To account for the imperfectness of the serological assays and the lack of a gold standard assay, we estimated *true* prevalence in a Bayesian framework, taking into account external information on diagnostic sensitivity and specificity of both assays. We assumed a Beta (1, 1) prior for the true prevalence, and derived prior Beta distributions for the sensitivity and specificity of the IFA and ELISA from Shaapan et al. [4] (Table 1). A Uniform (− 0.25, 0.25) prior was used for the covariance between both tests.

Table 1 Prior information on diagnostic sensitivity and specificity of the applied assays, based on Shaapan et al. [4]

Test	Sensitivity			Specificity		
	Distribution	Mean	95% UI	Distribution	Mean	95% UI
IFA	Beta(82,20)	0.80	(0.72–0.87)	Beta(181,17)	0.91	(0.87–0.95)
ELISA	Beta(92,10)	0.90	(0.84–0.95)	Beta(170,28)	0.86	(0.81–0.90)

UI uncertainty interval, *IFA* indirect immunofluorescence assay, *ELISA* enzyme-linked immunosorbent assay

Models were implemented independently for the five datasets. For each model, we simulated two chains of 20,000 iterations, of which the first 10,000 were discarded as burn-in. Convergence of the models was visually assessed using trace and density plots and numerically using the multivariate potential scale reduction factor [16]. The models were implemented in R 3.4. 0 [12] using the prevalence package version 0.4.0 [11].

Results

Apparent prevalence and between-test agreement
Table 2 shows the apparent test results per population and per diagnostic test, while Table 3 shows the cross-classification of samples based on both diagnostic tests. Across sheep sera, a positive agreement of 0.78 (95%CI 0.71–0.83) and a negative agreement of 0.92 (0.90–0.94) was found. Across goat sera, a positive agreement of 0. 94 (95%CI 0.88–0.97) and a negative agreement of 0.92 (0.85–0.97) was found.

True prevalence of anti-toxoplasma gondii antibodies in sheep and goats
The true prevalence of anti-*T. gondii* antibodies in the sera from the Egyptian goats was 62% (95% uncertainty interval [UI] 52–73%). The true prevalence in sheep varied between 4.1% and 26% in the four governorates. Specifically, Cairo had the lowest true prevalence (4.1%; 95%UI 0.2–11%), while higher and more similar prevalences were noted in Dakahlia (26%; 95%UI 16–36%) and Giza (23%; 95%UI 14–33%). An intermediate value was found for Sharkia governorate (12%; 95%UI 3.0–21%).

Discussion

T. gondii is widely prevalent in Egyptian sheep and goats, which may indicate an important role of small ruminants in the transmission of human toxoplasmosis in Egypt, especially when considering the Egyptian people's habit of eating undercooked grilled mutton and goat meat (*Kabab* and *Kofta*) [17]. The true prevalence was approximately two-fold higher in goats than in sheep in Dakahlia. The higher prevalence in goats was in line with Barakat et al. [6], who found prevalences of 55% and 44% in 306 goats and 320 sheep, respectively, from the Giza governorate; however, Ghoneim et al. [5] found prevalences of 42% and near 100% in 10 goats and 61 sheep from the Faiyum governorate, but this could have been due to the very low sample size. A higher *T. gondii* infection in goats than in sheep may be due to differences in herding practices [18], although this requires further investigation. The observed difference may also be confounded by a different age at slaughter; unfortunately, information on the age of the sampled animals was not available. Nonetheless, this high prevalence underlines the need to give extra attention to goats as a source for infection of humans, which is estimated to be 51% among pregnant women in Egypt [19]. Indeed, in most of the Egyptian rural areas, goat meat and milk are important human food products, which could facilitate the spread of toxoplasmosis to humans.

Our sheep study was carried out over four Egyptian Provinces: Cairo, Sharkia, Dakahlia, and Giza. The latter two governorates showed a nearly similar true prevalence (26%, and 23%, respectively), while an intermediate (12%) and very low value (4.1%) was found for Sharkia and Cairo, respectively. This may be attributed to the fact that Cairo (Capital of Egypt) is an urban area, with a lower cat density, unlike the three other rural governorates where cats are kept by the villagers as natural predators. Previous studies found 44% positive sheep sera in the Giza governorate [6], versus 46% and 42% in the Cairo governorate [4, 20]; thus not suggesting a lower prevalence in the urbanized Cairo governorate than in the rural governorates. However, no comparison with the other rural governorates was performed within the same study. More extensive sampling of sheep over time in different slaughterhouses in Cairo and in parallel in one of the three rural governorates will be needed to confirm our observation.

Our prevalences in sheep were lower than in other studies. This could be due to the specific pre-treatment

Table 2 Positive samples (*x*) and apparent prevalence (*AP*, %) with 95% exact confidence interval (CI) for *Toxoplasma gondii* infection by population and diagnostic test

Diagnostic test	Goats, Dakahlia (n = 100)		Sheep, Cairo (n = 100)		Sheep, Dakahlia (n = 100)		Sheep, Giza (n = 99)		Sheep, Sharkia (n = 99)	
	x	AP (95%CI)	x	AP (95%CI)	x	AP (95%CI)	x	AP (95%CI)	x	AP (95%CI)
IFA	54	54 (44–64)	20	20 (13–29)	38	38 (28–48)	32	32 (23–42)	34	34 (25–45)
ELISA	59	59 (49–69)	12	12 (6.4–20)	27	27 (19–37)	26	26 (18–36)	17	17 (10–26)

IFA indirect immunofluorescence assay, *ELISA* enzyme-linked immunosorbent assay

Table 3 Classification of samples according to indirect immunofluorescence assay (IFA) and enzyme-linked immunosorbent assay (ELISA)

Diagnostic test		Goats, Dakahlia	Sheep, Cairo	Sheep, Dakahlia	Sheep, Giza	Sheep, Sharkia
IFA	ELISA					
1	1	80	12	27	24	17
1	0	44	8	11	8	17
0	1	2	0	0	2	0
0	0	272	80	62	65	65

1 = positive; 0 = negative

of the serum samples. Nonetheless, previous studies demonstrated a close relation between results of serological tests on blood and serum in comparison with serum elutes from filter paper [8, 21, 22]; furthermore, we clearly showed that the sample treatment still allowed demonstrating *T. gondii*-specific antibodies. However, filter paper elutes have not been used before for demonstrating antibodies against *T. gondii* in sheep and goats. Whether the treatment decreases antibody concentrations and consequently also sensitivity should therefore be assessed. Nonetheless, since all samples were treated in a similar way in the present study, comparison between samples of different governorates remained possible.

A problem of comparing different seroprevalence studies on *T. gondii* infection remains the wide range of serological assays used [23–25]. In humans the Sabin-Feldman test is considered the gold standard, but is expensive and holds risks for the user because of the use of live tachyzoites [26]; the test is therefore rarely used in sheep [27]. More often, IFA, ELISA, an indirect hemagglutination assay or a modified agglutination test have been used [25]. IFA and ELISA are fast techniques with high sensitivity and specificity, suitable for epidemiological studies [26]. Previous studies applying the TLA-ELISA on sheep serum samples in the Netherlands [28] and in Cairo Egypt [4] reported sensitivities of 97. 8% and 90.1%, respectively, and specificities of 96.4% and 85.9%, respectively. Furthermore, Verhelst et al. [25] found a 100% agreement between the Toxo-spot IF® and the TLA-ELISA when applied to Belgian sheep sera. This was not the case in the present study, in which the positive agreement was 78% and 94% for sheep and goat sera, respectively, and the negative agreement was 92% for both sheep and goat sera. Several factors could be responsible for this difference, such as the sampled sheep breeds, the *T. gondii* strains infecting sheep in the field, or the presence of cross-reacting protozoa.

Even though the TLA-ELISA and IFA have been used in several studies, they have not been truly validated for use in sheep and goats. Therefore, in most studies several serological assays were performed and results combined

using Bayesian modeling [25, 28]. Nonetheless, there is a high need for validation and standardization of serological assays for seroprevalence studies on *T. gondii* in different species throughout the world. The accurate measurement of the seroprevalence in a comparative way would be an invaluable help to identify risk factors for infections in different regions and at a flock level. Furthermore, it is important to note that serological responses are only indicative of exposure to parasite antigens, but are not a definitive indication that the animals that tested positive are actually infected and capable of transmitting the infection to humans. For any definitive statement on the human health risk to be made samples of meat or milk would need to be tested by mouse bioassay to prove infectivity.

Conclusion

For the first time, we compared the prevalence of *T. gondii* infection in sheep across different Egyptian governorates, and demonstrated the prevalence of *T. gondii* infection in small ruminants from the Dakahlia governorate, one of the Nile delta governorates. The high prevalence of antibodies to *T. gondii* may indicate an important role of sheep and goat in the transmission of human toxoplasmosis in Egypt. Further validation of serological assays and methods, as well as molecular characterization studies, may provide further evidence on the public health risk of small ruminant toxoplasmosis in Egypt.

Abbreviations
AP: Apparent prevalence; CI: Confidence interval; ELISA: Enzyme-linked immunosorbent assay; IFA: Indirect immunofluorescence assay; NA: Negative agreement index; PA: Positive agreement index; PBS: Phosphate buffered saline; PBS-Tw: Phosphate buffered saline supplemented with 0.2% Tween®20; TLA: Total lysate antigen; UI: Uncertainty interval

Acknowledgements
Not applicable.

Funding
Yara Al-Kappany is the recipient of a scholarship from Erasmus Mundus Action 2 Welcome project funded by the European commission. The funding body had no role in the design of the study and collection, analysis, and interpretation of data and in writing the manuscript.

Authors' contributions
IEA, PD, EC conceived and designed the study. YMA, IEA collected the samples. MJ performed the laboratory analyses. BD analyzed the data. All authors critically appraised and interpreted the results. IEA drafted the first version of the manuscript. All authors provided feedback on the manuscript, and read and approved the final version.

Consent for publication
Not applicable.

Competing interests

BD is a member of the editorial board (Associate Editor) of BMC Veterinary Research. All other authors declare that they have no competing interests.

Author details

[1]Parasitology Department, Faculty of Veterinary Medicine, Mansoura University, Mansoura, Egypt. [2]Department of Public Health and Surveillance, Scientific Institute of Public Health (WIV-ISP), Rue Juliette Wytsmanstraat 14, 1050 Brussels, Belgium. [3]Laboratory of Parasitology, Faculty of Veterinary Medicine, Ghent University, Merelbeke, Belgium. [4]Department of Biomedical Sciences, Institute of Tropical Medicine, Antwerp, Belgium. [5]Laboratory of Immunology, Faculty of Veterinary Medicine, Ghent University, Merelbeke, Belgium.

References

1. Dubey JP. Toxoplasmosis in sheep—the last 20 years. Vet Parasitol. 2009a;163:1–4.
2. Buxton D, Maley SW, Wright SE, Rodger S, Bartley P, Innes EA. *Toxoplasma gondii* and ovine toxoplasmosis: new aspects of an old story. Vet Parasitol. 2007;149:25–8.
3. Boyer KM, Holfels E, Roizen N, Swisher C, Mack D, Remington J, et al. Risk factors for *Toxoplasma gondii* infection in mothers of infants with congenital toxoplasmosis: implications for prenatal management and screening. Am J Obstet Gynecol. 2005;192:564–71.
4. Shaapan RM, El-Nawawi FA, Tawfik MA. Sensitivity and specificity of various serological tests for the detection of *Toxoplasma gondii* infection in naturally infected sheep. Vet Parasitol. 2008;153:359–62.
5. Ghoneim NH, Shalaby SI, Hassanain NA, Zeedan GSG, Soliman YA, Abdalhamed AM. Comparative study between serological and molecular methods for diagnosis of toxoplasmosis in women and small ruminants in Egypt. Foodborne Pathog Dis. 2010;7:17–22.
6. Barakat AMA, Elaziz MMA, Fadaly HA. Comparative diagnosis of toxoplasmosis in Egyptian small ruminants by indirect hemagglutination assay and Elisa. Global. Veterinaria. 2009;3:9–14.
7. Devleesschauwer B, Pruvot M, Joshi DD, De Craeye S, Jennes M, Ale A, et al. Seroprevalence of zoonotic parasites in pigs slaughtered in the Kathmandu Valley of Nepal. Vector Borne Zoonotic Dis. 2013;13:872–6.
8. Turner C, Williams SM, Cumby TR. The inactivation of foot and mouth disease, Aujeszky's disease and classical swine fever viruses in pig slurry. J Appl Microbiol. 2000;89:760–7.
9. Bellini S, Rutili D, Guberti V. Preventive measures aimed at minimizing the risk of African swine fever virus spread in pig farming systems. Acta Vet Scand. 2016;58:82.
10. Verhelst D, De Craeye S, Dorny P, Melkebeek V, Goddeeris B, Cox E, et al. IFN-γ expression and infectivity of *Toxoplasma* infected tissues are associated with an antibody response against GRA7 in experimentally infected pigs. Vet Parasitol. 2011;179:14–21.
11. Devleesschauwer B, Torgerson P, Charlier J, Levecke B, Praet N, Roelandt S, et al. Prevalence: tools for prevalence assessment studies. R package version 0.4.0. 2015. http://cran.r-project.org/package=prevalence .
12. R Core Team. R: A language and environment for statistical computing. Vienna, Austria: R Foundation for Statistical Computing; 2017. http://www.R-project.org/.
13. Graham P, Bull B. Approximate standard errors and confidence intervals for indices of positive and negative agreement. J Clin Epidemiol. 1998;51:763–71.
14. Pouillot R, Delignette-Muller M-L. Evaluating variability and uncertainty in microbial quantitative risk assessment using two R packages. Int J Food Microbiol. 2010;142:330–40.
15. Speybroeck N, Devleesschauwer B, Joseph L, Berkvens D. Misclassification errors in prevalence estimation: Bayesian handling with care. Int J Public Health. 2013;58(5):791.
16. Brooks SP, Gelman A. General methods for monitoring convergence of iterative simulations. J Comput Graph Stat. 1998;7:434–55.
17. Hassan-Wassef H. Food habits of the Egyptians: newly emerging trends. East Mediterr Health J. 2004;10:898–915.
18. Tegegne D, Abdurahaman M, Yohannes M. Seroepidemiology and associated risk factors of *Toxoplasma gondii* in sheep and goats in southwestern Ethiopia. BMC Vet Res. 2016;12:280.
19. Ibrahim HM, Huang P, Salem TA, Talaat RM, Nasr MI, Xuan X, et al. Short report: prevalence of *Neospora caninum* and *Toxoplasma gondii* antibodies in Northern Egypt. Am J Trop Med Hyg. 2009;80:263–7.
20. El-Ghaysh AA, Mansour MM. Detection of antibodies to *Toxoplasma gondii* in an Egyptian sheep-herd using modern serological techniques. J Egypt Assoc Immunol. 1994;1:117–21.
21. Uggla A, Nilsson LA. Evaluation of a solid-phase immunoassay (DIG-ELISA) for the serodiagnosis of ovine toxoplasmosis. Vet Immunol Immunopathol. 1987;14:309–18.
22. Patel B, Holliman RE. Antibodies to *Toxoplasma gondii* in eluates from filter paper blood specimens. Br J Biomed Sci. 1994;51:104–8.
23. Dubey JP. Toxoplasmosis of animals and man. 2nd ed. Boca Raton, FL: CRC Press; 2009b.
24. Sakata FB, Bellato V, Sartor AA, de Moura AB, de Souza AP, Farias JA. *Toxoplasma gondii* antibodies sheep in Lages, Santa Catarina, Brazil, and comparison using IFA and ELISA. Rev Bras Parasitol Vet. 2012;21:196–200.
25. Verhelst D, De Craeye S, Vanrobaeys M, Czaplicki G, Dorny P, Cox E. Seroprevalence of *Toxoplasma gondii* in domestic sheep in Belgium. Vet Parasitol. 2014;205:57–61.
26. Su C, Dubey JP. Toxoplasma gondii. In: Liu D, editor. Molecular detection of foodborne pathogens. Boca Raton, FL: CRC Press; 2010. p. 741–51.
27. Cicek H, Babur C, Eser M. Seroprevalence of *Toxoplasma gondii* in Pirlak sheep in the Afyonkarahisar Province of Turkey. Turkiye Parazitol Derg. 2011;35:137–9.
28. Opsteegh M, Teunis P, Mensink M, Zuchner L, Titilincu A, Langelaar M, et al. Evaluation of ELISA test characteristics and estimation of *Toxoplasma gondii* seroprevalence in Dutch sheep using mixture models. Prev Vet Med. 2010;96:232–40.

18

Allogeneic mesenchymal stem cells improve the wound healing process of sheep skin

T. Martinello[1], C. Gomiero[1], A. Perazzi[2], I. Iacopetti[2], F. Gemignani[2], G. M. DeBenedictis[2], S. Ferro[1], M. Zuin[3], E. Martines[3], P. Brun[4], L. Maccatrozzo[1], K. Chiers[5], J. H. Spaas[6] and M. Patruno[1]* iD

Abstract

Background: Skin wound healing includes a system of biological processes, collectively restoring the integrity of the skin after injury. Healing by second intention refers to repair of large and deep wounds where the tissue edges cannot be approximated and substantial scarring is often observed. The objective of this study was to evaluate the effects of mesenchymal stem cells (MSCs) in second intention healing using a surgical wound model in sheep. MSCs are known to contribute to the inflammatory, proliferative, and remodeling phases of the skin regeneration process in rodent models, but data are lacking for large animal models. This study used three different approaches (clinical, histopathological, and molecular analysis) to assess the putative action of allogeneic MSCs at 15 and 42 days after lesion creation.

Results: At 15 days post-lesion, the wounds treated with MSCs showed a higher degree of wound closure, a higher percentage of re-epithelialization, proliferation, neovascularization and increased contraction in comparison to a control group. At 42 days, the wounds treated with MSCs had more mature and denser cutaneous adnexa compared to the control group. The MSCs-treated group showed an absence of inflammation and expression of CD3+ and CD20+. Moreover, the mRNA expression of hair-keratine (hKER) was observed in the MSCs-treated group 15 days after wound creation and had increased significantly by 42 days post-wound creation. Collagen1 gene (Col1α1) expression was also greater in the MSCs-treated group compared to the control group at both days 15 and 42.

Conclusion: Peripheral blood-derived MSCs may improve the quality of wound healing both for superficial injuries and deep lesions. MSCs did not induce an inflammatory response and accelerated the appearance of granulation tissue, neovascularization, structural proteins, and skin adnexa.

Keywords: Wound healing, Mesenchymal stem cells, Cell therapy, Regenerative medicine

Background

Skin is a multilayer organ that primarily functions as a protective barrier against the external environment, preventing dehydration and the penetration of external microorganisms [1]. Loss of the integrity of large portions of the skin, as a result of injury, may result in health issues, and poor quality of life [2]. Wound healing is a complex process that begins after injury and proceeds through three phases: hemostasis and inflammation, proliferation, and remodeling [3–5]. These phases are regulated by various cells, cytokines, and growth factors regulate these phases [3–5].

Wound healing re-establishes the skin's tensile strength and natural barrier function [6, 7]. Dysfunctional healing can lead to lifelong disability and an economic impact on breeding [8, 9]. To optimize wound healing, cell therapy may be an option for treating extensive and chronic wounds. The presence of mesenchymal stem cells (MSCs) in normal skin [10, 11] and their role in natural wound healing [11, 12] indicates that the use

* Correspondence: marco.pat@unipd.it
[1]Department of Comparative Biomedicine and Food Science, University of Padua, Viale dell'Università 16, 35020, Legnaro – Agripolis, Padua, Italy

of exogenous MSCs might be a means to treat wounds. MSCs are self-renewing, expandable, and able to differentiate into different cell lineages such as osteoblasts, adipocytes, chondrocytes, tenocytes, and myocytes [13–15]. Although bone marrow is one of the most common sources used to obtain MSCs [16, 17], other less invasive sources were used, such as peripheral blood, adipose tissue, and skin [11, 13, 14, 18–22].

The involvement of MSCs in the wound-healing process is significant. MSCs may regulate and improve the three phases of wound healing [23], contribute to the reduction of inflammation [7, 24], promote angiogenesis, reduce excessive wound contraction, attenuate scar formation [7, 25], and stimulate cell movement during epithelial remodeling [8]. Moreover, the immunosuppressive properties of MSCs allow for their potential use in allogeneic therapy. Although stem cell involvement in cutaneous wound healing has been studied in rodent models [22, 25, 26], this process has not been evaluated extensively in large animal models.

The aim of this study was to evaluate the specific effects of allogeneic MSCs in a sheep surgical wound model based on clinical, histopathological and molecular analyses. Moreover, macroscopic and microscopic study were carried out for testing the improvement of the regenerate tissue in the presence of MSCs, in the context of natural regeneration.

Methods
Animal model
Six female Bergamasca sheep of similar size and age, provided by a local farm, were acclimated to a stall (MAPS Department, University of Padua, Legnaro, Italy) for 2 weeks. Parasitological and biochemistry examinations were performed. The experiment was approved by The Body for the Protection of Animals (OPBA), ministerial decree n° 51/2015-PR released by the Health Department of Italy. Sheep were chosen because they are less neurologically developed than carnivores and equines and have sufficient superficial space on their backs for creation of experimental lesions. Moreover, sheep are also considered a possible research animal model for human medicine too [27–29]. The number of sheep was chosen based on sample size calculation and the "3Rs" principle (replacement-reduction-refinement) [30]. At the end of project, the animals were not sacrificed but located in a teaching farm.

Isolation of peripheral blood derived MSCs (PB-MSCs) from sheep
MSCs were isolated from the peripheral blood (PB) of six sheep that were not part of the wound model experimental design. From each animal, 100 ml of blood was taken from the jugular vein using a vacutainer containing the anticoagulant Li-heparin. The mononuclear cells were isolated using the protocol of Martinello et al. [13]. Briefly, the blood was diluted 1:1 with PBS (phosphate-buffered saline) and placed on Ficoll-paque solution (Amersham Biosciences) to obtain mononuclear cells in interphase after centrifugation. Cultures were maintained at 37 °C with 5% CO_2 in growth medium (DMEM 5671, Sigma-Aldrich) with 10% FCS (fetal bovine serum, Euroclone). On the day of application, PB-MSCs were trypsinized with 0.25% trypsin-EDTA (Euroclone, Italy) and resuspended in hyaluronic acid (Hyalgan®, Fidia).

Experimental design
In respect of the 3Rs principle [30], six lesions were performed according to the protocol established by Broeckx et al. [31], equidistant and symmetrical with each other on the back of six sheep to analyze the effect of five different therapeutic treatments (three conventional topical cream or gel, cold ionized plasma and MSCs). The distance of each lesion did not influence the result of trials. Six full-thickness square wounds (4×4 cm) were created on the back of six sheep. The lesions were created using a scalpel and a square guide model under sterilize surgical condition while the animals were under general anesthesia with analgesia [31]. In the present study, only the PB-MSCs treatment was evaluated and compared to phosphate saline buffer (PBS), used as placebo treatment.

In all six sheep, 1×10^6 PB-MSCs diluted in 1 ml of PBS were injected in the margins of the lesion dedicated to the MSCs study, and 1×10^6 PB-MSCs diluted in 1 ml of hyaluronic acid were topically applied at the center of the same lesion. In the control lesions of all six sheep, PBS only was administered topically to the wounds. After the application of the treatments, the lesions were bandaged with sterile gauze using the "wet-to-dry" method. The wounds were cleaned daily with PBS and the bandages were changed daily.

At two different time points (15 and 42 days from the induction of the lesions), two samples for each lesion treated with PB-MSCs and two samples for each control lesion were collected by means of a 6-mm punch biopsy with appropriate sedation and analgesic drug administration. Of the two collected samples for each time point, one was used for histopathological and immunohistochemistry protocols and one for molecular analyses.

Clinical evaluation
Lesion appearance was documented daily with photographs, using a ruler to measure wound size. Every week, the same-blinded investigator performed a clinical evaluation of the study animals. The observations were catalogued using the scoring system developed by Hadley et al. [32] (Table 1). The percentages of re-epithelialization and

Table 1 Skin-healing parameters scored in the experiment

Parameter	Score
Presence of exudate	1 absent
	2 small
	3 moderate
	4 abundant
Color of exudate	1 clear
	2 pink/red
	3 brown
	4 yellow
	5 green
Character of exudate	1 serous
	2 serosanguineous
	3 sanguineous
	4 purulent +
	5 purulent ++
	6 purulent +++
Gauze	1 dry/clean
	2 dry/stained
	3 moist
	4 wet
Hydration	1 Normal
	2 Maceration +
	3 Maceration ++
	4 Desiccation +
	5 Desiccation ++

wound contraction were measured at 7, 14, 21, 28 and 42 days post-wound creation.

Histopathological analysis

All 24 biopsy samples (6 PB-MSC at day 15, 6 PB-MSC at day 42, 6 control at day 15, 6 control at day 42) were used for histological evaluation and were glowed in OCT (Kaltek) and frozen in isopentane and liquid nitrogen. Samples were cut with cryostat into 5 µm slices before being mounted on slides and stained with Hematoxylin and Eosin (H&E). In order to obtain a full thickness examination, all samples were examined at difference depth (six chosen points). The presence of dermal and subcutaneous infiltrates, (immature) granulation tissue, undifferentiated mesenchymal tissue, and the development of adnexa were evaluated and scored using a 0 to 4 scale (0 absence, 1 presence, 2 small amount, 3 moderate amount, 4 abundant amount). Data were presented as percentage of relative frequency of the assigned values and calculated for each subject and for each parameter.

Immunohistological evaluation

The serial slices used for histopathological analysis were immunostained with polyclonal rabbit anti-human CD3 (Dako, 1:100), polyclonal rabbit anti-human CD20 (Thermo Fisher, 1:100), monoclonal mouse anti-human MHCII (Dako, 1:40), monoclonal mouse anti-human Ki67 (Dako, 1:10), and monoclonal rabbit anti-human vWF (Dako; 1:3200) antibodies. Immunolabeling was achieved with a high-sensitive horseradish spell out (PO) mouse or rabbit diaminobenzidine kit, with blocking of endogenous PO (Envision DAB+kit; Dako) in an autoimmunostainer (Cytomation S/N S38–7410-01; Dako). An antibody diluent (Dako), with background-reducing components was used to block hydrophobic interactions. The average of three fields from each slice was used to quantitatively evaluate different immunohistological parameters and all measurements were performed with a computer-based program (Leica microscope DM LB2 with Leica Application Suite LAS V4.0) using 20× magnification.

Real-time PCR analysis of Col1α1 and hKER gene expression

All 24 biopsy samples (6 PB-MSC at day 15; 6 PB-MSC at day 42; 6 control at day 15; 6 control at day 42) were used for molecular biology. Total RNA extraction was performed using Trizol (Life Technologies) reagent and quantified on a Nanodrop spectrophotometer (Thermo Scientific). The complementary DNA (cDNA) was synthetized to perform Real-Time PCR using the ABI 7500 Real-Time PCR system (Applied Biosystems) to evaluate Collagen 1α1 (Col1α1) and hair keratin (hKER) gene expression. All samples were tested in triplicate and untreated skin was used as a calibrator sample. The 2-ΔΔct method was used to analyze and normalize the RNA expression of the target genes with respect to endogenous housekeeping genes.
RPS24 - forward 5′ TTTGCCAGCACCAACGTTG 3′, reverse 5′ AAGGAACGCAAGAACAGAATGAA 3′, 18S - forward 5′ AAACGGCTACCACATCCAAG 3′, reverse 5′ TCCTGTATTGTTATTTTTCGTCAC 3′.

PCR primers were designed using Primer Express 3.0 software (Applied Biosystems). The sequences for the forward and reverse primers specific for each mRNA were as follows:
COL1α1 – forward 5′ GTACCATGACCGAGACGTGT 3′, reverse 5′ AGATCACGTCATCGCACAGCA 3′;
hKER – forward 5′ TGGTTCTGTGAGGGCTCCTT 3′, reverse 5′ GGCGCACCTTCTCCAGGTA 3′.

Statistical analysis

Data on clinical, histological, molecular, and immunohistochemical parameters were analysed using PROC MIXED, with animal as a random effect and repeated effect. The statistical linear model included the fixed effect of treatment (MSCs vs Placebo), time (week1, 2, 3, 4, 5, 6) and their interaction. The assumptions of the linear model were graphically inspected using residuals plots. For data that were not normally distributed (Shapiro-Wilks test < 0.90), the Mann-Whitney test was used (wound closure time, % of re-epithelialization and contraction, presence of exudate). The level of statistical significance was set at $p < 0.05$.

Results
Assessment of the healing process
Wound closure time for the PB-MSCs treated wounds was slightly quicker than that of the control group, average of wound closure time of six sheep was respectively 30.05 and 31.80 days (Fig. 1a-b). However, this was not a significant difference. Two weeks after wound creation, all animals in both the PB-MSCs-treated group and the control group had less than 40% re-epithelialization. Between day 14 and 28, the PB-MSCs-treated lesions had a higher percentage of re-epithelialization in comparison with the control group (58.69% vs 49.89% at day 21 and 93.5% vs 87% at day 28). However, this was not a significant difference. After 42 days of treatment, all wounds had 100% re-epithelialization(Fig. 2a). After two weeks of treatment, the PB-MSCs-treated wounds showed 81% contraction compared to 78% for the control PBS group. However, this was not a significant difference. All lesions had 100% contraction after 42 days of treatment (Fig. 2b).

Evaluation of skin-healing parameters
The PB-MSCs-treated wounds had a slight, but not-significant increase, in exudate compared to the control group. By the second week, exudate was absent from all lesions in both groups. For all lesions, the color of the exudate was pink/red and changed from serosanguinous to sanguineous over the course of the study.

During the first week post-wound creation, the gauze from all PB-MSCs-treated wounds was dry and clean while those of the PBS control group were slightly moist. However, this was not a significant difference. The wounds, from both groups, showed a normal state of hydration.

Histopathological examination
Dermal inflammation: at day 15, 33% of PB-MSCs-treated wounds presented with a moderate amount of dermal inflammation, while 67% presented with a small amount. In comparison, after 15 days, 50% of the control group presented with a moderate amount of dermal inflammation and 50% presented with a small amount. After 42 days, inflammation was completely absent in the PB-MSCs treated group, while 60% of the control group presented with a small amount of inflammation.

Subcutaneous inflammation: at day 15, 83% of PB-MSCs-treated wounds contained a small amount of subcutaneous inflammation. In contrast, 17% of the control group presented with moderate and 67% presented with a small amount of inflammation. After 42 days, subcutaneous inflammation was absent in all samples.

Immature granulation tissue: at day 15, all of the wounds in both groups presented an abundant amount

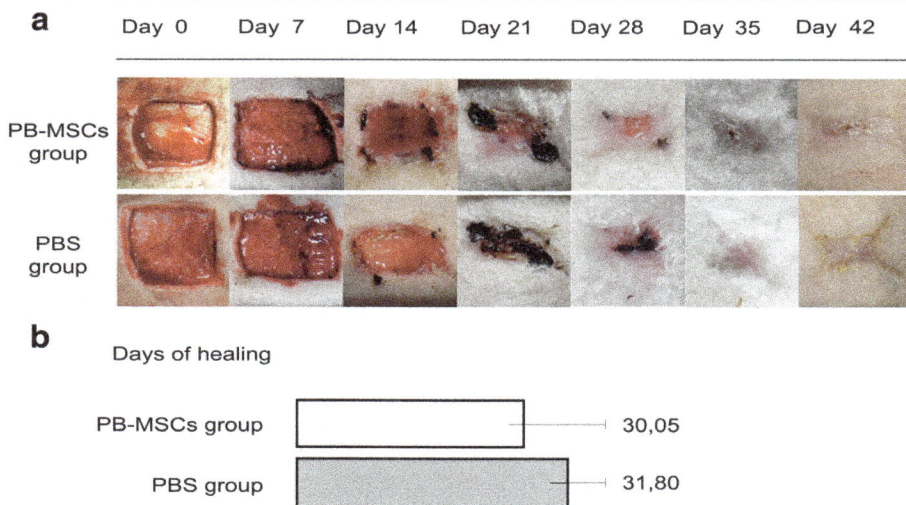

Fig. 1 Macroscopic analysis and the percentage of days of healing. **a** Serial macroscopic images of the wound site at different time points after PB-MSCs and PBS treatment. Between day 21 and 28, a smaller wound diameter and higher wound closure rate was observed in PB-MSCs-treated wounds. **b** The panel represents the percentage of days of healing. The wound closure time of the PB-MSC treated wounds (30,05 days) was slightly faster respect than the PBS-treated group (31,80 days)

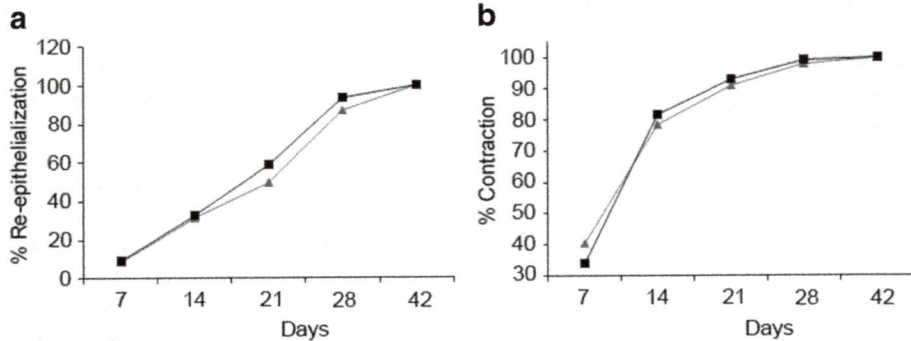

Fig. 2 Re-epithelialization and skin contraction. **a** The percentage of re-epithelization. **b** Percentage of contraction after 7, 14, 21, 28 and 42 days of treatment. PB-MSCs-treated wounds trend is represented by black lane, while PBS control group is indicated in grey lane

of immature granulation tissue (Fig. 3). Granulation tissue was absent from all wounds by day 42.

Undifferentiated mesenchymal tissue and cutaneous adnexa: undifferentiated mesenchymal tissue and cutaneous adnexa were observed only in samples collected at day 42. Hair follicles, sebaceous, and apocrine glands were present in all samples but the cutaneous adnexa observed in PB-MSCs-treated wounds appeared more mature and denser compare to the control group (Fig. 3).

After 15 days of treatment, ulceration was still present in all the samples. Complete re-epithelization was detected at day 42 in all samples.

Quantitative analysis of inflammatory, proliferative, vascular and structural factors

Quantitative immunohistochemical staining showed any increase of CD3+ and DC20+ positive cells was similar in both groups. A higher number of MHCII+ cells ($p < 0.5$) was observed after 15 days in PB-MSCs treated wounds (0.45 ± 0,03) compared to control group wounds (0.25 ± 0.02); this was not the case at day 42.

Within the newly formed dermis, the lesions treated with PB-MSCs had a higher Ki67 expression (0,661 ± 0,05) compared to the control group (0.313 ± 0,03). After 42 days, Ki67 expression, in both groups, began to

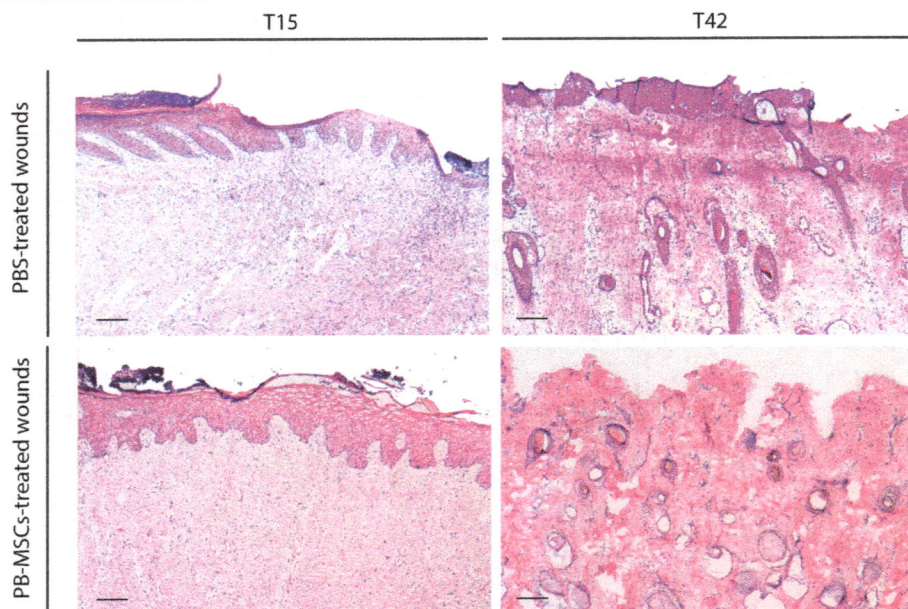

Fig. 3 Representative photomicrographs of PBS and PB-MSCs treated wounds (Hematoxylin-Eosin). Photomicrographs of PBS and PB-MSCs treated wounds analyzed at 15 and 42 days from treatments. The images show the presence of immature granulation tissue at 15 days, while mature connective tissue and developing cutaneous adnexa are present at 42 days. The lack of epidermis in representative image of PB-MSCs treated wounds at 42 days is an artefact. Scale bar 151,7 μm

decrease (Fig. 4). Using von Willebrand Factor (vWF) antibody staining, more dermal neovascularization was noticed in the PB-MSCs-treated wounds (4.15 ± 0,07) compared with the control lesions (3.32 ± 0,08) ($p < 0.5$). Neovascularization decreased in both groups during the wound healing process, showing the same protein expression values after day 42 (Fig. 4).

The molecular analysis (RT-PCR) of the Col1α1 gene indicated that at day 15 and 42, mRNA expression levels were statistically significant ($p < 0.5$) in the wounds treated with PB-MSCs (day 15: 75.09 ± 6,5, day 42: 87.65 ± 7,1) compared to the control group (day 15: 47.40 ± 3,6, day 42: 45.80 ± 5,3). PBS treatment did not influence the mRNA expression level of the Col1α1 gene (Fig. 5).

After 15 days, hKER mRNA expression (0.552 ± 0,05) was already present in the wounds treated with PB-MSCs. Furthermore, at day 42, the hKER expression level (5.016 ± 0,1) significantly ($p < 0.5$) increased in the PB-MSCs-treated lesions, but not in the control group's lesions. Control PBS alone did not stimulate cutaneous adnexa formation after 15 and 42 days (Fig. 5).

Discussion

MSCs represent a promising solution to promoting wound healing. The presence of these cells in normal skin [10] suggests their important role in maintaining skin homeostasis. There are different types of stem cells in the epidermis, dermis, and hair follicles [33], which

preserve the cellular state of the tissues. Several in vivo studies performed in small laboratory animals have demonstrated that stem cells accelerate wound healing. Many studies have hypothesizing that stem cells contribute to re-epithelization, vascularization, and extracellular remodeling [34–36]. The present study investigated the influence of allogeneic PB-MSCs treatment in a large animal experimental second intention wound healing model, evaluating their short and long-term effects on skin regeneration. Healing associated with a large and/or deep wound in which the tissue edges cannot be approximated is called secondary intention [37]. Wounds are left open to heal with the production of granulation tissue, followed by contraction and epithelialization [38]. Often, this type of healing can be associated with substantial scarring [37]. A previous study, using a murine model, showed that stem cells seeded on a nanostructured membrane helped primary intention healing, such as found with dermal burns [39]. Since MSCs are active in different phases of the healing process, it was hypothesized that they may also be used as a treatment for larger wounds that heal by second intention.

After skin injury, the inflammatory phase starts immediately. During this process, platelets aggregate at the injury site followed by the infiltration of neutrophils, macrophages, and T-lymphocytes [3]. The data presented in this paper show that there was no significant difference in level of inflammation between PBS-treated

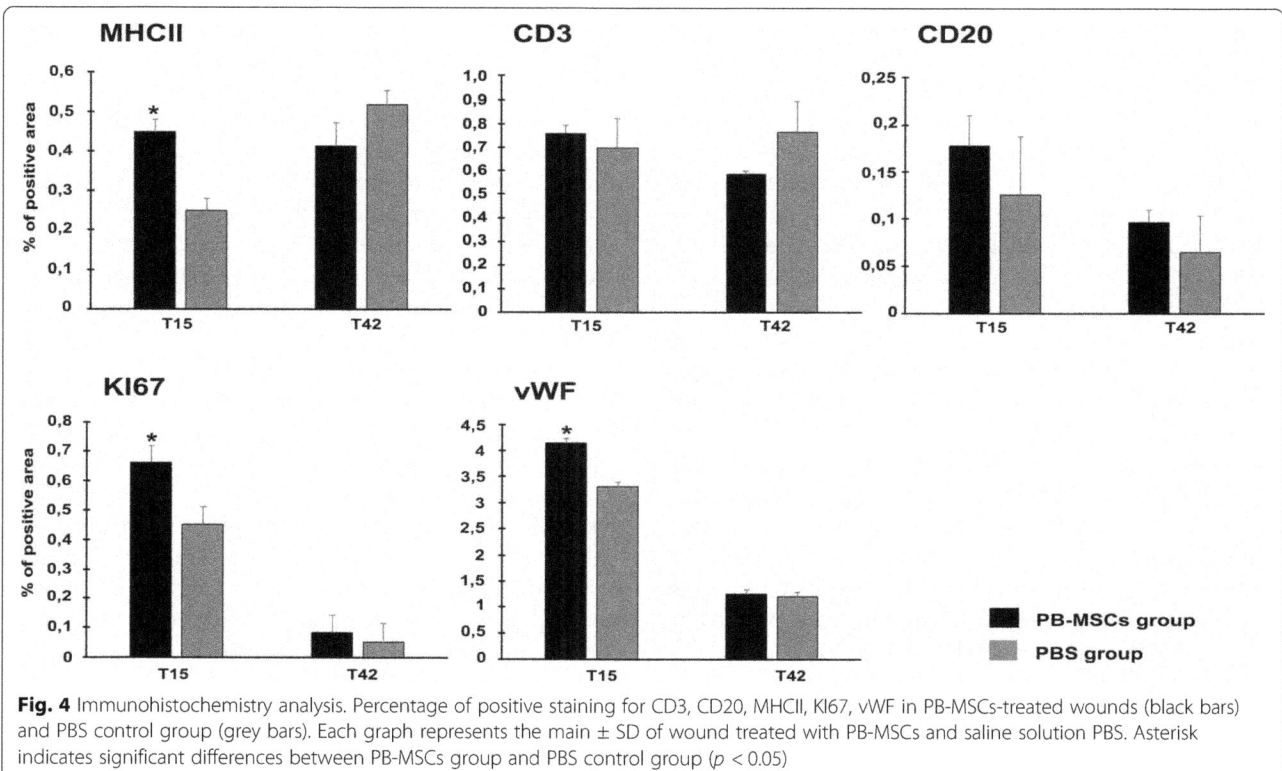

Fig. 4 Immunohistochemistry analysis. Percentage of positive staining for CD3, CD20, MHCII, KI67, vWF in PB-MSCs-treated wounds (black bars) and PBS control group (grey bars). Each graph represents the main ± SD of wound treated with PB-MSCs and saline solution PBS. Asterisk indicates significant differences between PB-MSCs group and PBS control group ($p < 0.05$)

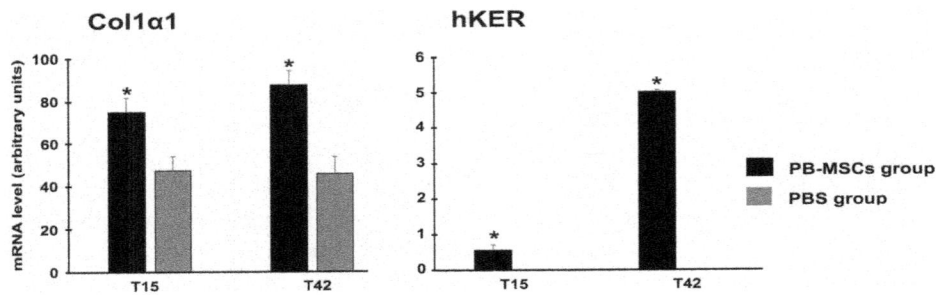

Fig. 5 Analyses of mRNA gene expression. mRNA expression of Col1α1 and hKER in PB-MSCs-treated wounds (black bars) and PBS control group (grey bars). Col1α1 and hKER were highly expressed in the treated wounds. Asterisk indicates significant differences between PB-MSCs and PBS control groups (p < 0.05)

and PB-MSCs-treated wounds. Microscopic evaluation indicated the presence of the inflammation phase 15 days post-injury in both the PBS control group and the PB-MSCs group at the dermal and subcutaneous levels. A notable result from the study was the complete absence of inflammation after 42 days in the PB-MSCs group whereas 60% of the PBS control group still presented with dermal inflammation. These results corroborate the findings of other studies. For example, Kim et al. [40] showed that experimental full-thickness wounds treated with topical allogeneic MSCs had increased healing and less inflammation, possibly due to the release of immunosuppressive factors in the wound bed that inhibit proliferation of immune cells such as B cells, T cells, and natural killers cells [41, 42]. This effect of allogenic MSCs was shown, in the current study, by absence of an increase of CD3+ and CD20+ cells (B lymphocytes) in MSCs-treated wounds. As discussed by Hussein et al. [43], CD3+ co-receptors helps to activate cytotoxic T lymphocytes, which constitute most of the mononuclear inflammatory cell infiltrate. Moreover, in the last decade, it has been found that MSCs also possess an antimicrobial effect, which helps to reduce excess inflammation from wound contaminants [44] and in the scar formation process [45]. The anti-inflammatory effect of PB-MSCs observed in the current study may result in a shortened inflammatory phase, thereby reducing myofibroblast and fibrocyte development and scar formation [46, 47].

After the inflammation phase, there is the proliferative phase with newly formed granulation tissue that covers the wound area to complete tissue repair. This phase is characterized by angiogenesis, which is important for attracting cytokines, sustaining the granulation tissue, and re-epithelization [48]. Histologically, the granulation tissue, evaluated in this study, was more abundant in wounds treated with PB-MSCs, although the amount of granulation tissue decreased for both cases and controls over time. The newly formed granulation tissue was seen at 15 days post-wound creation both in PBS and

PB-MSCs-treated wounds. Evidence of proliferative action by PB-MSCs was confirmed by an increase in Ki67 expression, with this protein present during all active phases of the cell cycle. The PB-MSCs treatment produced a significant increase in Ki67 expression compared to PBS treatment alone, which correlated with the presence of more abundant granulation tissue.

The increase in matrix and vessel formation, after MSCs treatment, may be attributed to the observed up-regulation of growth factors such as EGF, TGF-β1, and stromal-derived growth factor-1α [49]. The more active proliferation induced by PB-MSCs treatment was reflected by an increase in the percentage of re-epithelialization and contraction observed clinically. At 28 days, 93,5% of PB-MSCs-treated wounds were re-epithelized versus 87% of PBS treated wounds. In addition, wound contraction appeared earlier in the PB-MSCs-treated group. The histological data, obtained in this study, confirmed that MSCs might produce multiple pro-angiogenic factors at the lesion site, which stimulate endothelial cells and lead to new blood vessel formation in the wound bed. Revascularization of the wound bed is an important part of the normal wound healing process. Formation of new vessels is necessary to carry blood to the wound area, which requires oxygen and nutrients [50, 51].

The last phase of wound healing is maturation of the tissue. Collagen type 1 is the predominant collagen in normal skin and exceeds collagen type 3 by a ratio of 4:1. During wound healing, this ratio decreases to 2:1 because of an early increase in the deposition of collagen type 3. In this study, the expression of matrix protein collagen 1 was higher in PB-MSCs-treated wounds compared to only treatment with PBS at both 14 and 42 days, indicating an earlier process of wound healing. Moreover, in normal skin, a population of multipotent stem cells capable of generating all of the components of hair, as well as epithelial cells, is located in the hair follicle bulge [52]. These cells do not contribute to preservation of the interfollicular epidermis, but can differentiate into epidermal stem cells after a trauma [53]. In the current study, the treatment of wounds with allogeneic PB-MSCs resulted in the development of new hair follicles and probably also the activation of bulge cells.

Overall, the findings of this large animal study were similar to results from small animal studies. In fact, lesions created in rabbits and dogs [54, 55] demonstrated significantly earlier vascularization, fibroplasia, and maturation of collagen using autologous bone marrow-derived mononuclear cells compared to a control group. Formigli L et al. [56] demonstrated that MSCs seeded on bioengineering scaffolds induced enhanced re-epithelialization characterized by a multilayered epidermis, return of hair follicles, sebaceous glands, and enhanced blood vessel formation. The current study showed that treatment with PB-MSCs leads to a significant increase in the expression of hair keratin mRNA, with expression detectable at 14 days post-wound creation. Furthermore, after 42 days, microscopic evaluation showed an increased in hair follicles, sebaceous and apocrine glands in the PB-MSCs-treated group compared to the control group.

Conclusion

In the skin regeneration process, PB-MSCs play roles in different phases of wound healing, contributing to the healing process and, as it is confirmed from our paper, does not induce an inflammatory response. Despite some analyzed parameters did not show significant results the trend suggests a beneficial use of PB-MSCs not only for treating superficial injuries, but also for deeper lesions. PB-MSCs were able to speed up the appearance of granulation tissue, stimulate neovascularization, and increase structural proteins and skin adnexa.

Abbreviations
cDNA: complementary single strand DNA; Col1α1: Collagen 1α1; hKER: hair keratin; IHC: immunohistochemistry; MSCs: mesenchymal stem cells; OPBA: The Body for the Protection of Animals; PB: peripheral blood; PB-MSCs: MSCs isolated from peripheral blood; PBS: phosphate saline buffer; RPS24: ribosomal protein S24; RT-PCR: Real Time-PCR; vWF: von Willebrand factor

Acknowledgements
The authors are grateful to Prof. C. Budke (Texas A&M, USA) for the useful reading that improved the manuscript.

Funding
This work was supported by a grant from the University of Padova, Italy (BIRD161823/16.
DOR1683028/16 Dept. BCA, University of Padua). These grants were essential to provide costs for live animals and consumables for the laboratory analysis.

Authors' contributions
TM = contributions to conception and design, analysis and interpretation of data, involved in drafting the manuscript and revising. CG, LM = acquisition of data, analysis and interpretation of data. AP, FG, GMDB = acquisition of clinical data. SF, JHS, KC = analysis and interpretation of histopathological data. MZ, EM, PB = acquisition of plasma gas data, II, MP = contributions to conception and design, involved in drafting the manuscript and revising. All authors read and approved the final manuscript.

Consent for publication
Yes.

Competing interests
The authors declare that they have no competing interests.

Author details
[1]Department of Comparative Biomedicine and Food Science, University of Padua, Viale dell'Università 16, 35020, Legnaro – Agripolis, Padua, Italy. [2]Department of Animal Medicine, Production and Health, University of Padua, Padua, Italy. [3]Consorzio RFX, Padua, Italy. [4]Department of Molecular Medicine, University of Padua, Padua, Italy. [5]Department of Pathology, Bacteriology and Poultry Diseases, University of Gent, Ghent, Belgium. [6]Global Stem cell Technology-ANACURA group, Noorwegenstraat 4, 9940 Evergem, Belgium.

References
1. Pereira RF, Barrias CC, Granja PL, Bartolo PJ. Advanced biofabrication strategies for skin regeneration and repair. Nanomedicine. 2013;8:603–21.
2. Pereira RF, Bártolo PJ. Traditional therapies for skin wound healing. Advances in Wound Care. 2016;21:208–29.
3. Kondo T. Timing of skin wounds. Legal Med. 2007;9(2):109–14.
4. McGavin MD, Zachary JF. Pathologic basis of veterinary disease. Fourth edition: Mosby Elsevier; 2007. ISBN-10: 0323028705.
5. Borena BM, et al. Martens a, Broeckx SY, Meyer E, Chiers K, Duchateau L, Spaas JH. Regenerative skin wound healing in mammals: state-of-the-art on growth factor and stem cell based treatments. Cell Physiol Biochem. 2015; 36(1):1–23.
6. Singer AJ, Thode H Jr, McClain SA. Development of a histomorphologic scale to quantify cutaneous scars after burns. Acad Emerg Med. 2000;7:1083–8.
7. Cerqueira MT, Marques AP, Reis LR. Using stem cells in skin regeneration: possibilities and reality. Stem Cells Dev. 2012;20(21):1201–14.
8. Lipinski LC, Wouk AF, da Silva NL, Perotto D, Ollhoff RD. Effects of 3 topical plant extracts on wound healing in beef cattle. Afr J Tradit Complement Altern Med. 2012;9(4):542–7.
9. Arnold CE, Schaer TP, Baird DL, Martin BB. Conservative management of 17 horses with nonarticular fractures of the tibial tuberosity. Equine Vet J. 2003; 35(2):202–6.
10. Sellheyer K, Krahl D. Cutaneous mesenchymal stem cells. Current status of research and potential clinical applications. Hautarzt. 2010;61:429–34.
11. Maxson S, Lopez EA, Yoo D, Danilkovitch-Miagkova A, Leroux MA. Concise review: role of mesenchymal stem cells in wound repair. Stem Cells Transl Med. 2012;1(2):142–9.
12. Paquet-Fifield S, Schluter H, Li A, Aitken T, Gangatirkar P, Blashki D, Koelmeyer R, Pouliot N, Palatsides M, Ellis S, Brouard N, Zannettino A, Saunders N, Thompson N, Li J, Kaur P. A role for pericytes as microenvironmental regulators of human skin tissue regeneration. J Clin Invest. 2009;119:2795–806.
13. Martinello T, Bronzini I, Maccatrozzo L, Iacopetti I, Sampaolesi M, Mascarello F, Patruno M. Cryopreservation does not affect the stem characteristics of multipotent cells isolated from equine peripheral blood. Tissue Engineering Part C. 2010;16:771–81.
14. Martinello T, Bronzini I, Maccatrozzo L, Mollo A, Sampaolesi M, Mascarello F, Decaminada M, Patruno M. Canine adipose-derived-mesenchymal stem cells do not lose stem features after a long-term cryopreservation. Res Vet Sci. 2011;91:18–24.

15. Gomiero C, Bertolutti G, Martinello T, Van Bruaene N, Broeckx SY, Patruno M, Spaas JH. Tenogenic induction of equine mesenchymal stem cells by means of growth factors and low-level laser technology. Veterinary Research Communication. 2016;40(1):39–48.

16. Chu CR, Fortier LA, Williams A, Payne KA, McCarrel TM, Bowers ME, Jaramillo D. Minimally manipulated bone marrow concentrate compared with microfracture treatment of full-thickness chondral defects: a one-year study in an equine model. J Bone Joint Surg Am. 2018;100(2):138–46.

17. Sherman AB, Gilger BC, Berglund AK, Schnabel LV. Effect of bone marrow-derived mesenchymal stem cells and stem cell supernatant on equine corneal wound healing in vitro. Stem Cell Res Ther. 2017;8(1):120.

18. Cheng HY, Ghetu N, Wallace CG, Wei FC, Liao SK. The impact of mesenchymal stem cell source on proliferation, differentiation, immunomodulation and therapeutic efficacy. Journal of Stem Cell Research & Therapy. 2014;4(10)

19. Lyahyai J, Mediano DR, Ranera B, Sanz A, Remacha AR, Bolea R, Zaragoza P, Rodellar C, Martín-Burriel I. Isolation and characterization of ovine mesenchymal stem cells derived from peripheral blood. BMC Vet Res. 2012;8:169.

20. Heidari et al., Comparison of Proliferative and Multilineage Differentiation Potential of Sheep Mesenchymal Stem Cells Derived from Bone Marrow, Liver, and Adipose Tissue. Avicenna J Med Biotechnol Vol. 5, No. 2, April–June 2013.

21. Martinello, et al. Effects of in vivo applications of peripheral blood-derived mesenchymal stromal cells (PB-MSCs) and Platlet-rich plasma (PRP) on experimentally injured deep digital flexor tendons of sheep. J Orthop Res. 2013;31(2):306–14.

22. Hu MS, Rennert RC, McArdle A, Chung MT, Walmsley GG, Longaker MT, Lorenz HP. The role of stem cells during Scarless skin wound healing. Advance in Wound Care (New Rochelle). 2014;3(4):304–14. Otero-Viñas M, Falanga V. Mesenchymal stem cells in chronic wounds: the Spectrum from basic to advanced therapy. Advances in Wound Care, 2016; 5(4),149–163

23. Wu Y, Chen L, Scott PG, Tredget EE. Mesenchymal stem cells enhance wound healing through differentiation and angiogenesis. Stem Cells. 2007; 25:2648–59.

24. Liu P, Deng Z, Han S, Liu T, Wen N, Lu W, Geng X, Huang S, Jin Y. Tissue-engineered skin containing mesenchymal stem cells improves burn wounds. Artif Organs. 2008;32:925–31.

25. Jackson WM, Nesti LJ, Tuan RS. Concise review: clinical translation of wound healing therapies based on mesenchymal stem cells. Stem Cells Transl Med. 2012;1(1):44–50.

26. Cerqueira MT, Pirraco RP, Marques AP. Stem cells in skin wound healing: are we there yet? Advance in Wound Care (New Rochelle). 2016;5(4):164–75.

27. Music E, Futrega K, Doran MR. Sheep as a model for evaluating mesenchymal stem/stromal cell (MSC)-based chondral defect repair. Osteoarthr Cartil. 2018;26(6):730–40. https://doi.org/10.1016/j.joca.2018.03.006. Review

28. Chevrier A, Nelea M, Hurtig MB, Hoemann CD, Buschmann MD. Meniscus structure in human, sheep, and rabbit for animal models of meniscus repair. J Orthop Res. 2009;27(9):1197–203.

29. YR KJ, Evans RG, May CN. An ovine model for studying the pathophysiology of septic acute kidney injury. Methods Mol Biol. 2018;1717:207–18. https://doi.org/10.1007/978-1-4939-7526-6_16.

30. Russel WMD, Burch RL. The principles of Human Experimental Technique UFAW. London: Methuen & Co.; 1959. ISBN: 0900767782 9780900767784.

31. Broeckx SY, Borena BM, Van Hecke L, Chiers K, Maes S, Guest DJ, Meyer E, Duchateau L, Martens A, Spaas JH. Comparison of autologous versus allogeneic epithelial-like stem cell treatment in an in vivo equine skin wound model. Cytotherapy. 2015;17(10):1434–46.

32. Hadley HS, Stanley BJ, Fritz MC, Hauptman JG, Steficek BA. Effects of a cross-linked hyaluronic acid based gel in the healing of open wounds in dogs. Vet Surg. 2013;42:161–9.

33. Cui P, He X, Pu Y, Zhang W, Zhang P, Li C, Guan W, Li X, Ma Y. Biological characterization and pluripotent identification of sheep dermis-derived mesenchymal stem/progenitor cells. Biomed Res Int. 2014;2014:786234.

34. Yoshikawa T, Mitsuno H, Nonaka I, Sen Y, Kawanishi K, Inada Y, Takakura Y, Okuchi K, Nonomura A. Wound therapy by marrow mesenchymal cell transplantation. Plast Reconstr Surg. 2008;121:860–77.

35. Kwon DS, Gao X, Liu YB, Dulchavsky DS, Danyluk AL, Bansal M, Chopp M, McIntosh K, Arbab AS, Dulchavsky SA, Gautam SC. Treatment with bone marrow derived stromal cells accelerates wound healing in diabetic rats. Int Wound J. 2008;5:453–63.

36. Badillo AT, Redden RA, Zhang L, Doolin EJ, Liechty KW. Treatment of diabetic wounds with fetal murine mesenchymal stromal cells enhances wound closure. Cell Tissue Res. 2007;329:301–11.

37. Iocono JA, Ehrlich HP, Gottrup F, et al. The biology of healing. In: Leaper DL, Harding KG, editors. Wounds: biology and management. Oxford, England: Oxford University Press; 1998. p. 12–22.

38. You HJ, Han SK. Cell therapy for wound healing. J Korean Med Sci. 2014;29:311–9.

39. Souza CM, Mesquita LA, Souza D, Irioda AC, Francisco JC, Souza CF, Guarita-Souza LC, Sierakowski MR, Carvalho KA. Regeneration of skin tissue promoted by mesenchymal stem cells seeded in nanostructured membrane. Transplant Proc. 2014;46(6):1882–6.

40. Kim JW, Lee JH, Lyoo YS, Jung DI, Park HM. The effects of topical mesenchymal stem cell transplantation in canine experimental cutaneous wounds. Vet Dermatol. 2013;24:242–e53.

41. Matthay MA, Goolaerts A, Howard JP, Lee JW. Mesenchymal stem cells for acute lung injury: preclinical evidence. Critical Care Med. 2010; 38(Suppl 10):569–73.

42. Hass R, Kasper C, Bohm S, Jacobs R. Different populations and sources of human mesenchymal stem cells (MSC): a comparison of adult and neonatal tissue-derived MSC. Cell Communication and Signaling. 2011;9:12.

43. Hussein MR, Hassan HI. Analysis of the mononuclear inflammatory cell infiltrate in the normal breast, benign proliferative breast disease, in situ and infiltrating ductal breast carcinomas: preliminary observations. J Clin Pathol. 2006;59:972–7.

44. Mei SH, Haitsma JJ, Dos Santos CC, Deng Y, Lai PF, Slutsky AS, Liles WC, Stewart DJ. Mesenchymal stem cells reduce inflammation while enhancing bacterial clearance and improving survival in sepsis. American Journal of Respiratory Critical Care Medicine. 2010;182:1047–57.

45. Nuschke A. Activity of mesenchymal stem cells in therapies for chronic skin wound healing. Organ. 2014;10(1):29–37.

46. Rhett JM, Ghatnekar GS, Palatinus JA, O'Quinn M, Yost MJ, Gourdie RG. Novel therapies for scar reduction and regenerative healing of skin wounds. Trends Biotechnol. 2008;26:173–80.

47. Jackson WM, Nesti LJ, Tuan RS. Mesenchymal stem cell therapy for attenuation of scar formation during wound healing. Stem Cell Res Ther. 2012;3(3):20.

48. Burnouf T, Goubran HA, Chen T-M, Ou K-L, El- Ekiaby M, Radosevic M. Blood-derived biomaterials and platelet growth factors in regenerative medicine. Blood Rev. 2013;27:77–89.

49. Chen L, Tredget EE, Wu PY, Wu Y. Paracrine factors of mesenchymal stem cells recruit macrophages and endothelial lineage cells and enhance wound healing. PLoS One. 2008;3:e1886.

50. Morimoto N, Yoshimura K, Niimi M, et al. An exploratory clinical trial for combination wound therapy with a novel medical matrix and fibroblast growth factor in patients with chronic skin ulcers: a study protocol. Am J Transl Res. 2012;4:52–9.

51. Zhang Y, Wang T, He J, Dong J. Growth factor therapy in patients with partial-thickness burns: a systematic review and meta-analysis. Int Wound J. 2014;8:1–13.

52. Oshima H, Rochat A, Kedzia C, Kobayashi K, Barrandon Y. Morphogenesis and renewal of hair follicles from adult multipotent stem cells. Cell. 2001; 104:233–45.

53. Levy V, Lindon C, Zheng Y, Harfe BD, Morgan BA. Epidermal stem cells arise from the hair follicle after wounding. FASEB J. 2007;21:1358–66.

54. Borena BM, Pawde AAM, Aithal HP, Kinjavdekar P, Singh R, Kumar D. Evaluation of autologous bone marrow-derived nucleated cells for healing of full thickness skin wounds in rabbits. Int Wound J. 2010;7:249–60.

55. Borena BM, Pawde Amarpal AM, Aithal HP, Kinjavdekar P, Singh R, Kumar D. Evaluation of healing potential of autologous bone marrow-derived nucleated cells on incisional wounds in dogs. Indian Journal of Veterinary Surgery. 2009;30:85–9.

56. Formigli L, Paternostro F, Tani A, Mirabella C, Quattrini Li A, Nosi D, D'Asta F, Saccardi R, Mazzanti B, Lo Russo G, Zecchi-Orlandini S. MSCs seeded on bioengineered scaffolds improve skin wound healing in rats. Wound Repair Regen. 2015;23(1):115–23.

19

The influence of experimental inflammation and axotomy on leucine enkephalin (leuENK) distribution in intramural nervous structures of the porcine descending colon

Slawomir Gonkowski[*] [iD], Krystyna Makowska and Jaroslaw Calka

Abstract

Background: The enteric nervous system (ENS), located in the intestinal wall and characterized by considerable independence from the central nervous system, consists of millions of cells. Enteric neurons control the majority of functions of the gastrointestinal tract using a wide range of substances, which are neuromediators and/or neuromodulators. One of them is leucine–enkephalin (leuENK), which belongs to the endogenous opioid family. It is known that opioids in the gastrointestinal tract have various functions, including visceral pain conduction, intestinal motility and secretion and immune processes, but many aspects of distribution and function of leuENK in the ENS, especially during pathological states, remain unknown.

Results: During this experiment, the distribution of leuENK – like immunoreactive (leuENK-LI) nervous structures using the immunofluorescence technique were studied in the porcine colon in physiological conditions, during chemically-induced inflammation and after axotomy. The study included the circular muscle layer, myenteric (MP), outer submucous (OSP) and inner submucous plexus (ISP) and the mucosal layer. In control animals, the number of leuENK-LI neurons amounted to $4.86 \pm 0.17\%$, $2.86 \pm 0.28\%$ and $1.07 \pm 0.08\%$ in the MP, OSP and ISP, respectively. Generally, both pathological stimuli caused an increase in the number of detected leuENK-LI cells, but the intensity of the observed changes depended on the factor studied and part of the ENS. The percentage of leuENK-LI perikarya amounted to $11.48 \pm 0.96\%$, $8.71 \pm 0.13\%$ and $9.40 \pm 0.76\%$ during colitis, and $6.90 \pm 0.52\%$ $8.46 \pm 12\%$ and $4.48 \pm 0.44\%$ after axotomy in MP, OSP and ISP, respectively. Both processes also resulted in an increase in the number of leuENK-LI nerves in the circular muscle layer, whereas changes were less visible in the mucosa during inflammation and axotomy did not change the number of leuENK-LI mucosal fibers.

Conclusions: LeuENK in the ENS takes part in intestinal regulatory processes not only in physiological conditions, but also under pathological factors. The observed changes are probably connected with the participation of leuENK in sensory and motor innervation and the neuroprotective effects of this substance. Differences in the number of leuENK-LI neurons during inflammation and after axotomy may suggest that the exact functions of leuENK probably depend on the type of pathological factor acting on the intestine.

Keywords: Leucine-enkephalin (leuENK), Enteric nervous system (ENS), Inflammation, Axotomy, Colon, Pig

* Correspondence: slawomir.gonkowski@uwm.edu.pl
Department of Clinical Physiology, Faculty of Veterinary Medicine, University
of Warmia and Mazury, Oczapowski Str, 13 Olsztyn, Poland

Background

It is well-known that neural control of all functions of the gastrointestinal (GI) tract is performed by both extrinsic innervation [1–4] and the enteric nervous system (ENS) [5, 6]. Localization of extrinsic neurons supplying the stomach and intestine depend on the innervated regionof the GI tract. These neurons may be located in prevertebral sympathetic ganglia, sympathetic chain, dorsal root ganglia, sensory and parasympathetic ganglia of vagal nerve, as well as in parasympathetic nuclei of sacral spinal cord [1, 2, 4, 6]. In turn, the ENS is situated in the wall of the GI tract. It is composed of a large number of neurons, which are characterized by considerable independence from the central nervous system and, for this reason, it has also been called the "second" or "intestinal" brain [7]. The structure of the ENS depends on the species studied and the regionof the GI tract. In porcine, large intestine enteric neurons are grouped into three separate ganglionated plexuses, which are interconnected with a dense network of nerves (Fig. 1). There are three types of these

plexuses: myenteric plexus (MP) – located between the longitudinal and circular muscles, outer submucous plexus (OSP) – immediately adjacent to internal side of the circular muscle layer and inner submucous plexus (ISP) – positioned between muscularis mucosa and lamina propria [5, 8]. Neuronal cells within the above-mentioned plexuses belong to various functional classes, play different functions and show the presence of a broad range of active substances which, importantly, can play various roles as neuromediators and/or neuromodulators [6].

It should be pointed out that the ENS plays a key role in the regulation of intestinal functions not only in physiological conditions, but also during various pathological states. This fact may be expressed by structural, functional or chemical changes within enteric neurons, which may be a result of adaptive responses to operating factors. Till now, such changes in the ENS have been observed under physiological changes, such as development, aging or diet modification, but they are more visible during pathological processes, including intestinal and systemic diseases or nerve injury [8–11].

One of the many active substances occurring within enteric neurons is leucine-enkephalin (leuENK) which, along with methionine-enkephalin, is an endogenous opioid. These substances are pentapeptides derived from a precursor (proenkephalin) and described for the first time within the porcine brain in the 1970s [12]. Later investigations found that leuENK may also arise from α-neo-endorphin, dynorphin A or dynorphin B, which derived from prodynorphin [13]. Enkephalins, like as exogenous opioids (e.g. morphine) act via a different type of G-protein coupled receptors, including δ-opioid, κ-opioid and μ-opioid receptors [14, 15]. Although the main action of enkephalins is a mechanism connected with analgesia, previous studies have shown that these substances are also involved in other physiological processes, such as regulation of respiratory [16], urinary [17] or circulatory [18] systems.

To date, enkephalins have been described within the GI tract in various mammals species, including humans, in both the ENS and intestinal mucosal enterochromaffin cells [15, 19–21] as well as in the extrinsic innervation of intestine [1, 2]. Even so, the exact roles of opioids within the digestive system are not entirely clarified, although it is known that these substances mainly take part in reducing visceral pain [14]. Other functions of opioids within the GI tract include participation in intestinal immune processes, as well as inhibition of intestinal motility and secretion [14, 22, 23]. The inhibitory effects of enkephalins mentioned above cause an increase in food transit time in the GI tract and may contribute to constipation [14]. In addition, some previous studies have described changes in the number of enteric neurons immunoreactive to enkephalins under pathological factors, which suggests

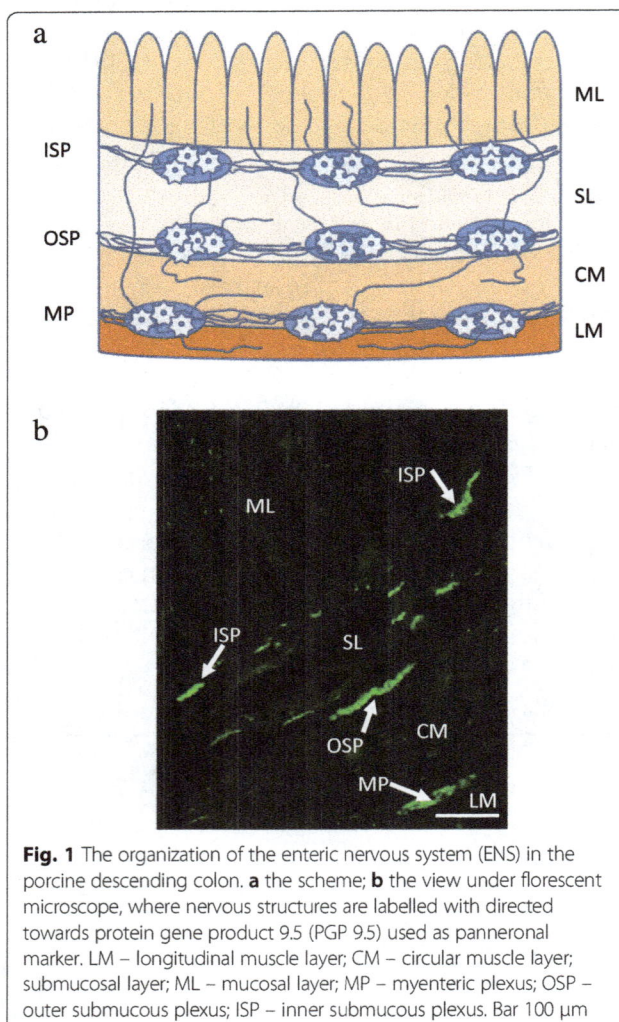

Fig. 1 The organization of the enteric nervous system (ENS) in the porcine descending colon. **a** the scheme; **b** the view under florescent microscope, where nervous structures are labelled with directed towards protein gene product 9.5 (PGP 9.5) used as panneronal marker. LM – longitudinal muscle layer; CM – circular muscle layer; submucosal layer; ML – mucosal layer; MP – myenteric plexus; OSP – outer submucous plexus; ISP – inner submucous plexus. Bar 100 μm

the participation of these substances in adaptive or neuroprotective processes within the intestine [1, 2, 21]. Nonetheless, many questions connected with the functions of leuENK in the ENS during pathological processes are not clear, especially in the GI tract of the pig, which is increasingly becoming an optimal laboratory animal due to well-established anatomical, biochemical and physiological resemblances to humans [24, 25]. These similarities primarily concern the organization of the ENS [26], which makes the pig a perfect animal model for experiments on changes in enteric neurons under various pathological factors.

The aim of the present investigation was to describe, for the first time, the influence of selected pathological factors, including experimental inflammation and axotomy on leucine enkephalin-like immunoreactive (leuENK-LI) enteric neurons in the porcine descending colon. It should be pointed out that the selection of this fragment of the GI tract as a subject of the present study was also not accidental, since various pathological processes often develop within the large intestine both in human and animals [8, 27–30].

Methods

Experimental animals

The present experiment was performed on 20 immature sows of the Large White Polish breed at the age of 8 weeks (about 18 kg body weight) bought on the commercial farm of pig production in Bałcyny (Poland). Animals were kept in normal laboratory conditions in the animal quarters of Faculty of Veterinary Medicine, University of Warmia and Mazury, Olsztyn (Poland) with feeding typical for this species and age of animal. The pigs were maintained in the pens (5 animals in each pen) with an area of about 4 m². All experimental procedures, as well as number of experimental animals were consistent with the instructions and agreements of the Local Ethical Committee in Olsztyn (Poland), with special attention paid to minimizing any stress reaction during and after surgery (agreement numbers: 90/2007 from 20 November 2007 and 85/2008 from 17 December 2008). Animals during unloading after the transport were randomly (without any determined method) divided into four groups, each of which consisted of five pigs located in separate pens. Then the assignment of animals from each pen to the particular procedures were performed by drawing lots. Experimental groups were as follows: control (C group) – without any experimental procedures, control 1 (C1) – animals subjected to "sham" operations, inflammatory (I group) - pigs with chemically-induced colitis and axotomy (A group), where animals were subjected to the cutting of specific nerves (see below).

After five-day adaptive period the experiment was started. The sows from C1, I and A groups were subjected to median laparotomy performed under general anaesthesia, which consisted of premedication with Stressnil (Janssen, Belgium, 75 μl/kg of body weight, given intramuscularly) 15 min. Before the administration of the main aesthetic – sodium thiopental (Thiopental, Sandoz, Kundl-Rakúsko, Austria; 20 mg/kg of body weight, given intravenously) prior to the surgery. Chemically-induced inflammation and axotomy were made using the methods previously described by Gonkowski et al. [31]. According to these methods, the animals of group I were injected with 80 μl of 10% formalin solution in saline (microinjections of 5–8 μl) into the wall of the descending colon (to a depth of 0,2–0,5 mm) in an area, where nerves from the inferior mesenteric ganglia supply the gut. Surgery on animals of group A consisted of the bilateral transection of caudal colonic nerves connecting to the inferior mesenteric ganglion with the descending colon (Fig. 2). Animals of group C1 were subjected to "sham" operations, which were aimed at exclusion of surgical manipulations on the enteric nervous system. Pigs of group C1were injected in the same manner as group I, but a pure saline solution was used instead of formalin. After 5 days, all animals were anaesthetized in the same way as described above and euthanized by an overdose of sodium thiopental and immediately perfused transcardially with freshly prepared 4% buffered paraformaldehyde (pH 7.4).

Double-labelling immunofluorescence technique

Segments of the descending colon (approximately 2 cm long) from the area where nerves from the inferior mesenteric ganglia supply the colon were collected from all pigs used in the experiment. Inflammation in the animals of group I was confirmed by a histopathological examination, which was made at the Laboratory of Histopathology (Faculty of Veterinary Medicine, University of Warmia and Mazury, Olsztyn, Poland). Immediately after collection, colonic fragments were post-fixed by immersion with 4% buffered paraformaldehyde (pH 7.4) for 20 min and rinsed in phosphate buffer (0.1 M, pH 7.4, at 4 °C) for 3 days with the exchange of buffer every day. The tissues were then put into 18% phosphate-buffered sucrose (at 4 °C) for 2 weeks. Finally, they were frozen at -22 °C. After freezing samples were coded and since then the assessors did know which samples belonged to which experimental group. Coding of samples was performed to avoid any bias in the evaluation of colonic fragments. Then tissues were cut perpendicular to the lumen of the colon into 14-μm-thick sections using microtome (Microm, HM 525, Walldorf, Germany) and fixed on glass slides.

The fragments of colon were subjected to standard double-labelling immunofluorescence, which has been described previously by Gonkowski et al. [31, 32]. During the present study, the combination of two kinds of primary antisera were used: mouse monoclonal

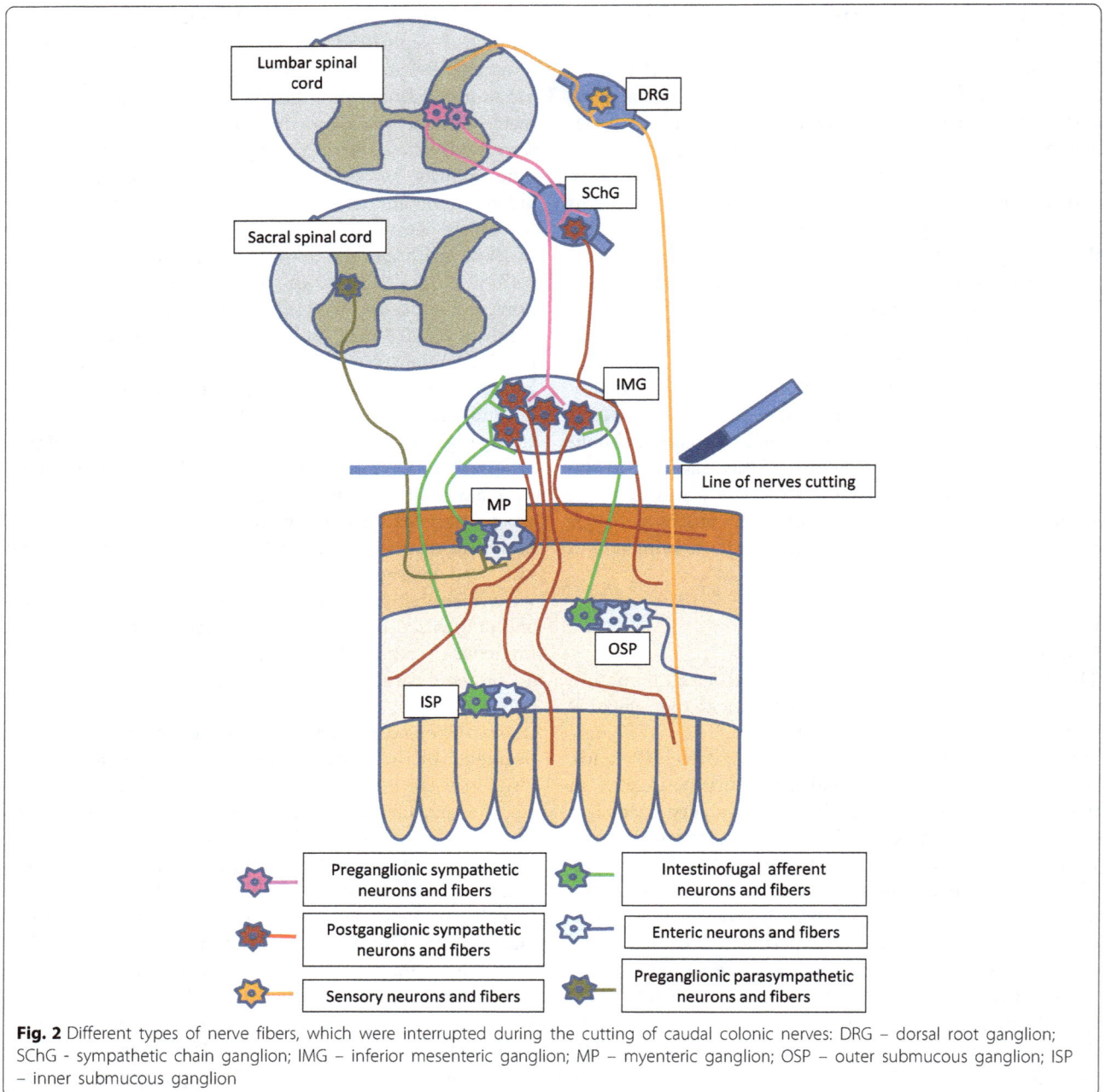

Fig. 2 Different types of nerve fibers, which were interrupted during the cutting of caudal colonic nerves: DRG – dorsal root ganglion; SChG - sympathetic chain ganglion; IMG – inferior mesenteric ganglion; MP – myenteric ganglion; OSP – outer submucous ganglion; ISP – inner submucous ganglion

antibody directed towards protein gene-product 9.5 (PGP 9.5, Biogenesis, UK, working dilution 1:2000, used here as a pan-neuronal marker) and rabbit polyclonal anti – leucine-enkephalin antibody (leuENK, Abcam, UK, working dilution 1:1000). The primary antisera mentioned above were visualized by a combination of species-specific secondary antibodies, i.e. Alexa fluor 488 donkey anti-mouse IgG and Alexa fluor 546 donkey anti-rabbit IgG (both from Invitrogen, Carlsbad, CA, USA, working dilution 1:1000).

Standard control probes were made to test the specificity of primary antibodies. These included pre-absorption, omission and replacement of primary antibodies by non–immune sera. The pre-absorption test consisted of incubation (for 18 h at 37 °C) of primary antibodies in working dilutions with the appropriate antigen. Antibodies directed towards PGP 9.5 were pre-incubated with native human protein gene product 9.5 (AbD Serotec, UK), and anti-leuENK antibody with synthetic human leucine-enkephalin peptide (Abcam, UK). The concentration of each antigen was 20 µg per 1 ml of diluted antibody. This procedure, as well as omission and replacement of primary antisera by non-immune sera, completely eliminated specific stainings.

Counting the nerve structures and statistical analysis

To delimit the percentage of enteric neurons immunoreactive to leuENK, at least 600 PGP-9.5- positive cells in a particular enteric plexus (MP, OSP and ISP) of each studied animal were examined for leuENK-like immunoreactivity. Double-labelled perikarya (only neurons with clearly visible nucleus) were evaluated using an Olympus BX51 microscope equipped with epi-fluorescence and appropriate filter sets. The obtained results were pooled and presented as mean ± SEM. To prevent double counting of leuENK-LI neurons, the sections were located at least 100 µm apart.

For semi-quantitative evaluation of the density of leuENK-LI nerves within the enteric ganglia, an arbitrary scale was used, where (–) indicated the absence of studied fibers, (+) - single fibers, (++) - rare fibers, (+++) – dense network of fibers and (++++) depicted a very dense meshwork of fibers studied.

In turn, denotation of the density of intramuscular and intramucosal nerves immunoreactive to leuENK was based on counting all leuENK-LI nerve fibers *per* microscopic observation field (0.55 mm^2). Nerves were counted in four sections per animal (in 5 fields per section) and the obtained data were pooled and presented as a mean. It should be pointed out that the nerves in the intestine can form very small bundles and it is not always possible to count exact numbers of fibers in fibre bundles. Therefore the present study includes only fibers that could have been counted.

All pictures were captured by a digital camera connected to a PC. Statistical analysis was carried out with an Anova test (Graphpad Prism v. 2.0; GraphPad Software Inc., San Diego, CA, USA). The differences were considered statistically significant at $p \leq 0.05$. No statistical power calculation was conducted prior to the study, and the determination of the size of animal experimental groups was based on data available in the previous studies, where the number of five animals in the case of pigs during neuro-immunofluorescence investigations are commonly accepted.

Results

During the present investigation, leuENK-positive nervous structures were observed within the descending colon in animals of all experimental groups and their number clearly depended on both the pathological factor studied and the part of the ENS (Table 1).

Under physiological conditions, a relatively small number of leuENK-LI neurons was noted in all "kinds" of enteric plexuses (Table 1). The percentage of such neuronal cells (in relation to number of all PGP 9.5-LI neurons) amounted to 4.86 ± 0.17%, 2.86 ± 0.28% and 1.07 ± 0.08% in MP, OSP and ISP, respectively. Regarding the view of individual enteric ganglia, only single neuronal cells immunoreactive to leuENK were noted in control animals (Fig. 3. I), but the majority of ganglia did not show the presence of leuENK-LI neurons (Figs. 4.I and 5.I). Moreover, intraganglionic leuENK-positive nerve fibers were observed in all kinds of enteric ganglia. The dense network (+++) of these nerves were encountered in OSP, whereas in MP they were rare (++), and in ISP only single (+) nerve processes immunoreactive to leuENK were noted (Table 1). A relatively dense network of clearly visible LeuENK-positive nerves (Fig. 6.Ia) was also observed in the circular muscle layer (15.30 ± 0.77 nerves/observation field). In turn, only single, thin and delicate leuENK-LI nerves were noted in the mucosal layer (Fig. 6.Ib).

Statistically significant differences were not observed between control and "sham" operated animals, but both pathological states studied changed the immunoreactivity to leuENK. These changes are generally expressed by an increase in the number of leuENK-positive neuronal structures, but their intensity depended on the pathological factor acting on the intestine (Table 1).

The highest percentage of neurons immunoreactive to leuENK were found within the myenteric plexus during the inflammatory process (Fig. 3.II), where the number of these cells increased by above 6.5 percentage points (pp). These changes were accompanied by an increase in the density of leuENK-LI intraganglionic nerve fibers. In animals suffering from inflammation, two "kinds" of myenteric ganglia were observed. One of them was characterized by the presence of leuENK-LI cells and a not very dense network of fibers immunoreactive to this peptide (Fig. 3.II). In the other MP, only very dense intraganglionic leuENK-LI nerves were visible, without leuENK-positive neurons (Fig. 3.II$_1$). Axotomy also caused an increase in the number of leuENK-LI neurons and intraganglionic fibers in the MP (Fig. 3.III), but these changes were less visible.

The pathological states studied caused an increase in the percentage of leuENK- positive neuronal cells located in OSP (Fig. 4). Contrary to MP, the differences between inflammation and axotomy were not observed. Both investigated states produced an increase in the number of leuENK-LI neurons by about 5.5 pp. (Table 1). These changes in OSP were accompanied by a decrease in the density of intraganglionic nerve fibers immunoreactive to leuENK (Fig. 4).

An increase in the leuENK-positive neuron number was also noted in the ISP (Fig. 5). As with myenteric plexus, more visible modifications were observed during the inflammatory process, where the percentage of leuENK-LI

Table 1 Leucine-enkephalin – like immunoreactive nervous structures in the porcine descending colon

Bowel part		No. of animal	C Group	C1Group	I Group	A Group
CML[1]		1	16.40	14.60	30.80	23.30
		2	17.15	17.15	28.90	26.20
		3	13.80	12.85	32.60	24.15
		4	13.15	14.50	29.05	23.15
		5	16.00	14.40	32.75	26.80
		average	15.30 ± 0.77[a]	14.70 ± 0.69[a]	30.82 ± 0.83[b]	24.72 ± 0.75[c]
MP	CB[2]	1	5.06	5.13	9.25	6.12
		2	5.23	5.15	12.06	8.63
		3	4.53	3.80	9.92	6.10
		4	4.38	4.30	11.44	7.60
		5	5.10	5.62	14.73	6.05
		average	4.86 ± 0.17[a]	4,80 ± 0.33[a]	11,48 ± 0.96[b]	6,90 ± 0.52[c]
	NF[3]	average	++	++	++++	+++
OSP	CB[2]	1	1.90	1.58	8.70	8.26
		2	3.05	3.38	8.76	8.68
		3	2.80	3.58	9.15	8.46
		4	2.90	1.38	8.60	8.15
		5	3.64	3.93	8.35	8.75
		average	2.86 ± 0.28[a]	2.77 ± 0.53[a]	8.71 ± 0.13[b]	8.46±12[b]
	NF[3]	average	+++	+++	+	+
ISP	CB[2]	1	0.84	0.58	7.52	3.48
		2	1.27	1.02	11.94	5.92
		3	1.07	0.60	9.61	4.64
		4	1.20	0.86	8.20	3.60
		5	0.98	1.34	9.73	4.76
		average	1.07 ± 0.08[a]	0.88 ± 0.14[a]	9.40 ± 0.76[b]	4.48 ± 0.44[c]
	NF[3]	average	+	+	+	+
S/ML[1]		1	0.85	1.20	3.00	0.85
		2	1.30	1.00	4.45	1.30
		3	1.15	1.30	4.20	1.20
		4	1.45	2.00	2.80	1.15
		5	1.50	1.15	3.65	1.95
		average	1.25 ± 0.12[a]	1.33 ± 0.17[a]	3.62 ± 0.32[b]	1.29 ± 0.18[a]

C group control animals, C1group "sham" operated animals, I group pigs suffering from inflammation, A group animals after axotomy
CML circular muscle layer, MP myenteric plexus, OSP outer submucous plexus, ISP inner submucous plexus, S/ML submucosal/mucosal layer, CB cell bodies, NF nerve fibers
[1]Average number of nerve profiles per area studied (mean ± SEM)
[2]Relative frequency of particular neuronal subclasses is presented as % (mean ± SEM) of all neurons counted within the ganglia stained for PGP 9.5
[3]The density of intraganglionic nerve fibers positive for leuENK is presented in arbitrary units
Statistically significant data ($p \leq 0.05$) between C group and group C1, I and A in the number of leuENK-LI nervous structures within particular part of the colon are marked by different letters and not significant data are marked by the same letters

cells grew nine-fold in relation to control animals and amounted to 9.40 ± 0.76% of all neurons immunoreactive to PGP 9.5. This was the clearest change of all parts of the ENS studied. Axotomy also caused an increase in the number of cells immunoreactive to leuENK, but changes were less visible (Table 1). Contrary to the number of

leuENK-positive neurons, the density of leuENK-LI intra-ganglionic nerves in the ISP was not changed during the pathological factors studied (Table 1).

Both inflammation and axotomy caused an increase in the density of leuENK-positive nerve fibers in the circular muscle layer (Fig. 6.I). In this case, as in the MP and

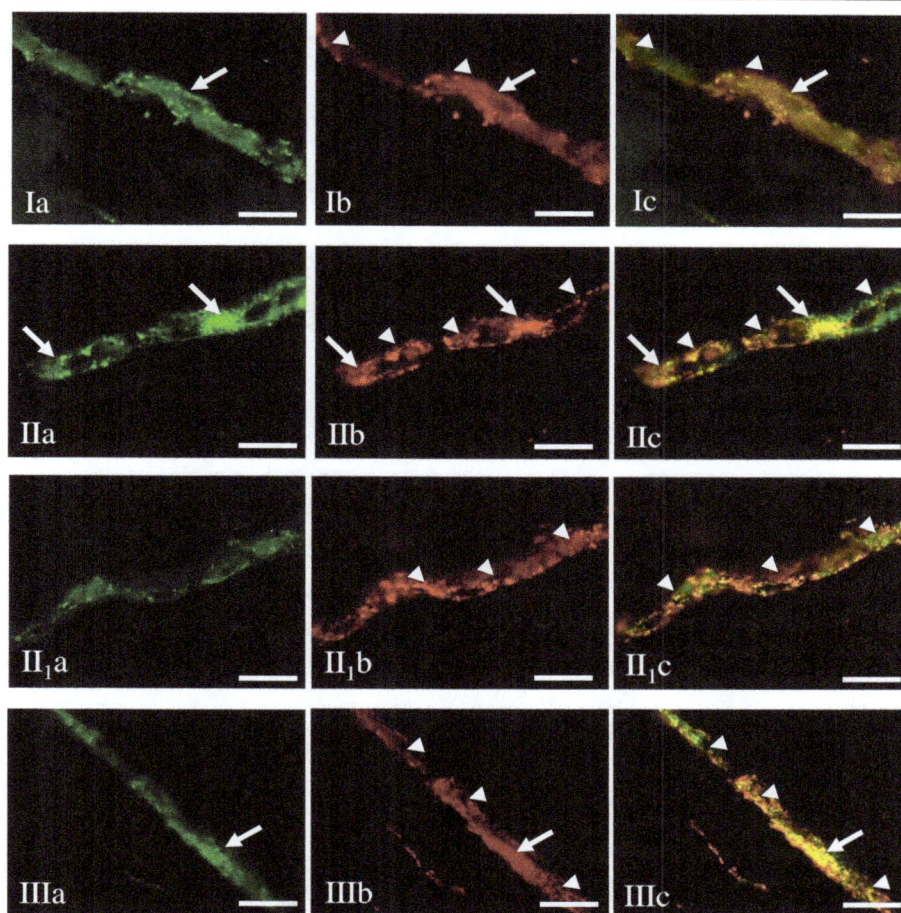

Fig. 3 Myenteric plexus of the porcine descending colon under physiological conditions (I), during inflammation (II, II1) and after axotomy (III) immunostained for PGP 9.5 (a) and leuENK (b). The right column of the pictures (c) shows the overlap of both stainings. Co-localisation of both antigens is indicated with arrows (perikarya) and arrow heads (nerve fibers). Bar, 20 μm

ISP, more visible changes were observed during inflammatory process, where the number of fibers immunoreactive to leuENK was doubled versus control animals (Table 1). In contrast, the number of leuENK-positive nerves localized in the mucosal layer only increased during inflammation, whereas axotomy did not change their number (Fig. 6.II, Table 1).

Discussion

During the present investigation, leuENK-positive nerve structures have been observed in all parts of the ENS studied, as well as in intramuscular and intramucosal nerve fibers. These structures under physiological conditions (in spite of nerves within circular muscle layer) were rather limited. This is in agreement with the majority of previous studies, where the presence of leuENK has been described both in the ENS and extrinsic innervation of the GI tract in various mammal species, including humans [15, 19, 21]. The results obtained during the present study show that distribution of leuENK-positive nervous structures in the descending colon of the pig and other studied species are similar. In the pig (like in rats, guinea pigs or cats) such structures were first observed in the myenteric plexus and the circular muscle layer, whereas in the mucosa and submucosal enteric ganglia they are rather rare [20, 33, 34]. Particular similarities in the distribution of leuENK-positive nervous structures in the intestinal wall may also be observed between humans [15, 35, 36] and pigs (this study). This suggests that relatively well-known resemblances in the ENS organization of both these species [26] also apply to the distribution of leuENK.

Moreover, it should be pointed out that there are some discrepancies in the localization of porcine enteric neurons immunoreactive to leuENK in the light of previous studies. For example, in the stomach, some investigations have noted the absence of this peptide in the ENS [37], other authors have described relatively numerous leuENK-LI neurons, especially in the MP [38]. The reason for these discrepancies remains unclear, although they

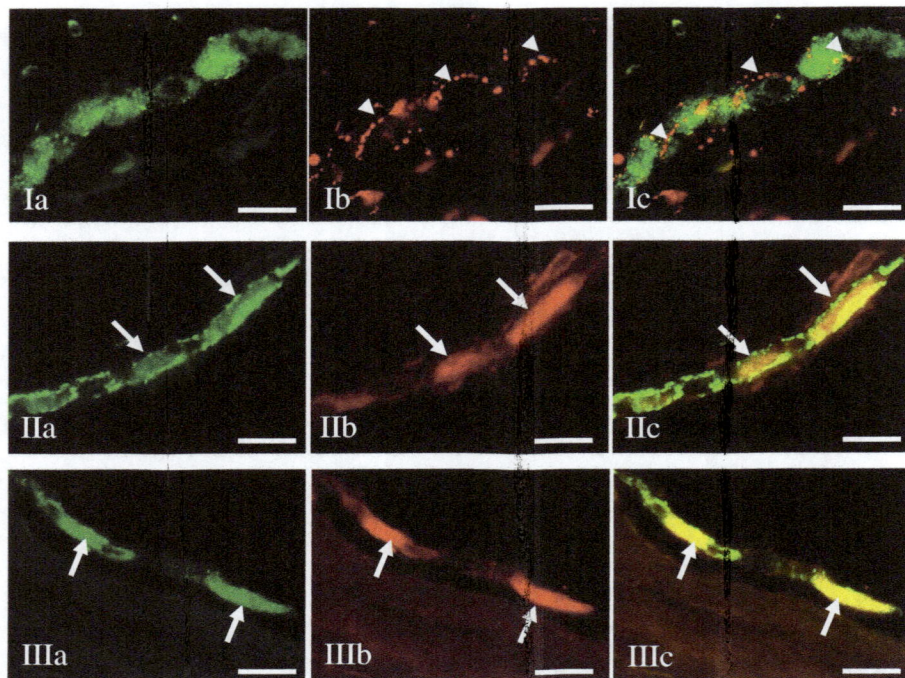

Fig. 4 Outer submucous plexus of the porcine descending colon under physiological conditions (I), during inflammation (II) and after axotomy (III) immunostained for PGP 9.5 (a) and leuENK (b). The right column of the pictures (c) shows the overlap of both stainings. Co-localisation of both antigens is indicated with arrows (perikarya) and arrow heads (nerve fibers). Bar, 20 μm

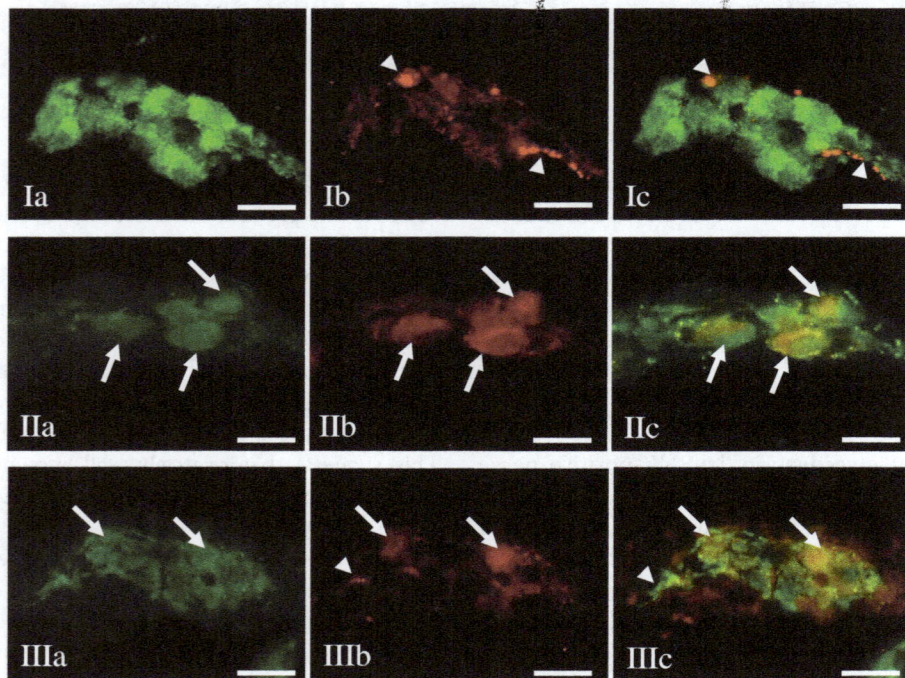

Fig. 5 Inner submucous plexus of the porcine descending colon under physiological conditions (I), during inflammation (II) and after axotomy (III) immunostained for PGP 9.5 (a) and leuENK (b). The right column of the pictures (c) shows the overlap of both stainings. Co-localisation of both antigens is indicated with arrows (perikarya) and arrow heads (nerve fibers). Bar, 20 μm

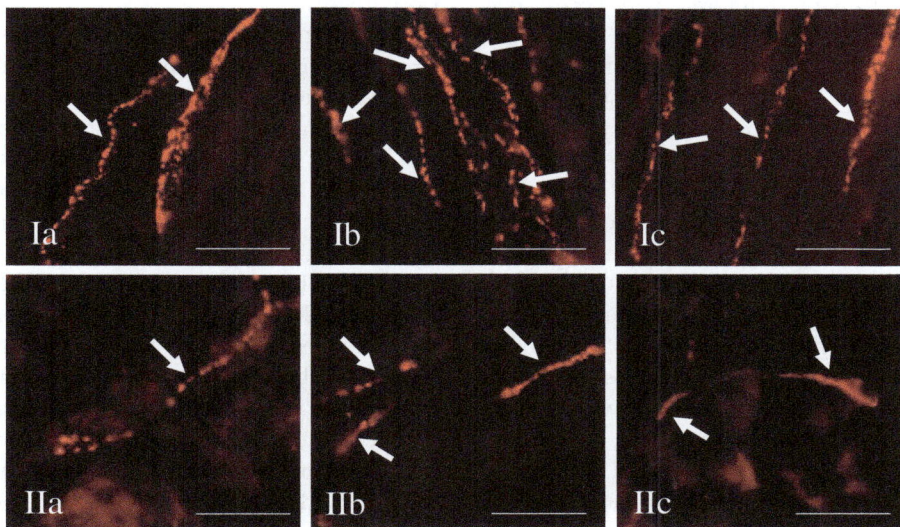

Fig. 6 Distribution pattern of nerve fibers (arrows) immunostained for leuENK in the circular muscle layer (I) and mucosal layer (II) of porcine descending colon under physiological conditions (a), during inflammation (b) and after axotomy (c). Bar, 20 μm

strongly suggest that enteric neurons can change leuENK expression under various physiological factors, such as nutritional differences, farming factors and/or environmental bacterial flora, which in all probability is associated with some currently unknown functions of opioids within the ENS.

Till now, opioids (including leuENK) in the digestive tract have been described as factors affecting the nociception, intestinal immune functions, motility and secretion [14] and their exact roles depends on both the type of opioid and receptors class. Opioid receptors, belonging to G-protein coupled receptors, have been described in the central and peripheral nervous structures, including the enteric nervous system [39–41] and three kinds of them (δ-opioid, κ-opioid and μ-opioid receptors) can participate in the regulation of gastrointestinal physiology. Previous studies showed that each class of opioid receptors has its preferred ligand and regulates specific processes. κ-opioid receptors, which bind mainly with dynorphin and μ-opioid receptors (their ligand is β-endorphin), are primarily responsible for visceral antinociception [14, 41]. In turn, δ-opioid receptors, preferred ligands of which are enkephalins, first of all exhibit an inhibitory effect on the intestinal motility, causing an increase in the transit time of digestive tract content [14]. Moreover, previous studies have described other gastrointestinal functions of opioids, such as suppression of intestinal secretion, participation in immune processes and protective roles during inflammation [14].

During the present study, both chemically-induced inflammation and axotomy caused changes in the number of leuENK-positive colonic nerve structures, and the influence of inflammatory processes has been more

visible. The differences in leuENK-like immunoreactivity have probably been connected with roles of opioids in intestinal immune functions. It is known that these substances, as well as their receptors, are present in macrophages, monocytes and some classes of lymphocytes and take part in mechanisms connected with cytokines production and T-cells proliferation [14, 42]. Another reason for the observed changes may be the participation of enkephalins in sensory stimuli conduction and anti-nociceptive mechanisms, which has been described within both the central nervous system and the enteric neurons [43–45].

Moreover, it is well-established that opioids can reduce the activity of intestinal muscles by various mechanisms, such as a reduction in the release of stimulating neurotransmitters, changes in the excitability of enteric neurons and/or neuromodulatory effects on intestinal motor neurons [14, 41, 46]. Enkephalins, acting through δ-opioid receptors, which are located mainly in the myenteric plexus of the ENS, are also a major factor reducing intestinal motility [14]. It is possible that changes observed during inflammation, where an often excessive increase of muscular contractility and diarrhoea are present, may result from compensatory activity of enteric neurons. In this case, the increase of leuENK expression would have to be aimed at peristalsis inhibition and return to physiological conditions. This assumption is all the more probable that the most numerous leuENK-LI nervous structures have been noted during the inflammatory process in the muscular plexus and circular muscle layer. Since enkephalins are also known as inhibitors of intestinal secretions, the mechanisms of changes noted in the ISP could be similar to those mentioned above.

Changes in leuENK-like immunoreactivity observed in animals suffering from inflammation may also result from the participation of opioids in mechanisms connected with typical symptoms of gastrointestinal diseases, including nausea and emesis [14] and/or from neuroprotective actions of these substances. Previous studies have shown that synthetic agonists of δ-opioid receptors may prolong the cell life of many organs, such as heart, lung or kidney [47, 48], as well as have a neuroprotective influence on neuronal cells in various regions of the brain [49–51] and spinal cord [52]. It is not yet known whether endogenous or synthetic enkephalins can prolong the life of neurons supplying the GI tract in the pig, but the results obtained during the present study, as well as previous investigations on extrinsic innervation of porcine digestive tract [2, 53, 54], where pathological factors also caused an increase in leuENK-like immunoreactivity strongly suggest this fact, because it is well-established that injury to neurons causes an increase in the expression of active substances which promote the regeneration of injured cells [55].

During the present study, changes in the number of leuENK-LI nervous structures after axotomy have been less visible than during inflammation. The cutting of nerves connecting the inferior mesenteric ganglia with the descending colon caused the destruction of various types of nerves (Fig. 2). The majority of these were postganglionic sympathetic nerves, axons of neurons located predominantly in the inferior mesenteric ganglia as well as in the sympathetic chain. Other nerves derived from afferent sensory cells located in dorsal root ganglia, parasympathetic preganglionic neurons in sacral spinal cord or intestinofugal afferent (also named "viscerofugal") neurons (IFANs). IFANs, belonging to the ENS, are situated in the wall of intestine, in all kinds of enteric plexuses, send axons to prevertebral sympathetic ganglia, including coeliac ganglion, cranial mesenteric ganglion or inferior mesenteric ganglion (depending on the segment of the GI tract) and take part in intestino – intestinal reflexes without the participation of the central nervous system [56–58]. Thus, changes in leuENK-like immunoreactivity noted after axotomy may result from compensatory activity of enteric neurons after the destruction of colonic extrinsic innervation or neuroprotective processes in IFANs, which is more likely in light of the relatively well-known neuroprotective effects of enkephalins, which were discussed above.

It should be pointed out that the mechanisms of changes observed during the present investigation may be variable. They can arise from an increase in leuENK synthesis in enteric neurons or modifications in intraneuronal transport of this substance from the cell body to nerve endings. In the case of fluctuations in leuENK synthesis, the changes may be due to different types of changes during transcription, translation or post-translational processes.

Conclusions

The present study shows that leuENK-positive nervous structures are present in the porcine descending colon and their distribution is similar to other species, especially to humans, which suggests the possibility of using the pig as a model animal in investigations of this peptide in the human ENS. Moreover, leuENK-like immunoreactivity in the porcine ENS undergoes fluctuations during chemically-induced inflammation and after axotomy. The increase in the number of leuENK-LI neurons and nerves can suggest neuroprotective roles of the described substance within the ENS. The present results are the first step toward exploitation of synthetic enkephalins as a medicine, not only during colitis, which has been suggested by previous studies [45], but also after intestinal innervation damage. On the other hand, the observed changes clearly depended on both the part of the ENS and the type of factor studied. This is probably caused by different roles of leuENK in intestinal innervation, according to the type of acting pathological stimulus. These roles are still not fully explained and require further study.

Abbreviations

A group: Animals after axotomy; C group: Control group; C1 group: "Sham" operated animals; ENS: Enteric nervous system; GI tract: Gastrointestinal tract; I group: Inflammatory group; ISP: Inner submucous plexus; leuENK: Leucine enkephalin; MP: Myenteric plexus; OSP: Outer submucous plexus; Pp: Percentage point; SEM: Standard error of the mean

Funding

Publication was supported by the grant No 15.610.003–300 (statutory research) of University of Warmia and Mazury in Olsztyn (Poland) and KNOW (Leading National Research Centre) Scientific Consortium "Healthy animal – Safe Food", decision of Ministry of Science and Higher Educiation, No. 05–1/KNOW 2/2015.

Authors' contributions

SG is the originator and main performer of experiment as well as the major contributor in writing the manuscript. KM takes part in the statistical analysis of data creation of figures. JC checked the manuscript linguistically and grammatically. All authors read and approved the final manuscript.

Competing interests

The authors declare that they have no competing interests.

References

1. Wojtkiewicz J, Rowniak M, Gonkowski S, Crayton R, Majewski M, Robak A, Bialkowska J, Barczewska M. Proliferative enteropathy (PE)-induced changes in the calbindin-immunoreactive (CB-IR) neurons of inferior mesenteric ganglion supplying the descending colon in the pig. J Mol Neurosci. 2012;48:757–65.
2. Wojtkiewicz J, Rowniak M, Crayton R, Gonkowski S, Robak A, Zalecki M, Majewski M, Klimaschewski L. Axotomy-induced changes in the chemical coding pattern of Colon projecting Calbindin-positive neurons in the inferior mesenteric ganglia of the pig. J Mol Neurosci. 2013;51:99–108.
3. Chen BN, Sharrad DF, Hibberd TJ, Zagorodnyuk VP, Costa M, Brookes SJ. Neurochemical characterization of extrinsic nerves in myenteric ganglia of the Guinea pig distal colon. J Comp Neurol. 2015;523:742–56.
4. Rytel L, Calka J. Acetylsalicylic acid-induced changes in the chemical coding of extrinsic sensory neurons supplying the prepyloric area of the porcine stomach. Neurosci Lett. 2016;23:218–24.

5. Gonkowski S, Substance P. As a neuronal factor in the enteric nervous system of the porcine descending colon in physiological conditions and during selected pathogenic processes. Biofactors. 2013;39:542–51.

6. Furness JB, Callaghan BP, Rivera LR, Cho HJ. The enteric nervous system and gastrointestinal innervation: integrated local and central control. Adv Exp Med Biol. 2014;817:39–71.

7. Avetisyan M, Schill EM, Heuckeroth RO. Building a second brain in the bowel. J Clin Invest. 2015;125:899–907.

8. Gonkowski S, Burlinski P, Calka J. Proliferative enteropathy (PE) – induced changes in galanin – like immunoreactivity in the enteric nervous system of the porcine distal colon. Acta Vet-Beogr. 2009;59:321–30.

9. Saffrey MJ. Ageing of the enteric nervous system. Mech Ageing Dev. 2004;125:899–906.

10. Di Giancamillo A, Vitari F, Bosi G, Savoini G, Domeneghini C. The chemical code of porcine enteric neurons and the number of enteric glial cells are altered by dietary probiotics. Neurogastroenterol Motil. 2010;22:e271–8.

11. Vasina V, Barbara G, Talamonti L, Stanghellini V, Corinaldesi R, Tonini M, De Ponti F, De Giorgio R. Enteric neuroplasticity evoked by inflammation. Auton Neurosci. 2006;126-127:264–72.

12. Hughes J, Smith TW, Kosterlitz HW, Fothergill LA, Morgan BA, Morris HR. Identification of two related pentapeptides from the brain with potent opiate agonist activity. Nature. 1975;258:577–9.

13. Dores RM, Lecaudé S, Bauer D, Danielson PB. Analyzing the evolution of the opioid/orphanin gene family. Mass Spectrom Rev. 2002;21:220–43.

14. De Schepper HU, Cremonini F, Park MI, Camilleri M. Opioids and the gut: pharmacology and current clinical experience. Neurogastroenterol Motil. 2004;16:383–94.

15. Brehmer A, Lindig TM, Schro F, Neuhuber W, Ditterich D, Rexer M, Rupprecht H. Morphology of enkephalin-immunoreactive myenteric neurons in the human gut. Histochem Cell Biol. 2005;123:131–8.

16. Greer JJ, Carter J, Al-Zubaidy EZ. Opioid depression of respiration in neonatal rats. J Physiol. 1995;485:845–55.

17. Klarskov P. Enkephalin inhibits presynaptically the contractility of urinary tract smooth muscle. Br J Urol. 1987;59:31–5.

18. Feldman PD, Parveen N, Sezen S. Cardiovascular effects of Leu-enkephalin in the nucleus tractus solitarius of the rat. Brain Res. 1996;709:331–6.

19. Herbrecht F, Bagnol D, Cucumel F, Julé Y, Cupo A. Biochemical characterization and quantification of enkephalins in brain, prevertebral ganglia and digestive tract: a comparative study on cats, Guinea-pigs and rats. Regul Pept. 1995;57:85–95.

20. Bagnol D, Henry M, Cupo A, Julé Y. Distribution of enkephalin-like immunoreactivity in the cat digestive tract. J Auton Nerv Syst. 1997;64:1–11.

21. Wojtkiewicz J, Równiak M, Crayton R, Majewski M, Gonkowski S. Chemical coding of zinc-enriched neurons in the intramural ganglia of the porcine jejunum. Cell Tissue Res. 2012;350:215–23.

22. Puig MM, Pol O. Peripheral effects of opioids in a model of chronic intestinal inflammation in mice. J Pharmacol Exp Ther. 1998;287:1068–75.

23. Pol O, Palacio JR, Puig MM. The expression of delta- and kappa-opioid receptor is enhanced during intestinal inflammation in mice. J Pharmacol Exp Ther. 2003;306:455–62.

24. Litten-Brown JC, Corson AM, Clarke L. Porcine models for the metabolic syndrome, digestive and bone disorders: a general overview. Animal. 2010;4:899–920.

25. Verma N, Rettenmeier AW, Schmitz-Spanke S. Recent advances in the use of Sus scrofa (pig) as a model system for proteomic studies. Proteomics. 2011;11:776–93.

26. Brown DR, Timmermans JP. Lessons from the porcine enteric nervous system. Neurogastroenterol Motil. 2004;16(Suppl 1):50–4.

27. Kosugi I, Tada T, Tsutsui Y, Sato Y, Mitsui T, Itazu I. Giant inflammatory polyposis of the descending colon associated with a Crohn's disease-like colitis. Pathol Int. 2002;52:318–21.

28. Gonkowski S, Burlinski P, Szwajca P, Calka J. Changes in cocaine- and amphetamine – regulated transcript – like immunoreactive (CART-LI) nerve structures of the porcine descending colon during proliferative enteropathy. Bull Vet Inst Pulawy. 2012;56:199–203.

29. Rychlik A, Gonkowski S, Nowicki M, Calka J. Cocaine- and amphetamine-regulated transcript immunoreactive nerve fibres in the mucosal layer of the canine gastrointestinal tract under physiological conditions and in inflammatory bowel disease. Vet Med Czech. 2015;60:361–7.

30. Kinugasa T, Akagi Y. Status of colitis-associated cancer in ulcerative colitis. World J Gastrointest Oncol. 2016;8:351–7.

31. Gonkowski S, Burliński P, Skobowiat C, Majewski M, Całka J. Inflammation- and Axotomy-induced changes in galanin-like immunoreactive (GAL-LI) nerve structures in the porcine descending colon. Acta Vet Hung. 2010;58:91–103.

32. Gonkowski S, Rychlik A, Nowicki M, Nieradka R, Bulc M, Calka J. A population of nesfatin 1-like immunoreactive (LI) cells in the mucosal layer of the canine digestive tract. Res Vet Sci. 2012;93:1119–21.

33. Schultzberg M, Hökfelt T, Nilsson G, Terenius L, Rehfeld JF, Brown M, Elde R, Goldstein M, Said S. Distribution of peptide- and catecholamine-containing neurons in the gastro-intestinal tract of rat and Guinea-pig: immunochemical studies with antisera to substance P, enkephalins, somatostatin, gastrin, cholecystokinin, neurotensin and dopamine b-hydroxylase. Neuroscience. 1980;5:689–744.

34. Furness JB, Costa M, Miller RJ. Distribution and projections of nerves with enkephalin-like immunoreactivity in the Guinea-pig small intestine. Neuroscience. 1983;8:653–64.

35. Keast JR, Furness JB, Costa M. Distribution of certain peptide-containing nerve fibres and endocrine cells in the gastrointestinal mucosa in five mammalian species. J Comp Neurol. 1985;236:403–22.

36. Wattchow DA, Furness JB, Costa M. Distribution and coexistence of peptides in nerve fibers of the external muscle of the human gastrointestinal tract. Gastroenterology. 1988;95:32–41.

37. Kaleczyc J, Klimczuk M, Franke-Radowiecka A, Sienkiewicz W, Majewski M, Łakomy M. The distribution and chemical coding of intramural neurons supplying the porcine stomach - the study on normal pigs and on animals suffering from swine dysentery. Anat Histol Embryol. 2007;36:186–93.

38. Rekawek W, Sobiech P, Gonkowski S, Zarczyńska K, Snarska A, Wasniewski T, Wojtkiewicz J. Distribution and chemical coding patterns of cocaine- and amphetamine-regulated transcript-like immunoreactive (CART-LI) neurons in the enteric nervous system of the porcine stomach cardia. Pol J Vet Sci. 2015;18:515–22.

39. Simon EJ. Opioid receptors and endogenous opioid peptides. Med Res Rev. 1991;11:357–74.

40. Minami M, Satoh M. Molecular biology of the opioid receptors: structures, functions and distributions. Neurosci Res. 1995;23:121–45.

41. Sternini C, Patierno S, Selmer IS, Kirchgessner A. The opioid system in the gastrointestinal tract. Neurogastroenterol Motil. 2004;16(Suppl. 2):3–16.

42. Stein C. The control of pain in peripheral tissue by opioids. N Engl J Med. 1995;332:1685–90.

43. Skinner K, Basbaum AI, Fields HL. Cholecystokinin and enkephalin in brain stem pain modulating circuits. Neuroreport. 1997;8:2995–8.

44. Cabot PJ, Carter L, Schäfer M, Stein C. Methionine-enkephalin-and Dynorphin A-release from immune cells and control of inflammatory pain. Pain. 2001;93:207–12.

45. Sobczak M, Pilarczyk A, Jonakowski M, Jarmuż A, Sałaga M, Lipkowski AW, Fichna J. Anti-inflammatory and antinociceptive action of the dimeric enkephalin peptide biphalin in the mouse model of colitis: new potential treatment of abdominal pain associated with inflammatory bowel diseases. Peptides. 2014;60:102–6.

46. Holzer P. Opioid receptors in the gastrointestinal tract. Regul Pept. 2009;155:11–7.

47. Wu G, Zhang F, Salley RK, Diana JN, Su TP, Chien S. Delta opioid extends hypothermic preservation time of the lung. J Thorac Cardiovasc Surg. 1996;111:259–67.

48. Su TP. Delta opioid peptide [d-Ala (2), d-Leu (5)] enkephalin promotes cell sur-vival. J Biomed Sci. 2000;7:195–9.

49. Zhang J, Gibney GT, Zhao P, Xia Y. Neuroprotective role of delta-opioid receptors in cortical neurons. Am J Phys Cell Phys. 2002;282:C1225–34.

50. Borlongan CV, Wang Y, Su TP. Delta opioid peptide (d-ala 2, d-Leu 5) enkephalin: linking hibernation and neuroprotection. Front Biosci. 2004;9:3392–8.

51. Iwata M, Inoue S, Kawaguchi M, Nakamura M, Konishi N, Furuya H. Effects of delta-opioid receptor stimulation and inhibition on hippocampal survival in a rat model of forebrain ischaemia. Br J Anaesth. 2007;99:538–46.

52. Liu H, Chen B, Zhang Y, Qiu Y, Xia Y, Li S, Yao J. Protective effect of delta opioid agonist [D-Ala2, D-Leu5] enkephalin on spinal cord ischemia reperfusion injury by regional perfusion into abdominal aorta in rabbits. Neurosci Lett. 2015;584:1–6.

53. Palus K, Całka J. Alterations of neurochemical expression of the coeliac-superior mesenteric ganglion complex (CSMG) neurons supplying the prepyloric region of the porcine stomach following partial stomach resection. J Chem Neuroanat. 2016;72:25–33.

54. Palus K, Całka J. The influence of prolonged acetylsalicylic acid supplementation-induced gastritis on the neurochemistry of the sympathetic neurons supplying prepyloric region of the porcine stomach. PLoS One. 2015;10:e0143661.

55. Arciszewski MB, Ekblad E. Effects of vasoactive intestinal petide and galanin on survival of cultured porcine myenteric neurons. Regul Pept. 2005;125:185–92.

56. Luckensmeyer GB, Keast JR. Immunohistochemical characterisation of viscerofugal neurons projecting to the inferior mesenteric and major pelvic ganglia in the male rat. J Auton Nerv Syst. 1996;61:6–16.

57. Furness JB. Types of neurons in the enteric nervous system. J Auton Nerv Syst. 2000;81:87–96.

58. Ermilov LG, Miller SM, Schmalz PF, Hanani M, Lennon VA, Szurszewski JH. Morphological characteristics and immunohistochemical detection of nicotinic acetylcholine receptors on intestinofugal afferent neurones in Guinea-pig colon. Neurogastroenterol Motil. 2003;15:289–98.

20

Genetic diversity of *Coxiella burnetii* in domestic ruminants in central Italy

M. Di Domenico[1]*(iD), V. Curini[1], V. Di Lollo[1], M. Massimini[1], L. Di Gialleonardo[1], A. Franco[2], A. Caprioli[2], A. Battisti[2] and C. Cammà[1]

Abstract

Background: As the epidemiology of human Q Fever generally reflects the spread of *Coxiella burnetii* in ruminant livestock, molecular characterization of strains is essential to prevent human outbreaks. In this study we report the genetic diversity of *C. burnetii* in central Italy accomplished by MST and MLVA-6 on biological samples from 20 goat, sheep and cow farms.

Results: Five MST and ten MLVA profiles emerged from the analysis establishing a part of *C. burnetii* strain world atlas. In particular, ST32 occurred on 12 farms (60%), prevalently in goat specimens, while ST12 (25%) was detected on 4 sheep and 1 goat samples. ST8 and a variant of this genotype were described on 2 different sheep farms, whereas ST55 was observed on a goat farm. Five complete MLVA profiles different from any other published genotypes were described in this study in addition to 15 MLVA incomplete panels. Despite this, polymorphic markers Ms23, Ms24 and Ms33 enabled the identification of samples sharing the same MST profile.

Conclusions: Integration of such data in international databases can be of further help in the attempt of building a global phylogeny and epidemiology of Q fever in animals, with a "One Health" perspective.

Keywords: *Coxiella burnetii*, Genotyping, MST, MLVA, Italy

Background

Coxiella burnetii is the causative agent of Query (Q) fever in humans, a zoonotic disease present throughout the world, except in New Zealand [1], and coxiellosis in animals [2]. It is an obligate intracellular bacterium, replicating in eukaryotic cells that shows a biphasic developmental cycle: the large cell variant (LCV) corresponding to the intracellular replicative form and the small cell variant (SCV) which represents the host cell free stable form that is highly resistant to different environmental stresses [3]. The SCVs may persist in the environment for years [1, 3]. *C. burnetii* is found in urine, faeces and milk of infected animals, although transmission to humans is most frequently due to the inhalation of aerosolized bacteria that are spread in the environment by infected animals after delivery or abortion. Amniotic fluid and placenta contain the highest concentration of bacteria [4]. In recent years, an increasing number of animals have been reported to shed the bacterium, including pets, reptiles, ticks, rodents and birds [5–8], however, the main reservoirs of *C. burnetii* are cattle, sheep, and goats [9, 10]. Because Q fever is a zoonosis, the epidemiology of human infections generally reflects the circulation of the bacterium in ruminant livestock and serological investigation. Prophylaxis and molecular characterization of the strains circulating is therefore essential in order to prevent human outbreaks. Several genotyping methods have been described thus far. Until 2005, these techniques were based on plasmid typing, restriction fragment length polymorphism followed by pulsed-field gel electrophoresis (RFLP-PFGE) analysis, and sequence analysis of individual genes (e.g. *16S*). Nevertheless, some of these methods exhibited limitations on inter- and intra-laboratory reproducibility and poor discrimination power that hindered their widespread use [11]. In 2005, typing via sequence analysis of multi-spacer regions (MST) was introduced by Glazunova et al. [12], who identified 10 highly variable intergenic spacers allowing the unambiguous characterization of the agent. MST

* Correspondence: m.didomenico@izs.it
[1]Istituto Zooprofilattico Sperimentale dell'Abruzzo e del Molise "G. Caporale", Campo Boario, 64100 Teramo, Italy
Full list of author information is available at the end of the article

genotyping has high levels of discrimination and helps to trace the spread of *C. burnetii* from one region to another and to define phylogenetic relationships [13, 14]. Otherwise, multi-locus Variable Number Tandem Repeat (VNTR) analysis (MLVA) was first established by Svraka et al. [15], and then improved by Arricau-Bouvery et al. [16]. Since their rapid evolution VNTR are extremely polymorphic, therefore MLVA usually provide a better discriminatory power, which is often suitable for epidemiological purposes [17]. MST and MLVA are both PCR-based techniques, and they have the potential to be used directly on non-cultured samples, avoiding the culture step of the pathogen that requires biosafety level 3 and long-time analyses. In this study, we report the genetic diversity of *C. burnetii* by MST and MLVA with the aim of describing the strains circulating in central Italy, taking into account the knowledge acquired after the Q Fever outbreaks in the Netherlands [18–20].

Methods
Biological samples and DNA extraction
During the period 2012–2015 20 farms were positive for *C. burnetii* by Real Time PCR [21], with slight modifications [14]. Farms were named G1-G10 for goat, S1-S9 for sheep, and C1 for cow (Table 2). Brain, spleen, lung and liver were sampled from aborted goat (N. 10) and sheep (N. 9) fetuses and vaginal swabs were collected from the relative parturient animals. Placenta was collected only in two sampling sessions, these samples referred to a goat (G2) and sheep (S9). A milk sample was collected from a serological positive cow (C1). Only one DNA sample per farm was used for genotyping purposes; the selection was driven by the lowest C_t values obtained by Real Time PCR. Details of the samples per geographic origin, specimen, year of collection and C_t value are reported in Table 1 and Fig. 1.

DNA from fetal organs was isolated using the Maxwell 16 Tissue DNA Purification Kit (Promega) following the manufacturer's instructions. Cells from milk were recovered by centrifuging 50 ml at 2000 g for 10 min; the pellet was resuspended with 300 μl of nuclease free water and then DNA was isolated using the Maxwell 16 Cell DNA Purification Kit (Promega) following the manufacturer's instructions.

MST and MLVA
Multispacer Sequence Typing of *C. burnetii* DNA was performed on 20 specimens as previously described by Glazunova et al. [12] with some modifications [14]. Raw sequence data were assembled using SeqMan Pro (DNASTAR Lasergene 10 Core Suite). The coded alleles were compared with the sequences in the reference database available on the website http://ifr48.timone.univ-mrs.fr/mst/coxiella_burnetii/groups.html. MLVA was performed on the same samples for the 6 loci panel [16] with some

Table 1 Detailed information of the specimens

ID[a]	Province	Year	Species	Specimen	Ct value
G1	Pescara	2015	goat	brain	31
G2	Pescara	2014	goat	placenta	8
G3	Caserta	2015	goat	lung	28
G4	L'Aquila	2015	goat	brain	28
C1	Chieti	2016	cattle	milk	23
S1	L'Aquila	2015	sheep	brain	29
S2	Pisa	2012	sheep	vaginal swab	21
S3	Rieti	2015	sheep	spleen	29
S4	Viterbo	2014	sheep	lung	30
G5	Roma	2013	goat	vaginal swab	20
G6	Roma	2012	goat	spleen	26
G7	Latina	2014	goat	liver	19
G8	Latina	2014	goat	spleen	24
G9	Roma	2014	goat	lung	22
S5	Livorno	2014	sheep	spleen	22
S6	Grosseto	2015	sheep	lung	23
S7	Grosseto	2015	sheep	spleen	24
S8	Firenze	2013	sheep	vaginal swab	22
S9	Grosseto	2015	sheep	placenta	28
G10	Rieti	2014	goat	lung	26

[a]Farm ID designation according to animal species host: *G* goat, *S* sheep, *C* cattle. Samples per farm with the lowest Ct values were reported only

modifications described by Tilburg et al. [22]. Furthermore, in order to obtain complete panels, new reverse primers were designed for the refractive markers Ms24 (5'-ACAAGCTATTTACTCCCTTTCTGC-3') and Ms34 (5'-GCGTTAGTGTGCTTATCTCTTG-3') by the Primer Express v3.0.1 software (Applied Biosystems), while for Ms23 and Ms33 primers from the website http://mlva.u-psud.fr/MLVAnet/spip.php?rubrique50 were selected. Three more DNA samples for which MST profile was previously reported [14], were included in the VNTR analysis for a total of 23 different strains. The amplification products were diluted 1:10 and 1:100 in distilled water and 1.2 μl of every single dilution was added to reaction mixture containing 10.5 μl formamide and 0.3 μl of Gene Scan 500 size standard marker (Applied Biosystems). PCR products were denatured for 30 s at 95 °C, cooled on ice for 2 min and then run on AB3130XL capillary sequencer with POP7 polymer. VNTR fragments were finally sized via GeneMapper software v.4.0 (Applied Biosystems). DNA from the Nine Mile strain (RSA 493) was used as a reference. According to the online database support site (http://mlva.u-psud.fr/mlvav4/genotyping/index.php), the MLVA profile of the Nine Mile strain is 9 27 4 6 9 5 for markers Ms23, Ms24, Ms27, Ms28, Ms33 and Ms34, respectively. For each marker the repeats number was determined inferring the sizes of the sample

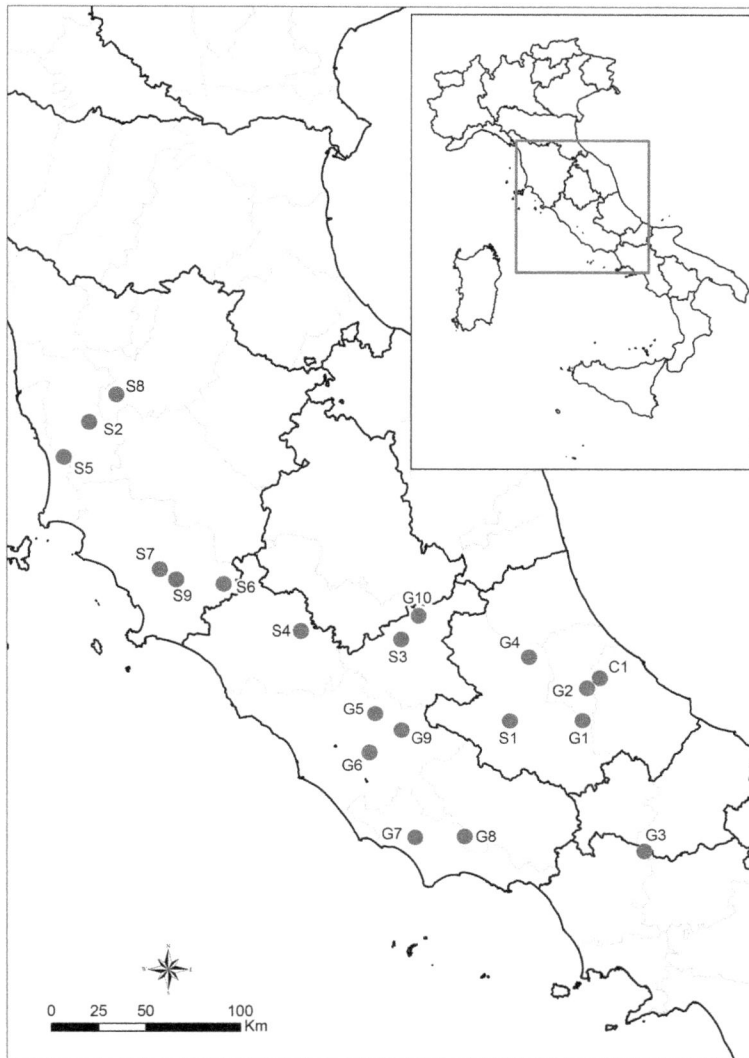

Fig. 1 Geographical localization of the farms in central Italy. Farm ID was designated according to animal species host: G = goat; S = sheep; C = cattle

fragments with those obtained using the reference strain, run in the same time. Ten samples were sequenced to confirm the repeats number and the presence of Insertion Sequences (IS).

Results

The MST analysis from the 10 goat, 9 sheep and 1 cattle PCR-positive farms revealed two dominant previously described Sequence Types: ST32, reported in 12 farms (60%), and ST12, reported in 5 farms (25%). In addition, ST8 and ST55 were reported in the farms S5 (5%) and G1 (5%) respectively. Finally, a variant of the ST8 (proposed ST62) previously described only in this geographic area [14], was described in farm S1 (5%). This new allelic combination is 5 4 2 5 1 5 3 2 4 4 for the spacers Cox2, Cox5, Cox18, Cox20, Cox22,

Cox37, Cox51, Cox56, Cox57, and Cox61 respectively. A complete panel was accomplished for all samples through MST analysis (Table 2, Fig. 2), while incomplete results were obtained for MLVA (Table 2). Although only partial panels were gained for the samples G1, G3-G6, G8-G10, S1-S5, S9 and C1, the high resolution power of MLVA allowed differentiation of strains sharing the same MST profile such as the case of the sample from farm G3, with ST32 like eight other farms, but showing a unique allele (12 tandem repeats) in the Ms24 marker. The case of strains from goat farms G7 and G8, located in the same geographic area and having the same ST32, is analogous. The variation in the number of tandem repeats in the locus Ms24 allowed the differentiation of the two genotypes showing 11 and 10 tandem repeats respectively, as well as the strains from

Table 2 MST and MLVA genotyping of *C. burnetii* DNA from domestic ruminants in central Italy

Farm ID[a]	Province	Specimen	MST	MLVA 6 panel						Study MLVA genotype
				Ms23	Ms24	Ms27	Ms28	Ms33	Ms34	
G1	Pescara	brain	ST55			4	5	6		1
G2	Pescara	placenta	ST32	9	10	2	3	5 (IS1111)[b]	3	2
G3	Caserta	lung	ST32		12	2	3		3	3
G4	L'Aquila	brain	ST32		10	2	3		3	
C1	Chieti	individual milk	ST32	5	10	2	3		3	
S1	L'Aquila	brain	ST8[c]			4	5	8	2	4
S2	Pisa	vaginal swab	ST32		10	2	3		3	
S3	Rieti	spleen	ST32	10		2	3		3	
S4	Viterbo	lung	ST32			2	3		3	
G5	Roma	vaginal swab	ST32		10	2	3		3	
G6	Roma	spleen	ST32		10	2	3		3	
G7	Latina	liver	ST32	8	11	2	3	5 (IS1111)[b]	3	5
G8	Latina	spleen	ST32		10	2	3	9	3	6
G9	Roma	lung	ST32		10	2	3	5 (IS1111)[b]	3	
S5	Livorno	spleen	ST8		7	4	3	8	2	7
S6	Grosseto	lung	ST12							
S7	Grosseto	spleen	ST12							
S8	Firenze	vaginal swab	ST12							
S9	Grosseto	placenta	ST12		11	2	3		3	
G10	Rieti	lung	ST12		12				3	
Di Domenico et al. 2014 (C2) [14]	Pescara	individual milk	ST20[c]	6	12	2	7	6	9	8
Di Domenico et al. 2014 (G11) [14]	L'Aquila	individual milk	ST55	9	22	4	1	3	1	9
Di Domenico et al. 2014 (G12) [14]	L'Aquila	brain	ST8[c]	9	8	4	1	5	2	10

[a]Farm ID designation according to animal species host: *G* goat, *S* sheep, *C* cattle
[b]The presence of the Insertion Sequence IS1111 was confirmed by sequencing
[c]New genotypes described by Di Domenico et al. [14]

farms S9 and G10 (11 and 12 tandem repeats, respectively), while showing the same ST12. Moreover, the polymorphism of the marker MS33 allows the discrimination of the strain G8 (9 tandem repeats) from G2 and G9 (5 tandem repeats + IS1111), all of these sharing the same ST32. Ten different MLVA genotypes emerged from the analysis (Table 2). Three samples are omitted from the MLVA description because of the limited amount of DNA available. The five complete MLVA profiles compared with those deposited in the public MLVA database (http://mlva.u-psud.fr/mlvav4/genotyping/index.php) did not match with any of the published genotypes. Sequence analysis of the samples G2, G7, and G9 confirmed the presence of the IS1111.

Discussion

Discrimination of bacterial strains based on the analysis of their genetic content has become widely used due to its high resolution. Genetic variability can explain most of the phenotypic heterogeneity within bacterial population, such as host specificity, pathogenicity, antibiotic resistance, and virulence [23]. Currently, two main discriminant methods are commonly used for *Coxiella burnetii* genotyping: MST and MLVA [4]. MST has succeeded in establishing reliable correlations of STs with geographic distribution, clinical manifestations, and epidemiology of strains. Despite some STs being found worldwide, many others are restricted to specific areas. As a result, it has also been qualified as a "geotyping method" [4]. For example, to date ST17 has been isolated only from French Guiana, where it causes severe forms of the disease [24]. Acute Q Fever patients from French Guiana demonstrate significantly higher phase I IgG titers, whereas phase II antibodies are generally revealed at this stage of the disease [4]. Of note, pneumonia is observed in 83% of the patients representing the genotype with the highest prevalence of community-acquired pneumonia in the world [4, 25]. Analogously, ST33 is distributed in different small areas

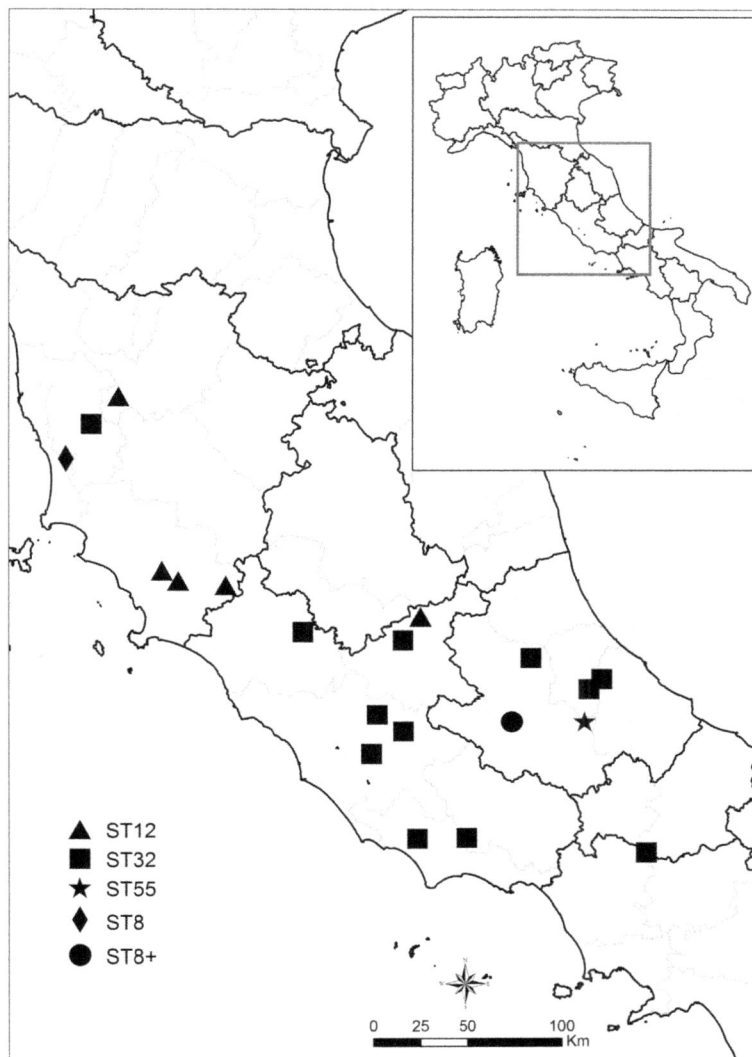

Fig. 2 Genetic diversity. Sequence types are expressed accordingly to MST database. ST8+ refers to a variant of the ST8 as previously described (Di Domenico et al. [14])

of Europe including Germany, where it was first described, and the Netherlands where it spread via France, causing the largest Q Fever outbreak ever described [26, 27]. Strains belonging to the ST23 group were reported in ticks, birds, ruminants and humans only in a restricted area between Eastern Europe (Czech Republic and Slovakia) and Asian countries (Russia, Kazakhstan, Mongolia and Uzbekistan), (http://ifr48.timone.univ-mrs.fr/mst/coxiella_burnetii/strains.html). Another interesting case is that of ST21. Despite two isolates from France and one from the United States, it is mainly reported in Nova Scotia [12]. On the other hand, some sequence types are distributed worldwide, such as ST16.

In Italy, sequence types already described are ST16, ST18 and ST29 (http://ifr48.timone.univmrs.fr/mst/coxiella_burnetii/strains.html), a novel sequence type similar to ST20, an additional sequence type related to ST8 and ST55 discovered in bovine milk, goat fetus and goat milk, respectively [14]. ST12 and ST32 are closely related on the basis of phylogenetic analysis [13, 14] and largely distributed along the area considered in the present study. Indeed, these two genotypes recur in 85% of the specimens (17/20) in cow, sheep and goat. These findings confirmed the spread of these genotypes in Tuscany as previously reported [28]. Interestingly, ST12 has been detected in clinical human samples from France, Switzerland and Senegal (http://ifr48.timone.univmrs.fr/mst/coxiella_burnetii/strains.html), whereas in animals it has been only found in Italy [28]. This result suggests that goat and sheep could represent an important source of human Q fever in this country. Although the oral exposure is still controversial [9], the transmission of *C. burnetii* to humans through inhalation of contaminated aerosol is widely recognized, especially for certain at-risk categories, such as farmers, veterinarians, or people living close to or exposed to infected flocks [29–32]. Conversely, the zoonotic origin of ST32 and the transmission to human have been already assessed. Indeed, as reported in the database http://ifr48.timone.univ-mrs.fr/mst/coxiella_burnetii/strains.html, it was identified in a goat placenta in Austria, and detected in human heart valve and aortic biopsy in Germany and France, respectively. Our study confirms ST32 detection in sheep specimens and it firstly describes the occurrence of this zoonotic genotype in cows. Three additional MST profiles have been reported in the present study: ST8, ST55 and a novel ST closely related to the ST8 previously identified in the same area [14]. Notably, ST8 was responsible for two human chronic Q Fever cases in Portugal [33].

The Q Fever outbreak in the Netherlands pushed toward the molecular investigation by both MST and MLVA of *C. burnetii* strains in different countries, not only in Europe. As VNTRs are important source of genetic polymorphisms for strain typing due to their rapid evolution, MLVA approach is particularly useful for epidemiological purposes. Unfortunately, PCR amplification is not always successful for all markers, so that partial genotypes are frequently obtained causing underestimation of the genotypic diversity [33–36]. Moreover, insertions or deletions may impair the estimation of the number of tandem repeats [37]. In particular, Ms23 and Ms33 both harbour a recognition site for the IS1111 insertion sequence upstream the repeat units and therefore may constitute preferred targets for insertions [37].

In the present study, five complete MLVA profiles were gained, all of which were different from the previously reported genotypes, including those recently found in Italy [35], while 15 panels were incomplete. Despite this, MLVA enabled 10 genotypes to be identified, instead of the 5 obtained by MST. Moreover, by means of sequencing we detected the presence of the IS1111 within the repetitive region of the Ms33 marker in three different goat samples sharing the same sequence type (ST32).

Conclusions

Our study based on MST and MLVA-6, established a part of *C. burnetii* strain atlas in central Italy. Integration of such data with international databases can be of further help in the attempt of building a global phylogeny and epidemiology of Q fever in animals, with a "One Health" perspective.

Abbreviations

LCV: Large cell variant; MLVA: Multi Locus VNTR Analysis; MST: Multispacer Sequence Typing; RFLP-PFGE: Restriction fragment length polymorphism followed by pulsed-field gel electrophoresis; SCV: Small cell variant; ST: Sequence Type; VNTR: Variable Number Tandem Repeat

Acknowledgements

The authors wish to thank Luigi Sorbara and Fiorentino Stravino for outstanding technical assistance.

Funding

Funding was provided by the Italian Ministry of Health. The funding organization played no role in the design of study, choice of specimens, review and interpretation of data, or preparation or approval of manuscript.

Authors' contributions

All authors have made substantial contributions to conception and design of the study and have been involved in revising the manuscript critically for important intellectual content. Samples collection and acquisition of data for diagnosis purposes have been accomplished by LDG, AF, AC and AB. MDD, VC, CC, VDL and MM have made substantial contributions to acquisition and analysis of genotyping data. MDD have made significant contributions to interpretation of data and together with all other authors have been involved in drafting the manuscript. All authors have given final approval of the version to be published. Finally, all authors agree all aspects of the work ensuring that questions related to the accuracy or integrity of any part of the work are appropriately investigated and resolved.

Competing interests

The authors declare that they have no competing interests.

Author details

[1]Istituto Zooprofilattico Sperimentale dell'Abruzzo e del Molise "G. Caporale", Campo Boario, 64100 Teramo, Italy. [2]Istituto Zooprofilattico Sperimentale del Lazio e della Toscana "M. Aleandri", Via Appia Nuova 1411, 00178 Roma, Italy.

References

1. Raoult D, Marrie TJ, Mege JL. Natural history and pathophysiology of Q fever. Lancet Infect Dis. 2005;5:219–26.
2. Pearson T, Hornstra HM, Hilsabeck R, Gates LT, Olivas SM, Birdsell DM, Hall CM, German S, Cook JM, Seymour ML, Priestley RA, Kondas AV, Clark Friedman CL, Price EP, Schupp JM, Liu CM, Price LB, Massung RF, Kersh GJ, Keim P. High prevalence and two dominant host-specific genotypes of *Coxiella burnetii* in U.S. milk. BMC Microbiol. 2014;14:41.
3. Tissot-Dupont H, Raoult D. Q Fever. Infect Dis Clin N Am. 2008;22:505–14.
4. Eldin C, Mélenotte C, Mediannikov O, Ghigo E, Million M, Edouard S, Mege JL, Maurin M, Raoult D. From Q fever to *Coxiella burnetii* infection: a paradigm change. Clin Microbiol Rev. 2017;30:115–90.
5. Pascucci I, Di Domenico M, Dall'Acqua F, Sozio G, Cammà C. Detection of Lyme disease and Q fever agents in wild rodents in central Italy. Vector-Borne Zoonotic Dis. 2015;15(7):404–11.
6. D'amato F, Million M, Edouard S, Delerce J, Robert C, Marrie T, Raoult D. Draft genome sequence of *Coxiella burnetii* Dog Utad, a strain isolated from a dog-related outbreak of Q fever. New Microbes New Infect. 2014;2:136–7.
7. Langley JM, Marrie TJ, Covert A, Waag DM, Williams JC. Poker players' pneumonia. An urban outbreak of Q fever following exposure to a parturient cat. N Engl J Med. 1988;319(6):354–6.
8. Socolovschi C, Reynaud P, Kernif T, Raoult D, Parola P. Rickettsiae of spotted fever group, *Borrelia valaisiana*, and *Coxiella burnetii* in ticks on passerine birds and mammals from the Camargue in the south of France. Ticks Tick-borne Dis. 2012;3:354–9.
9. Scientific Opinion on Q fever. EFSA panel on Animal Health and Welfare (AHAW) EFSA panel on Biological Hazards (BIOHAZ) (Chapter 4 on Food Safety). EFSA J. 2010;8(5):1595.
10. Agerholm J. *Coxiella burnetii* associated reproductive disorders in domestic animals-a critical review. Acta Vet Scand. 2013;55:13.
11. Massung RF, Cutler SJ, Frangoulidis D. Molecular typing of *Coxiella burnetii* (Q fever). Adv Exp Med Biol. 2012;984:381–96.
12. Glazunova O, Roux V, Freylikman O, Sekeyova Z, Fournous G, Tyczka J, Tokarevich N, Kovacova E, Marrie TJ, Raoult D. *Coxiella burnetii* genotyping. Emerg Infect Dis. 2005;11(8):1211–7.
13. Hornstra HM, Priestley RA, Georgia SM, Kachur S, Birdsell DN, Hilsabeck R, Gates LT, Samuel JE, Heinzen RA, Kersh GJ, Keim P, Massung RF, Pearson T. Rapid typing of *Coxiella burnetii*. PLoS One. 2011;6(11):e26201.

14. Di Domenico M, Curini V, De Massis F, Di Provvido A, Scacchia M, Camma C. *Coxiella burnetii* in Central Italy: novel genotypes are circulating in cattle and goats. Vector-Borne Zoonotic Dis. 2014;14(10):710–5.

15. Svraka S, Toman R, Skultety L, Slaba K, Homan WL. Establishment of a genotyping scheme for *Coxiella burnetii*. Microbiol Lett. 2006;254(2):268–74.

16. Arricau-Bouvery N, Hauck Y, Bejaoui A, Frangoulidis D, Bodier C, Souriau A, Meyer H, Neubauer H, Rodolakis A, Vergnaud G. Molecular characterization of *Coxiella burnetii* isolates by infrequent restriction site-PCR and MLVA typing. BMC Microbiol. 2006;6:38.

17. van Belkum A. Tracing isolates of bacterial species by multilocus variable number of tandem repeat analysis (MLVA). FEMS Immunol Med Microbiol. 2007;49:22–7.

18. Roest HIJ, Ruuls RC, Tilburg JJHC, Nabuurs-Franssen MH, Klaassen CHW, Vellema P, van den Brom R, Dercksen D, Wouda W, Spierenburg MAH, van der Spek AN, Buijs R, de Boer AG, Willemsen PTJ, van Zijderveld FG. Molecular epidemiology of *Coxiella burnetii* from ruminants in Q fever outbreak, the Netherlands. Emerg Infect Dis. 2011;17(4):668–75.

19. Roest HIJ, Tilburg JJHC, van der Hoek W, Vellema P. The Q fever epidemic in the Netherlands: history, onset, response and reflection. Epidemiol Infect. 2011;139:1):1–12.

20. Tilburg JJHC, Rossen JWA, van Hannen EJ, Melchers WJG, Hermans MHA, van de Bovenkamp J, Roest HJIJ, de Bruin A, Nabuurs-Franssen MH, Horrevorts AM, Klaassen CHW. Genotypic diversity of *Coxiella burnetii* in the 2007-2010 Q fever outbreak episodes in the Netherlands. J Clin Microbiol. 2012;50(3):1056–8.

21. Panning Panning M, Kilwinski J, Greiner-Fischer S, Peters M, Kramme S, Frangoulidis D, Meyer H, Henning K, Drosten C. High throughput detection of *Coxiella burnetii* by real-time PCR with internal control system and automated DNA preparation. BMC Microbiol. 2008;8:77.

22. Tilburg JJHC, Roest HJIJ, Nabuurs-Franssen MH, Horrevorts AM, Klaassen CHW. Genotyping reveals the presence of a predominant genotype of Coxiella burnetii in consumer milk products. J Clin Microbiol. 2012;50(6):2156–8.

23. Li W, Raoult D, Fournier PF. Bacterial strain typing in the genomic era. FEMS Microbiol. 2009;33:892–916.

24. Mahamat A, Edouard S, Demar M, Abboud P, Patrice JY, La Scola B, Okandze A, Djossou F, Raoult D. Unique clone of *Coxiella burnetii* causing severe Q fever, French Guiana. Emerg Infect Dis. 2013;19(7):1102–4.

25. D'Amato F, Eldin C, Georgiades K, Edouard S, Delerce J, Labas N, Raoult D. Loss of TSS1 in hypervirulent *Coxiella burnetii* 175, the causative agent of Q fever in French Guiana. Comp Immunol Microbiol Infect Dis. 2015;41:35–41.

26. Tilburg JJHC, Roest HJIJ, Buffet S, Nabuurs-Franssen MH, Horrevorts AM, Raoult D, Klaassen CHW. Epidemic genotype of *Coxiella burnetii* among goats, sheep, and humans in the Netherlands. Emerg Infect Dis. 2012;18(5):887–8.

27. D'Amato F, Rouli L, Edouard S, Tyczka J, Million M, Robert C, Nguyen TT, Raoult D. The genome of *Coxiella burnetii* Z3055, a clone linked to the Netherlands Q fever outbreaks, provides evidence for the role of drift in the emergence of epidemic clones. Comp ImmunolMicrobiol Infect Dis. 2014;37:281–8.

28. Galiero A, Fratini F, Cammà C, Di Domenico M, Curini V, Baronti I, Turchi B, Cerri D. Occurrence of *Coxiella burnetii* in goat and ewe unpasteurized cheeses: screening and genotyping. Int J Food Microbiol. 2016;237:47–54.

29. Manfredi Selvaggi T, Rezza G, Scagnelli M, Rigoli R, Rassu M, De Lalla F, Pellizzer GP, Tramarin A, Bettini C, Zampieri L, Belloni M, Pozza ED, Marangon S, Marchioretto N, Togni G, Giacobbo M, Todescato A, Binkin N. Investigation of a Q-fever outbreak in northern Italy. Eur J Epidemiol. 1996;12(4):403–8.

30. Starnini G, Caccamo F, Farchi F, Babudieri S, Brunetti B, Rezza G. An outbreak of Q fever in a prison in Italy. Epidemiol Infect. 2005;133:377–80.

31. Van den Brom R, van Engelen E, Roest HI, van der Hoek W, Vellema P. *Coxiella burnetii* infections in sheep or goats: an opinionated review. Vet Microbiol. 2015;181(1–2):119–29.

32. Brooke RJ, Mutters NT, Péter O, Kretzschmar ME, Teunis PF. Exposure to low doses of *Coxiella burnetii* caused high illness attack rates: insights from combining human challenge and outbreak. Epidemic. 2015;11:1–6.

33. Santos AS, Tilburg JJHC, Botelho A, Barahona MJ, Núncio MS, Nabuurs-Franssen MH, Klaassen CHW. Genotypic diversity of clinical *Coxiella burnetii* isolates from Portugal based on MST and MLVA typing. Int J Med Microbiol. 2012;302:253–6.

34. Račić I, Spičić S, Galov A, Duvnjak S, Zdelar-Tuk M, Vujnović A, Habrun B, Cvetnić Ž. Identification of *Coxiella burnetii* genotypes in Croatia using multi-locus VNTR analysis. Vet Microbiol. 2014;173(3–4):340–7.

35. Ceglie L, Guerrini E, Rampazzo E, Barberio A, Tilburg JJHC, Hagen F, Lucchese L, Zuliani F, Marangon S, Natale A. Molecular characterization by MLVA of *Coxiella burnetii* strains infecting dairy cows and goats of North-Eastern Italy. Microbes Infect. 2015;17(11–12):776–81.

36. Sulyok KM, Kreizinger Z, Hornstra HM, Pearson T, Szigeti A, Dán Á, Balla E, Keim PS, Gyuranecz M. Genotyping of *Coxiella burnetii* from domestic ruminants and human in Hungary: indication of various genotypes. BMC Vet Res. 2014;10:107.

37. Sidi-Boumedine K, Duquesne V, Prigent M, Yang E, Joulié A, Thiéry R, Rousset E. Impact of IS1111 insertion on the MLVA genotyping of *Coxiella burnetii*. Microbes Infect. 2015;17(11–12):789–94.

21

Effects of dietary supplementation with polyphenols on meat quality in Saanen goat kids

Roberta Cimmino[1]* ⓘ, Carmela M. A. Barone[2], Salvatore Claps[3], Ettore Varricchio[4], Domenico Rufrano[3], Mariangela Caroprese[5], Marzia Albenzio[5], Pasquale De Palo[6], Giuseppe Campanile[7] and Gianluca Neglia[7]

Abstract

Background: Diet supplementation with polyphenols is a novel strategy to improve meat quality in livestock, by preventing oxidative deterioration of lipids and protein. Polyphenols have beneficial effects on both human and animal health and can be obtained from several sources, such as olive mill wastewaters (OMWW). These are severe environmental pollutants and therefore may be recycled and utilized in other sectors. The aim of this study was to evaluate growth performance, meat characteristics, fatty acid composition, antioxidant status, different forms of myoglobin and malondialdehyde formation in kids who received a diet supplemented with polyphenols obtained from OMWW. Weaned goat kids ($n = 18$) were divided into two homogenous groups: control (C) group ($n = 9$) received a fattening standard diet while the other group ($n = 9$) received the same diet, supplemented with 3.2 mg/day of polyphenols powder extract (PE group). Average daily gain (ADG) was calculated 10 days apart throughout the study. After 78 days, the kids were slaughtered and pH and carcass yield were evaluated. *Longissimus thoracis et lumborum* muscle was collected and utilized for chemical analysis, meat quality evaluation and oxidative stability.

Results: No differences were recorded in ADG, carcass weight, pH and dressing between the two groups. Furthermore a similar meat proximate composition, texture and color was observed. Dietary polyphenols supplementation significantly ($P < 0.01$) decreased short chains (<C12:0) (2.93 + 0.50 and 0.35 + 0.40 g/100 g of fatty acids, for C and PE Group, respectively), and saturated (49.22 ± 2.39 and 39.51 ± 1.95 g/100 g, in C and PE Group, respectively) fatty acids. Furthermore, a higher ($P < 0.05$) proportion of monounsaturated (34.35 ± 2.84 and 42.22 ± 2.32 g/100 g, in C and PE Group, respectively) fatty acids was recorded. Malondialdehyde formation was significantly ($P < 0.05$) lower in PE compared to C Group (0.25 ± 0.005 and 0.15 ± 0.005, in C and PE Group, respectively).

Conclusions: Polyphenols dietary supplementation has positive effects on kid meat, improving fatty acid profile and reducing malondialdehyde contents. Furthermore the utilization of OMWW as the source of polyphenols may represent an innovative strategy to re-utilize agri-food industry wastes.

Keywords: Polyphenols, Kids, Fatty acids, Sensory evaluation, Malondialdehyde

Background

An increased awareness of the consumers was recently observed regarding red meat's production and consumption. A typical example was the latest report of the World Health Organization (WHO), which classified the consumption of red meat as *"probably carcinogenic for man"* [1], although, as outlined by other authors [2], there are still a number of

gaps in the current knowledge about this topic. Furthermore, eating red meat has been correlated with higher incidence of chronic diseases [3]. For this reason there has been an increased interest for new breeding techniques which confer potential health benefits to the consumers [4].

The supplementation of ruminants diet with antioxidants is considered an effective strategy for changing and ameliorating the fatty acid composition of meat in response to consumer demands [4]. In fact, it has been observed that

---
* Correspondence: r.cimmino@anasb.it
[1]Italian Buffalo Breeders Association, V. Petrarca 42/44, 81100 Caserta, Italy

the inclusion of linseed and fish oil on the lamb diet may have a negative impact on the oxidative stability and on physicochemical properties of the meat [5]. In addition, the changes that occur in the muscle during the post-mortem period may create an unbalanced proportion between the antioxidant and pro-oxidant capability, increasing the risks of oxidative damage [6]. The antioxidant status of the muscle is the main factor affecting oxidative deterioration in meat [7]. Oxidative deterioration of lipids and proteins in meat could adversely affect its nutritional quality and shelf life, reducing flavor, color and quality of meat, with a negative impact on meat consumption.

Thus, it is essential to preserve the quality and the safety of the meat by attenuating oxidative deterioration. Recently, the interest of food processing industries in the use of natural antioxidants rather than synthetic counterparts was increased, for either low environmental impact and economical reasons [8]. Furthermore natural antioxidants are well accepted by the consumers, because they are considered safe and healthy [9]. For these reasons, many studies have been carried out to develop new natural antioxidants, especially from plants [8], among which polyphenols. Polyphenols are a large family of more than 8000 natural compounds derived from plants and characterized by the presence of a phenol ring in their structure [10, 11]. Epidemiological, clinical and nutritional studies strongly support the evidence that dietary phenolic compounds are effective in the prevention of common diseases, including cancer, neurodegenerative diseases and gastrointestinal disorders [12]. Furthermore, a large number of studies suggest several immunomodulatory and anti-inflammatory properties of these compounds in humans [13, 14]. Some studies, performed in livestock, reported the positive effects of diet supplementation with polyphenols on pigs and chicken health. [11, 15, 16]. In addition, it has been hypothesized that dietary supplementation with polyphenols would increase beneficial lipids and oxidative stability of myoglobin and reduce the content of malondialdehyde (MDA) [see 11 for review].

Polyphenols can be obtained from several sources. One of these is the olive oil sector, particularly important in Mediterranean countries, such as Spain (the leading producer), Italy, Greece, Turkey, Syria and Tunisia. Olive mill wastewaters (OMWW) are the main pollutant by-phase and traditional suction systems Mills. The management of OMWW is a serious environmental issue for the presence of organic compounds that turn OMWW into phytotoxic materials [17]. Nevertheless, OMWW contains valuable resources, such as polyphenols, which can represent about 10% of OMWW dry weight [18]. Some olive oil derived-compounds, such as hydroxytyrosol, have an important role in preventing cardiovascular diseases [19]. Thus OMWW could be recycled and utilized in other

sectors, such as animal feeding, to reduce its environmental impact.

Therefore, the objective of this study was to investigate the effects of the supplementation with polyphenols obtained from OMWW in the diet of Saanen goat kids on growth performance and meat quality. In particular, meat proximate composition, texture and colorimetric properties, fatty acid composition, lipid oxidation and the level of the different forms of myoglobin of *Longissimus thoracis et lumborum* (LTL) muscle were evaluated.

Methods
Animals, experimental design and diet composition
All experimental procedures were approved by the Ethical Committee on Animal Research of the University of Naples (Protocol number: 2014/0105988 of 1st December 2014), and the study was carried out in accordance with EU Directive 2010/63/EU for animal experiments.

The trial was performed on 18 Saanen female kids maintained in the experimental farm of CREA (Research Centre for Animal Production and Aquaculture) located in Bella (PZ), in the South of Italy, at 40°75′ latitude and 15°67′ W longitude, and 802 m above sea level. The kids were allowed natural suckling until weaning, that occurred at 69 days of age. After weaning the animals were divided into 2 homogeneous groups, according to age and live body weight (LW), recorded at both birth and weaning: Control (C) group ($n = 9$): received a fattening standard diet (Table 1); Polyphenols extract (PE) group ($n = 9$):

Table 1 Ingredients (%) and proximate composition (% of dry matter) of the basal diet administered to the kids throughout the experimental period

Ingredients (%) of concentrate administered		
Maize meal	60	
Faba bean (*Viccia faba minor*)	15	
Alfalfa pellets	5	
Soybean meal	5	
Wheat bran	10	
Oat	5	
	Concentrate	Dehydrated alfalfa
Dry Matter (g/100 g weight)	87.5	85.6
Ash	2.7	8.3
Crude protein	11.8	19.5
Crude fiber	6.19	25.6
Neutral detergent fiber	18.2	37.5
Acid detergent fiber	8.9	34.3
Lignin	2.1	9.8
Ether extract	3.8	1.9
Non structural carbohydrates	63.5	25.6
Starch	56.8	–

received the same diet of control group, supplemented with a powder of polyphenols extract from OMWW (3.2 mg/day, see below). The amount of polyphenols powder was defined according to antioxidant activity (see below) and some studies performed in growing lambs [5].

The kids of each group were individually penned in boxes (1 m²) throughout the study that lasted 78 days. All kids had free access to water and received the same ration, consisting of alfalfa hay ad libitum and a starter concentrate administered in increasing amount according to the growth following the recommendation of National Research Council [20]. Feed was administered twice a day, at 8.00 and 16.00 and polyphenols powder was mixed on daily bases with the concentrate in the PE Group. The inclusion level of polyphenols extract was variable throughout the study according the amount of administered concentrate from 1.70% at the start of the trial to 0.83% in the last week (1.16% on average). Feed intake was determined from orts (refusals) collected daily in the morning (when present) before the next feed administration. The amount and the composition of orts were utilized to calculate the dry matter (DM) intake and the composition of the ingested diet. Individual feedstuff and orts were sampled every 15 days and the analyses were carried out as per AOAC (Association of Official Analytical Chemists) procedures [21] after drying at 65 °C and mechanical reduction of the samples (granulometry 1 mm for all analyses). The chemical composition of individual feedstuff is reported in Table 1.

Each animal was individually weighed 10 days apart and average daily gain (ADG) was calculated by dividing the difference between two consecutive LW measurements (LW1 and LW2) with the number of days elapsed (LW2 – LW1/days).

Polyphenols determination

Polyphenols extract from OMWW was obtained and characterized as reported by Parrillo et al. [22]. Briefly, the polyphenolic extract was obtained by membrane separation of OMWW according to previous studies [23] and the main phenolic compounds were identified by HPLC (LC-4000 Series Integrated HPLC Systems, JASCO, Japan) according to Azaizeh et al. [18] (Table 2). Total phenols content in OMWW extract was determined by the Folin-Ciocalteu method [24]. The antioxidant activity of the OMWW extract was evaluated by using the free radical ABTS (2,2-Azino-bis-3-ethylbenzothiazoline-6-sulfonic acid), according to the procedures described by Re et al. [25].

Slaughtering procedure and muscle sampling

At the end of the experimental period (147 days of age), all the kids were left overnight with ad libitum access to

Table 2 Main phenolic compounds of olive mill wastewaters (OMWW) extracts

Main compounds	mg/kg	Percent
Hydroxytyrosol	20,829	21.27
Flavonoid	3278	3.35
Tyrosol	3947	4.03
Caffeic acid	9991	10.20
Verbascoside	17,449	17.82
Hydroxytyrosol derivatives (OHTY-glycol, OHTY-glucoside, 3,4-diidrossifenilethanol – 3,4-DHPEA Elenolic acid mono-Aldehyde, 3,4-DHPEA– AC hydroxytyrosol acetate)	26,826	27.40
Verbascoside derivatives (isoverbascoside, β-hydroxyverbascoside,β-hydroxyisoverbascoside)	6498	6.64
Other derivatives of cinnamic acid (cinnamic acid, o-, p- coumaric acid, ferulic acid)	3372	3.44
Caffeic acid derivatives (chlorogenic acid, neochlorogenic acid, 1-O-caffeoylquinic acid, 3,5-O-dicaffeoylquinic acid)	4681	4.78
Other polyphenols	1055	1.08
Total polyphenols (TPC) (mg/kg)	97,926	
Antioxidant activity (mmolitrolox/kg sample)	8521	

water and slaughtering procedures were carried out in accordance to the EU Regulation 2009/1099/EC on the protection of animals at the time of killing. The animals were stunned by captive bolt and the exsanguination from the jugular vein was carried out. After slaughtering, evisceration and dressing, each carcass was weighed and the pH was measured 45 min post-mortem in the LTL muscle (between 11th and 13th thoracic vertebra), using an automatic digital pH-metertest-205 (TestoInc, Sparta, NJ, USA), equipped with a penetrating electrode. The probe was calibrated with pH 4 and 7 standard buffer solutions. The dressed carcass comprises the body after removing skin, head, fore feet (at the carpal–metacarpal joint), hind feet (at the tarsal–metatarsal joint), lung, heart, liver, spleen, kidneys, kidney fat, and gastrointestinal tract fat. Furthermore, the stomachs (rumen, reticulum, omasum, and abomasum) and the postruminal tract (intestine and caecum) were removed. Dressing percentage (DP) was calculated according to the following formula:

$$DP = (\text{hot carcass weight}/\text{live weight}) \times 100$$

All carcasses were stored at 2 °C for 24 h. Ultimate pH was assessed 24 h *post-mortem* on *LTL* muscle and the carcasses were weighed again. After that, a professional butcher removed the *LTL* muscle, between the sixth thoracic and 5th lumbar vertebra, from the right side of each carcass. *LTL* samples were stored at 4 °C and used for chemical analysis, meat quality evaluation and oxidative stability. Furthermore, pH was also assessed 48 h post-mortem on LTL muscle as described above.

Chemical analyses

All analyses were carried out in the Laboratories of the Department of Veterinary Medicine and Animal Production (DMVPA), and Department of Agriculture Sciences (DIA) of University of Naples Federico II (Italy). The chemical composition of *LTL* muscle was determined on refrigerated (4 °C) samples according to the AOAC procedures [26] by using a Foodscan equipment (Food Scan™Lab 78,810). Each sample was analyzed in duplicate.

Texture measurements

Tenderness was evaluated on meat cooked in a thermo-statically controlled water bath at 90 °C on day 1 after slaughtering. To monitor temperature achieved in the middle of the sample (70 °C), a portable digital thermometer (TEMP7 thermometer digital microprocessor for Pt100 probes, TECNAFOOD MO, IT) was used. After cooking the samples were dried with paper, in order to eliminate the moisture on the surface, and held in a cooler at 4 °C before coring. A minimum of four cores of 1.27 cm^2 from each LTL muscle were obtained parallel to the longitudinal orientation of the muscle fibers. The *Instron 5565* with a *Warner–Bratzler shear* (WBS) device and crosshead speed set at 100 mm/min and a load cell of 500 kg [27] was used. According to Girard et al. [28], the measured parameters were *Shear myofibrillar force* (SMF) and *Warner-Bratzler Shear Force* (WBSF) both expressed in kg. Indeed in a shear force curve some peaks of less importance may be observed, before and after maximum positive peak shear force (WBSF). Bouton & Harris [29] related these first small peaks to the myofibrillar component of shear force coinciding with initial yield (SMF). WBSF minus SMF estimates the connective tissue contribution.

Instrumental color

The color of the meat was determined on the surface of samples using a U3000 spectrophotometer, equipped with integrating sphere (Hitachi, Tokyo, Japan). Although illuminant D$_{65}$ is largely utilized, the use of Illuminant A is recommended by AMSA [30]. For this reason color coordinates, employing the CIEL*a*b* system with two illuminants, D$_{65}$ (6500 K) and A (2856 K) and 10° standard observer, were Lightness (L*); redness (a*), and yellowness (b*) were also calculated, according to AMSA [30]. The color analyses were carried out in duplicate every day, from 24 h post-mortem until day 7, on samples (diameter 2.54 cm, thick 2 cm) stored at 4 °C. The samples were placed on white trays and wrapped with oxygen-permeable film.

Color difference (ΔE*) between each day of storage and the day 0 was calculated as follows:

$$\Delta E^* = \left(\Delta L^{*2} + \Delta a^{*2} + \Delta b^{*2}\right)^{1/2}$$

Where AL*, Δa*and Δb*are the differences between L*, a* and b* values at time 0 and the individual readings each day.

Metmyoglobin (MMb), deoxymyoglobin (DMb) and oxymyoglobin (OMb) percentages were estimated according to [30] on the basis of the *Reflex Attenuance* (A) at the isobestic points 572, 525, 473 and 730 nm (nm). The *Reflex Attenuance* (A) was identified as:

$$A = \log\left(1/R\right)$$

where R expresses the reflectance at a specific wavelength in decimal (0.30 rather than 30%).

Therefore from the *Reflex Attenuance* (A), it was possible to estimate the three forms of myoglobin:

$$\%MMb = [1.395 - (A572 - A730/A525 - A730)]\, x100$$

$$\%DMb = [2.375^*(1 - ((A473 - A730)./(A525 - A730)]x100$$

$$\%OMb = [(100 - (\%MMb + \%DMb)]$$

Fatty acid analysis

Fatty acid composition of the meat was assessed on fresh samples on day 1 post-slaughtering. The muscle was blended in a food processor and the lipids were extracted from 5 g samples in duplicate, using chloroform:methanol (2:1, *v/v*) [31]. The extracted lipids were transmethylated to their fatty acid methyl esters (FAME) according to Christie [32]. The amount of fatty acid (g/100 g of FAME) was determined by gas chromatography using a chromatograph (DANI fast GC, Italy), with a flame ionization detector and equipped with a capillary column (TR-CN 100) (60 m × 0.25 mm diameter × 20 μm). Helium was used as carrier gas at a flow rate of 1.2 mL/min. The split ratio was 50:1, the injector was set at 280 °C and the detector at 240 °C. The oven temperature was programmed and held at 80 °C for 5 min, then increased to 165 °C at 5 °C/min, held at 165 °C for 1 min, and then increased to 260 °C at 3 °C/min, and then held at 260 °C for 1 min.

Identification of each fatty acid was obtained by comparing the chromatogram with the reference standard mixture of Supelco 37 component series FAME MIX (Supelco Bellefonte, PA, USA) and a mixture of CLA isomers (Nu-Chek-Prep, Inc., Elysian, MN, USA). Retention time and area of each peak were calculated using the Clarity software (Clarity v.2.4.1.77, Data Apex Ltd., 2005). The atherogenic index (AI) and the thrombogenic index (TI) were obtained by using the following equations [33].

$$AI : [(4 \times C14 : 0) + C16 : 0 + C18 : 0]$$

$$/[\Sigma MUFA + \Sigma PUFA-n6 + \Sigma PUFA-n3]$$

$$TI : [(C14 : 0 + C16 : 0 + C18 : 0)$$

$$/(0.5 \times MUFA) + (0.5 \times PUFA-n6)$$

$$+(3 \times PUFA-n3) + (PUFA-n3/PUFA-n6)]$$

Evaluation of lipid oxidation

The evaluation of lipid oxidation in meat was based on the determination of malondialdehyde (MDA), a secondary lipid oxidation product (nmol/µg of meat). MDA was determined by a specific kit (Lipid Peroxidation (MDA) Assay Kit, Sigma-Aldrich, USA) according to the manual instructions. MDA was quantified by spectrophotometric analysis (Model 680 Microplate Reader, Biorad, Italy), at a wave length of 532 nm. MDA content was evaluated at 1, 3 and 7 days *post-slaughtering*.

Statistical analysis

The experimental data were subjected to one way analysis of variance (ANOVA) using the linear model of the SAS software [34], following confirmation of normality and homogeneity of variance. The influence of dietary treatment (polyphenols supplemented diet or control diet),storage (day 1, 3 and 7) and their interaction (dietary treatment x storage) were used as main factors to analyze color parameters, MDA levels and the different forms of myoglobin. Results are presented as mean values with standard error of mean (SEM). The mean were compared using the Student's *t*-test. The differences were considered significant when P values were lower than 0.05, whereas a tendency was considered for $P < 0.10$.

Results

Animal performance and carcass characteristics

The effects of polyphenols supplementation on growth performance and carcass properties are shown in Table 3. A similar ADG from weaning to slaughter and a similar weight at slaughter were observed in the two Groups. Furthermore, carcass weights and yields were similar between groups at 0 and 24 h post mortem (Table 3). No significant differences were observed in the pH decline values between the groups both at 45′ minutes and 24 h post-mortem (Table 3).

Meat quality

A similar meat chemical composition was recorded in C and PE group (Table 4).

The texture assay highlights a double peak (SMF = 1st peak and WBSF = 2nd peak)in 9/9 animals of C samples

Table 3 Age and weight at weaning and at slaughter, average daily gain (ADG), feed consumption, carcass characteristics and dressing percentage in Control (C) and polyphenols extract (PE) group (mean values ± SEM)

	Groups		Significance
	C	PE	P
Age at weaning (days)	69.8 ± 1.30	68.9 ± 1.50	0.65
Age at slaughtering (days)	146.8 ± 1.30	145.9 ± 1.50	0.65
Weaning body weight (kg)	11.0 ± 0.75	10.7 ± 0.86	0.78
Final body weight (kg)	17.9 ± 0.99	18.6 ± 1.28	0.69
Weight gain (kg)	6.94 ± 0.60	7.93 ± 0.61	0.26
ADG (kg/g)	0.09 ± 0.01	0.10 ± 0.01	0.26
Concentrate intake (kg/day)	0.30 ± 0.00	0.29 ± 0.00	0.33
Hay intake (kg/day)	0.68 ± 0.01	0.67 ± 0.01	0.39
DM intake (kg/day)	0.84 ± 0.02	0.85 ± 0.03	0.57
Carcass weight at slaughtering (kg)	10.0 ± 0.51	10.0 ± 0.77	0.98
Carcass weight at 24 h (kg)	9.7 ± 0.50	9.7 ± 0.74	0.99
pH at slaughtering	6.8 ± 0.11	6.8 ± 0.12	0.90
pH at 24 h	5.8 ± 0.03	5.8 ± 0.03	0.64
Dressing at slaughtering (%)	55.8 ± 1.41	54.0 ± 0.48	0.15
Dressing at 24 h (%)	54.6 ± 1.39	52.3 ± 0.45	0.14

compared to 3/9 animals of PE counterparts. SMF values obtained by deformation curves where a double peak was evident, tended to be higher ($P = 0.06$) in PE compared to C group, whereas similar WBSF values were recorded (Table 5).

The color coordinates detected by using the two illuminants (A and D65) were similar between the two groups. In fact, although the illuminant A would give more emphasis on the proportion of red wave lengths, no significant differences were recorded in lightness (L*); redness (a*), yellowness (b*) between the two groups and throughout the storage. Furthermore, no significant interaction storage x treatment was assessed. Slight differences were observed in color differences (ΔE*) during the

Table 4 Proximate meat composition of LTL muscle in control (C) and polyphenols extract (PE) group (mean values ± SEM)

	Groups		Significance
	C	PE	P
Moisture (%)	66.6 ± 0.46	64.1 ± 0.35	0.11
Fat (%)	13.7 ± 0.46	16.5 ± 0.31	0.09
Protein (%)	18.3 ± 0.30	18.7 ± 0.40	0.43
Collagen (%)	2.3 ± 0.28	3.0 ± 0.16	0.14
Ash (%)	0.91 ± 0.12	0.84 ± 0.05	0.58
pH*	6.1 ± 0.01	6.2 ± 0.01	0.51

*pH was assessed on LTL sample 48 h after slaughtering

Table 5 *Shear myofibrillar force* (SMF) and *Warner-Bratzler Shear Force* (WBSF) recorded in the LTL muscle of control (C) and polyphenols extract (PE) group (mean values ± SEM)

	Groups		Significance
	C	PE	P
WBSF, kg	4.5 ± 0.10	4.6 ± 0.10	0.86
SMF, kg	0.53 ± 0.07	1.14 ± 0.33	0.06

first three days of storage in the meat of PE Group compared to the control when analyzed by illuminant A, while an opposite trend was recorded with the illuminant D65. In any case, no statistical differences were recorded (Fig. 1) and subsequently the changes showed the same trend with both illuminants.

Neither dietary treatment ($P = 0.77$ and $P = 0.99$, for OMb and MMb, respectively) nor storage period ($P = 0.33$ and $P = 0.10$, for OMb and MMb, respectively) and their interaction ($P = 0.98$ and $P = 0.95$, for OMb and MMb, respectively) influenced the estimated levels of myoglobin forms. As expected, during the storage period, the estimated level of MMb tended to increase in both groups, while the levels of OMb decreased until 3 days and increased after 7 days of storage (Table 6).

Fatty acid composition and oxidative stability

The fatty acids composition of LTL intramuscular lipids in the two groups is shown in Table 7. Dietary polyphenols supplementation determined a decrease of short chain fatty acids (<C12:0) as well as of Myristic (C14:0) Myristoleic (C14:1), Pentadecanoic (C15:0), Palmitic (C16:0) and Palmitoleic (C16:1) acids.

A significant lower ($P < 0.01$) percentage of saturated (SFA) and higher ($P < 0.05$) proportion of monounsaturated (MUFA) fatty acids were recorded in PE group compared to C group. However, no differences were recorded in the

total amount of polyunsaturated fatty acids (PUFA), while the unsaturated fatty acids (UFA):SFA ratio was influenced by the treatment ($P < 0.01$). Furthermore, the PUFA:SFA (P:S) ratio tended to be higher ($P = 0.09$) in PE group, compared to C group. Therefore, the meat of PE group was also characterized by a significant lower AI index ($P < 0.05$) and TI index ($P < 0.01$), compared to the control counterparts.

In relation to lipid oxidation, a significant effect of dietary treatment was observed. In particular, lower ($P < 0.05$) MDA values were recorded in PE group, compared to C group (0.25 ± 0.005 and 0.15 ± 0.005 in Group C and PE, respectively), as showed in Fig. 2, whereas neither the storage ($P = 0.96$) nor the interaction storage x dietary treatment ($P = 0.97$) influence the results.

Discussion

This study aimed to ascertain the influence of dietary polyphenols administration on growth performance and meat quality in kids. Contrasting results have been reported on the effects of plant extracts administration on live weight gain, carcass weight and dressing percentage either in ruminants [35, 36], and not ruminants [11, 37]. According to some authors [37] no adverse effect on growth performance or protein and aminoacids digestibility were recorded in broilers after 2.5 g/kg of grape seed extracts administration, while a delayed growth rate was observed after 5 g/kg of supplementation. On the contrary, other authors [38] report an increase of final body weight and DWG in broilers fed dietary polyphenol-rich grape seed. Few studies have been performed in ruminants till now. Similar growth performance were observed in growing lambs fed a diet supplemented with grape pomace, vitamin E or grape seeds extract [39]. Similarly, the supplementation of the diet with pomegranate seed pulp [40] or with vitamin E, Turmeric powder or *Andrographis paniculata* powder [35] did not affect the growth of kids.

Also in our study, growth performance was not influenced by dietary polyphenols supplementation, but some

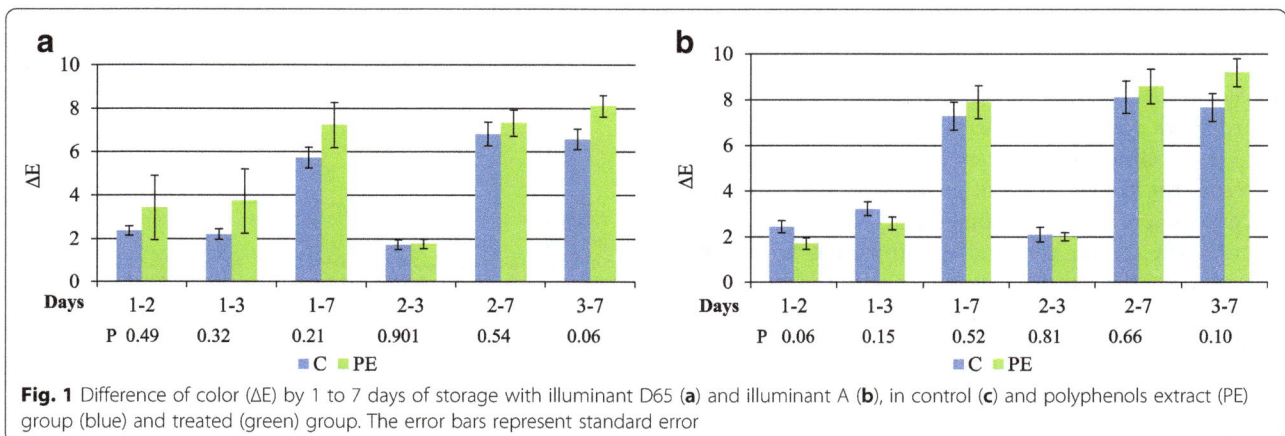

Fig. 1 Difference of color (ΔE) by 1 to 7 days of storage with illuminant D65 (**a**) and illuminant A (**b**), in control (**c**) and polyphenols extract (PE) group (blue) and treated (green) group. The error bars represent standard error

Table 6 Estimated levels (mean percentage ± SEM) of metmyoglobin (MMb) and oxymyoglobin (OMb), in the LTL muscle of control (C) and treated (PE) group, throughout the storage at 4 °C

	Groups	1	2	3	7
MMb	C	40.73 ± 1.88	45.28 ± 1.18	53.57 ± 1.67	61.78 ± 3.60
	PE	40.99 ± 1.94	47.67 ± 1.52	54.41 ± 1.88	63.37 ± 3.94
OMb	C	68.93 ± 2.70	34.04 ± 3.04	15.07 ± 3.24	43.87 ± 9.42
	PE	62.66 ± 7.02	38.13 ± 4.83	21.42 ± 5.69	51.48 ± 10.28

Table 7 Mean fatty acid composition (g/100 g of FAME) of LTL muscle in control (C) and polyphenols extract (PE) group (mean values ± SEM)

Fatty Acids	Groups C	PE	Significance P
< C12:0	2.93 ± 0.50 [B]	0.35 ± 0.40 [A]	0.001
C12:0	0.45 ± 0.36	0.20 ± 0.07	0.06
C14:0	4.00 ± 0.58 [b]	2.28 ± 0.47 [a]	0.04
C14:1	0.46 ± 0.06 [B]	0.23 ± 0.05 [A]	0.01
C15:0	0.58 ± 0.06 [b]	0.41 ± 0.05 [a]	0.04
C16:0	23.81 ± 0.85[B]	20.22 ± 0.70[A]	0.01
C16:1	3.05 ± 0.41 [b]	1.63 ± 0.33 [a]	0.02
C17:0	1.12 ± 0.17	1.17 ± 0.14	0.79
C17:1	1.33 ± 0.17	0.91 ± 0.14	0.08
C18:0	15.60 ± 1.25	14.31 ± 1.02	0.43
C18:1 cis	27.16 ± 3.15 [a]	36.01 ± 2.57 [b]	0.05
C18:1 trans	0.08 ± 0.73	1.64 ± 0.60	0.12
C18:2 cis	12.22 ± 1.07	13.64 ± 0.88	0.32
C18:2 trans n6	0.48 ± 0.05	0.37 ± 0.04	0.09
C18:3 n6 γ-linolenic	0.10 ± 0.03	0.07 ± 0.02	0.43
C20:0	0.05 ± 0.04	0.03 ± 0.03	0.74
C20:1	1.62 ± 0.15	1.59 ± 0.12	0.89
C20:2 n6	0.10 ± 0.03	0.05 ± 0.02	0.31
CLA cis 9–trans 11	1.22 ± 0.12	1.02 ± 0.10	0.24
C20:3 n3	1.00 ± 0.40	1.22 ± 0.33	0.67
C20:4 n6	0.72 ± 0.75	1.63 ± 0.61	0.36
C22:0	0.32 ± 0.12	0.27 ± 0.10	0.76
C22:6 n3	0.63 ± 0.40	0.24 ± 0.33	0.47
C24:0	0.33 ± 0.20	0.26 ± 0.16	0.77
C24:1	0.63 ± 0.38	0.20 ± 0.31	0.39
PUFA n6	13.62 ± 0.96	15.77 ± 1.30	0.31
PUFA n3	1.63 ± 0.75	1.47 ± 0.61	0.86
n6/n3	8.36 ± 0.86	10.73 ± 0.91	0.41
ΣSFA	49.22 ± 2.39[B]	39.51 ± 1.95[A]	0.01
ΣMUFA	34.35 ± 2.84[a]	42.22 ± 2.32[b]	0.05
ΣPUFA	16.47 ± 2.02	18.26 ± 1.65	0.51
UFA:SFA	1.08 ± 0.13 [A]	1.55 ± 0.07 [B]	0.01
PUFA:SFA	0.34 ± 0.04	0.47 ± 0.05	0.09
AI	1.28 ± 0.19 [b]	0.69 ± 0.16 [a]	0.03
TI	2.13 ± 0.32[B]	1.38 ± 0.26 [A]	0.01

Different superscript letters within the same row indicate significant differences ([a,b]: $P < 0.05$; [A,B]: $P < 0.01$)
Note: Σ SFA = (<C12:0 + C12:0 + C14:0 + C15:0 + C16:0 + C17:0 + C18:0 + C20:0 + C22:0 + C24:0), Σ MUFA = (C14:1 + C16:1 + C17:1 + C18:1cis + C18:1trans + C20:1 + C24:1), Σ PUFA = (ΣCLA+Σn3 + Σn6),
UFA:SFA = [(ΣMUFA+ΣPUFA)/ΣSFA],
PUFA:SFA = (ΣPUFA/ΣSFA),
AI: [(4 x C14:0) + C16:0 + C18:0] / [ΣMUFA +ΣPUFA-n6 + ΣPUFA-n3],
TI: [(C14:0 + C16:0 + C18:0) / (0.5xMUFA) + (0.5xPUFA-n6) + (3xPUFA-n3) + (PUFA-n3 / PUFA-n6)]

aspects need to be considered. First of all about 8000 phenolic compounds have been identified in different plant species [11] and in different amounts, creating serious difficulties in making a comparison among the studies. Furthermore, fewer information is present on bioavailability of phenolic compounds in ruminants and their effects on bacterial rumen population, fermentation and absorption [10] compared to non-ruminants [11]. It can not be ruled out that different compounds and/or different amounts of polyphenols may lead to different results.

Also the proximate composition of the meat in our trial was not influenced by dietary polyphenols supplementation. This result is in agreement with previous studies carried out in goat, in which dietary herbal antioxidants [35] or vitamin E [36] were administered. Similarly, vitamin E supplementation did not affect carcass traits and dressing percentage in lambs [41]. Regarding the texture, the meat of control kids showed the occurrence in all samples of an initial yield putatively related to the myofibrillar component of shear force and classified as shear myofibrillar force [29]. On the contrary, a single peak was recorded in 6/9 animals of the PE group. For this reason, SMF values tended to be higher in PE group compared to C group. However, no differences were recorded for WBSF. This last result may indicate a similar collagen content of the meat recorded in the two groups. The evaluation of the contribution of either myofibrillar and connective tissue to meat texture after the cooking process can not be obtained without performing additional measurements [28]. Some authors [42, 43] suggested that the connective tissue contribution may be recorded by subtracting the initial yield from the shear force peak: although the shear force does not accurately represent the contribution of connective tissue to muscle tenderness [27, 44, 45], it may be hypothesized a larger contribution of the connective component to the final hardness in the control group, compared to the PE counterparts. To our knowledge, few studies have been performed to evaluate texture parameters in small ruminants after polyphenols diet supplementation. Carnosic acid supplementation seems to be an useful tool to improve meat sensory characteristics in fattening lambs,

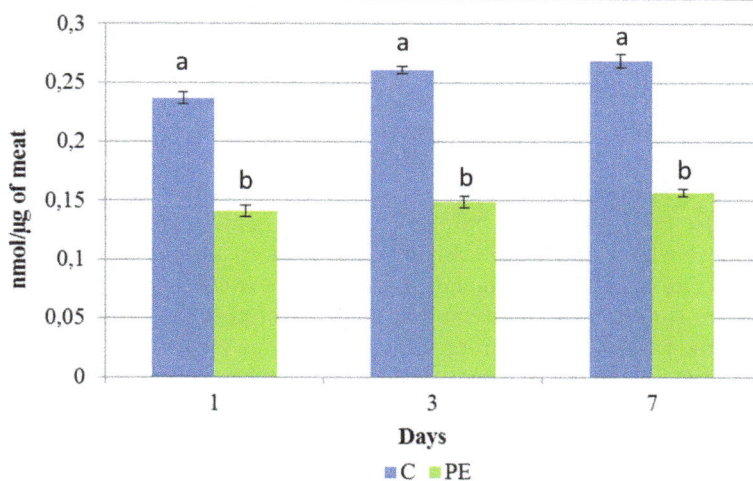

Fig. 2 Lipid peroxidation in meat during the days of storage (expressed in MDA nmol/μg of meat), the error bars represent standard error

similarly to Vitamin E [46]. On the contrary, quercetin dietary supplementation seemed to worsen texture in fattening lambs [47].

No differences were observed in our study on color stability between the two groups, in accordance with previous reports carried out in lambs fed naringin supplementation [48]. The color is an important meat quality attribute because it is the first aspect which attracts consumers when choosing fresh meat. Color stability is mainly due to the increase in myoglobin oxidation and consequent metmyoglobin formation and accumulation. Kid's meat is generally characterized by lower a* values compared to beef and lamb meat, outlining the difference in myoglobin concentration that reduces the discoloration. The effects of antioxidants on color stability in small ruminants have been analyzed in several studies in both kids [36, 40, 49] and lambs [39, 50] with contrasting results. According to some authors [40] no differences were recorded in kids fed pomegranate seed pulp on L*, while an increase of a* together with a decrease of b* was assessed. Similarly, no influence of vitamin E supplementation on color parameters was recorded in kids reared both in pen groups or maintained on pastures, although the rearing system significantly influenced meat color [36]. Dietary supplementation with antioxidants (vitamin E and flavonoids in particular) influences color parameters in some [39, 51], but not all [50] studies performed in growing lambs. Other authors reported a significantly lighter meat after polyphenols administration in lambs, probably because of their action in chelating iron and consequent low haemoglobin concentration [52]. Definitely, the results obtained in this study suggest that the concentration of MMb and OMb in meat, and consequently meat color, was not affected by dietary polyphenols supplementation. Only a slight lower, although

not significant, ΔE* was observed in PE group, when detected by illuminant A. The latter determined higher value placing more emphasis on the proportion of red wave lengths. This aspect is important when the differences between treatments have to be highlighted, as outlined by AMSA [30]. Similar results were obtained by other authors [7] by using a dietary supplementation with a blend of 80% canola oil and 20% palm oil, and were justified by physical changes that occur during the storage, and not directly related to the oxidative process [49], since no significant effects on color and oxidative stability of myoglobin was recorded. However, it can not be ruled out that the amount of dietary polyphenols was not able to highlight significant effects, thus further studies are needed to evaluate different dosages.

In our study polyphenols extract supplementation affected the fatty acid profile of the meat. In particular, a significantly higher concentration of MUFA, together with a reduction of SFA was observed in PE Group compared to control, but no effects were recorded on PUFA. The acidic composition of meat in ruminants is generally different from that of non-ruminants. The PUFA/SFA ratio is lower because of the hydrogenation of UFA in the rumen, while this process does not occur in monogastrics that absorb UFA without any transformation in the gastrointestinal tract. Therefore, several UFA that are normally present in the diet of ruminants, are saturated and can not be present in derived food [53]. Thus, the diet plays the main role in the modification of fatty acid composition and rumen microbial population, as demonstrated by the evidence that rumen population activity on dietary UFA hydrogenation depends by the administered diet [54]. Several studies carried out in vitro [55–57] demonstrated that some phenolic compounds can modulate rumen fermentation, reducing the rate of biohydrogenation of fatty acids. In particular, some plants with high polyphenols content would be able to lower the biohydrogenation of PUFA in the rumen, increasing their by-pass and, consequently, the formation of CLA [56]. Similarly, an increase of C18 PUFA was recorded after 24 h in vitro incubation of an experimental diet with leaf fractions of papaya, extracted with hexane or chloroform [57]. However, only an increase of C18:1 (oleic acid) was recorded in our study, together with a similar n6 and n3 PUFA and a similar concentration of CLA, although a slight increase of C18:1 trans was observed. It is known that the presence of CLA in the tissues depends from the desaturation of C18:1 *trans-11*, catalyzed through Δ9 desaturase [58]. However, the amount of C18:1 *trans-11* in ruminant meat and milk is influenced by its hydrogenation to stearic acid (C18:0) or other C18:1 isomers at the rumen level by the microbial population [59]. It can not be ruled out that the amount of polyphenols extract utilized in our study might need to be increased to further improve the fatty acid profile of the meat. In a recent study [36], a lower amount of SFA, together with an increase of MUFA and lower AI and TI indexes were observed in kids after 450 mg/kg of vitamin E diet supplementation compared to control counterparts. Adeyemi et al. [4], reported a lower concentration of C16:0 and C16:1 n7 in goats fed 4 and 8% oil blend compared with control, justified by displacement or dilution effects of other fatty acids and by the decrease in the activity of lipogenic enzymes responsible for the synthesis of medium chain FA or the preferential incorporation of long chain FA from diet and/or adipose tissues. In a recent trial [5] dietary supplementation with red wine extract significantly increased the levels of C20:5 n3 in lambs, but no influence on oxidative stability was recorded. In any case, the P:S ratio, recognized as a healthy index, tended to be higher in PE group compared to control. Because of the hydrogenation of PUFA by the microbial population, the meat of ruminants is usually characterized by a low P:S ratio [60], leading to an increased risk of cardiovascular diseases. Therefore, the modification of fatty acid profile obtained in the present study suggests that dietary polyphenols administration may result in kid meat with beneficial effects on human health.

A further positive effect of polyphenols administration was recorded on oxidative stability of the meat, as demonstrated by the reduction of MDA, an aldehyde commonly utilized as marker of secondary lipid oxidation in meat and characterized by mutagenic and carcinogenic properties. In fact, MDA changes the link between lipoproteins and scavenger receptors on the surface of macrophages causing cholesterol ester buildup and foam cell formation and may also react with deoxyadenosine and deoxyguanosine, nitrogenous DNA bases, potentially causing mutations [61].

Although the values recorded in our study may be considered a limiting point from where rancid flavor overpowers flavor in beef [62] and MDA concentration is probably overestimated by spectrophotometric method [63], a significant decrease was observed in PE group compared to C group.

The use of antioxidant substances, such as high polyphenols content in the diet, reduces lipid peroxidation and the formation of DNA additives, by inhibiting the formation of MDA. A possible explanation of the antioxidant activity of polyphenols can be sought in the potential of chelating metal ions [64]: they are able to bind iron, forming insoluble compounds at the gastrointestinal lumen level and making it unavailable for absorption. Iron has been recognized as the most likely catalyst to promote lipid peroxidation by the formation of free radicals formed following the Fenton reaction [65]. This is supported by several studies carried out in chicken poultry [37], rats [66] and pigs [67], in which the inclusion in the diet of polyphenols obtained by different sources significantly reduces plasma iron concentration. Therefore, the reduction of iron absorption may explain the lower MDA content recorded in PE group compared to C group. However, it can not be ruled out that this phenomenon may be due to an enhancement of muscle oxidative stability, via an increased expression of Δ9 desaturase enzyme, as demonstrated in lambs fed tannins supplementation [68], or to the increase of endogenous antioxidant defense [69]. In any case, our results are in agreement with other studies in which a significant reduction in MDA formation following polyphenols administration was recorded in goat [40], chicken [70] and turkey meat [71].

Conclusions

In conclusion, although no effects were observed on meat proximate composition, texture and colorimetric properties, the results of this study demonstrated that the supplementation of the diet with polyphenols extracts by OMWW is able to improve the fatty acid profile of kid meat, together with a reduction of MDA content. The utilization of this source of polyphenols may represent a novel strategy to re-utilize agri-food industry wastes as additives for livestock, meeting either meat industry and consumers requests of natural additives in animal farming. In any case, further studies are needed to test different doses of polyphenols and to better understand the role that the different categories of these compounds may have on meat quality.

Abbreviations

ΔE*: Color difference; A: Reflex attenuance; a*: Redness; ADG: Average daily gain; AI: Atherogenic index; ANOVA: analysis of variance; AOAC: Association of official analytical chemists; b*: Yellowness; C group: Control group; CREA: Research centre for animal production and aquaculture; DIA: Department of agriculture sciences; DM: Dry matter; DMb: Deoxymyoglobin; DMVPA: Department of veterinary medicine and animal production; DP: Dressing percentage; FAME: Fatty acid methyl esters; L*: Lightness; LTL: Longissimus thoracis et lumborum; LW: Live body weight; MDA: Malondialdehyde; MMb: Metmyoglobin; MUFA: Monounsaturated fatty acid; OMb: Oxymyoglobin; OMWW: Olive mill wastewaters; PE group: Polyphenols extract group; PUFA: Polyunsaturated fatty acid; SEM: Standard error of mean; SFA: Saturated fatty acid; SMF: Shear myofibrillar force; TI: Thrombogenic index; TPC: Total polyphenols; UFA: unsaturated fatty acid; WBS: Warner–bratzler shear; WBSF: Warner-bratzler Shear Force

Acknowledgements

The authors gratefully acknowledge Mr. Roberto Di Matteo for his technical support during meat quality analyses.
The authors gratefully acknowledge Dr. Emiliana Capurro, PhD, Dr. Trevor Cullen (Dublin, Ireland) and Dr. Claire Teter for language editing.

Funding

This research did not receive any specific grant from funding agencies in the public, commercial, or not-for-profit sectors.

Authors' contributions

RC, GC and GN designed the experiment. EV, SC and MC prepared the polyphenols and determined their antioxidant activity. DR, SC, CMAB, MA and RC carried out the experiments. CMAB, RC and PDP performed the statistical analysis of the results. RC, GN, PDP and MC prepared the draft of the manuscript. All authors contributed to the analysis of the data, discussion of results and implications and commented on the manuscript at all stages. All authors read and approved the final manuscript.

Competing interests

The authors declare that they have no competing interests.

Author details

[1]Italian Buffalo Breeders Association, V. Petrarca 42/44, 81100 Caserta, Italy. [2]Department of AgriculturalSciences, Federico II University, Via Università 133, 80055 Portici, Naples, Italy. [3]Research Centre for Animal Production and Aquaculture (CREA, S.S. 7 Appia, 85051, Bella Muro, PZ, Italy. [4]Department of Sciences and Technologies, University of Sannio, V. Port'Arsa 11, 82100 Benevento, Italy. [5]Department of Agricultural Food and Environmental Sciences, University of Foggia, Via Napoli 25, 71122 Foggia, Italy. [6]Department of Veterinary Medicine, University "Aldo Moro" of Bari, S.P. per Casamassima, km 3, Valenzano, 70010 Bari, Italy. [7]Department of Veterinary Medicine and Animal Production, Federico II University, V. F. Delpino 1, 80137 Naples, Italy.

References

1. International Agency for Research on Cancer. Carcinogenicity of consumption of red and processed meat. Lyon France: Report of the International Agency for Research on cancer. Press release no 240; 2015.

2. Domingo JL, Nadal M. Carcinogenicity of consumption of red meat and processed meat: a review of scientific news since the IARC decision. Food Chem Toxicol. 2017;105:256–61.

3. Wolk A. Potential health hazards of eating red meat. J Intern Med. 2017; 281:106–22.

4. Adeyemi KD, Shittu RM, Sabow AB, Ebrahimi M, Sazili AQ. Influence of diet and postmortem ageing on oxidative stability of lipids, myoglobin and myofibrillar proteins and quality attributes of gluteus medius muscle in goats. PLoS One. 2016;11:e0154603.

5. Muíño I, Apeleo E, de la Fuente J, Pérez-Santaescolástica C, Rivas-Cañedo A, Pérez C, et al. Effect of dietary supplementation with red wine extract or

6. Sabow AB, Sazili AQ, Aghwan AA, Zulkifli I, Goh YM, Ab-Kadir MZA, et al. Changes of microbial spoilage lipid-protein oxidation and physicochemical properties during postmortem refrigerated storage of goat meat. Anim Sci J. 2016;87:816–26.

7. Adeyemi KD, Sabow AB, Shittu RM, Karim R, Karsani SA, Sazili AQ. Impact of chill storage on antioxidant status, lipid and protein oxidation, color, drip loss and fatty acids of semimembranosus muscle in goats. CyTA-J Food. 2015; https://doi.org/10.1080/19476337.2015.1114974.

8. Jiang J, Xiong Y. Natural antioxidants as food and feed additives to promote health benefits and quality of meat products: a review. Meat Sci. 2016;120:107–17.

9. Naveena BM, Sen AR, Vaithiyanathan S, Babji Y, Kondaiah N. Comparative efficacy of pomegranate juice, pomegranate rind powder extract and BHT as antioxidants in cooked chicken patties. Meat Sci. 2008;80:1304–8.

10. Gessner DK, Ringseis R, Eder K. Potential of plant polyphenols to combat oxidative stress and inflammatory processes in farm animals. J Anim Physiol An Nutr (Berl). 2017;101:605–28.

11. Lipiński K, Mazur M, Antoszkiewicz Z, Purwin C. Polyphenols in monogastric nutrition: a review. Ann Anim Sci. 2017;17:41–58.

12. Zhang PY. Polyphenols in health and disease. Cell Biochem Biophys. 2016; 73:649–64.

13. Pandey KB, Rizvi SI. Plant polyphenols as dietary antioxidants in human health and disease. Oxidative Med Cell Longev. 2009;2:270–8.

14. Zhang H, Tsao R. Dietary polyphenols, oxidative stress and antioxidant and anti-inflammatory effects. Curr Opin Food Sci. 2016;8:33–42.

15. Brenes A, Viveros A, Chamorro S, Arija I. Use of polyphenol-rich grape by-products in monogastric nutrition. A review Anim Feed Sci Tech. 2016;211:1–17.

16. Laudadio V, Ceci E, Lastella NM, Tufarelli V. Dietary high-polyphenols extra-virgin olive oil is effective in reducing cholesterol content in eggs. Lipids Health Dis. 2015;14:5.

17. Federici F, Fava F, Kalogerakis N, Mantzavinos C. Valorisation of agro-industrial by-products effluents and waste: concept opportunities and the case of olive mill wastewaters. J Chem Technol Biot. 2009;84: 895–900.

18. Azaizeh H, Halahlih F, Najami N, Brunner D, Faulstich M, Tafesh A. Antioxidant activity of phenolic fractions in olive mill wastewater. Food Chem. 2012;134:2226–34.

19. Fernández-Bolaños JG, López O, López-García MA, Marset A. Biological properties of hydroxytyrosol and its derivatives. In: Boskou D, editor. Olive oil - constituents. Quality. Health properties and bioconversions. Croatia: InTech open; 2012. p. 375–96.

20. NRC (National Research Council). Nutrient requirements of small ruminants: sheep, goats, cervids and new world camelids. Washington, DC: The National Academies Press; 2007. https://doi.org/10.17226/11654

21. AOAC. Association of official analytical chemists: official methods of analysis. 18[th]ed. 2004, Arlington, VA, USA: Association of Official Analytical Chemists.

22. Parrillo L, Coccia E, Volpe MG, Siano F, Pagliarulo C, Scioscia E, Varricchio E, Safari O, Eroldogan T, Paolucci M. Olive mill wastewater-enriched diet positively affects growth, oxidative and immune status and intestinal microbiota in the crayfish. Astacusleptodactylus Aquaculture. 2017;473:161–8.

23. Cassano A, Conidi C, Giorno L, Drioli E. Fractionation of olive mill wastewaters by membrane separation techniques. J Hazard Mat. 2013;248-249:185–93.

24. Singleton VL, Orthofer R, Lamuela-Raventos RM. Analysis of total phenols and other oxidation substrates and antioxidants by means of Folin–Ciocalteu reagent. Methods Enzymol. 1999;299:152–78.

25. Re R, Pellegrini N, Proteggente A, Pannala A, Yang M, Rice-Evans C. Antioxidant activity applying and improved ABTS radical action decolorization assay. Free Radical Bio Med. 1999;26:1231–7.

26. AOAC. Determination of Fat, Moisture, and Protein in Meat and Meat Products: official Methods of Analysis. 18th ed. Gaithersburg: Method; 2007.

27. American Meat Science Association (AMSA). Research guidelines for cookery, sensory evaluation and instrumental tenderness measurements of meat. 2nd Edition Version 1.02. Chicago: Illinois: American Meat Science Association; 2016.

28. Girard I, Bruce HL, Basarab JA, Larsen IL, Aalhus JL. Contribution of myofibrillar and connective tissue components to the Warner–Bratzler shear force of cooked beef. I Meat Sci. 2012;92:775–82.

29. Bouton PE, Harris PV. Factors affecting tensile and Warner–Bratzler shear values of raw and cooked meat. J Texture Stud. 1978;9:395–413.

30. AMSA. Meat color measurements guidelines. Chicago: IL, USA: American Meat Science Association; 2012.

31. Folch J, Lees M. Sloane Stanley GH. A simple method for the isolation and purification of total lipides from animal tissues. J Biol Chem. 1957;226:497–509.

32. Christie WW. Preparation of methyl esters-part 2. Lipid Technol. 1990;2:79–80.

33. Ulbricht TLV, Southgate DAT. Coronary hearth disease: seven dietary factors. Lancet. 1991;338:985–92.

34. SAS Institute. SAS procedures guide version 6. 3rd ed. SAS Institute Inc: Cary, NC; 1990.

35. Karami M, Alimon AR, Goh YM, Sazili AQ, Ivan M. Effects of dietary herbal antioxidants supplemented on feedlot growth performance and carcass composition of male goats. Am J Anim Vet Sci. 2010;5:33–9.

36. Yakan A, Ates CT, Alasahana S, Odabasioglua F, Unal N, Ozturkc OH, Gungor OF, Ozbeyaz C. Damascus kids' slaughter, carcass and meat quality traits in different production systems using antioxidant supplementation. Small Ruminant Res. 2016;136:43–53.

37. Chamorro S, Viveros A, Centeno C, Romero C, Arija I, Brenes A. Effects of dietary grape seed extract on growth performance, amino acid digestibility and plasma lipids and mineral content in broiler chicks. Animal. 2013;7:555–61.

38. Abu Hafsa SH, Ibrahim SA. Effect of dietary polyphenol-rich grape seed on growth performance, antioxidant capacity and ileal microflora in broiler chicks. J Anim Physiol An Nutr (Berl). 2017; https://doi.org/10.1111/jpn.12688.

39. Guerra-Rivas C, Vieira C, Rubio B, Martínez B, Gallardo B, Mantecón AR, Lavín P, Manso T. Effects of grape pomace in growing lamb diets compared with vitamin E and grape seed extract on meat shelf life. Meat Sci. 2016;116:221–9.

40. Emami A, FathiNasri MH, Ganjkhanlou M, Zali A, Rashidi L. Effects of dietary pomegranate seed pulp on oxidative stability of kid meat. Meat Sci. 2015; 104:14–9.

41. Macit M, Aksakal V, Emsen E, Esenbu¨ga N, Aksu MI. Effect of vitamin E supplementation on fattening performance: non carcass components and retail cut percentages and meat quality traits of Awassi lambs. Meat Sci. 2003;64:1–6.

42. Bouton PE, Harris PV, Shorthose WR. Changes in shear parameters of meat associated with structural changes produced by aging, cooking and myofibrillar contraction. J Food Sci. 1975;40:1122–6.

43. Lawrie RA, Ledward DA. Lawrie's meat science. 7th edition. Cambridge England: Woodhead Publishing limited p. 2006;442

44. Bouton PE, Ford AL, Harris PV, Shorthose WR, Ratcliff D, Morgan JHL. Influence of animal age on the tenderness of beef: muscle differences. Meat Sci. 1978;2:301–11.

45. Harris PV, Shorthose WR. Meat texture. In: Lawrie RA, editor. Developments in meat science-4. London, England: Elsevier Applied Science Publishers; 1988. p. 245–86.

46. Morán L, Andrés S, Bodas R, Prieto N, Giráldez FJ. Meat texture and antioxidant status are improved when carnosic acid is included in the diet of fattening lambs. Meat Sci. 2012;91:430–4.

47. Andrés S, Huerga L, Mateo J, Tejido ML, Bodas R, Morán L, Prieto N, Rotolo L, Giráldez FJ. The effect of quercetin dietary supplementation on meat oxidation processes and texture of fattening lambs. Meat Sci. 2014;96:806–11.

48. Bodas R, Prieto N, Jordán MJ, López-Campos O, Giráldez FJ, Morán L, et al. The liver antioxidant status of fattening lambs is improved by naringin dietary supplementation at 0.15% rates but not meat quality. Animal. 2012; 6:863–70.

49. Morales-delaNuez A, Moreno-Indias I, Falcón A, Argüello A, Sánchez-Macias D, Capote J, Castro N. Effects of various packaging systems on the quality characteristic of goat meat. Asian-Austral J Anim. 2009;22:428–32.

50. Muela E, Alonso V, Campo MM, Sañudo C, Beltrán JA. Antioxidant diet supplementation and lamb quality throughout preservation time. Meat Sci. 2014;98:289–95.

51. Inserra L, Priolo A, Lanza M, Bognanno M, Gravador R, Luciano G. Dietary citrus pulp reduces lipid oxidation in lamb meat. Meat Sci. 2014;96:1489–93.

52. Samman S, Sandström B, Toft MB, Bukhave K, Jensen M, Sørensen SS, Hansen M. Green tea or rosemary extract added to foods reduces non heme-iron absorption. Am J Clin Nutr. 2001;73:607–12.

53. Chilliard Y, Glasser F, Ferlay A, Bernard L, Rouel J, Doreau M. Diet, rumen biohydrogenation and nutritional quality of cow and goat milk fat. Eur J Lipid Sci Technol. 2007;109:828–55.

54. Bessa RJ, Alves SP, Jeronimo E, Alfaia CM, Prates JA, Santos-Silva J. Effect of lipid supplements on ruminal biohydrogenation intermediates and muscle fatty acids in lambs. Eur J Lipid Sci Technol. 2007;109:868–78.

55. Cabiddu A, Salis L, Tweed JK, Molle G, Decandiaa M, Lee M. The influence of plant polyphenols on lipolysis and biohydrogenation in dried forages at different phenological stages: in vitro study. J Sci Food Agric. 2010;90:829–35.

56. Jayanegara A, Kreuzer M, Wina E, Leiber E. Significance of phenolic compounds in tropical forages for the ruminal bypass of polyunsaturated fatty acids and the appearance of biohydrogenation intermediates as examined in vitro. Anim Prod Sci. 2011;51:1127–36.

57. Jafari S, Meng GY, Rajion MA, Jahromi MF, Ebrahimi M. Manipulation of rumen microbial fermentation by polyphenol rich solvent fractions from papaya leaf to reduce green-house gas methane and biohydrogenation of C18 PUFA. J Agric Food Chem. 2016;64:4522–30.

58. Palmquist DL, St-Pierre N, McClure KE. Tissue fatty acid profiles can be used to quantify endogenous rumenic acid synthesis in lambs. J Nutr. 2004;134: 2407–14.

59. Griinari JM, Bauman DE. Biosynthesis of conjugated linoleic acid and its incorporation into meat and milk in ruminants. In: Yurawecz MP, Mossoba MM, JKG K, Pariza MW, Nelson GJ, editors. Advances in conjugated linoleic acid research. Champaign: AOCS press; 1999. p. 180–200.

60. Nieto G, Ros G. Modification of fatty acid composition in meat through diet: effect on lipid peroxidation and relationship to nutritional quality – a review. Lipid peroxidation, Dr. angel Catala: In Tech; 2012. https://doi.org/10. 5772/51114. Available from: https://www.intechopen.com/books/lipid-peroxidation/modification-of-fatty-acid-composition-in-meat-through-diet-effect-on-lipid-peroxidation-and-relatio

61. Del Rio D, Stewart AJ, Pellegrini N. A review of recent studies on malondialdehyde as toxic molecule and biological marker of oxidative stress. Nutr Metab Cardiovasc Dis. 2005;15:316–28.

62. Campo MM, Nute GR, Hughes SI, Enser M, Wood JD, Richardson RI. Flavour perception of oxidation in beef. Meat Sci. 2006;72:303–11.

63. Reitznerová A, Šuleková M, Nagy J, Marcin´cák S, Semjon B, C´ertík M, Klempová T. Lipid peroxidation process in meat and meat products: a comparison study of malondialdehyde determination between modified 2-Thiobarbituric acid spectrophotometric method and reverse-phase high-performance liquid chromatography. Molecules. 2017;22

64. Shahidi F, Janitha PK, Wanasundara PD. Phenolic antioxidants. Critical Reviews in Food Sci Nutr. 1992;32:67–103.

65. Liochev SI, Hausladen A, Beyer WF Jr, Fridovich I. NADPH: ferredoxin oxidoreductase acts as a paraquatdiaphorase and is a member of the soxrs regulon. P Natl AcadSci USA. 1994;91:1328–31.

66. Marouani N, Chahed A, Hedhili A, Hamdaoui MH. Both aluminium and polyphenols in green tea decoction (Camelia sinensis) affect iron status and haematological parameters in rats. Eur J Nutr. 2007;46:453–9.

67. Lee SH, Shinde PL, Choi JY, Kwon IK, Lee JK, Pak SI, Cho WT, Chae BJ. Effects of tannic acid supplementation on growth performance, blood haematology, iron status and faecal microflora in weanling pigs. Livestock Sci. 2010;131:281–6.

68. Vasta V, Priolo A, Scerra M, Hallet KG, Wood JD, Doran O. Δ(9) desaturase protein expression and fatty acid composition of longissimus dorsi muscle in lambs fed green herbage or concentrate with or without added tannins. Meat Sci. 2009;82:357–64.

69. López-Andrés P, Luciano G, Vasta V, Gibson TM, Biondi L, Priolo A, Mueller-Harvey I. Dietary quebracho tannins are not absorbed, but increase the antioxidant capacity of liver and plasma in sheep. Br J Nutr. 2013;110:1–8.

70. Goñi I, Brenes A, Centeno C, Viveros A, Saura-Calixto F, Rebole´ A, Arija I, Esteve R. Effect of dietary grape pomace and vitamin E on growth performance, nutrient digestibility and susceptibility to meat lipid oxidation in chickens. Poultry Sci. 2007;86:508–16.

71. Juskiewicz J, Jankowski J, Zielinski H, Zdunczyk Z, Mikulski D, Antoszkiewicz Z, Kosmala M, Zdunczyk P. The fatty acid profile and oxidative stability of meat from turkeys fed diets enriched with n-3 polyunsaturated fatty acids and dried fruit pomaces as a source of polyphenols. PLoS One. 2017;12: e0170074. https://doi.org/10.1371/journal.pone.0170074.

22

Efficacy of live attenuated porcine reproductive and respiratory syndrome virus 2 strains to protect pigs from challenge with a heterologous Vietnamese PRRSV 2 field strain

Tatjana Sattler[1,2]*, Jutta Pikalo[1], Eveline Wodak[1], Sandra Revilla-Fernández[1], Adi Steinrigl[1], Zoltán Bagó[1], Ferdinand Entenfellner[3], Jean-Baptiste Claude[4], Floriane Pez[4], Maela Francillette[4] and Friedrich Schmoll[1]

Abstract

Background: Effective vaccines against porcine reproductive and respiratory syndrome virus (PRRSV), especially against highly pathogenic (HP) PRRSV are still missing. The objective of this study was to evaluate the protective efficacy of an experimental live attenuated PRRSV 2 vaccine, composed of two strains, against heterologous challenge with a Vietnamese HP PRRSV 2 field strain. For this reason, 20 PRRSV negative piglets were divided into two groups. The pigs of group 1 were vaccinated with the experimental vaccine, group 2 remained unvaccinated. All study piglets received an intranasal challenge of the HP PRRSV 2 on day 0 of the study (42 days after vaccination). Blood samples were taken on days 7 and 21 after vaccination and on several days after challenge. On day 28 after challenge, all piglets were euthanized and pathologically examined.

Results: On days 7 and 21 after vaccination, a PRRSV 2 viraemia was seen in all piglets of group 1 which remained detectable in seven piglets up to 42 days after vaccination. On day 3 after challenge, all piglets from both groups were positive in PRRSV 2 RT-qPCR. From day 7 onwards, viral load and number of PRRSV 2 positive pigs were lower in group 1 than in group 2. All pigs of group 1 seroconverted after PRRSV 2 vaccination. PRRSV antibodies were detected in serum of all study pigs from both groups from day 14 after challenge onwards. In group 2, moderate respiratory symptoms with occasional coughing were seen following the challenge with HP PRRSV 2. Pigs of group 1 remained clinically unaffected. Interstitial pneumonia was found in four piglets of group 1 and in all ten piglets of group 2. Histopathological findings were more severe in group 2.

Conclusions: It was thus concluded that the used PRRSV 2 live experimental vaccine provided protection from clinical disease and marked reduction of histopathological findings and viral load in pigs challenged with a Vietnamese HP PRRSV 2 field strain.

Keywords: HP PRRSV 2, Vaccine, Efficacy, Viral replication, Immune response

* Correspondence: tasat@vetmed.uni-leipzig.de
[1]Institute for Veterinary Disease Control, AGES, Robert-Koch-Gasse 17, 2340 Mödling, Austria
[2]Clinic for Ruminants and Swine, University of Leipzig, An den Tierkliniken 11, 04103 Leipzig, Germany
Full list of author information is available at the end of the article

Background

The porcine reproductive and respiratory syndrome (PRRS), caused by PRRS virus (PRRSV), is of great importance in the pig industry worldwide. Recently [1] the PRRSV has been divided into PRRSV 1, the former genotype 1 (European strain, Lelystad virus) [2] and PRRSV 2, the former genotype 2 (North American strain) [3], both of which are of high genetic variability [4]. Highly pathogenic (HP) PRRSV 2 strains have caused great economic losses in Asia, beginning with an especially extensive outbreak with a high mortality not only in piglets but also in sows 2006 in China [5]. Since then, a lot of different subtypes of HP PRRSV 2 have been described [6–8]. Less virulent (non-HP) PRRSV 2 variants, some of them having already been detected in 1996, were reported to occur in Asian countries as well [9].

Because of the wide distribution and the high morbidity and mortality caused by HP PRRSV 2 strains in Asia, efficient immunization strategies are necessary to minimize problems in affected farms. Modified live vaccines often proved to be effective in controlling the infection with PRRSV 1 or (non-HP) PRRSV 2 by reducing the viral shedding and protection against re-infection [10–12]. In many cases, however, commercial vaccines are not as effective as necessary. This can on one hand be caused by the ability of PRRSV to modulate the immune response [13] and is on the other hand due to the high genetic variability of the virus [14]. Unsatisfactory results were especially seen after infection with heterologous virus, where only partial protection could be achieved [15]. PRRSV 2 vaccination with a homologous vaccine conferred better protection, especially against HP PRRSV 2 [16, 17]. Until now, the most effective protection against infection with HP PRRSV 2 was provided by attenuated HP PRRSV 2 vaccines in experimental challenge studies [18]. It is assumed that the highest benefit from vaccination occurs when the vaccine virus is genetically as close to the field virus as possible, as was reported in a study on a homologous attenuated PRRSV 2 live vaccine in China [19]. Another aspect would be the special induction of cellular immunity which has been tried with a homologous DNA vaccine [20].

There is, however, no commercially available vaccine on the market that is able to protect efficiently against infection with HP PRRSV 2 in Vietnam. For any live vaccine it is necessary to identify a batch which is both safe and highly effective in inducing a protective immune response. The aim of this study was to evaluate the potential suitability efficacy of an experimental vaccine containing two live attenuated PRRSV 2 strains in protecting pigs from challenge with a low-passage Vietnamese HP PRRSV 2 field isolate by studying the clinical symptoms, growth parameters, the viral replication and development of antibodies against PRRSV 2.

The response to the challenge was compared to not pre-vaccinated pigs.

Methods

Experimental design, animals and housing

Twenty male piglets (landrace and large white crossbreds) from a PRRSV 1 and 2 negative farm were selected during the suckling period, marked with an individual ear tag and randomly divided into two groups of 10 piglets each. All piglets were routinely vaccinated twice against Mycoplasma hyopneumoniae (2 ml i.m., Hyoresp, Merial, Halbergmoos, Germany) at the age of 5 and 21 days and against PCV-2 (1 ml i.m., Ingelvac Circoflex, Boehringer Ingelheim, Germany) at the age of six weeks. At the age of 21 days, ten piglets (group 1) were housed in the experimental stable, sized 12 m^2. Another ten piglets (group 2) were housed in a separate room of the experimental unit with the same size. The units were cleaned daily by qualified personnel. The piglets had permanent free access to drinking water, playing and nuzzling material and were fed ad libitum by an automatic feeder with commercial nursery piglet diet containing colistin sulfate (10 mg/kg body weight, Colistin Mix, AniMed Service AG, Dobl, Austria), amoxicillintrihydrate (20 mg/kg body weight, Amoxi-Mix 10%, AniMed Service AG) and 100 mg zinc oxide/kg body weight (Vetzink®, approved special import from Denmark by Chevita, Wels, Austria) per day. After an adaptation period of five days (day – 42 of the experiment), all piglets of group 1 received an intramuscular injection of 2 ml of a re-suspended experimental vaccine made of two PRRSV 2 strains, containing 10^5 50% tissue culture infective dose ($TCID_{50}$) of each strain per dose (strains kindly provided by Kyoto Biken Laboratories, Inc., Kyoto, Japan). This corresponds to a viral load of $1.38E + 09$ copies/ml, as determined by reverse transcription quantitative real-time polymerase chain reaction (RT-qPCR). At approximately ten weeks of age (42 days after vaccination, day 0 of the experiment) all piglets of both groups received an intranasal challenge of 2 ml of the challenge virus, an HP PRRSV 2 field strain as described below.

All piglets underwent a daily clinical examination (through visual examination). Blood samples were taken from the piglets of group 1 on days – 42, – 35, – 7 and from piglets of both groups on days 0, 3, 7, 10, 14 and 28 of the experiment. Rectal body temperature of each piglet was measured on the blood sampling days. On days 0 and 28 the pigs were weighed and the weight gain was calculated. Housing, animal care and experimental protocol of the study were approved by the local ethics committee (Agency of the Government in Lower Austria, Department of Agrarian Law).

Virus strain, titration, calculation of TCID$_{50}$

The HP PRRSV 2 strain Vietnam_PRRSV_AGES/568-30FC/13 (GenBank accession number KM588915, in the following called "challenge virus") was isolated from serum of a naturally infected pig from a Vietnamese farm, in which severe clinical symptoms of PRRS and a high mortality among pigs were evident. This strain had been identified as HP PRRSV 2 field strain, based on an Nsp2 specific RT-PCR and sequencing [7]. To produce a sufficient quantity of the test virus, the virus was pooled from three consecutive passages in MARC-145 cells over 4 days.

To calculate the infectious PRRSV 2 titer, the Spearman-Karber method was used. PRRSV 2 titers were expressed as TCID$_{50}$/mL. The infectious titer of the virus stock was calculated to be 10^5 TCID$_{50}$/mL. The PRRSV 2 RNA concentration in the virus stock was 7.28E + 08 copies/ml, as quantified by RT-qPCR.

RNA extraction and PRRSV ORF7 RT-qPCR

Nucleic acid extraction from serum and tissue samples (lung tissue, pulmonary lymph nodes and tonsillar scrapings) was conducted using the Nucleospin® Virus Core kit and the Nucleospin 96® RNA kit (Macherey-Nagel, GenXpress, Wiener Neudorf, Austria), respectively, on the automated platform Freedom EVO® 150 (Tecan, Grödig, Austria), following the instructions of the manufacturer.

To detect PRRSV 1 and 2 RNA, the samples were analysed by a commercial ORF7 RT-qPCR assay that allows the simultaneous detection and differentiation of PRRSV 1 and 2 (TaqMan® PRRSV Reagents and Controls, Life Technologies, Brunn am Gebirge, Austria) on the ABI 7500 Fast Real-Time PCR System (Life Technologies). For absolute quantification, a PRRSV 2 RNA dilution series with known copy numbers ranging from 1.0E + 00−1.0E + 07 copies/µl was assayed in parallel.

PRRSV 2 ORF5 amplification and sequencing

The challenge virus stock, the experimental vaccine as well as representative RT-qPCR positive samples collected during the animal experiment (group 1: two serum samples on day − 21, three serum samples on day 7 and nine tonsillar scraping samples on day 28; group 2: four serum samples on day 7 and three tonsillar scrapings on day 28) were subjected to conventional ORF5 RT-PCR, sequencing and phylogenetic analysis. Due to the genetic diversity of some newly emerged Asian HP PRRSV 2 strains [8], specific primers were applied [6; 7]. The corresponding ORF5 RT-PCR products were separated by gel electrophoresis in 1.5% agarose gels stained with ethidium bromide and DNA bands of the expected sizes were excised from the agarose gel and recovered using the QIAquick® Gel Extraction Kit (Qiagen, Hilden, Germany). Sequencing reactions were performed in

both directions using the same primers as for ORF5 RT-PCR and the BigDye® Terminator v3.1 Cycle Sequencing Kit (Life Technologies). Sequencing reactions were purified with the DyEx® 2.0 Spin kit (Qiagen). Purified sequencing reactions were resolved on the 3130xl Genetic Analyzer (Life Technologies) and sequence raw data was created with the Data Collection Software (version 2.0, Applied Biosystems, Life Technologies). The raw sequence data was assembled and the consensus sequences were generated using SeqScape Software (version 2.5, Applied Biosystems, Life Technologies). A multiple sequence alignment was done in BioEdit [21], followed by Neighbour joining tree construction (Maximum Composite substitution model, complete deletion of gaps, 1000 bootstrap iterations) using MEGA5 [22].

PRRSV 2 next generation sequencing (NGS)

Two serum samples from group 1 pigs taken 21 days after vaccination and four samples (two serum and two tonsillar scrapings) from the same group taken 3, 7 and 28 days, respectively, after challenge were selected for NGS. Additionally, the experimental vaccine and the challenge strain were tested.

Prior to NGS, RNA samples were again tested for PRRSV 1 and 2 using the real-time PCR diagnostic assay Bio-T kit® PRRSV (Biosellal, Lyon, France). Total RNA was converted to cDNA and amplified with a combination of one-step and two-step reverse transcription polymerase chain reaction (RT-PCR). One-step RT-PCRs were performed with the One Step RT-PCR kit (Qiagen). Two-step RT-PCRs were performed with the SuperScript® III First strand kit (Invitrogen, Carlsbad, USA) and the Kapa LongRange HotStart PCR kit (Kapa, Wilmington, USA).

Libraries were prepared using the Ion Xpress™ Plus Fragment Library Kit for AB Library Builder™ System (Life Technologies) according to manufacturer's instructions. The obtained libraries were sequenced by the Ion Torrent PGM sequencer using the 316v2 chip (Life Technologies). Fastq files were analyzed with CLC Genomics Workbench 7.5.1 software (Qiagen). Briefly, reads were trimmed (default parameters) then mapped to the PRRSV strain VR2332 sequence (GenBank No. EF536003.1) with the NGS Reference Assembly tool (default parameters). Alignments and phylogenetic analysis (Neighbour joining, Kimura80, 1000 bootstraps) were all performed with CLC.

PRRSV antibody ELISA

The presence of PRRSV antibodies in serum from all piglets on each sampling day was assessed by ELISA (IDEXX PRRS X3, IDEXX, Westbrook, USA) following the instructions of the manufacturer.

Necropsy and histopathology

On day 28 of the experiment, all piglets were narcotized by intramuscular application of Azaperone (2 mg/kg body weight) and Ketamine (20 mg/kg body weight) and then euthanized by intracardial application of 5 ml T61®. Necropsy was performed on all 20 pigs with the main focus on pulmonary lesions and pulmonary lymph nodes. Gross pulmonary lesions were semi-quantified using a scoring scheme after Halbur et al. [23].

For histologic investigation, tissue samples from lungs (cranial and caudal lobe) and pulmonary lymph nodes were taken and fixed in 7,5% neutral buffered formalin. After embedding in paraffin, 4 µm sections were cut and routinely stained with hematoxylin and eosin (HE) and evaluated by light microscopy. Histopathological lung alterations were clustered/quantified according to the scoring scheme as previously described [23] using the following criteria: 0 = no histological alterations, 1 = mild interstitial pneumonia, 2 = moderate multifocal interstitial pneumonia, 3 = moderate diffuse interstitial pneumonia and 4 = severe interstitial pneumonia. Tonsillar scrapings and tissue samples of lung and pulmonary lymph nodes from each piglet were prepared for detection of PRRSV 2 RNA.

Statistical analysis

Data were tested for normal distribution with the Kolmogorov-Smirnov-test. Since most parameters were not normally distributed, differences between the groups were tested with the Mann-Whitney-U-test. Differences between the sampling times were assessed with the Friedman's variance analysis test followed by the Wilcoxon test. In cases with more than two sampling times (as was the case in viral load tested by PRRSV 2 RT-qPCR and body temperature) a correction of the alpha error of the significance value was done. Differences of the outcomes of PRRSV 2 RT-qPCR and ELISA and the occurrence of histologic lesions between the groups on each time point were tested with the Fisher's exact test. Differences with a $P < 0.05$ were considered significant. Correlations between parameters were tested with the rank correlation after Spearman. The correlation coefficient r was indicated in the text if a correlation was found.

Results

Clinical data

At the beginning of the experiment, all piglets appeared clinically healthy. After PRRSV 2 vaccination, a slight decrease of appetite was observed for a few days in most of the piglets of group 1. After challenge, pigs of group 1 remained clinically unaffected. Piglets of group 2 showed decreased appetite for a few days after challenge. In most piglets of group 2, occasional coughing and slightly increased lacrimation were observed from day 14 onwards. In two cases, cyanoses on ears, tail and scrotum were seen. These symptoms disappeared after two days. The rectal body temperature did neither increase after vaccination nor after challenge and did not differ between the groups. No significant differences in body weight and weight gain from day 0 to day 28 were detected between the groups.

PRRSV 2 RT-qPCR and ORF5 sequencing

All study piglets tested negative by PRRSV 1 and 2 RT-qPCR at the beginning of the experiment. The piglets of group 1 tested positive on day – 35 (day 7 after vaccination), which was the first sampling day after vaccination. On day 3 after challenge with HP PRRSV 2, all piglets of both groups tested positive in PRRSV 2 RT-qPCR. Viral loads and number of positive piglets per group on the respective days is shown in Fig. 1a. From day 7 onwards, viral load in the serum of the positive piglets was significantly higher in group 2 than in group 1. The number of PRRSV2 positive pigs was significantly higher in group 2 than in group 1 from day 10 onwards. The PRRSV 2 load in the tissue samples and the number of positive samples are shown in Fig. 1b. In both groups, median viral loads were highest in tonsillar scrapings, followed by lung lymph nodes and lung tissue. In group 1, fewer piglets tested positive in lung tissue than in group 2, although this difference was not statistically significant. Although all piglets tested positive by PRRSV 2 RT-qPCR in tonsillar scrapings on day 28, the viral load in tonsillar scrapings was significantly lower in group 1 than in group 2. There was a positive correlation between PRRSV 2 loads in serum on day 28 and lung and lung lymph nodes. Viral load in tonsillar scrapings was positively correlated with that in serum on days 7, 10, 14, 21 and 28.

Both strains of the experimental vaccine (sequences kindly provided by Kyoto Biken) group within the same cluster as AGES 1048, which was amplified directly from the vaccine (Fig. 2). Sequencing of the ORF5 amplified from four serum samples of group 1 collected on day – 21 showed 100% nucleotide sequence identity to the experimental vaccine strain AGES 1048. On days 7 and 28 after challenge, ORF5 sequences in all sequenced samples from both groups were identical or almost identical to the challenge virus sequence (fig. 2). Sample 1308–5, taken 7 days after challenge, was most distant to the challenge virus (a difference of four nucleotides, equal to 98% sequence identity). In contrast, experimental vaccine and challenge virus only showed 91% sequence identity in the ORF5 region used for comparison (218 bp).

PRRSV NGS

Samples subjected to NGS are listed in Table 1. A graphical view of a multiple alignment between all tested samples

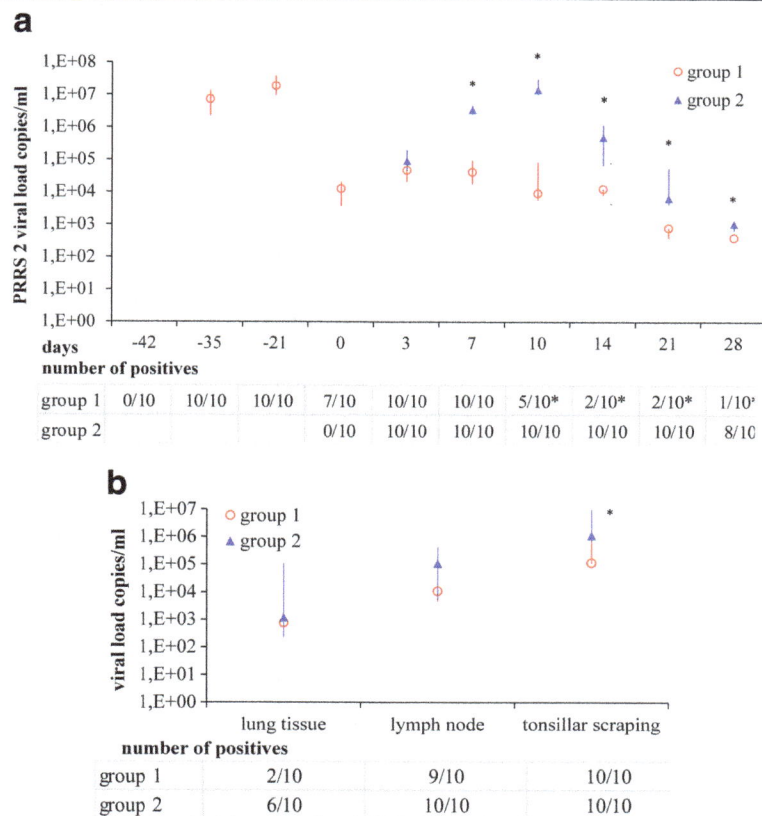

Fig. 1 PRRSV loads (copies/ml) in serum (**a**) and tissue samples (**b**) of the study piglets. Blood sampling points were: before vaccination (day − 42), days − 35, − 21, 0, 3, 7, 10, 14, 21 and 28 after challenge with an HP PRRSV 2 field strain; tissue sample were collected on day 28. Data are given as median, 1st and 3rd quartile. Group 1: pre-vaccinated with a new PRRSV 2 live vaccine, group 2: not pre-vaccinated. On time points marked with *, differences between the groups were significant

and three reference sequences is shown in Fig. 3a. Not all samples could be sequenced over the entire ORF2 to ORF7 region. In sample 1171–04 (21 days after vaccination), the ORFs 5–6 are missing. In sample 1414-3c (tonsillar scraping 28 days after vaccination), ORFs 3–6 are missing. From samples 1171–06 and 1302–01 no sequence could be obtained. Figure 3b and c show the distance tree and the nucleotide sequence identity in the tested samples. Nucleotide sequence identity between the experimental vaccine and the challenge virus was 91.57%. The viral sequences generated from the sample taken before challenge was > 99.9% identical to the experimental vaccine virus, whereas all sequences obtained from post-challenge samples were > 99.8% identical to the challenge virus (Fig. 3). Thus, NGS confirmed the results obtained by partial sequencing of the ORF5 region and corroborated that all viral sequences recovered from post-challenge samples were derived from the challenge virus (Fig. 3).

PRRSV antibody ELISA

All piglets were PRRSV antibody negative at the beginning of the experiment. Nine out of the ten piglets of group 1 had seroconverted by day 21 after vaccination. The S/P value of the remaining piglet was slightly beneath the test cut-off. On day 0, PRRSV antibodies were present in all piglets of group 1 and in no piglet of group 2. All piglets were PRRSV antibody positive on day 14 after challenge (Table 2).

Gross pathology and histopathology

An induration of the pulmonary parenchyma was found in eight piglets of group 1 and all piglets of group 2. In all piglets, pulmonary lymph nodes were at least moderately enlarged, however, piglets of group 2 had a more pronounced and generalized lymph node enlargement. Histologically, lymphatic hyperplasia was found in all piglets, which was again more pronounced in the piglets of group 2. An overview of the lung histology results is shown in Table 3. An interstitial pneumonia (intralobular as well as peribronchial) occurred significantly more often in group 2. Gross and histological pulmonary lesions due to lymphohistiocytic interstitial pneumonia were significantly more severe in group 2 as can be seen in Table 4. In Fig. 4, representative microphotos from

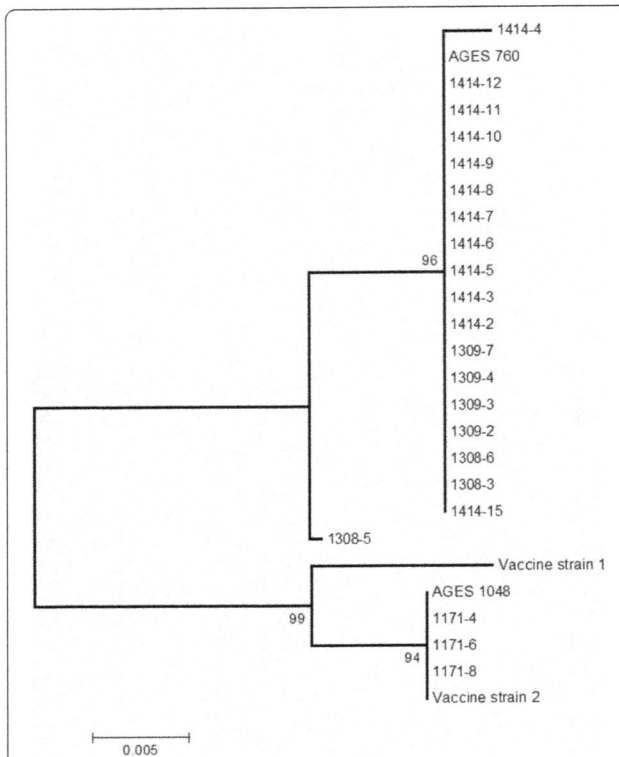

Fig. 2 Neighbour joining tree based on partial ORF5 sequences. Obtained from samples of the study piglets, the challenge virus (AGES 760) and the tested experimental PRRSV 2 live vaccine (AGES 1048). Sequences from the two virus strains included in the experimental vaccine (Vaccine strain 1 and 2) were kindly provided by Kyoto Biken. The size of the alignment was 218 bp. Numbers along the branches show the percentage of 1000 bootstrap iterations

the lungs of three affected pigs are shown. Overall, the histopathological findings indicated more severe lesions in group 2.

Discussion

In this study we assessed the efficacy of two live attenuated PRRSV 2 strains to protect piglets from challenge with a heterologous HP PRRSV 2 field strain (Vietnam_PRRSV_AGES/568-30FC/13; GenBank: KM588915) that

had initially been isolated in 2013 from pig serum from a Vietnamese farm.

Although the challenge strain was molecularly typed as highly pathogenic based on a deletion in the Nsp2 region [7], the clinical symptoms upon challenge of unvaccinated piglets (group 2) were moderate. One reason for this could be intrinsic low pathogenicity of the challenge virus. HP PRRSV 2 strains with different pathogenicity in animal experiments have been described [24, 25]. A recent study referred about different pathogenicity of HP PRRSV type 2 strains isolated from Northern and Southern Vietnam, with higher pathogenicity of the Northern strain [26]. However, the challenge virus likely caused severe clinical symptoms and a high mortality in the Vietnamese farm where it was isolated from. The relatively mild symptoms observed in unvaccinated piglets might also be due to the fact that the animals were healthy at the time of challenge and bacterial infections were prevented by the use of antibiotics throughout the study. Furthermore, tests for antibodies against classical swine fever, Aujeszky's disease, swine influenza virus and *Actinobacillus pleuropneumoniae* as well as *Mycoplasma hyopneumoniae* DNA at day 28 were negative in all study piglets (data not shown). Since the challenge virus remained genetically stable during the three cell culture passages, it is unlikely that the mild clinical symptoms are due to a genetic attenuation of the virus. Attenuation can only be expected after several passages [27]. In other studies low passages were successfully used for challenge as well [28].

In piglets of group 1 (vaccinated), no clinical symptoms were seen after PRRSV challenge. This is in line with the significantly milder gross pulmonary lesions and histopathological findings compared to group 2 and proves the efficacy of the tested vaccine to prevent clinical symptoms and diminish pathological lesions after infection with the heterologous HP PRRSV 2 challenge virus. As determined by NGS, experimental vaccine viral strains and challenge virus only shared 91.57% nucleotide sequence identity over the entire ORF2 – ORF7

Table 1 Samples selected for next generation sequencing

Name	Sample identity	Sample type	RT-qPCR cq
AGES 1048	PRRSV 2 live vaccine	vaccine	19
AGES 760	HP PRRSV 2 challenge strain	cell culture	20
1171–04	day 21 after vaccination, ear tag 104	serum	29
1171–06	day 21 after vaccination, ear tag 106	serum	26
1302–01	day 3 after challenge, ear tag 101	serum	32
1308–05	day 7 after challenge, ear tag 106	serum	30
1414-3c	day 28 after challenge, ear tag 104	tonsillar scraping	32
1414-5c	day 28 after challenge, ear tag 106	tonsillar scraping	30

All of the selected piglets were pre-vaccinated with an experimental PRRSV 2 live vaccine

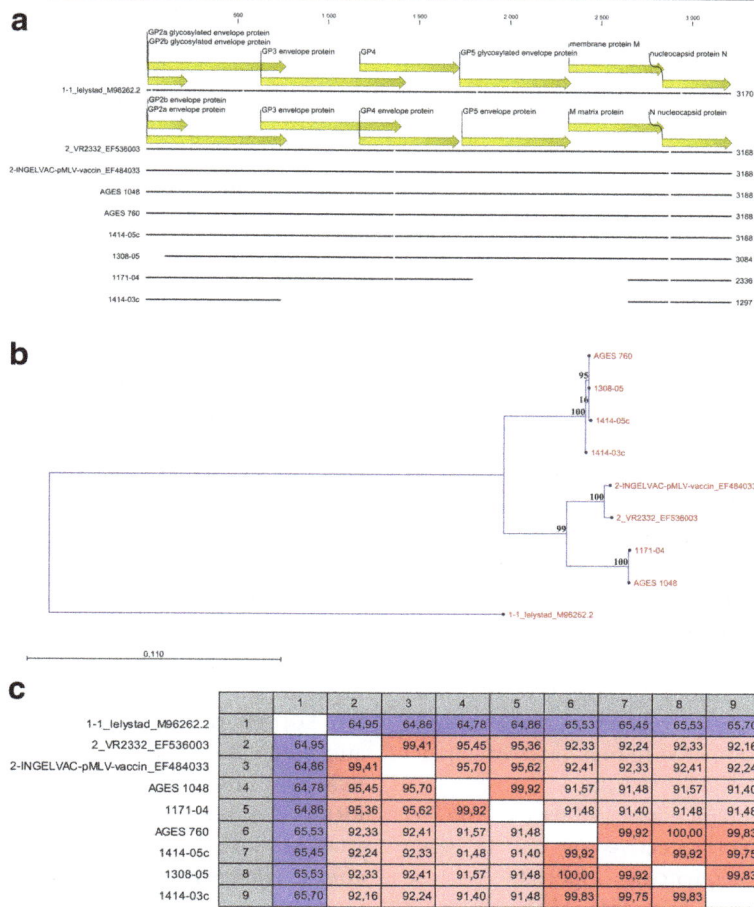

Fig. 3 Multiple alignment (**a**), neighbour joining tree (**b**) and nucleotide sequence identity matrix (**c**). Graphical view for selected references, sequenced samples of the study piglets, the challenge virus and the tested PRRSV 2 live vaccine. Reference strains: Lelystad virus for PRRSV 1, INGELVAC pMLV and VR2332 for PRRSV 2. The size of the alignment was 1201 bp. Numbers along the branches in (**b**) show the bootstrap values (%) after 1000 bootstrap iterations

region. Studies testing the efficacy of the vaccine strains used in this study after challenge with HP PRRSV 2 are not available. In a study of Wei et al. [16], clinical symptoms after HP PRRSV 2 challenge could not be prevented but diminished using a PRRSV 2 attenuated live vaccine. The control group in their study, however, developed severe clinical symptoms after challenge. Similar results were found in other studies [17, 29, 30].

Intranasal challenge with the defined PRRSV dose resulted in detectable virus replication in all study piglets on the first day of sampling after challenge (day 3). In the unvaccinated group 2, high viral loads were detected in serum, lung, pulmonary lymph nodes and tonsillar scrapings. This proves a rapid virus replication in the unvaccinated piglets. The viral loads were comparable to those found by Hu et al. [31] and Han et al. [32] after challenge with HP PRRSV type 2 strains isolated in

Table 2 Results of PRRSV antibody ELISA in the study piglets

Study day	No. of positive piglets								
	−42	−21	0	3	7	10	14	21	28
Group 1	0	9	10	10	10	10	10	9	9
Group 2	0		0	0	0	6	10	10	10

Group 1 – pre-vaccinated with an experimental PRRSV 2 live vaccine
Group 2 – not pre-vaccinated

Table 3 Histopathological pulmonary findings in pigs challenged with an HP PRRSV 2 field strain ($n = 10$ per group)

	No. of piglets group 1	No. of piglets group 2
Interstitial pneumonia	4[a]	10[b]
Alveolar histiocytosis	4	7
Desquamative/purulent bronchitis	3	7
Dystelectasis	8	10

Group 1 – pre-vaccinated with an experimental PRRSV 2 live vaccine
Group 2 – not pre-vaccinated
Significant differences ($P < 0.05$) between the groups are indicated with different letters

Table 4 Score of gross and histological pulmonary lesions due to interstitial pneumonia modified after Halbur et al. [23] in pigs challenged with an HP PRRSV 2 field strain (n = 10 per group) (Median (1st; 3rd quartile))

	Group 1	Group 2
Gross pulmonary lesions	14.5 (9.0; 20.8)[a]	36.5 (26.5; 57.3)[b]
Interstitial pneumonia	1.5 (1.0; 2.0)[a]	3.0 (2.0; 3.0)[b]

Group 1 – pre-vaccinated with an experimental PRRSV 2 live vaccine
Group 2 – not pre-vaccinated
Significant differences (P < 0.05) between the groups are indicated with different letters

China. In the vaccinated pigs (group 1), the viral load in serum as well as the number of viraemic piglets were significantly lower than in group 2. Lager et al. [17] describe similar levels of protection conferred by a homologous HP PRRSV 2 vaccine, using virus isolation as readout instead of PCR. This further underlines the efficacy of the experimental vaccine tested in our study. In a study using different dosages of an HP PRRSV 2 vaccine, the experimental vaccine was able to protect the study pigs from viraemia after homologous challenge when administered at least at the two-fold dosage used in our study [28]. Other studies, using heterologous PRRSV vaccines, also referred about partial protection (fewer clinical symptoms and viraemia compared to non-vaccinated pigs) against challenge with HP PRRSV [29, 30].

NGS was chosen to sequence a larger part of the experimental samples in a cost-effective way and to verify the results from Sanger sequencing, especially because one post-challenge sequence from group1 (sample No 1308–05) differed from the remaining post-challenge sequences. To investigate whether this sample might be a recombinant between vaccine and field virus, NGS was applied to obtain (a) a longer stretch of sequence to improve identification of potential recombination breakpoints;

(b) to obtain a higher coverage of the sequence in question. In some cases, not the entire ORF2 – ORF7 sequence could be obtained by NGS, or sequencing failed completely. This is probably due to the relatively high Cq value in some of these samples. In cases of samples with cq values above 25, the success of NGS diminishes as has been described in another study [33].

The HP PRRSV 2 challenge strain used in this study remained genetically stable not only during replication in cell culture but also during the animal experiment as shown by both partial sequencing of the ORF5 and NGS of the ORFs 2 to 7. The experimental vaccine strain found in the pigs remained genetically stable as well during the animal experiment. Furthermore, there was no evidence of recombination between experimental vaccine and challenge virus, as all viral sequences obtained from experimental animals before and after challenge were more than 99% identical to the experimental vaccine and challenge virus sequence, respectively.

The humoral immune response to the tested vaccine is shown by detection of PRRSV antibodies by ELISA in all vaccinated piglets (group 1) on day 21 after vaccination. Lager et al. [17] obtained similar results, whereas others [16, 30] report that on day 21 after PRRSV 2 live vaccination, only a part of the pigs was PRRSV antibody positive by ELISA. In group 2, all piglets had developed PRRSV antibodies on day 14 after challenge. In other studies, PRRSV antibodies were found on this time point after challenge as well [34, 35].

Conclusions

Vaccination with new live attenuated PRRSV 2 strains induced an immune response as shown by timely production of PRRSV antibodies. Experimental infection with the heterologous HP PRRSV 2 challenge virus resulted in

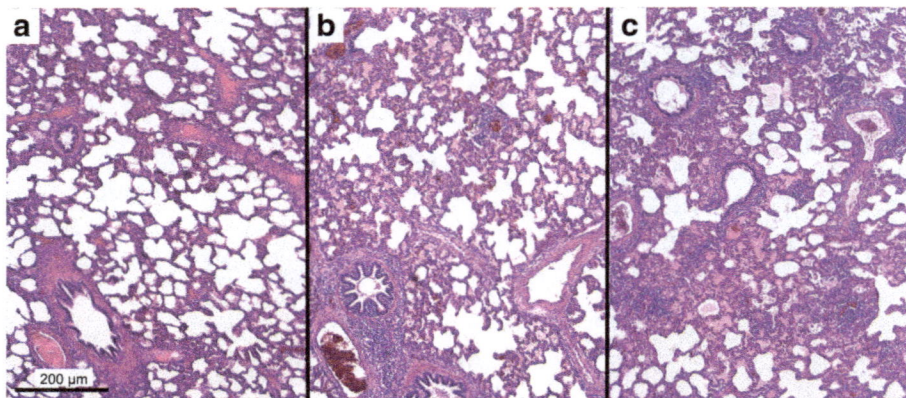

Fig. 4 Pulmonary lesions in pigs challenged with an HP PRRSV 2 field strain. **a**: pig from group 1 (vaccinated) without inflammatory alterations (score 0); **b**: pig from group 2 (non-vaccinated) showing moderate multifocal lymphohistiocytic interstitial pneumonia with peribronchial and perivascular accentuation (score 2); **c**: pig from group 2 displaying moderate diffuse lymphohistiocytic interstitial pneumonia with peribronchial, perivascular and intralobular accentuation (score 3). Microphoto; H&E-staining; Bar = 200 μm

viraemia in all study piglets that was significantly lower in animals vaccinated with the experimental vaccine. Although PRRSV loads in serum and tissues of the unvaccinated study piglets were high, the development of clinical symptoms was moderate. Nevertheless, histological findings indicated interstitial pneumonia and/or other pulmonary lesions in all of the unvaccinated piglets. No clinical symptoms and less severe pathological findings were seen in the vaccinated piglets. Thus, the tested live attenuated PRRSV 2 strains were able to provide an efficient partial protection against heterologous challenge with a Vietnamese HP PRRSV 2 field strain.

Abbreviations
HP: Highly pathogenic; NGS: Next generation sequencing; PRRSV: Porcine Reproductive and respiratory disease virus; RT-qPCR: Reverse transcription quantitative real-time polymerase chain reaction; TCID$_{50}$: 50% tissue culture infective dose

Acknowledgements
The project was funded by Sanphar Asia. Many thanks to Kyoto Biken for providing the attenuated PRRSV strains. The authors want to thank the teams of the departments of Molecular Biology, Serology/Virology and Pathology Center East of the Institute for Veterinary Disease Control Mödling, Austria, as well as the technical personnel for perfect preparing of the experimental stable.

Author contributions
Conception and design of the study: TS, JP, FE, FS; Acquisition of data: TS, JP, ZB, SRF; Analysis and interpretations of data: TS, AS, JBC, FP, MF; Drafting of the article TS, AS, JBC; Revision of the manuscript JP, FE, SRF, ZB, FP, MF, FS; Final approval: all.

Funding
The study was funded by SANPHAR Division Holding GmbH. Design of the study, collection, analysis, and interpretation of data and the writing of the manuscript were not influenced by the funding body.

Competing interests
The authors declare that they have no competing interest.

Author details
[1]Institute for Veterinary Disease Control, AGES, Robert-Koch-Gasse 17, 2340 Mödling, Austria. [2]Clinic for Ruminants and Swine, University of Leipzig, An den Tierkliniken 11, 04103 Leipzig, Germany. [3]Veterinary Practice Entenfellner, Bonnleiten 8, 3073 Stössing, Austria. [4]BioSellal, Bâtiment Accinov, 317 avenue Jean Jaurès, 69007 Lyon, France.

References
1. Adams MJ, Lefkowitz EL, King AMQ, Harrach B, Harrison RL, Knowles AM, et al. Ratification vote on taxonomic proposals to the international committee on taxonomy of viruses (2016). Arch Virol. 2016;161:2921–49.
2. Wensvoort G. Lelystad virus and the porcine epidemic abortion and respiratory syndrome. Vet Res. 1993;24:117–24.
3. Nelson EA, Christopher-Hennings J, Drew T, Wensvoort G, Collins JE, Benfield DA. Differentiation of U.S. and European isolates of porcine reproductive and respiratory syndrome virus by monoclonal antibodies. J Clin Microbiol. 1993;31:3184–9.
4. Stadejek T, Stankevicius A, Murtaugh MP, Oleksiewicz MB. Molecular evolution of PRRSV in Europe: current state of play. Vet Microbiol. 2013; 165:21–8.
5. Tian K, Yu X, Zhao T, Feng Y, Cao Z, Wang C, et al. Emergence of fatal PRRSV variants: unparalleled outbreaks of atypical PRRS in China and molecular dissection of the unique hallmark. PLoS One. 2007;2:e526.
6. Feng Y, Zhao T, Nguyen T, Inui K, Ma Y, Nguyen TH, et al. Porcine respiratory and reproductive syndrome virus variants, Vietnam and China, 2007. Emerg Infect Dis. 2008;14:1774–6.
7. Li B, Fang L, Guo X, Gao J, Song T, Bi J, et al. Epidemiology and evolutionary characteristics of the porcine reproductive and respiratory syndrome virus in China between 2006 and 2010. J Clin Microbiol. 2011;49:3175–83.
8. Thuy NT, Thu NT, Son NG, Ha LTT, Hung VK, Nguyen NT, et al. Genetic analysis of ORF5 porcine reproductive and respiratory syndrome virus isolated in Vietnam. Microbiol Immunol. 2013;57:518–26.
9. An TQ, Zhou YJ, Liu GQ, Tian ZJ, Li J, Qiu HJ, et al. Genetic diversity and phylogenetic analysis of glycoprotein 5 of PRRSV isolates in mainland China from 1996 to 2006: coexistence of two NA subgenotypes with great diversity. Vet Microbiol. 2007;123:43–52.
10. Martelli P, Gozio S, Ferrari L, Rosina S, De Angelis E, Quintavalla C, et al. Efficacy of a modified live porcine reproductive and respiratory syndrome virus (PRRSV) vaccine in pigs naturally exposed to a heterologous European (Italian cluster) field strain: clinical protection and cell-mediated immunity. Vaccine. 2009;27:3788–99.
11. Linhares DCL, Cano JP, Wetzell T, Nerem J, Torremorell M, Dee SA. Effect of modified-live porcine reproductive and respiratory syndrome virus (PRRSv) vaccine on the shedding of wild-type virus from an infected population of growing pigs. Vaccine. 2012;30:407–13.
12. Park C, Seo HW, Han K, Kang I, Chae C. Evaluation of the efficacy of a new modified live porcine reproductive and respiratory syndrome virus (PRRSV) vaccine (Fostera PRRS) against heterologous PRRSV challenge. Vet Microbiol. 2014;172:432–42.
13. Renukaradhya GJ, Alekseev K, Jung K, Fang Y, Saif L. Porcine reproductive and respiratory syndrome virusinduced immunosuppression exacerbates the inflammatory response to porcine respiratory coronaviru in pigs. Viral Immunol. 2010;23:457–66.
14. Hu J, Zhang C. Porcine reproductive and respiratory syndrome virus vaccines: current status and strategies to a universal vaccine. Transbound Emerg Dis. 2014;61:109–20.
15. Roca M, Gimeno M, Bruguera S, Segalés J, Díaz I, Galindo-Cardiel IJ, et al. Effects of challenge with a virulent genotype II strain of porcine reproductive and respiratory syndrome virus on piglets vaccinated with an attenuated genotype I strain vaccine. Vet J. 2012;193:92–6.
16. Wei Z, Thang J, Zhuang J, Sun Z, Gao F, Yuan S. Immunization of pigs with a type 2 modified live PRRSV vaccine prevents the development of a deadly long lasting hyperpyrexia in a challenge study with highly pathogenic PRRSV JX143. Vaccine. 2013;31:2062–6.
17. Lager KM, Schlink SN, Brockmeier SL, Miller LC, Henningson JN, Kappes MA, et al. Efficacy of type 2 PRRSV vaccine against Chinese and Vietnamese HP-PRRSV challenge in pigs. Vaccine. 2014;32:6457–62.
18. Leng X, Li Z, Xia M, He Y, Wu H. Evaluation of the efficacy of an attenuated live vaccine against highly pathogenic porcine reproductive and respiratory syndrome virus in young pigs. Clin Vacc Immunol. 2012;19:1199–206.
19. Tian ZJ, An TQ, Zhou YJ, Peng JM, Hu SP, Wei TC, et al. An attenuated live vaccine based on highly pathogenic porcine reproductive and respiratory syndrome virus (HP-PRRSV) protects piglets against HP-PRRS. Vet Microbiol. 2009;138:34–40.
20. Suradhat S, Wongyanin P, Kesdangsakonwut S, Teankum K, Lumyai M, Triyarach S, et al. A novel DNA vaccine for reduction of PRRSV-induced negative immunomodulatory effects: a proof of concept. Vaccine. 2015;33: 3997–4003.
21. Hall TA. BioEdit: a user-friendly bilogical sequence alignment editor and analysis program for windows 97/98/NT. Nucleic Acids Symp Ser. 1999;41:95–8.
22. Tamura K, Peterson D, Peterson N, Stecher G, Nei M, Kumar S. MEGA5: molecular evolutionary genetics analysis using maximum likelihood, evolutionary distance, and maximum parsimony methods. Mol Biol Evol. 2011;28(10):2731–9.
23. Halbur PG, Paul GS, Frey ML, Landgraf J, Eernisse K, Meng XJ, et al. Comparison of the pathogenicity of two US porcine reproductive and respiratory syndrome virus isolates with that of Lelystad virus. Vet Pathol. 1995;32:648–60.
24. Li Y, Xue C, Wang L, Chen X, Chen F, Cao Y. Genomic analysis of two Chinese strains of porcine reproductive and respiratory syndrome viruses with different virulence. Virus Genes. 2010;40:374–81.
25. Guo B, Lager KM, Schlink SN, Kehrli ME Jr, Brockmeier SL, Miller LC, et al. Chinese and Vietnamese strains of HP-PRRSV cause different pathogenic outcomes in United States high health swine. Virology. 2013;446:238–50.

26. Do TD, Park C, Choi K, Jeong J, Nguyen TT, Nguyen DQ, et al. Comparison of experimental infection with northern and southern Vietnamese strains of highly pathogenic porcine reproductive and respiratory syndrome virus. J Comp Pathol. 2015a;152:227–37.

27. Chen Y, He S, Sun L, Luo Y, Sun Y, Xie J, et al. Genetic variation, pathogenicity, and immunogenicity of highly pathogenic porcine reproductive and respiratory syndrome virus strain XH-GD at different passage levels. Arch Virol. 2016;161:77–86.

28. Yu X, Zhou Z, Cao Z, Wu J, Zhang Z, Xu B, et al. Assessment of the safety and efficacy of an attenuated live vaccine based on highly pathogenic porcine reproductive and respiratory syndrome virus. Clin Vaccine Immunol. 2015;22:493–502.

29. Do TD, Park C, Choi K, Jeong J, Nguyen TT, Nguyen DQ, et al. Comparison of two genetically distant type 2 porcine reproductive and respiratory syndrome virus (PRRSV) modified live vaccines against Vietnamese highly pathogenic PRRSV. Vet Microbiol. 2015b;179:233–41.

30. Charoenchanikran P, Kedkovid R, Sirisereewan C, Woonwong Y, Arunorat J, Sitthichareonchai P, et al. Efficacy of Fostera® PRRS modified live virus (MLV) vaccination strategy against a Thai highly pathogenic porcine reproductive and respiratory syndrome virus (HP-PRRSV) infection. Trop Anim Health Prod. 2016;48:1351–9.

31. Hu SP, Zhang Z, Liu YG, Tian ZJ, Wu DL, Cai XH, et al. Pathogenicity and distribution of highly pathogenic porcine reproductive and respiratory syndrome virus in pigs. Transbound Emerg Dis. 2013;60:351–9.

32. Han Z, Liu Y, Wang G, He Y, Hu S, Li Y, et al. Comparative analysis of immune responses in pigs to high and low pathogenic porcine reproductive and respiratory syndrome viruses isolated in China. Transbound Emerg Dis. 2013;195:313–8.

33. Zhang J, Zheng Y, Xia XQ, Chen Q, Bade SA, Yoon KJ, et al. High-throughput whole genome sequencing of porcine reproductive and respiratory syndrome virus from cell culture materials and clinical specimens using next-generation sequencing technology. J Vet Diagn Investig. 2017;29:41–50.

34. Cong Y, Huang Z, Sun Y, Ran W, Zhu L, Yang G, et al. Development and application of a blocking enzyme-linked immunosorbent assay (ELISA) to differentiate antibodies against live and inactivated porcine reproductive and respiratory syndrome virus. Virology. 2013;444:310–6.

35. Gerber PF, Gimenez-Lirola LG, Halbur PG, Zhou L, Meng XJ, Oppriessnig T. Comparison of commercial enzyme-linked immunosorbent assays and fluorescent microbead immunoassays for detection of antibodies against porcine reproductive and respiratory syndrome virus in boars. J Virol Meth. 2014;197:63–6.

23

Non-typhoidal *Salmonella* serovars in poultry farms in central Ethiopia: prevalence and antimicrobial resistance

Tadesse Eguale (ID)

Abstract

Background: Poultry is one of the common sources of non-typhoidal *Salmonella* and poultry products are the major sources of human infection with non-typhoidal *Salmonella*. In spite of flourishing poultry industry in the country, data on prevalence and antimicrobial susceptibility of non-typhoidal *Salmonella* serovars at farm level is not available in Ethiopia. This study investigated prevalence, serotype distribution and antimicrobial resistance of non-typhoidal *Salmonella* in poultry farms in Addis Ababa and its surrounding districts.

Results: A total of 549 fresh pool of fecal droppings (*n* = 3 each) were collected from 48 poultry farms and cultured for *Salmonella* using standard laboratory technique and serotyped using slide agglutination technique. Susceptibility of *Salmonella* isolates to18 antimicrobials was tested according to CLSI guideline using Kirby-Bauer disk diffusion assay. *Salmonella* was recovered in 7 (14.6%) of the farms and 26 (4.7%) of the samples. *Salmonella* was more common in poultry farms with larger flock size than in the smaller ones and in Ada'a district as compared to other districts. All isolates were obtained from farms containing layers. Two out of 6 (33.3%) farms that kept birds in cage were positive for *Salmonella* while only 5 (11.9%) of the 42 farms who used floor system were positive. Oxytetracycline was used widely in 40 (83.3%) of the farms, followed by amoxicillin 14 (29.2%) and sulfonamides 11 (22.9%). *Salmonella* Saintpaul was the dominant serotype detected accounting for 20 (76.9%) of all isolates. Other serovars, such as *S.* Typhimurium3 (11.5%), *S.* Kentucky 2 (7.7%) and *S.* Haifa 1 (3.8%) were also detected. Of all the *Salmonella* isolates tested, 24 (92.3%) were intermediately or fully resistant to sulfisoxazole and streptomycin, 12 (46.2%) to cephalothin, while 11 (42.3%) were resistant to ampicillin, amoxicillin+clavulanic acid, kanamycin and chloramphenicol. Multidrug resistance (MDR) to several drugs was common in *S.* Kentucky and *S.* Saintpaul.

Conclusion: Despite low prevalence of *Salmonella* in poultry farms in the study area, circulation of MDR strains in some farms warrant special biosecurity measures to hinder dissemination of these pathogens to other farms and the public. Moreover, awareness creation on prudent use of antimicrobials is recommended.

Keywords: Poultry, Non-typhoidal *Salmonella*, Antimicrobial resistance, Prevalence, Ethiopia

Background

Salmonella is one of the major causes of food-borne diseases worldwide [1]. Poultry and other food animals are considered the common reservoirs of *Salmonella enterica* and undercooked poultry products are the major sources of human infection with non-typhoidal *Salmonella* [2, 3]. Several host unrestricted *S. enterica* serovars frequently isolated from poultry without

showing any clinical signs usually infect a wider range of hosts and cause disease in humans as well [4].

It has been shown that some of the most commonly detected serovars in chickens in a given geographic area are also among the top serovars associated with human infections indicating the role of *Salmonella* colonization of poultry farms to public health [5]. Knowledge on distribution of *Salmonella* serovars in food animals and humans is useful to understand the trends of *Salmonella* epidemiology and to identify serovars that cluster over time and space. Temporal and spatial variation in rate

Correspondence: tadesse.eguale@aau.edu.et
Aklilu Lemma Institute of Pathobiology, Addis Ababa University, P.O. Box 1176, Addis Ababa, Ethiopia

and distribution of *Salmonella* serovars in poultry industry has been reported [2, 6].

Developed countries conduct routine surveillance of *Salmonella* in poultry farms to understand the level of colonization by *Salmonella*, serovars involved and drug resistance profile with the aim of designing ways of reducing public health salmonellosis of poultry origin [7, 8]. However, in developing countries like Ethiopia, little effort is made to monitor *Salmonella* in poultry farms and information on prevalence and serotype distribution as well as phenotypic and genotypic relatedness of *Salmonella* isolated from poultry and humans is not well documented. Local knowledge on prevalence of *Salmonella*, serotype distribution and associated risk factors is important to implement appropriate control strategy to reduce wider dissemination of important zoonotic serovars [2].

There is little available literature on farm level prevalence and serotype distribution of non-typhoidal *Salmonella* in poultry farms in Ethiopia. Previous studies conducted on retail raw chicken products reported 17.9% prevalence of *Salmonella*, the dominant serovars being *S.* Braenderup (31.5%), *S.* Anatum (25.9%), *S.* Saintpaul (14.8%) and *S.* Uganda (11.1%) [9]. Another study also reported that 14% of chicken carcass from supermarkets in Addis Ababa were positive for *Salmonella.* *S.* Braenderup (41.4%), *S.* Hadar (20.7%), *S.* Newprt (13.8%) and *S.* Typhimurium (10.3%) were the dominant serovars detected in poultry products in Addis Ababa [10]. However, source of *Salmonella* contamination in these poultry products could be either from farm or due to cross contamination during slaughter, transportation or storage. Recent study in southern Ethiopia showed that 16.7% of samples from poultry and the environment of three poultry farms were positive for *Salmonella* although this study did not show whether *Salmonella* isolates were host specific *Salmonella* serovars or host unrestricted non-typhoidal *Salmonella* serovars [11].

Majority of the *Salmonella* isolates from poultry products and poultry farms in the previous studies were found to be resistant to several antimicrobials. Information on farm level prevalence and antimicrobial susceptibility status of isolates can explain the level of public health risk associated with poultry products. The aim of this study was therefore to determine the prevalence, serotype distribution and antimicrobial resistance of *salmonella* in poultry farms in central Ethiopia. The type of antimicrobials and disinfectants commonly employed in poultry farms were also assessed.

Methods
Study design, study area and study animals
A cross-sectional study was conducted in Addis Ababa and 3 districts of Oromia region located at the outskirt of Addis Ababa from July 2013–January 2014. A total of 549 pooled fresh fecal droppings (from 3 chicken each) were collected in 48 farms (Ada'a district *n* = 33, Addis Ababa *n* = 6, Sebeta *n* = 6, Barake *n* = 3). Inclusion of farms in the sampling was based on representation of the area under study, willingness of the owners, availability of poultry farms in the study area, and the flock having a minimum of 50 birds. Most of the poultry farms investigated in the current study were those from Ada'a district due to large number of poultry farms in this district.

Data and sample collection
Information such as type of poultry farm, whether it is broiler or layer, flock size, birds housing system, age of birds, purpose and types of antimicrobials and disinfectants commonly used in the farm during the last 6 months were recorded using a purposively designed questionnaire. Collection of data was performed at the time of fecal sample collection from each farm. Pooled fresh fecal droppings(from 3 chickens) were collected using clean disposable gloves in to sterile zippered plastic bags which were transported to microbiology laboratory of Aklilu Lemma Institute of Pathobiology, Addis Ababa University in an ice box within 3–4 h of collection.

Salmonella isolation, identification, serotyping and phage typing
Salmonella isolation and identification was carried out using conventional methods [12, 13]. Briefly, fresh fecal droppings from three chicken was thoroughly mixed of which 10 g of feces was suspended in 90 ml of buffered peptone water (BPW) (Becton Dickinson, Sparks, MD) and incubated overnight at 37 °C. Enrichment, culturing on selective media, and biochemical analysis of presumptive *Salmonella* colonies was conducted as shown previously [14]. Genus specific PCR was used to confirm isolates suspected to be *Salmonella* by biochemical tests [15]. *Salmonella* Typhimurium (ATCC 14028) was used as a positive control during biochemical analysis and PCR. Confirmed *Salmonella* isolates were stored at − 80 °C in 20% glycerol till further investigation.

Serotyping and phage typing of *Salmonella* isolates was conducted at the World Organization for Animal Health (OIÉ) Reference Laboratory for salmonellosis, Public Health Agency of Canada's National Microbiology at Guelph. Determination of serovars was conducted using serum agglutination technique as shown previously [16, 17], based on identification of somatic (O) antigens [18] and flagellar (H) antigens [19].

Antimicrobial susceptibility testing

Isolates were investigated for susceptibility to 18 antimicrobials using the Kirby-Bauer disk diffusion method according to Clinical and Laboratory Standards Institute guidelines [20]. Antimicrobials used in the current study were amikacin (30 μg), amoxicillin + clavulanic acid (20/10 μg), ampicillin (10 μg), cefoxitin (30 μg), ceftriaxone (30 μg), cephalothin (30 μg), chloramphenicol (30 μg), ciprofloxacin (5 μg), gentamicin (10 μg), kanamycin (30 μg), nalidixic acid (30 μg), neomycin (30 μg), nitrofurantoin (100 μg), streptomycin (10 μg), sulfisoxazole (1000 μg), sulfamethoxazole + trimethoprim (23.75/1.25 μg), trimethoprim (5 μg) and tetracycline (30 μg). All of them were from Sensi-Discs, Becton, Dickinson and Company, Loveton, USA. The interpretation cut off points for susceptibility status of isolates was based on the CLSI guidelines [20]. For the purpose of analysis, all readings classified as intermediate were considered as resistant unless indicated. *E. coli* ATCC 25922 was used as a quality control organism.

Statistical analysis

Sample level prevalence of *Salmonella* was calculated as percentage of *Salmonella* culture positive fecal samples among total number of samples examined. Farm level prevalence was calculated as the percentage of farms with one or more *Salmonella* culture positive pooled fecal sample among the total farms sampled. Association of *Salmonella* detection with various factors was tested using exact test and *p*-value < 0.05 was considered significant.

Results

Farm level *Salmonella* occurence with respect to various factors

Salmonella was isolated from 14.6% (7/48) of poultry farms with individual sample level prevalence of 4.7% (Table 1). *Salmonella* was more common in poultry farms with larger flock size and in age group of 2–6 months (Table 2). Majority of the farms studied contained layers or young pullets grown for egg production (*n* = 43, 89.6%); whereas only (*n* = 5; 11.4%) were keeping broilers. *Salmonella* was not detected from the broiler

farms. *Salmonella* isolation was also more common in farms of the Ada'a district as compared to other districts. Majority of the farms (*n* = 42; 87.5%) keep their birds on floor system and 12.5% (6/48) use cage system. Out of the farms that use cage system 33.3% (2/6) were positive for *Salmonella* whereas 11.9% (5/42) of farms that use floor system were found positive for *Salmonella*.

Antimicrobials used in poultry farms

Among the common antimicrobials, oxytetracycline was used widely in 40 (83.3%) of the farms, followed by amoxicillin (29.2%) and sulfonamides (22.9%). Other antimicrobials such as fluoroquinolones (enrofloxacin and ciprofloxacin), and florfenicol were also used in 11 (22.9%) and 7 (14.6%) of the farms respectively, whereas 6(12.5%) of the farms reported that they did not use any antimicrobials during last 6 months. None of the farms reported use of antimicrobials as feed additive. All of the farms use antimicrobials for therapeutic or prophylactic purposes when there is one or more sick birds in the flock. Interestingly, in one of the poultry farms in Adaa district, the use of human preparation of ciprofloxacin tablet was observed. *Salmonella* was recovered more frequently in farms which use only amoxicillin, sulfadimidine and oxytetracycline than those farms which use fluoroquinolones and florfenicol. Recent use of antimicrobials and occurrence of *Salmonella* in farms is shown in Table 3. All samples from six farms with no history of use of antimicrobials were also not culture positive for *Salmonella*. Twenty-three (47.9%) of the farms reported use of sodium hypochlorite disinfectant as foot bath, for cleaning poultry houses before introduction of new stock and to clean feeding utensils, while 4(8.3%) of the farms used copper sulfate. The remaining 21(43.8%) of the poultry farms were not using any disinfectant.

Salmonella serotype distribution and antimicrobial susceptibility

Salmonella Saintpaul was the dominant serotype detected in poultry farms accounting for 20 (76.9%) of all isolates. Other serotypes, such as *S.* Typhimurium (*n* = 3), *S.* Kentucky (*n* = 2) and *S.* Haifa (*n* = 1) were also detected. Rate of resistance to antimicrobials tested and

Table 1 Prevalence of *Salmonella* in poultry farms in Addis Ababa and surrounding districts

	No. of farms	Average no. of birds /farm	No. of samples[a]	No. positive samples	% positive samples	(%) positive farms
Ada'a	33	4638	464	25	5.4	18.2
Addis Ababa	6	1075	45	1	2.2	16.7
Barake	3	395	18	0	0	0
Sebeta	6	627	22	0	0	0
Total	48	1684	549	26	4.7	14.6

[a]Samples were pool of fecal droppings from 3 chicken

Table 2 Occurrence of *Salmonella* in poultry farms stratified by selected factors

Selected Factors	No. of farms	No. of *Salmonella* positive farms	% of farms positive for *Salmonella*	*p*-value*
Commodity type				
Layers	43	7	16.3	1.000
Broilers	5	0	0	
Use of disinfectants				
Yes	26	6	23.1	0.106
No	22	1	4.5	
Age of birds in months				
< 2	8	0	0	0.608
2–6	17	4	24	
7–12	13	2	15.4	
> 12	10	1	10	
Flock size				
< 1000(Small)	22	2	9.1	0.648
1000–5000(Medium)	17	3	17.7	
> 5000(Large)	9	2	22.2	
Poultry housing system				
Floor	42	5	11.9	0.206
Cage	6	2	33.3	

*Exact test was used to obtain *p*-value

resistance patterns of the isolates are shown in Tables 4 and 5 respectively. Of all the *Salmonella* isolates tested, (*n* = 24, 92.3%) were resistant to sulfisoxazole and streptomycin, (*n* = 12, 46.2%) of the isolates were resistant to cephalothin, while (*n* = 11, 42.3%) were resistant to ampicillin, amoxicillin + clavulanic acid, kanamycin and chloramphenicol (Table 4).

Overall, multidrug resistance was commonly detected in *Salmonella* isolates in the current study particularly in strains belonging to *S.* Saintpaul and the two *S.* Kentucky isolates. All *S.* Saintpaul strains in the current study were isolated from farms in Ada'a district. However, there was wide diversity in their antimicrobial susceptibility pattern even among isolates obtained from

the same farm. Some of them were resistant to only few antimicrobials while others were MDR to several antimicrobials. The two *S.* Kentucky isolates were resistant to 9 of the 18 antimicrobials tested (Table 5).

Discussion

Colonization of poultry with *Salmonella* without detectable clinical signs at farm level followed by contamination of poultry products with subsequent access to human food chain has been considered as the major sources of human salmonellosis [21, 22]. *Salmonella* in healthy poultry is the main risk factor for possible outbreak of human salmonellosis and epidemiological studies have shown the huge contribution of contaminated

Table 3 Recent use of antimicrobials and occurrence of *Salmonella* in poultry farms

Type of Antimicrobials used during the last 6 months	No. of farms	No. of *Salmonella* positive farms	% of farms positive for *Salmonella*
Amoxicillin only	2	0	0
Oxytetracyline only	18	4	22.2
Oxytetracycline + ciprofloxacin	3	0	0
Oxytetracycline + florfenicol + enrofloxacin	4	0	0
Oxytetracycline + sulfonamides	3	1	33.3
Oxytetracycline + amoxicillin	4	1	25
Oxytetracycline + sulfonamides + amoxicillin	8	1	12.5
Did not use antimicrobial agent	6	0	0

Table 4 *Salmonella* serovar distribution and rate of resistance to antimicrobial agents

Antimicrobial agents	*Salmonella* serovars and resistance rate[a]				Total No. (%) resistant
	S. Saintpaul (*n* = 20)	*S.* Typhimurium (*n* = 3)	*S.* Kentucky (*n* = 2)	*S.* Haifa (*n* = 1)	
	No. resistant (%)	No. resistant (%)	No. resistant (%)	No. resistant (%)	
Ampicillin	9 (45)	0	2 (100)	0	11 (42.3)
Amoxicillin+clavulanic acid	9 (45)	0	2 (100)	0	11 (42.3)
Chloramphenicol	10 (50)	0	1 (50)	0	11 (42.3)
Cephlothin	10 (50)	0	2 (100)	0	12 (46.2)
Ciprofloxacin	0	0	2 (100)	0	2 (7.7)
Cefoxitin	0	0	0	0	0
Gentamicin	0	0	2 (100)	0	2 (7.7)
Kanamycin	8 (40)	2 (66.7)	0	1 (100)	11 (42.3)
Sulfamethoxazole+trimethoprim	0	0	0	1 (100)	1 (3.9)
Trimethoprim	0	0	0	1 (100)	1 (3.9)
Tetracycline	4 (20)	1 (33.3)	2 (100)	1 (100)	8 (30.8)
Sulfisoxazole	18 (90)	3 (100)	2 (100)	1 (100)	24 (92.3)
Streptomycin	18 (90)	3 (100)	2 (100)	1 (100)	24 (92.3)
Nitrofurantoin	5 (25)	1 (33.3)	0	1 (100)	7 (26.7)
Nalidixic acid	2 (10)	0	2 (100)	1 (100)	5 (19
Neomycin	3 (15)	0	0	0	3 (11.5)

[a]Isolates with intermediate susceptibility were also considered resistant for this analysis

poultry products to human salmonellosis [23, 24]. In fact, some countries have shown that successful control measures involving surveillance, improved biosecurity and vaccination targeting specific serovars in poultry can result in reduction of human salmonellosis cases [21, 24].

In the current study, 7(14.6%) of the 48 examined poultry farms were positive for *Salmonella*. This is very much low compared to studies conducted in Morocco and Nigeria where 76.7% and [25], 43.6% [26] of the poultry farms were contaminated by *Salmonella,* respectively. Sample level prevalence of *Salmonella* was also low in the current study (4.7%) compared to previous studies conducted elsewhere. For instance, *Salmonella* prevalence in fecal samples from conventional poultry farms in USA was reported to be 38.8% while it was 5.6% in organic farms [27]. *Salmonella* prevalence in conventional poultry is usually very high in different countries [28–31]. The possible reason for low prevalence of *Salmonella* in the current study could be due to the fact that most of the poultry farms in the current study were small scale farms holding small number of birds unlike most of the large commercial poultry farms where they keep thousands of birds and the feeding and management activities associated with intensification allows easy dissemination of the pathogen within the farm. This finding is in agreement with previous report where large farms were significantly associated

with high prevalence of *Salmonella* compared to medium and small farms [32].

Both farm level and pooled sample level prevalence of *Salmonella* was high in farms from Ada'a district compared to other areas, which could be due to larger number of poultry farms examined from this district compared to others as well as difference in agroecology. Ada'a district is highly concentrated with large number of poultry farms and is located in rift valley which is relatively warm region compared to Addis Ababa, Sebeta and Barake districts. The fact that most of the large poultry farms in the country including the parent stocks are located in Ada'a district and most of the farms from this area shared a single serotype, *S.* Saintpaul implies the possibility of transmission of *Salmonella* from farm to farm in this town. *Salmonella* Saintpaul is not frequently isolated from poultry in other previous studies elsewhere. *Salmonella* Kentucky was the dominant serovar in studies conducted in Nigeria [26] and Bangladesh [31] and *S.* Entertidis was dominant in Spain [33]; while *S.* Typhimurium was dominant in China [34]. Although there is no serotype data on *Salmonella* isolates from poultry at farm level in Ethiopia, previous study from poultry food items in Addis Ababa did not report *S.* Saintpaul [35]. As most of the farms obtain their day old chickens or pullets from a few parent stock farms located in this district, there is likelihood of contamination of poultry from source farms. In addition, *S.* Saintpaul

Table 5 *Salmonella* serotypes isolated from poultry farms and their antimicrobial resistance pattern

No.	Study site	Farm Code	Isolate code	Serotype	Resistance pattern	
					Intermediate	Resistant
1	Adaa	DZP-20	DP-213 T	Kentucky	C	Amp,Amc,Cf,Cip,Gm,Te,Su,S,Na
2	Adaa	DZP-20	DP-220 T	Kentucky	–	Amp,Amc,Cf,Cip,Gm,Te,Su,S,Na
3	Addis Ababa	AAP-08	AP-H2O	Haifa	K,S	Sxt,Tmp,Te,Su,Nitro,Na
4	Adaa	DZP-03	DP-23 T	Saintpaul	Su	–
5	Adaa	DZP-03	Dp-24 T	Saintpaul	Su,S	–
6	Adaa	DZP-03	Dp-25R	Saintpaul	SuS	–
7	Adaa	DZP-03	DP-26R	Saintpaul	Cip,Su,SNitro,N	–
8	Adaa	DZP-03	DP-27R	Saintpaul	Su,S	–
9	Adaa	DZP-11	DP-116 T	Typhimurium	K	Te,Su,S
10	Adaa	DZP-08	DP-70 T	Typhimurium	SuS	–
11	Adaa	DZP-08	DP-71 T	Typhimurium	K,Su,S,Nitro	–
12	Adaa	DZP-33	DP-107	Saintpaul	Amc,Cf,K,S	Amp,C,Te,Su
13	Adaa	DZP-33	DP-117	Saintpaul	Amc,Cf,K,S	Amp,C,Su
14	Adaa	DZP-33	DP-128	Saintpaul	-	-
15	Adaa	DZP-33	DP-131	Saintpaul	K,S	–
16	Adaa	DZP-33	DP-110	Saintpaul	Amc,Cf,S	Amp,C,Te,Su
17	Adaa	DZP-33	DP-114	Saintpaul	SuS	
18	Adaa	DZP-33	DP-126	Saintpaul	Cf,S	Amp,Amc,C,Su
19	Adaa	DZP-12	DP-313	Saintpaul	S,K	Amp,Amc,C,Cf,Te,Su
20	Adaa	DZP-12	DP-325	Saintpaul	Amc,Cf,Su,S,Nitro	Amp,Amc,C,Cf,Su,S,Nitro,Na
21	Adaa	DZP-12	DP-327	Saintpaul	K,S,Nitro	Su
22	Adaa	DZP-12	DP-328	Saintpaul	Amc,Cip,S,N	Amp,C,Cf,Te,Su
23	Adaa	DZP-12	DP-339	Saintpaul	K,Su,S	–
24	Adaa	DZP-12	DP-322	Saintpaul	Amc,Cf, Su,S,Nitro	Amp,Amc,C,Cf,Su,S,Nitro,Na
25	Adaa	DZP-12	DP-326	Saintpaul	Cf,S	Amp,Amc,C,Su
26	Adaa	DZP-12	DP-308	Saintpaul	Amc,Cip,K,Su,S,Na,N	Amp,C,Cf,Nitro

Amp ampicillin, *Amc* amoxicillin and clavulanic acid, *Cf* cephalothin, *Cip* ciprofloxacin, *Gm* gentamicin, *K* kanamycin, *Tmp* trimethoprim, *Te* tetracycline, *Su* sulfisoxazole, *S* streptomycin, *Nitro* nitrofurantoin, *Na* nalidixic acid, *N* neomycin, -sensitive

was the major serotype detected in dairy farms in this study area which suggests possibility of transmission between dairy and poultry farms [14].

The high rate of resistance to sulfixazole and streptomycin (92.3%) is not concordant with the current rate of use of antimicrobials in farms investigated. However, previously, different sulfonamide drugs and streptomycin together with penicillin were the common antimicrobials frequently used in the country for treatment of various infectious diseases in veterinary medicine and recent studies showed that sulfonamides and streptomycin are the 2nd and 3rd most prescribed veterinary medications respectively in the study area next to oxytetracycline [36]. Similarly, high resistance rate to ampicillin and tetracycline could be due to long term use of these antimicrobials in

veterinary medicine including poultry. Interestingly, the two *S.* Kentucky isolates resistant to several drugs including nalidixic acid and ciprofloxacin were isolated from one of a few farms which reported use of fluoroquinolones for therapeutic purposes in the farm suggesting possible contribution of use of these drugs in the farm for selection of these strains. Eleven (42.3%) of the isolates in the current study, most of which belonging to *S.* Saintpaul from farms in Ada'a district were resistant to chloramphenicol unlike previous study where all of the isolates obtained from food of animal origin including poultry products were fully susceptible to chloramphenicol [35]. Unlike previous study in south Ethiopia [11] where extremely high proportion of *Salmonella* isolates (97.8%) were resistant to second generation cephalosporin (cefoxitin), in this study, none of the isolates were resistant to this drug. This could be due to over use of betalactam drugs in the previous farms.

Conclusion

Despite low prevalence of *Salmonella* in poultry farms in the study area, circulation of MDR strains in some farms warrant special biosecurity measures to hinder dissemination of these pathogens to other farms and the public. Moreover, awareness creation on prudent use of antimicrobials is recommended.

Abbreviations

BPW: Buffered peptone water; MDR: Multi-drug resistance; RVB: Rappaport-vassiliadis broth; TTB: Tetrathionate broth; XLT-4: Xylose lysine tergitol 4

Acknowledgments

The author would like to thank Mr. Haile Alemayehu and Mr. Nega Nigussie for their support during sample collection and laboratory analysis. Dr. Roger P. Johnson, Dr. Linda Cole, Shaun Kernaghan, Ketna Mistry, Ann Perets and Betty Wilkie of the Public Health Agency of Canada, National Microbiology Laboratory at Guelph are also acknowledged for serotyping of *Salmonella* isolates.

Funding

This study was supported by WHO Advisory Group on Integrated Surveillance of Antimicrobial Resistance. The funding agency was not involved in design of study, data collection, analysis of data and manuscript writing.

Author's contributions

TE was involved in conception of the study, laboratory work, data analysis and preparation of the manuscript.

Consent for publication

Not applicable.

Competing interests

The author declares that he has no competing interests.

References

1. Zhao S, Datta AR, Ayers S, Friedman S, Walker RD, White DG. Antimicrobial-resistant Salmonella serovars isolated from imported foods. Int J Food Microbiol. 2003;84(1):87–92.
2. Foley SL, Nayak R, Hanning IB, Johnson TJ, Han J, Ricke SC. Population dynamics of Salmonella enterica serotypes in commercial egg and poultry production. Appl Environ Microbiol. 2011;77(13):4273–9.
3. Braden CR. Salmonella enterica serotype Enteritidis and eggs: a national epidemic in the United States. Clin Infect Dis. 2006;43(4):512–7.
4. Gast RK. Serotype-specific and serotype-independent strategies for preharvest control of food-borne Salmonella in poultry. Avian Dis. 2007; 51(4):817–28.
5. Foley SL, Lynne AM, Nayak R. Salmonella challenges: prevalence in swine and poultry and potential pathogenicity of such isolates. J Anim Sci. 2008; 86(14 Suppl):E149–62.
6. Sivaramalingam T, McEwen SA, Pearl DL, Ojkic D, Guerin MT. A temporal study of Salmonella serovars from environmental samples from poultry breeder flocks in Ontario between 1998 and 2008. Can J Vet Res. 2013;77(1):1–11.
7. Wegener HC, Hald T, Lo Fo Wong D, Madsen M, Korsgaard H, Bager F, Gerner-Smidt P, Mølbak K. Salmonella control programs in Denmark. Emerg Infect Dis. 2003;9(7):774–80.
8. Hendriksen RS, Vieira AR, Karlsmose S, Lo Fo Wong DM, Jensen AB, Wegener HC, Aarestrup FM. Global monitoring of Salmonella serovar distribution from the World Health Organization global foodborne infections network country data Bank: results of quality assured laboratories from 2001 to 2007. Foodborne Pathog Dis. 2011;8(8):887–900.
9. TibaiJuka B, B M, G H, J K. Occurrence of salmonellae in retail raw chicken products in Ethiopia. Berl Munch Tierarztl Wochenschr. 2003;116(1–2):55–8.
10. Endrias Z, Poppe C. Antimicrobial resistance pattern of Salmonella serotypesisolated from food items and personnel in AddisAbaba, Ethiopia. Trop Anim Prod. 2009;41:241–9.
11. Abdi RD, Mengstie F, Beyi AF, Beyene T, Waktole H, Mammo B, Ayana D, Abunna F. Determination of the sources and antimicrobial resistance patterns of Salmonella isolated from the poultry industry in southern Ethiopia. BMC Infect Dis. 2017;17(1):352.
12. Molla B, Sterman A, Mathews J, Artuso-Ponte V, Abley M, Farmer W, Rajala-Schultz P, Morrow WE, Gebreyes WA. Salmonella enterica in commercial swine feed and subsequent isolation of phenotypically and genotypically related strains from fecal samples. Appl Environ Microbiol. 2010;76(21):7188–93.
13. WHO. Who Global Foodborne Infections Network Laboratory Protocol, Isolation of Salmonella spp From Food and Animal Feaces. 5th ed; 2010. http://antimicrobialresistance.dk/CustomerData/Files/Folders/6-pdf-protocols/63_18-05-isolation-of-salm-220610.pdf.
14. Eguale T, Engidawork E, Gebreyes AW, Asrat D, Alemayehu H, Medhin G, Johnson RP, Gunn JS. Fecal prevalence, serotype distribution and antimicrobial resistance of salmonellae in dairy cattle in Central Ethiopia. BMC Microbiol. 2016;16(1):1–11.
15. Cohen ND, Neibergs HL, McGruder ED, Whitford HW, Behle RW, Ray PM, Hargis BM. Genus-specific detection of salmonellae using the polymerase chain reaction (PCR). J Vet Diagn Investig. 1993;5(3):368–71.
16. Grimont PAD, Weill FX. Antigenic Formulae of the Salmonella Serovars. 9th ed. Paris: Institut Pasteur; 2007.
17. Issenhuth-Jeanjean S, Roggentin P, Mikoleit M, Guibourdenche M, de Pinna E, Nair S, Fields PI, Weill FX. Supplement 2008-2010 (no. 48) to the white-Kauffmann-Le minor scheme. Res Microbiol. 2014;165(7):526–30.
18. Ewing WH. (1986) Edwards and Ewing's identification of Enterobacteriaceae, Elsevier Science Publishing Co. Inc. New York, N.Y, 4th ed.
19. Shipp CR, Rowe B. A mechanised microtechnique for salmonella serotyping. J Clin Pathol. 1980;33(6):595–7.
20. CLSI: Performance Standards for Antimicrobial Susceptibility Testing; Twenty-Third Informational SupplementM100-S23. In., vol. 33; 2013.
21. Cosby DE, Cox NA, Harrison MA, Wilson JL, Buhur RJ, Fedorka-Cray PJ. Salmonella and antimicrobial resistance in broilers:A review. J Appl Poult Res. 2015;24:408–26.
22. Butaye P, Michael GB, Schwarz S, Barrett TJ, Brisabois A, White DG. The clonal spread of multidrug-resistant non-typhi Salmonella serotypes. Microbes Infect. 2006;8(7):1891–7.
23. Antunes P, Mourao J, Campos J, Peixe L. Salmonellosis: the role of poultry meat. Clin Microbiol Infect. 2016;22(2):110–21.
24. Hugas M, Beloeil P. Controlling Salmonella along the food chain in the European Union - progress over the last ten years. Euro Surveill. 2014;19(19)
25. Ziyate N, Karraouan B, Kadiri A, Darkaoui S, Soulay A. Prevalence and antimicrobial resistance of Salmonella isolates in Moroccan laying hens farms. J Appl Poult Res. 2016;25:539–46.
26. Fagbamila IO, Barco L, Mancin M, Kwaga J, Ngulukun SS, Zavagnin P, Lettini AA, Lorenzetto M, Abdu PA, Kabir J, et al. Salmonella serovars and their distribution in Nigerian commercial chicken layer farms. PLoS One. 2017; 12(3):e0173097.
27. Alali WQ, Thakur S, Berghaus RD, Martin MP, Gebreyes WA. Prevalence and distribution of Salmonella in organic and conventional broiler poultry farms. Foodborne Pathog Dis. 2010;7(11):1363–71.
28. Gaffga NH, Barton Behravesh C, Ettestad PJ, Smelser CB, Rhorer AR, Cronquist AB, Comstock NA, Bidol SA, Patel NJ, Gerner-Smidt P, et al. Outbreak of salmonellosis linked to live poultry from a mail-order hatchery. N Engl J Med. 2012;366(22):2065–73.

29. Basler C, Forshey TM, Machesky K, Erdman MC, Gomez TM, Nguyen TA, Behravesh CB. Multistate outbreak of human Salmonella infections linked to live poultry from a mail-order hatchery in Ohio–march-September 2013. MMWR Morb Mortal Wkly Rep. 2014;63(10):222.

30. Taylor M, Leslie M, Ritson M, Stone J, Cox W, Hoang L, Galanis E, Bowes V, Byrne S, de With N, et al. Investigation of the concurrent emergence of Salmonella enteritidis in humans and poultry in British Columbia, Canada, 2008-2010. Zoonoses Public Health. 2012;59(8):584–92.

31. Barua H, Biswas PK, Olsen KE, Christensen JP. Prevalence and characterization of motile Salmonella in commercial layer poultry farms in Bangladesh. PLoS One. 2012;7(4):e35914.

32. Adesiyun A, Webb L, Musai L, Louison B, Joseph G, Stewart-Johnson A, Samlal S, Rodrigo S. Survey of Salmonella contamination in chicken layer farms in three Caribbean countries. J Food Prot. 2014;77(9):1471–80.

33. Alvarez-Fernandez E, Alonso-Calleja C, Garcia-Fernandez C, Capita R. Prevalence and antimicrobial resistance of Salmonella serotypes isolated from poultry in Spain: comparison between 1993 and 2006. Int J Food Microbiol. 2012;153(3):281–7.

34. Kuang X, Hao H, Dai M, Wang Y, Ahmad I, Liu Z, Zonghui Y. Serotypes and antimicrobial susceptibility of Salmonella spp. isolated from farm animals in China. Front Microbiol. 2015;6:602.

35. Zewdu E: Prevalence, distribution and antimicrobial resistance profile of Salmonella isolated from food items and personnel in Addis Ababa, Ethiopia 2004.

36. Beyene T, Endalamaw D, Tolossa Y, Feyisa A. Evaluation of rational use of veterinary drugs especially antimicrobials and anthelmintics in Bishoftu, Central Ethiopia. BMC Res Notes. 2015;8:482.

24

Clinical impact of deoxynivalenol, 3-acetyl-deoxynivalenol and 15-acetyl-deoxynivalenol on the severity of an experimental *Mycoplasma hyopneumoniae* infection in pigs

Annelies Michiels[1]* [iD], Ioannis Arsenakis[1], Anneleen Matthijs[1], Filip Boyen[2], Geert Haesaert[3], Kris Audenaert[3], Mia Eeckhout[3], Siska Croubels[4], Freddy Haesebrouck[2] and Dominiek Maes[1]

Abstract

Background: The mycotoxin deoxynivalenol (DON) is highly prevalent in cereals in moderate climates and therefore pigs are often exposed to a DON-contaminated diet. Pigs are highly susceptible to DON and intake of DON-contaminated feed may lead to an altered immune response and may influence the pathogenesis of specific bacterial diseases. Therefore, the maximum guidance level in feed is lowest in this species and has been set at 900 µg/kg feed by the European Commission. This study aimed to determine the effect of in-feed administration of a moderately high DON concentration (1514 µg/kg) on the severity of an experimental *Mycoplasma hyopneumoniae* (*M. hyopneumoniae*) infection in weaned piglets. Fifty *M. hyopneumoniae*-free piglets were assigned at 30 days of age [study day (D)0] to four different groups: 1) negative control group (NCG; *n* = 5), 2) DON-contaminated group (DON; *n* = 15), 3) DON-contaminated and *M. hyopneumoniae*-inoculated group (DONMHYO; *n* = 15), 4) *M. hyopneumoniae*-inoculated group (MHYO; n = 15). The piglets were fed the experimental diets ad libitum for five weeks and were monitored during this period and euthanized at day 35 [27 days post infection (DPI)] or 36 (28 DPI). The main parameters under investigation were macroscopic lung lesions (MLL) at euthanasia, respiratory disease score (RDS) from day 8 until day 35, histopathologic lesions and log copies of *M. hyopneumoniae* DNA detected by qPCR, determined at the day of euthanasia.

Results: No significant difference was obtained for MLL at euthanasia, RDS (8–35), histopathologic lung lesions and log copies of *M. hyopneumoniae* DNA in the DONMHYO and MHYO group and consequently, no enhancement of the severity of the *M. hyopneumoniae* infection could be detected in the DONMHYO compared to the MHYO group.

Conclusions: Under present conditions, the findings imply that feed contaminated with DON (1514 µg/kg) provided to weaned pigs for five weeks did not increase the severity of an experimental *M. hyopneumoniae* infection. Further research is needed to investigate the impact of DON on *M. hyopneumoniae* infections in a multi-mycotoxin and multi-pathogen environment.

Keywords: *Mycoplasma hyopneumoniae*, Experimental, Challenge, Clinical, Impact, Deoxynivalenol, Feed, Weaned piglets

* Correspondence: Annelies.Michiels@UGent.be
[1]Department of Reproduction, Obstetrics and Herd Health, Faculty of Veterinary Medicine, Ghent University, Salisburylaan 133, 9820 Merelbeke, Belgium
Full list of author information is available at the end of the article

Background

The mycotoxin deoxynivalenol (DON) is a fungal metabolite produced mainly by *Fusarium graminearum* and *Fusarium culmorum* [1, 2]. *Fusarium*-produced mycotoxins, of which by toxicological viewpoint the trichothecenes DON and T-2 toxin, zearalenone and fumonisins are the most important [3], have been reported worldwide in many cereal-based cropping systems. *Fusarium* species have traditionally been associated with temperate cereals, as these fungi require lower temperatures for growth and mycotoxin production [4]. Indeed, DON is one of the most common natural mycotoxin contaminants of wheat and other small cereal grains harvested in moderate climate zones [5–7]. Extensive data on global mycotoxin occurrence showed that 59% of 5819 samples of animal feed tested positive on DON presence [3, 4, 8]. In low doses, DON causes anorexia, decreased weight gain and immune stimulation [2]. Pigs are known to be a very sensitive animal species to DON, mainly because of the high oral bioavailability and differences in metabolism of this mycotoxin compared to other species [2]. In moderate to high doses (above 840 µg/kg feed), decreased feed intake or feed refusal, vomiting and immune suppression are seen [2, 9, 10]. In fact, the EU (European Union) recommended maximum pig feed guidance level for DON is 900 µg/kg, which is the lowest one compared to other farm animal species for total diets and compounds of total diets. For adult ruminants and poultry a maximum of 5000 µg/kg and 2000 µg/kg for calves and lambs are set as guidance levels [11]. On top of that, the pig consumes a cereal rich diet and DON is frequently detected in wheat, barley, corn and by-products [9, 12].

It is known that DON can have an impact on the pathogenesis of several bacterial diseases [3, 13, 14]. Exposing porcine ileal loop tissue to a DON-concentration of 1 µg/ml, potentiated the inflammatory response and significantly enhanced *Salmonella* Typhimurium invasion in and passage of the bacterium across the intestinal epithelium [15]. Furthermore, DON (0.025 µg/ml) induced an enhanced uptake of *Salmonella* Typhimurium in porcine macrophages, indicating the capacity of DON to modulate the innate immune system, and thus to increase the susceptibility of the pig to *Salmonella* Typhimurium infections [16]. Deoxynivalenol might, due to the immunomodulatory effect on the host and/or the immediate impact on the pathogen, have an impact on the course of respiratory infectious diseases in swine. However only limited in vivo information is available. A three-week ingestion period of feed contaminated with high levels of 2500 µg/kg and 3500 µg/kg DON resulted in a higher viremia and lung viral load in case of a

Porcine Circovirus type 2 (PCV2) infection, and a lower body weight gain, more lung lesions and mortality in porcine reproductive and respiratory syndrome virus (PRRSv)-infected pigs, respectively [17, 18]. Pigs receiving fumonisin B1 in a concentration of 10,000 µg/kg feed, and dually infected with *Bordetella bronchiseptica* (*B. bronchiseptica*) and *Pasteurella multocida*, (*P. multocida*) were at greater risk to develop pneumonia and had an increase of the extent and severity of the pathological changes compared to dually infected pigs that did not receive fumonisin B1 [19]. Oral gavage of *P. multocida*-infected pigs with a crude extract of fumonisin B1 in a concentration of 500 µg/kg body weight (BW) per day for a period of seven days resulted in the pigs coughing more, in increased bronchoalveolar lavage fluid total cells, macrophages and lymphocytes, and resulted in an increased occurrence of lung lesions compared to the pigs only infected with *P. multocida* [20]. Dietary exposure to fumonisin B1 in a concentration of 12,000 µ/kg feed increased the risk on PRRSv–associated disease [21] and induced pulmonary edema which may aggravate *M. hyopneumoniae* infection [22].

Mycoplasma hyopneumoniae is causing tremendous economic losses in all intensive pig producing countries worldwide [23], despite many attempts to control the disease (enzootic pneumonia) through vaccination strategies and control measures. Consequently, there is a high prevalence of both *M. hyopneumoniae* infections and a high contamination rate of feed with mycotoxins, more specific DON, in Europe [8]. Therefore, the odds for a pig to ingest feed contaminated with DON, whilst simultaneously being infected with *M. hyopneumoniae* is high. The present study aimed to investigate the effect of in-feed administration of DON at a moderately high level of 1540 µg/kg feed, on the clinical course of an experimental *M. hyopneumoniae* infection with two genetically different *M. hyopneumoniae* strains in weaned piglets.

Methods

Study animals and experimental design

The study was compliant with all relevant institutional and European standards for animal care and experimentation. The experiment was approved by the Ethics Committee for Animal Experiments of the Faculty of Veterinary Medicine and Faculty of Bioscience Engineering, Ghent University (approval number EC2015/112). Fifty *M. hyopneumoniae*-free Rattlerow-Seghers piglets (RA-SE Genetics NV, Ooigem, Belgium) were included in the study. The herd of origin has been free of *M. hyopneumoniae* and PRRSv since 2012 based on repeated serological testing, absence of clinical signs and pneumonia lesions, and nested polymerase chain reaction (nPCR) testing on tracheobronchial swabs as previously described [24]. The gilts and sows in the herd were vaccinated against *Erysipelothrix rhusiopathiae* and Parvovirus before insemination. No vaccinations were administered to the piglets. The piglets were weaned on average at 26 days of age and moved four days later to the experimental facilities of the Faculty of Veterinary Medicine, Ghent University, Belgium. The piglets were individually identified by means of an ear tag. The study design, the different parameters and timing are summarized in Table 1. Upon arrival (D0) the piglets were randomly allocated to four different groups: 1) negative control group (NCG; $n = 5$): sham-inoculated D8, D9 + control diet, 2) DON-contaminated group (DON; $n = 15$): sham-inoculated D8, D9 and DON-diet (1514 μg/kg), 3) DON-contaminated and *M. hyopneumoniae*-inoculated group (DONMHYO; n = 15): experimentally inoculated with *M. hyopneumoniae* D8, D9 and DON-diet (1514 μg/kg), 4) *M. hyopneumoniae*-inoculated group (MHYO; n = 15): experimentally inoculated with *M. hyopneumoniae* D8, D9 + control diet. The number of 15 animals in the treatment groups enabled to find a difference of 5.26 ± 4.7 in the main parameter, namely macroscopic lung lesions (two-sided test) with 95% certainty and a statistical power of 80%. This difference is biologically relevant and was based on previous research in our research group [25]. The NCG (5 pigs) was used to verify whether the purchased piglets remained *M. hyopneumoniae* negative throughout the study. The different groups were housed in four different facilities equipped with absolute filtered chambers (HEPA U15) in order to avoid cross-infection of *M. hyopneumoniae* between the different groups. The pigs had free access to drinking water and were fed ad libitum.

Mycoplasma hyopneumoniae strains and challenge infection

The pigs were inoculated with two different strains of *M. hyopneumoniae*: a highly virulent strain F7.2C and low virulent strain F1.12. Both strains had been differentiated and characterized at proteomic level with Sodium-Dodecyl-Sulphate Polyacrylamide gelelectrophoresis (SDS-page) [26] and at genomic level with Random Amplified Polymorphic DNA (RAPD), Amplified Fragment Length Polymorphism (AFLP), PCR-Random Fragment Length Polymorphism (PCR-RFLP) of the p146 gene, Variable Number of Tandem Repeats (VNTR) analysis of p97, with Multiple-Locus of VNTR Analysis (MLVA) [27–29] and used in previous studies [24, 25, 30]. Previous research has shown that pigs are often infected with two or even three genetically different *M. hyopneumoniae* strains [29, 31, 32], and in slaughter pigs infected with different strains, more lung lesions can be detected [32]. Therefore, the pigs in the present study were inoculated with two genetically different *M. hyopneumoniae* strains, as performed previously by this research group [18], to mimic the situation in the field. All pigs were anesthetised via the intramuscular route with 0.22 ml/kg BW of a mixture of tiletamine, zolazepam (Zoletil 100®, Virbac, Louvain-la-Neuve, Belgium) and xylazine (Xyl-M®

Table 1 Experimental design, sample collections and timing in the different experimental groups

Study day, D	Groups			
	NCG (*n* = 5)	DON (*n* = 15)	DONMHYO (*n* = 15)	MHYO (*n* = 15)
D0[a]	Arrival			
	Weight			
	Randomisation			
D0-D35/36[b]	Commercial feed ad libitum	DON-contaminated feed ad libitum	DON-contaminated feed ad libitum	Commercial feed ad libitum
D1-D35[c]	RDS			
D8	Sham-inoculation[d]	Sham-inoculation[d]	F7.2C[e]-inoculation	F7.2C[e]-inoculation
	Weight			
	Blood			
D9	Sham-inoculation[d]	Sham-inoculation[d]	F1.12A[f]-inoculation	F1.12A[f]-inoculation
D21	BALF			
	Blood			
D35/36[b]	Weight			
	Necropsy			
	Lung sample			
	BALF			

[a]the average age of the pigs at arrival was 26 days, [b]all pigs of the NCG and DON group, and five pigs of MHYO were necropsied at D35. Ten pigs of MHYO and all pigs of DONMHYO were necropsied at D36, [c]RDS was not determined at D0 and D36 because the piglets arrived later than 8 a.m. (hour of performing coughing score every day) and part of the piglets were already euthanized on D36, respectively, [d]sham-inoculation was performed with sterile Friis medium, [e]highly virulent strain of *M. hyopneumoniae*, [f]low virulent strain of *M. hyopneumoniae*

NCG negative control group, DON deoxynivalenol contaminated group, DONMHYO deoxynivalenol contaminated +*M. hyopneumoniae*-inoculated group, MHYO *M. hyopneumoniae*-inoculated group, RDS respiratory disease score, BALF bronchoalveolar lavage fluid

2%, VMD, Arendonk, Belgium) and the pigs of the DONMHYO and MHYO groups were endotracheally inoculated with 7 ml of inoculum containing 10^7 CCU/ml of strain F7.2C on D8 and 7 ml of inoculum containing 10^7 CCU/ml of strain F1.12A on D9. On both inoculation days, the pigs of the NCG and DON group were endotracheally sham-inoculated with 7 ml of sterile Friis medium. At D35 or D36 of the experiment, the pigs were euthanized using deep anaesthesia by intramuscularly administering 0.3 ml/kg BW of a mixture of tiletamine, zolazepam (Zoletil 100®, Virbac, Louvain-la-Neuve, Belgium) and xylazine (Xyl-M® 2%, VMD, Arendonk, Belgium), followed by exsanguination. For practical reasons and to avoid *M. hyopneumoniae* contamination of the samples, all pigs of the NCG and DON group were euthanized at D35, followed by five animals of the MHYO group. All the other animals of the MHYO and all animals of the DONMMHYO group were necropsied on D36.

Deoxynivalenol contaminated diet

A commercial antibiotic-free diet for weaned piglets was purchased (Leievoeders N.V., Waregem, Belgium). Before purchasing the feed, a sample of the batch was tested with liquid chromatography-tandem mass spectrometry (LC-MS/MS) according to Monbaliu et al. [33, 34] for the presence of DON, 3-Acetyldeoxynivalenol (3-ADON), zearalenone and fumonisin B1+ B2, the levels were below the reporting limit of 50 µg/kg, 50 µg/kg, 10 µg/kg and 50 µg/kg, respectively. The piglets of the NCG and MHYO groups were fed this commercial diet from D0 until D35/36. A part of the purchased feed (1300 kg) was transported to the laboratory of the Department of Applied Biosciences (Faculty of Bioscience Engineering, Ghent University) to add the target concentration of 1800 µg/kg feed DON. This procedure was followed: the reference strain *Fusarium graminearum (F. graminearum)* MUCL 42841 (Mycothèque de l'Université catholique de Louvain) was used to produce the DON-culture. The strain was grown in liquid mineral (MIN) medium supplemented with L-arginin as a selective nitrogen source, as previously described by Gardiner et al. [35]. After 14 days of cultivation, the culture was filtered and centrifuged. The obtained concentration of DON was determined with LC-MS/MS by adding 150 µl of the resulting undiluted MIN medium to 5 g of certified blank wheat standard (Sigma Aldrich, Overijse, Belgium). In total, 7535 mg/kg DON was quantified and 1076 mg/kg acetylated DON (3-ADON + 15-ADON) in the grown DON-culture. Next to the presence of DON, the inoculum was tested with LC-MS/MS for the presence of other *F. graminearum* trichothecenes such as nivalenol, neosolaniol, fusarenon-X, diacetoxyscirpenol, HT-2 toxin and T-2 toxin, and results were below the detection limit. Also no zearalenone was detected. Eight l of inoculum were obtained for preparation of 1300 kg of DON-contaminated feed in a concentration aimed at twice the recommended maximum pig feed level of 900 µg/kg or 1800 µg/kg DON. First, 2.67 l of the inoculum was mixed with 10 kg of feed to obtain a thoroughly mixed premix of the DON-contaminated feed. Subsequently, the premix with inoculum was added to 433.3 kg of feed in a feed mill and thoroughly mixed for at least 40 min. The same procedure was repeated twice, to obtain the total amount of 1300 kg of DON-contaminated feed in a concentration of 1800 µg/kg and the feed was collected again in the original 25 kg bags of the feeding company. After preparation of the contaminated feed, a mixed sample originating from three DON-contaminated feed bags was taken and was submitted for LC-MS/MS to obtain the true DON-concentration of the contaminated feed. The results of the LC-MS/MS of the contaminated feed were 407 ± 120 µg/kg DON, 280 ± 100 µg/kg 3-ADON and 827 ± 300 µg/kg 15-ADON, resulting in a total DON and acetylated DON (3-ADON + 15-ADON)-concentration of 1514 µg/kg in the contaminated feed sample. This feed was used in the DON and DONMHYO-groups from D0 until D35/36 of the study.

Clinical and performance parameters

From D0 until D35/36 onwards, the piglets were observed daily at 8 a.m. for at least half an hour by the same researcher to assess appetite, faecal consistency and presence of dyspnea and tachypnea. A faecal consistency score was used to evaluate the faeces found on the pen floor before cleaning [36]: 1 (firm and shaped), 2 (soft and shaped), both addressed as a normal faecal consistency in pigs, 3 (loose) and 4 (watery), with scores 3 and 4 considered as abnormal. Daily, from D1 until D35 a respiratory disease score (RDS), ranging from 0 to 6 was recorded according to Halbur et al. (1996) [37]. Score 0 was obtained when a pig did not cough. Score 1, 3 and 5 were respectively designated as mild, moderate and severe couging after encouraged move. Score 2, 4, 6 were respectively obtained when mild, moderate and severe coughing in rest was present.The RDS was not determined at D0 and D36, as the pigs arrived later than 8 a.m. at the facilities and already part of the animals was euthanized on D35. The daily RDS values were averaged for the following periods: D1–7, D8–35/36 and D1–35/36. All pigs were weighed (kg) at the day of arrival (D0), the first inoculation day (D8) and the day of euthanasia (D35/36). The average daily gain (ADG, kg/pig/day) was calculated from D0–7, D8–35/36 and D0–35/36 by subtracting the starting weights from the final weights, divided by the number of days during that period.

Macroscopic and histopathologic lung lesions

The lungs were removed and macroscopic lung lesions (MLL) (D35/36) were determined according to Hannan et al. (1982) [38] from each pig. Consequently, the lungs were transported to the laboratory of the Department of Pathology, Bacteriology and Avian Diseases, Faculty of Veterinary Medicine, Ghent University from the necropsy rooms from the experimental facilities in the same department and from each lung in each pig, samples from the right apical, cardiac and diaphragmatic lung lobes were collected. In case a lesion was present, a sample was collected including both healthy and affected lung tissue, at the border of the lesion. The 10% neutral formalin fixed and paraffin embedded samples were stained with hematoxylin and eosin. Using light microscopy the samples were investigated and scored for the degree of peribronchiolar and perivascular lymphohistiocytic infiltration and nodule formation (cuffing) [39]. The scoring system ranged from 1 to 5, with score 1 and 2 considered not to be related with *M. hyopneumoniae* infection as previously described [24, 39, 40].The percentage of air (percentage of lung area occupied by air) was examined by means of an automated image analysis system (Leica application suite AF Lite (Diegem, Belgium) and image J (Bethesda, Maryland, USA) [41]. This parameter is inversely proportional to the lymphohistiocytic infiltration in the lung tissue and the intrabronchiolar-and bronchial exudate [25].

Quantitative PCR for *M. hyopneumoniae*

Two weeks post inoculation (PI) (D21) of the high virulent strain F7.2C (D8), bronchoalveolar lavage (BAL) fluid from all pigs, while conscious was collected. After snaring the pigs and opening their mouth with a gag, a catheter (Portex® Dog Catheter with Female Luer Mount, Smiths Medical International Ltd. Kent, United Kingdom) was inserted allowing to flush the lungs with 10 ml of sterile phosphate buffered saline (PBS). The PBS fluid was subsequently aspirated. At the day of necropsy (D35/36) the head bronchus of the left part of the lung was flushed with 10 ml of sterile PBS before collection of the histopathological samples. After collection, the BAL fluids were stored at − 70 °C until they were analysed. The DNA was extracted with the DNeasy Blood & Tissue kit (QIAGEN, Qiagen Benelux, B.V., Antwerp, Belgium) with the DNA Purification protocol for bloods or bloody fluids (spin protocol) on 200 µl of BAL fluid according to the manual instructions and quantitative PCR (qPCR) was performed as previously described to detect the number of *M. hyopneumoniae* organisms [42]. Briefly, after DNA-extraction, qPCR was performed with primers p102f (5′GTCAAAGTCAAAGT CAGCAAAC 3′) and p102r (5′AGCTGTTCAAATGC TTGTCC 3′) using SensiMixTM SYBR (Bioline GmbH, Luckenwalde, Germany) in the CFX384 real-time PCR detection system (Bio-Rad, Nazareth, Belgium). A tenfold dilution series of *M. hyopneumoniae* DNA of strain

F7.2C was used to convert the threshold values to the number of *M. hyopneumoniae* organisms. Values below the dilution of 1.50×10^1 (1.18 log copies) were considered as negative [31, 42].

Serology

A blocking ELISA (IDEIA™ *Mycoplasma hyopneumoniae* EIA kit, Oxoid Limited, Hampshire, UK) was performed according to the instructions in the protocol manual and as previously described [24] to detect antibodies against *M. hyopneumoniae* in the blood collected at D28, D21 and at necropsy (D35/36). Sera with optical density < 50% of the average value of the OD-buffer control were considered to be positive. All values above or equal to 50% of the average value of the OD-buffer control were classified as negative .

Routine bacteriological culture on bronchoalveolar lavage fluid

Ten µl of BAL fluid collected at necropsy (D35/36) of each pig was inoculated on Columbia agar supplemented with 5% sheep blood (Oxoid, Hampshire, UK) with a *Staphylococcus pseudintermedius* streak for bacteriological examination [43]. Plates were incubated for 48 h in a 5% CO_2-enriched environment at 35 °C for identification of respiratory bacteria in the lungs. All macroscopically different colonies were identified to the species level (score value > 2.000) with a Bruker Daltonic Microflex LT Biotyper Biotyper MALDI-TOF mass spectrometer by using the direct transfer method and α-cyano-4-hydroxycinnamic acid (HCCA) as matrix, according to the manufacturer's guidelines. The spectra were obtained and analysed with the MBT Compass software version 3.1. (Bruker Daltonik), which included a database of 6903 mean spectra projections.

Statistical analysis

The independent variable in the statistical analyses was 'group': NCG, DON, DONMHYO and MHYO-group. The dependent variables were RDS, weight, ADG, MLL, histopathology and percentage of air, qPCR-results, percentage of *M. hyopneumoniae* qPCR-positive samples, *M. hyopneumoniae* specific AB expressed in OD values and percentage of ELISA *M. hyopneumoniae* positive samples. These variables were all run in separate models with 'pig' as statistical unit and no additional factors included into the model. The normality of the data was investigated by means of descriptive statistics, except for the binary data (*M. hyopneumoniae* qPCR positive samples and ELISA *M. hyopneumoniae* positive samples). The parameters BW, ADG and percentage air analysis were normally distributed and a one-way analysis of variance (ANOVA) test was used. In case of the RDS, a repeated measures ANOVA was performed. Scheffé's post-hoc test was used to make pair-wise comparisons. The qPCR-results and *M. hyopneumoniae* specific antibodies were not normally distributed and therefore, a non-parametric Kruskal-Wallis test was used, with the Dunn-Bonferroni approach to make pair-wised comparisons, as well for MLL and histopathology results. In case of the normally distributed data, the mean and standard deviation (SD) were reported, in case of the non-parametric data, the median and the interquartile range were reported. All analyses for these parameters were performed with SPSS 23 for Windows (SPSS inc. Illinois, USA). Percentage of seropositive pigs and percentage of pigs testing positive with qPCR in each group were analyzed using binomial logistic regression (R version 3.3.1) [44]. The results were considered to be statistically significant when $P < 0.05$.

Results

Clinical and performance parameters

In none of the groups, feed refusals or vomiting were observed. No tachypnea, nor dyspnea were observed in any of the groups. Post-weaning diarrhoea was observed from D0 onwards until D3 in all groups, therefore all pigs were treated IM with Colistin sulphate (Colivet 'S', Prodivet,

Eynatten, Belgium) from D0 onwards for 5 days, according to the product leaflet. The post weaning diarrhoea lasted until D8, D3, D3 and D7 in NCG, DON, DONMHYO and MHYO, respectively. In three groups some pigs with faecal consistency score 3 were noticed throughout the study: in NCG two pigs at D20 and D23 respectively, in DON one pig at D13 and in DONMHYO two pigs at D20 and D23, respectively. In MHYO, normal faecal consistency was observed throughout the study. Coughing was not observed in the NCG. All results of the different time periods for this parameter in the study are shown in Table 2 (RDS $_{1-7}$, RDS $_{8-35}$ and RDS $_{1-35}$). The RDS from the first inoculation day onwards until euthanasia (RDS $_{8-35}$) were 0 ± 0, 0.0071 ± 0.028, 1.04 ± 0.82, 1.14 ± 0.92 for NCG, DON, DONMHYO and MHYO, respectively ($P < 0.001$). No statistically significant differences were obtained between all groups for D1–7. For D8–35 and D1–35 a statistically significant difference was obtained between the experimentally infected (DONMHYO and MHYO) and non-infected pigs (NCG and DON) ($P < 0.001$). However, no statistically significant differences were obtained for DONMHYO and MHYO groups in each time period of the study (Table 2). The daily course of RDS $_{1-35}$ for each group is shown in Fig. 1.

Table 2 Results of the clinical parameters, macroscopic and microscopic lung lesions in the different experimental groups

Parameter	Groups				
	NCG (n = 5)	DON (n = 15)	DONMHYO (n = 15)	MHYO (n = 15)	P-value
RDS					
D1–7	0 ± 0	0 ± 0	0 ± 0	0 ± 0	1.00
D8–35	0 ± 0^a	0.01 ± 0.03^a	1.04 ± 0.82^b	1.14 ± 0.92^b	< 0.001
D1–35	0 ± 0^a	0.01 ± 0.03^a	0.83 ± 0.84^b	0.91 ± 0.94^b	< 0.001
Weight ± SD (kg)					
D0	6.37 ± 1.09	6.39 ± 0.92	6.38 ± 1.04	6.42 ± 1.23	1.00
D8	7.70 ± 1.13	8.04 ± 1.27	8.20 ± 1.08	8.11 ± 1.90	0.92
D35/36	20.10 ± 2.33	19.45 ± 3.83	21.02 ± 3.18	21.75 ± 4.36	0.38
ADG (kg/pig/day)					
D0–8	0.52 ± 0.14	0.53 ± 0.19	0.56 ± 0.15	0.67 ± 0.17	0.12
D0–35/36	0.39 ± 0.06	0.37 ± 0.11	0.41 ± 0.08	0.43 ± 0.10	0.41
D8–35/36	0.46 ± 0.08	0.42 ± 0.13	0.46 ± 0.10	0.49 ± 0.10	0.37
MLL, histopathology and percentage of air D35/36					
MLL	0 ± 0^a	0 ± 0^a	2.77 ± 3.22^b	5.87 ± 7.32^b	< 0.001
Histopathology	1.70 ± 0.20^a	2.00 ± 0.30^a	2.40 ± 0.80^b	2.40 ± 0.90^b	< 0.001
Percentage of air (%)	45.94 ± 6.54	46.85 ± 6.86	49.75 ± 9.90	46.96 ± 8.51	0.25

Respiratory disease score (RDS), bodyweight, average daily weight gain (ADG), macroscopic lung lesions (MLL), histopathology score of the lungs and percentage of air in the lungs for NCG, DON group, DONMHYO group, MHYO group
The parameters bodyweight, ADG, percentage of air analysis were analysed by means of one way analysis of variance, the parameter RDS with a repeated measures analysis of variance, in both cases with Scheffé post-hoc test to make pair-wised comparisons, therefore means ± SD are reported. The parameter MLL and histopathology were analysed with Kruskal-Wallis test and the Dunn-Bonferroni approach to make pair-wised comparison, therefore medians and interquartile range are reported. Different superscripts in one row are statistically different ($P < 0.05$)
NCG: negative control group, *DON*: deoxynivalenol contaminated group, *DONMHYO*: deoxynivalenol contaminated +*M. hyopneumoniae*-inoculated group, *MHYO*: *M. hyopneumoniae*-inoculated group, *SD*: standard deviation, *n* number, *D* Day of the study, *ADG* average daily gain, *RDS* respiratory disease score, *MLL* macroscopic lung lesions

There were no significant differences between the groups for the parameter BW and the ADG during the different time periods (Table 2).

Macroscopic and histopathologic lung lesions

The MLL of the NCG, DON, DONMHYO and MHYO groups were 0 ± 0, 0 ± 0, 2.77 ± 3.22 and 5.87 ± 7.32, respectively ($P < 0.001$). The histopathological lung lesions were 1.70 ± 0.20, 2.00 ± 0.30, 2.40 ± 0.80 and 2.40 ± 0.90 for the NCG, DON, DONMHYO and MHYO group, respectively ($P < 0.001$). The percentage of air was 45.94 ± 6.54, 46.85 ± 6.86, 49.75 ± 9.90 and 46.96 ± 8.51 for NCG, DON, DONMHYO and MHYO, respectively ($P = 0.25$). There were no statistically significant differences between the DONMHYO and MHYO groups for these above-mentioned parameters, however there were statistically significant differences between the experimentally infected (DONMHYO and MHYO) and non-infected pigs (NCG and DON) ($P < 0.001$), except for the parameter percentage of air analysis ($P = 0.25$) (Table 2).

Quantitative PCR for *M. hyopneumoniae*

The samples of the NCG and DON group remained negative throughout the study. The qPCR results at D21 were 0.05 ± 0.37, 0.63 ± 0.81, 3.92 ± 2.60 and 3.24 ± 1.91 ($P < 0.001$) and at D35/36 0.80 ± 0.70, 0.29 ± 1.09, 4.05 ± 1.38 and 4.20 ± 1.00 ($P < 0.001$) for NCG, DON, DONMHYO and MHYO, respectively. There were no significant differences between the DONMHYO and MHYO group, however there were statistically significant differences between the experimentally infected (DONMHYO and MHYO) and non-infected pigs (NCG and DON) ($P < 0.001$) for D21 and D35/36 (Table 3).

Serology

The serological results are presented in Table 3. All pigs of the NCG and DON remained serologically negative throughout the study. The OD-values of the serological results at D35/36 were 1.15 ± 0.27, 1.16 ± 0.18, 0.25 ± 0.18 and 0.23 ± 0.09 for the NCG, DON, DONMHYO and MHYO, respectively. At necropsy (D35/36), all pigs of DONMHYO and MHYO were serologically positive.

Routine bacteriological culture on bronchoalveolar lavage fluid

Few colonies of *Streptococcus suis* were isolated from one pig of both the DONMHYO (1/15) and MHYO (1/15) groups. In addition, few colonies of *B. bronchiseptica* were isolated from one pig of DON (1/15) and four pigs of MHYO (4/15).

Discussion

In the present study, the outcome of the main parameters (RDS, MLL, histopathological lesions and log copies of *M. hyopneumoniae* DNA in bronchoalveolar lavage

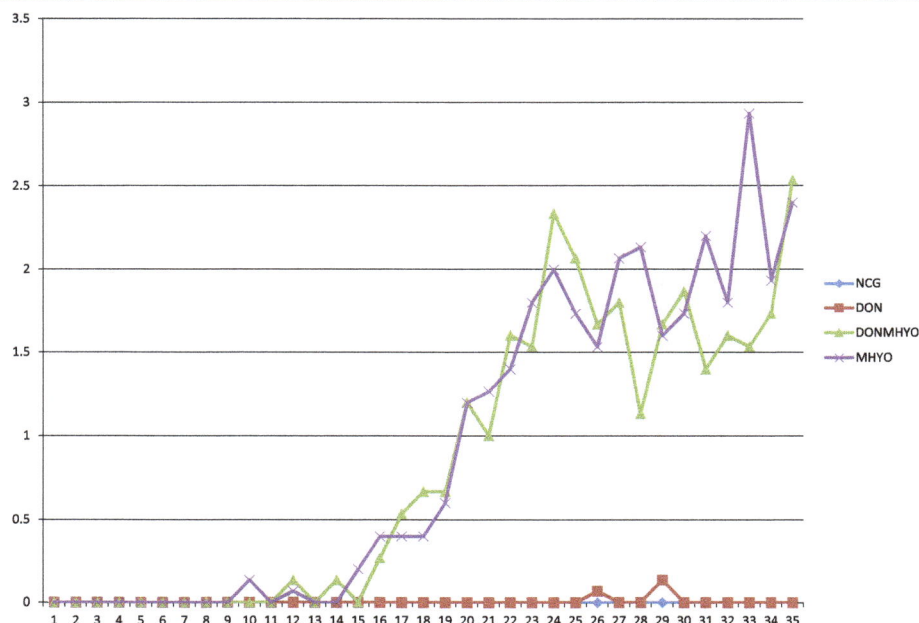

Fig. 1 Course of average respiratory disease score (RDS) from day 1 until the day of necropsy (D 35). Average RDS from D1 until D35 in the negative control group (NCG), the DON-contaminated group (DON), DON-contaminated and *M. hyopneumoniae*-inoculated group (DONMHYO), and *M. hyopneumoniae*-inoculated group (MHYO). The challenge infections were performed on D8 (a highly virulent *M. hyopneumoniae* strain F7.2C) and D9 (a low virulent strain F1.12) in DONMHYO and MHYO. The NCG and DON were sham-inoculated with sterile Friis medium on D8 and D9

Table 3 Results of *M. hyopneumoniae*-DNA detection in the BALF and serology in the different experimental groups

Parameter	Groups				
	NCG ($n = 5$)	DON ($n = 15$)	DONMHYO ($n = 15$)	MHYO ($n = 15$)	P-value
qPCR (log copies of *M. hyopneumoniae* DNA /ml BALF) ± SD					
D21	0.05 ± 0.37[a]	0.63 ± 0.81[a]	3.92 ± 2.60[b]	3.24 ± 1.91[b]	< 0.001
D35/36	0.80 ± 0.70[a]	0.29 ± 1.09[a]	4.05 ± 1.38[b]	4.20 ± 1.00[b]	< 0.001
Percentage of *M. hyopneumoniae* qPCR positive samples (n of positive samples/total n of samples tested)					
D21 (%)	0 (0/5)[a]	0 (0/15)[a]	87 (13/15)[b]	93 (14/15)[b]	< 0.001
D35/36 (%)	0 (0/5)[a]	0 (0/15)[a]	100 (15/15)[b]	100 (15/15)[b]	< 0.001
M. hyopneumoniae specific AB expressed in OD-values ± SD					
D8	1.68 ± 0.23	1.83 ± 0.34	1.71 ± 0.19	1.70 ± 0.19	0.27
D21	1.15 ± 0.22[a]	1.29 ± 0.23[a]	1.16 ± 0.36[ab]	0.94 ± 0.26[b]	0.001
D35/36	1.15 ± 0.27[a]	1.16 ± 0.18[a]	0.25 ± 0.18[b]	0.23 ± 0.10[b]	< 0.001
Percentage of ELISA *M. hyopneumoniae* positive samples (n of positive samples/total n of samples tested)					
D8 (%)	0 (0/5)	0 (0/15)	0 (0/15)	0 (0/15)	1.00
D21 (%)	0 (0/5)	0 (0/15)	0 (0/15)	13 (2/15)	0.17
D35/36 (%)	0 (0/5)[a]	0 (0/15)[a]	100 (15/15)[b]	100 (15/15)[b]	< 0.001

Respiratory disease score (RDS), bodyweight, average daily weight gain (ADG), macroscopic lung lesions (MLL), histopathology score of the lungs and percentage of air in the lungs for NCG, DON group, DONMHYO group, MHYO group
The parameter qPCR and *M. hyopneumoniae* specific AB were analysed with a non-parametric Kruskal-Wallis and the Dunn-Bonferroni approach to make pair-wised comparisons, therefore median and interquartile range are reported. The prevalence of *M. hyopneumoniae* qPCR positive samples and the prevalence of ELISA *M. hyopneumoniae* positive samples were analysed with binomial logistic regression. Different superscripts in one row are statistically different ($P < 0.05$)
NCG: negative control group, DON: deoxynivalenol contaminated group, DONMHYO: deoxynivalenol contaminated + *M. hyopneumoniae*-inoculated group, MHYO: *M. hyopneumoniae*-inoculated group, Different superscripts in one row are statistically different ($P < 0.05$), n number, SD standard deviation, D Day of the study, M. hyopneumoniae: Mycoplasma hyopneumoniae, AB antibodies, OD optical densities

fluid) was not statistically different in *M. hyopneumoniae* infected pigs that received or did not receive DON-contaminated feed. This indicates that ingestion of DON-contaminated feed at a concentration of 1514 μg/kg, exceeding the maximum guidance level for pigs according to EU regulation, did not aggravate the severity of an experimental *M. hyopneumoniae*-infection in weaned piglets.

The challenge infection was successful as all animals in the *M. hyopneumoniae* inoculated groups (MHYO and DONMHYO group) coughed, showed lung lesions (except for one animal in the MHYO group), seroconverted and *M. hyopneumoniae*-DNA was detected in BAL fluid at necropsy. The results obtained in the *M. hyopneumoniae* challenged pigs (DONMHYO and MHYO group) differed significantly from the non-challenged pigs (DON group, NCG). The latter pigs remained serologically negative throughout the entire study period and no *M. hyopneumoniae*- DNA was detected in BAL fluid two weeks post-inoculation, nor at necropsy. The obtained data in the challenged groups were comparable with a previous study performed by our research group with the same challenge model [24], demonstrating the repeatability of the model. The two *M. hyopneumoniae*-strains were not administered on the same day, as the infection model practiced at our research group uses seven ml inoculum containing 10[7]

CCU of the strain/ml [31]. It is not known whether piglets of four to five weeks of age (on average 26 days, D0 at arrival and inoculation at D8) are able to cope with twice the inoculum volume (14 ml for both strains). In addition, it is not known whether both strains will grow next to each other, or whether one strain will overgrow the other strain consuming most of the nutrients, when these *M. hyopneumoniae*-strains are grown in the same culture flask. Consequently, it was decided to challenge the piglets on two consecutive days.

In the present study, DON was aimed to be administered at two times the maximum guidance level of DON advised by the European commission for pigs (1800 μg/kg feed) [11]. This dose was selected as Madson et al. [45] stated that moderate (1000–5000 μg/kg feed) to high concentrations (> 5000 μg/kg feed) are associated with delayed or supressed immune responses due to leukocyte apoptosis and resulting in increased disease susceptibility [45, 46]. It was demonstrated that a dose-related decrease in daily feed intake is observed when administering DON-supplemented feed to pigs, and that this effect was mostly significant to the control feed in the range of 2000 μg/kg – 4000 μg/kg [47–49]. Therefore, a dose below this range was chosen to be administered to the pigs in this study to limit feed refusals and to avoid hampering of DON possibly influencing the investigated parameters. The commercial, un-spiked feed was tested before the start of the study, not only for the presence of DON, but also for zearalenone and fumonisin B1 + B2, to avoid these components having an effect in the feed of the control groups (MHYO group, NCG) and avoiding an additional or synergistic effect to the DON-toxicity in the DON-spiked groups (DON group, DONMHYO group) [46]. After preparation of the contaminated feed, a mixed feed sample from three DON-contaminated feed bags was submitted to LC-MS/MS to detect the DON-concentration and to ensure thorough mixing of the DON-inoculum in the feed. The results of the LC-MS/MS was 1514 µg/kg of DON and acetylated forms and deviated only slightly from the target concentration of 1800 µg/kg. Cytotoxicity of the produced DON in the study was not tested on beforehand, as the deleterious effects of DON in the pigs are known and the same DON inoculation method in pig feed was already successfully used in a study by Goossens et al. [50]. In the latter study, the in vivo effect of DON on intestinal damage influencing the resorption of doxycycline was determined in pigs.

It remains to be elucidated why no impact of DON in the pigs was observed in the present study. The degree of susceptibility of certain breeds/lines to the effects of DON might be one explanation [51]. For instance a lower severity in porcine circovirus type 2 associated histopathological lesions has been shown for Piétrain compared to landrace pigs [52], but so far no studies have assessed the impact of genetic differences in sensitivity to the effects of DON.

No feed-refusals were observed in the pigs administered the DON-contaminated feed. The minimum emetic dose for orally distributed DON in pigs is 100 µg/kg [53], yet vomiting was not observed in this study. The reason why this did not occur is unclear. It must be noted that the observation period of the pigs (30 min. of RDS-scoring, 15 min. Feeding and cleaning in the morning, 10 min. of observation and cleaning in the evening) was fairly short compared to the time the pigs spent in the facilities. However, the main researcher was always present after providing the feed to identify possible sick pigs, as most healthy pigs start eating immediately after providing the feed and Young et al., [54] observed that vomiting occurs within minutes after ingesting the DON-contaminated feed. Pestka et al. [53] stated that vomiting due to DON-contaminated feed, is more likely if the feed is ingested at once and not via smaller portions throughout the day. The contaminated feed and hence the DON in the DONMHYO and DON pigs, was consumed throughout the day as the pigs were fed ad libitum and had freely access to the feed.

The *M. hyopneumoniae*-infected pigs administered the DON-supplemented feed did not have a higher RDS compared to the pigs only infected with *M. hyopneumoniae*.

No other studies are available investigating the impact of DON on respiratory tract disease signs in *M. hyopneumoniae*-infected pigs. However the effect of fumonisin B1 in combination with *M. hyopneumoniae* or the effect of DON on other pathogens has been studied. Pósa et al., [22] studied the effect of 20,000 µg/kg fumonisin B1 on *M. hyopneumoniae*-infected pigs, however no firm conclusions could be drawn regarding the difference in coughing between the *M. hyopneumoniae*-infected pigs with or without fumonisin B1 supplementation in the feed. Halloy et al. [20] investigated the impact of fumonisin B1 administered orally (500 µg/kg BW per day, seven days) in *P. multocida*-infected pigs and concluded that these pigs coughed more compared to the pigs only infected with *P. multocida*.

It is known that *M. hyopneumoniae* infections can negatively influence production parameters, such as ADG [55]. Deoxynivalenol ingestion in pigs may result in reduced feed intake, and subsequently decrease ADG [51, 56, 57]. In the present study, DON did not decrease ADG in the *M. hyopneumoniae* inoculated animals. This result is in agreement with Accensi et al. [9] (840 µg DON/kg feed), Gerez et al. [46] (1500 µg DON/kg feed), and Savard et al. [17]. The latter authors studied the impact of DON (2500 and 3000 µg DON/kg feed) in pigs that were simultaneously infected with PCV2. Pósa et al., [22], did not find an effect of fumonisin B1 (20, 000 µg/kg) on ADG in *M. hyopneumoniae*-infected pigs. Rotter et al. [58] observed an adaptation of pigs to oral ingestion of *Fusarium* mycotoxins from one week onwards. In that study, pigs were fed a diet mixed with naturally contaminated corn (28,700 µg/kg DON, 8600 µg/kg 15-ADON and1,100 µg/kg ZEA) to obtain DON-concentrations of 750, 1500 or 3000 µg DON /kg feed during 28 days. During the first week, the exposed pigs had lower weight gains than the pigs fed a non-contaminated control feed. At the end of the study, however, the overall weight gains did not differ anymore between these groups. This adaptation might be one of the reasons why we did no see an effect of DON on daily growth. The relative low number of pigs, followed up during a limited period of time, which is inherent in experimental infection studies like ours, also makes it difficult to obtain a statistical significant difference in daily growth [31, 59, 60]. No effect of DON ingestion on the macroscopic and histopathological lung lesions was observed in the *M. hyopneumoniae* infected group. This finding is in agreement with Pósa et al. [19], who neither saw a statistical difference in macroscopic lung lesions between the *B. bronchiseptica* and *P. multocida* dually-infected groups with or without supplementation of fumonisin B1 in the feed (10,000 µg/kg) of three-day-old piglets. Savard et al. (2015) on the other hand, did observe an effect of DON-supplementation in the feed (3500 µg/kg, three weeks) on macroscopic lung lesions in PRRSv-infected pigs [51]. In this study, DON-ingestion did not influence the number of log copies detected in the *M. hyopneumoniae* infected animals. Similarly, Savard et al. [61] did not observe an effect of DON-contamination on the presence of viral RNA, measured with qPCR, in PRRSv-infected pigs. It is not clear why the obtained effects of dietary DON vary among experiments, It might be explained by different factors such as starting weight or age of the pigs, the contamination source of DON (natural versus artificial contamination), presence of other known or unknown undetected fungal metabolites or pathogens, duration of the study (adaptation), number of pigs used in the study, gender of the pigs, health status, nutritional balance of the pig and statistical design of the study [10, 62, 63]. It is not known why DON did not influence a *M. hyopneumoniae* infection under the circumstances in this study. Deoxynivalenol has a good distribution in the lung of the pig [64]. However, as *M. hyopneumoniae* is attached to the cilia of the upper respiratory tract and does not invade the parenchyma of the lung [23, 65, 66], it might be that DON is not able to exert its effect on the pathogen. More research is needed to investigate this relationship in pigs, for instance by in vitro tests on pig tracheal explants. On the other hand discrepancies have been reported between in vivo (higher viremia in pigs exposed to DON in the feed) and in vitro (decreased PRRSv replication in MARC-cells) effects of DON on PRRSv [18, 61], thus extrapolation of in vitro results to in vivo effects has to be done cautiously.

Conclusions

No effect was observed of DON contamination in a moderately high dose in the feed on the severity of an experimental *M. hyopneumoniae* infection in weaned piglets. In the field, however, the impact of DON-contaminated feed on a *M. hyopneumoniae*-infection might be more expressed, because mostly multi-mycotoxin contamination of the feed occurs [3, 52–54], the pigs can be exposed to DON-contaminated feed for a longer period than the five week exposure period in this study, often suboptimal housing and climate conditions may prevail and other pathogens may be present [55, 56]. Further research should assess the impact of DON on *M. hyopneumoniae* infections under these multi-pathogen and multi-mycotoxins circumstances and investigating the impact of DON in vitro on *M. hyopneumoniae* in tracheal explant cells. More research could also focus on factors influencing the effect on DON such as health status, gender, age, and possible genetic resistance of the pigs.

Abbreviations
3-ADON: 3-Acetyldeoxynivalenol; ADG: average daily gain; ANOVA: one way analysis of variance; BAL: bronchoalveolar lavage; BW: body weight; D: study day; DON: deoxynivalenol; DON: DON-contaminated; DONMHYO: DON-contaminated and *M. hyopneumoniae*- inoculated; EU: European Union; F.

graminearum: *Fusarium graminearum*; LC-MS/MS: liquid chromatography-tandem mass spectrometry; MHYO: *M. hyopneumoniae*- inoculated; MIN: liquid mineral; MLL: macroscopic lung lesions; NCG: negative control group; nPCR: nested polymerase chain reaction; PBS: phosphate buffered saline; PI: post inoculation; qPCR: quantitative PCR; RDS: respiratory disease score

Acknowledgements
The authors are grateful to Marleen Foubert for the assistance with preparing the inoculum, to Pieter-Jan De Temmerman for the help with the binary data analysis and to the colleagues for assistance during the necropsy. Thank you to the feeding company Leievoeders N.V. for the flexibility in re-testing and delivering the feed and thank you to RA-SE for wanting to provide the *M. hyopneumoniae*-negative pigs.

Funding
The study was partially financed by ZOETIS, Belgium. The role of the funding body was merely financial. ZOETIS, Belgium did not have any role in the design of the study, collection, analysis of the data, and writing of the manuscript.

Authors' contributions
AMi designed the study protocol, wrote the ethical application, grew the *M. hyopneumoniae* strains for the experimental infections, mixed the feed with the DON-inoculum, performed the trial, performed the laboratory analyses, performed the statistical analyses, interpreted the data and wrote the manuscript. IA reviewed the study protocol, assisted with performing the trial, assisted with interpretation of the data and reviewed the manuscript. AMa reviewed the study protocol, assisted with performing the trial, assisted with the lab work, assisted with interpretation of the data and reviewed the manuscript. FB reviewed the study protocol, the ethical application dossier, assisted with interpretation of the data and reviewed the manuscript and helped with the laboratory analysis. GH and KA were responsible for producing the DON inoculum from the reference strain of *F. graminearum* and supervised the production of the DON-contaminated feed. ME and SC assisted with the interpretation of the mycotoxin testing of the commercial feed and the DON-inoculated feed. GH, KA, ME, SC, FH and DM designed the study protocol, reviewed the ethical application dossier and the manuscript and assisted with interpretation of the data. All authors gave their final approval for publication of the study.

Competing interests
The authors declare that they have no competing interests.

Author details
[1]Department of Reproduction, Obstetrics and Herd Health, Faculty of Veterinary Medicine, Ghent University, Salisburylaan 133, 9820 Merelbeke, Belgium. [2]Department of Pathology, Bacteriology and Avian Diseases, Faculty of Veterinary Medicine, Ghent University, Salisburylaan 133, 9820 Merelbeke, Belgium. [3]Department of Applied Biosciences, Faculty of Bioscience Engineering, Ghent University, Campus Schoonmeersen, Valentin Vaerwyckweg 1, 9000 Ghent, Belgium. [4]Department of Pharmacology, Toxicology and Biochemistry, Faculty of Veterinary Medicine, Ghent University, Salisburylaan 133, 9820 Merelbeke, Belgium.

References
1. Whitlow WM, Hagler JR, Diaz DE. Mycotoxins in feed. Feedstuffs. 2010:74–84.
2. Bracarense AP, Lucioli J, Grenier B, Drociunas Pacheco G, Moll WD, Schatzmayr G, et al. Chronic ingestion of deoxynivalenol and fumonisin, alone or in interaction, induces morphological and immunological changes in the intestine of piglets. Br J Nutr. 2012;107:1776–86.
3. Antonissen G, Martel A, Pasmans F, Ducatelle R, Verbrugghe E, Vandenbroucke V, et al. The impact of *Fusarium* mycotoxins on human and animal host susceptibility to infectious diseases. Toxins. 2014;6:430–52.

4. Placinta CM, D'Mello JPF, Macdonald AMC. A review of worldwide contamination of cereal grains and animal feed with *Fusarium* mycotoxins. Anim Feed Sci Technol. 1999;78:21–37.

5. Kostelanska M, Dzuman Z, Malachova A, Capouchova I, Prokinova E, Skerikova A, et al. Effects of milling and baking technologies on levels of deoxynivalenol and its masked form deoxynivalenol-3-glucoside. J Agric Food Chem. 2011;59:9303–12.

6. Miller JD. Mycotoxins in small grains and maize: old problems, new challenges. Food Addit Contam. 2008;25:219–30.

7. Rasmussen PH, Nielsen KF, Ghorbani F, Spliid NH, Nielsen GC, Jørgensen LN. Occurrence of different trichothecenes and deoxynivalenol-3-β-D-glucoside in naturally and artificially contaminated Danish cereal grains and whole maize plants. Mycotoxin Res. 2012;28:181–90.

8. Rodrigues I, Naehrer K. A three-year survey on the worldwide occurrence of mycotoxins in feedstuffs and feed. Toxins. 2012;4:663–75.

9. Accensi F, Pinton P, Callu P, Abella-Bourges N, Guelfi J-F, Grosjean F, Oswald IP. Ingestion of low doses of deoxynivalenol does not affect hematological, biochemical, or immune responses of piglets. J Anim Sci. 2006;84:1935–42.

10. Dersjant-Li Y, Verstegen MW, Gerrits WJ. The impact of low concentrations of aflatoxin, deoxynivalenol or fumonisin in diets on growing pigs and poultry. Nutrition Res Rev. 2003;16:223–39.

11. Commission recommendation of 17 August 2006 on the presence of deoxynivalenol, zearalenone, ochratoxin A, T-2, HT-2 and fumonisin in products intended for animal feeding. In: Ojeu. vol. L229;2006;7–9.

12. Pierron A, Alassane-Kpembi I, Oswald I. Impact of mycotoxin on immune response and consequences for pig health. Anim Nutr. 2016;2:63–8.

13. Antonissen G, Van Immerseel F, Pasmans F, Ducatelle R, Haesebrouck F, Timbermont L, et al. The mycotoxin deoxynivalenol predisposes for the development of *Clostridium perfringens*-induced necrotic enteritis in broilers. PLoS One. 2014;9:1–8.

14. Antonissen G, Croubels S, Pasmans F, Ducatelle R, Eeckhaut V, Devreese M, et al. Fumonisins affect the intestinal microbial homeostasis in broiler chickens. predisposing to necrotic enteritis Vet Res. 2015;46:1–11.

15. Vandenbroucke V, Croubels S, Martel A, Verbrugghe E, Goossens J, Van Deun K, et al. The mycotoxin Deoxynivalenol potentiates intestinal inflammation by *Salmonella* Typhimurium in porcine Ileal loops. PLoS One. 2011;6:1–8.

16. Vandenbroucke V, Croubels S, Verbrugghe E, Boyen F, De Backer P, Ducatelle R, et al. The mycotoxin deoxynivalenol promotes uptake of *Salmonella* Typhimurium in porcine macrophages, associated with ERK1/2 induced cytoskeleton reorganization. Vet Res. 2009;40:1–12.

17. Savard C, Provost C, Alvarez F, Pinilla V, Music N, Jacques M, et al. Effect of deoxynivalenol (DON) mycotoxin on in vivo and in vitro porcine circovirus type 2 infections. Vet Microbiol. 2015;176:257–67.

18. Savard C, Pinilla V, Provost C, Gagnon CA, Chorfi Y. In vivo effect of deoxynivalenol (DON) naturally contaminated feed on porcine reproductive and respiratory syndrome virus (PRRSV) infection. Vet Microbiol. 2014;147:419–26.

19. Pósa R, Donkó T, Bogner P, Kovács M, Repa I, Magyar T. Interaction of *Bordetella bronchiseptica*, *Pasteurella multocida*, and fumonisin B1 in the porcine respiratory tract as studied by computed tomography. Can J Vet Res. 2011;75:176–82.

20. Halloy DJ, Gustin PG, Bouhet S, Oswald I. Oral exposure to culture material extract containing fumonisins predisposes swine to the development of pneumonitiscaused by *Pasteurella multocida*. Toxicology. 2005;213:34–44.

21. Ramos CM, Martinez EM, Carrasco AC, Puente JHL, Quezada F, Perez JT, et al. Experimental trial of the effect of Fumonisin B1 and the PRRS virus in swine. J Anim Vet Adv. 2010;9:1301–10.

22. Pósa R, Magyar T, Stoev SD, Glávits R, Donkó T, Repa I, et al. Use of computed tomography and histopathologic review for lung lesions produced by the interaction between *Mycoplasma hyopneumoniae* and Fumonisin mycotoxins in pigs. Vet Pathol. 2013;50:971–9.

23. Maes D, Sibila M, Kuhnert P, Segalés J, Haesebrouck F, et al. Update on *Mycoplasma hyopneumoniae* infections in pigs: knowledge gaps for improved disease control. Transbound Emerg Dis. 2017;1:1–15.

24. Michiels A, Arsenakis I, Boyen F, Krecjci R, Haesebrouck F, Maes D. Efficacy of one dose vaccination against experimental infection with two *Mycoplasma hyopneumoniae* strains. BMC Vet Res. 2017;13:1–10.

25. Vicca J. 2005. Virulence and antimicrobial susceptibility of *Mycoplasma hyopneumoniae* isolates from pigs. PhD thesis. Faculty of Veterinary Medicine, Ghent University, ISBN 90-5864-086-8, 219 pp.

26. Calus D, Baele M, Meyns T, de Kruif A, Butaye P, Decostere A, et al. Protein variability among *Mycoplasma hyopneumoniae* isolates. Vet Microbiol. 2007; 120:284–91.

27. Stakenborg T, Vicca J, Butaye P, Maes D, Peeters J, de Kruif A, et al. The diversity of *Mycoplasma hyopneumoniae* within and between herds using pulsed-field gel electrophoresis. Vet Microbiol. 2005;109:29–35.

28. Stakenborg T, Vicca J, Maes D, Peeters J, de Kruif A, Haesebrouck F, et al. Comparison of molecular techniques for the typing of *Mycoplasma hyopneumoniae* isolates. J Microbiol Methods. 2006;66:263–75.

29. Vranckx K, Maes D, Calus D, Villarreal I, Pasmans F, Haesebrouck F. Multiple locus variable number of tandem repeats analysis is a suitable tool for the differentiation of *Mycoplasma hyopneumoniae* strains without cultivation. J Clin Microbiol. 2011;49:2020–3.

30. Villarreal I, Maes D, Meyns T, Gebruers F, Calus D, Pasmans F, et al. Infection with a low virulent *Mycoplasma hyopneumoniae* isolate does not protect piglets against subsequent infection with a highly virulent *M. hyopneumoniae* isolate. Vaccine. 2009;27:1875–9.

31. Vranckx K, Maes D, Del Pozo Sacristán R, Pasmans F. Haesebrouck F. A longitudinal study of the diversity and dynamics of *Mycoplasma hyopneumoniae* infections in pig herds. Vet Microbiol. 2011;156:315–21.

32. Michiels A, Vranckx K, Piepers S, Del Pozo Sacristán R, Arsenakis I, Boyen F, et al. Impact of diversity of *Mycoplasma hyopneumoniae* strains on lung lesions in slaughter pigs. Vet Res. 2017;48:1–14.

33. Monbaliu S, Van Poucke C, Van Peteghem C, Van Poucke K, Heungens K, De Saeger S. Development of a multi-mycotoxin liquid chromatography/ tandem mass spectrometry method for sweet pepper analysis. Rapid Commun Mass Spectrom. 2009;23:3–11.

34. Monbaliu S, Van Poucke C, DeTavernier C, Dumoulin F, Van De Velde M, Schoeters E, et al. Occurrence of mycotoxins in feed as analyzed by a multi-mycotoxin LC-MS/MS method. J Agric Food Chem. 2010;58:66–71.

35. Gardiner DM, Kazan K, Manners JM. Nutrient profiling reveals potent inducers of trichothecene biosynthesis in fusarium graminearum. Fungal Genet Biol. 2009;46:604–13.

36. Pedersen KS, Toft N. Intra- and inter-observer agreement when using a descriptive classification scale for clinical assessment of faecal consistency in growing pigs. Prev Vet Med. 2011;98:288–91.

37. Halbur PG, Paul PS, Meng XJ, Lum MA, Andrews JJ, Rathje JA. Comparative pathogenicity of nine US porcine reproductive and respiratory syndrome virus (PRRSV) isolates in a five-week-old cesarean-derived, colostrum-deprived pig model. J Vet Diagn Investig. 1996;8:11–20.

38. Hannan PC, Bhogal BS, Fish JP. Tylosin tartrate and tiamutilin effects on experimental piglet pneumonia induced with pneumonic pig lung homogenate containing mycoplasmas, bacteria and viruses. Res Vet Sci 1982;33:76–88.

39. Morris C, Gardner I, Hietala S, Carpenter T, Anderson R, Parker K. Seroepidemiologic study of natural transmission of *Mycoplasma hyopneumoniae* in a swine herd. Prev Vet Med. 1995;21:323–37.

40. Del Pozo Sacristán R. Treatment and control of *Mycoplasma hyopneumoniae* infections. Ghent: Ghent University; 2014.

41. Rasband WS. ImageJ In.: U. S. National Institutes of Health, Bethesda, Maryland, USA; 1997-2016.

42. Marois C, Dory D, Fablet C, Madec F, Kobisch M. Development of a quantitative real-time TaqMan PCR assay for determination of the minimal dose of *Mycoplasma hyopneumoniae* strain 116 required to induce pneumonia in SPF pigs. J Appl Microbiol. 2010;108:1523–33.

43. Villarreal I, Maes D, Vranckx K, Calus D, Pasmans F, Haesebrouck F. Effect of vaccination of pigs against experimental infection with high and low virulence *Mycoplasma hyopneumoniae* strains. Vaccine. 2011;29:1731–5.

44. R: A language and environment for statistical Computing URL https://www.R-project.org/, Accessed 18 Nov 2017.

45. Madson DM, Ensley SM, Patience JF, Gauger PC, Main RG. Diagnostic assessment and lesion evaluation of chronic deoxynivalenol ingestion in growing swine. Swine Health Prod. 2014;22:78–83.

46. Pestka JJ. Mechanisms of deoxynivalenol-induced gene expression and apoptosis. Food Addit Contam Part A. 2008;25:1128–40.

47. Bergsjø B, Matre T, Nafstad I. Effects of diets with graded levels of deoxynivalenol on performance in growing pigs. Zentralbl Veterinarmed A. 1992;39:752–8.

48. Dänicke S, Goyaerts T, Valenta H, Razzazi E, Böhm J. On the effects of deoxynivalenol (DON) in pig feed on growth performance, nutrients utilization and DON metabolism. J Anim Feed Sci. 2004;13:539–56.

49. Prelusky DB, Gerdes RG, Underhill KL, Rotter BA, Jui PY, Trenholm HL. Effects of low-level dietary deoxynivalenol on haematological and clinical parameters of the pig. Nat Toxins. 1994;2:97–104.

50. Goossens J, Vandenbroucke V, Pasmans F, De Baere S, Devreese M, Osselaere A, et al. Influence of mycotoxins and a mycotoxin adsorbing agent on the oral bioavailability of commonly used antibiotics in pigs. Toxins. 2012;4:281–95.

51. Eriksen GS, Pettersson H. Toxicological evaluation of trichothecenes in animal feed. Anim Feed Sci Technol. 2004;114:205–39.

52. Opriessnig T, Patterson AR, Madson DM, Pal N. Difference in severity of porcine circovirus type two-induced pathological lesions between landrace and Pietrain pigs. J Anim Sci. 2009;87:1582–90.

53. Pestka JJ. Deoxynivalenol: toxicity, mechanisms and animal health risks. Anim Feed Sci Technol. 2007;137:283–98.

54. Young LG, McGirr L, Valli VE, Lumsden JH, Lun A. Vomitoxin in maize fed to young pigs. J Anim Sci. 1983;57:655–64.

55. Thacker EL, Minion FC. Mycoplasmosis. In: Zimmerman JJ, Karriker LA, Ramirez A, Schwartz KJ, Stevenson GW, editors. Diseases of Swine, 9th edition. Ames, Iowa: Wiley-Blackwell;2012.

56. Friend DW, Trenholm HL, Elliot JI, Thompson BK, Hartin KE. Effect of feeding vomitoxin-contaminated wheat to pigs. Can J Anim Sci. 1982;62:1211–22.

57. Trenholm HL, Cochrane WP, Cohen H, Elliot JI, Farnworth ER, Friend DW, et al. Survey ofvomitoxin contamination of 1980 Ontario white winter wheat crop: results of survey and feeding trials. J Assoc Off Anal Chem. 1983;66: 92–7.

58. Rotter BA, Thompson BK, Lessard M, Trenholm HL, Tryphonas H. Influence of low-level exposure to fusarium mycotoxins on selected immunological and hematological parameters in young swine. Toxicol Sci. 1994;23:117–24.

59. Arsenakis I, Panzavolta L, Michiels A, Del Pozo Sacristán R, Boyen F, Haesebrouck F, et al. Efficacy of Mycoplasma hyopneumoniae vaccination before and at weaning against experimental challenge infection in pigs. BMC Vet Res. 2016;12:1–7.

60. Jensen CS, Ersbøll AK, Nielsen JP. A meta-analysis comparing the effect of vaccines against Mycoplasma hyopneumoniae on daily weight gain in pigs. Prev Vet Med. 2002;54:265–78.

61. Savard C, Gagnon CA, Chorfi Y. Deoxynivalenol (DON) naturally contaminated feed impairs the immune response induced by porcine reproductive and respiratory syndrome virus (PRRSV) live attenuated vaccine. Vaccine. 2015;33:3881–6.

62. Goyarts T, Dänicke S, Rothkötter HJ, Spilke J, Tiemann U, Schollenberger M. On the effects of a chronic deoxynivalenol intoxication on performance, haematological and serum parameters of pigs when diets are offered either for ad libitum consumption or fed restrictively. J Vet Med A Physiol Pathol Clin Med. 2005;52:305–14.

63. Rotter BA, Thompson BK, Lessard M. Effects of deoxynivalenol-contaminated diet on performance and blood parameters in growing swine. Can J Anim Sci. 1995;75:297–302.

64. Prelusky DB, Trenholm HL. Tissue distribution of Deoxynivalenol in swine dosed intravenously. J Agric Food Chem. 1991;39:748–51.

65. Blanchard B, Vena MM, Cavalier A, Le Lannic J, Gouranton J, Kobisch M. Electron microscopic observation of the respiratory tract of SPF piglets inoculated with Mycoplasma hyopneumoniae. Vet Microbiol. 1992;30:329–41.

66. Jaques M, Blanchard B, Foiry B, Kobisch M. In vitro colonization of porcine trachea by Mycoplasma hyopneumoniae. Ann Rech Vét, INRA Editions. 1992; 23:239–47.

PERMISSIONS

The contributors of this book come from diverse backgrounds, making this book a truly international effort. This book will bring forth new frontiers with its revolutionizing research information and detailed analysis of the nascent developments around the world.

We would like to thank all the contributing authors for lending their expertise to make the book truly unique. They have played a crucial role in the development of this book. Without their invaluable contributions this book wouldn't have been possible. They have made vital efforts to compile up to date information on the varied aspects of this subject to make this book a valuable addition to the collection of many professionals and students.

This book was conceptualized with the vision of imparting up-to-date information and advanced data in this field. To ensure the same, a matchless editorial board was set up. Every individual on the board went through rigorous rounds of assessment to prove their worth. After which they invested a large part of their time researching and compiling the most relevant data for our readers.

The editorial board has been involved in producing this book since its inception. They have spent rigorous hours researching and exploring the diverse topics which have resulted in the successful publishing of this book. They have passed on their knowledge of decades through this book. To expedite this challenging task, the publisher supported the team at every step. A small team of assistant editors was also appointed to further simplify the editing procedure and attain best results for the readers.

Apart from the editorial board, the designing team has also invested a significant amount of their time in understanding the subject and creating the most relevant covers. They scrutinized every image to scout for the most suitable representation of the subject and create an appropriate cover for the book.

The publishing team has been an ardent support to the editorial, designing and production team. Their endless efforts to recruit the best for this project, has resulted in the accomplishment of this book. They are a veteran in the field of academics and their pool of knowledge is as vast as their experience in printing. Their expertise and guidance has proved useful at every step. Their uncompromising quality standards have made this book an exceptional effort. Their encouragement from time to time has been an inspiration for everyone.

The publisher and the editorial board hope that this book will prove to be a valuable piece of knowledge for researchers, students, practitioners and scholars across the globe.

LIST OF CONTRIBUTORS

Christian P. Bertholle, Ellen Meijer, Arjan Stegeman and Arie van Nes
Department of Farm Animal Health, Faculty of Veterinary Medicine, Utrecht University, Yalelaan 7, NL-3584 CL Utrecht, The Netherlands

Willem Back
Department of Equine Sciences, Faculty of Veterinary Medicine, Utrecht University, Yalelaan 112-114, NL-3584 CM Utrecht, The Netherlands
Department of Surgery and Anaesthesia of Domestic Animals, Ghent University, 17 Salisburylaan 133, B-9820 Merelbeke, Belgium

P. René van Weeren
Department of Equine Sciences, Faculty of Veterinary Medicine, Utrecht University, Yalelaan 112-114, NL-3584 CM Utrecht, The Netherlands

Hu Suk Lee and Hung Nguyen-Viet
International Livestock Research Institute, Regional Office for East and Southeast Asia, Room 301-302, B1 Building, Van Phuc Diplomatic Compound, 298 Kim Ma Street, Ba Dinh District, Hanoi, Vietnam

Johanna Lindahl
International Livestock Research Institute, Nairobi, Kenya
Swedish University of Agricultural Sciences, Uppsala, Sweden

Nguyen Viet Khong, Vuong Bui Nghia and Huyen Nguyen Xuan
National Institute of Veterinary Research, 86 Truong Chinh, Phuong Mai, Dong Da, Hanoi, Vietnam

Delia Grace
International Livestock Research Institute, Nairobi, Kenya

K. Domańska-Blicharz, A. Lisowska, A. Jacukowicz, A. Pikuła and Z. Minta
Department of Poultry Diseases, National Veterinary Research Institute, Al. Partyzantów 57, 24-100 Puławy, Poland

Ł. Bocian
Department of Epidemiology and Risk Assessment, National Veterinary Research Institute, Al. Partyzantów 57, 24-100 Puławy, Poland

Sungwon Kim
The Roslin Institute and R(D)SVS, University of Edinburgh, Easter Bush, Midlothian EH25 9RG, UK

Myeongseon Park and Rami A. Dalloul
Avian Immunobiology Laboratory, Department of Animal and Poultry Sciences, Virginia Tech, Blacksburg, VA 24061, USA

Ariel E. Leon and Dana M. Hawley
Department of Biological Sciences, Virginia Tech, Blacksburg, VA 24061, USA

James S. Adelman
Department of Natural Resource Ecology and Management, Iowa State University, Ames, IA 50011, USA

Antje Römer, Jürgen Wallmann and Heike Kaspar
Federal Office of Consumer Protection and Food Safety, Berlin, Germany

Gesine Scherz, Saskia Reupke, Jessica Meißner and Manfred Kietzmann
University of Veterinary Medicine Hannover, Foundation, Institute of Pharmacology, Toxicology and Pharmacy, Hanover, Germany

Beata Dolka and Piotr Szeleszczuk
Department of Pathology and Veterinary Diagnostics, Faculty of Veterinary Medicine, Warsaw University of Life Sciences-SGGW, Nowoursynowska 159c St., Warsaw 02-776, Poland

Dorota Chrobak-Chmiel
Department of Preclinical Sciences, Faculty of Veterinary Medicine, Warsaw University of Life Sciences-SGGW, Ciszewskiego 8 St., Warsaw 02-786, Poland

László Makrai
Department of Microbiology and Infectious Diseases, Faculty of Veterinary Science, Szent István University, Hungária krt. 23-25, Budapest H-1143, Hungary

Tsegaw Fentie, Wassie Molla, Birhanu Ayele, Seleshe Nigatu and Ashenafi Assefa
College of Veterinary Medicine and Animal Sciences, University of Gondar, Gondar, Ethiopia

Nigusie Fenta
Livestock and Fisheries Development Office, Dembia District, North Gondar, Ethiopia

Samson Leta
College of Veterinary Medicine and Agriculture, Addis Ababa University, Bishoftu, Ethiopia

Yechale Teshome
Faculty of Agriculture, Debre Markos University, Debre Markos, Ethiopia

Kebede Amenu
School of Veterinary Medicine, Hawassa University, Hawassa, Ethiopia
International Livestock Research Institute, Addis Ababa, Ethiopia
Department of Microbiology, Immunology and Veterinary Public Health, College of Veterinary Medicine and Agriculture, Addis Ababa University, Bishoftu, Ethiopia

Barbara Szonyi and Barbara Wieland
International Livestock Research Institute, Addis Ababa, Ethiopia

Delia Grace
International Livestock Research Institute, Nairobi, Kenya

T. Tekle and G. Gari
National Animal Health Diagnostic and Investigation Center-Protozoology unit, Addis Ababa, Ethiopia

G. Terefe and H. Ashenafi
Department of Pathology & Parasitology, Addis Ababa University College of Veterinary Medicine and Agriculture, Bishoftu, Ethiopia

T. Cherenet
Minstry of Livestock and Fisheries, Addis Ababa, Ethiopia

K. G. Akoda and A. Teko-Agbo
Ecole Inter- Etats des Sciences et Médecine vétérinaires de Dakar, Dakar, Fann, Senegal

J. Van Den Abbeele
Department of Biomedical Sciences Veterinary Protozoology, Institute of Tropical Medicine, Unit 155 Nationalestraat, B-2000 Antwerp, Belgium

P.-H. Clausen and A. Hoppenheit
Institute for Parasitology and Tropical Veterinary Medicine, Freie Universitaet Berlin, Robert-von-Ostertag Str. 7-13, 14163 Berlin, Germany

R. C. Mattioli
Food and Agriculture Organization of the United Nations, Viale delle Terme di Caracalla, 00153 Rome, Italy

R. Peter
Global Alliance for Livestock Veterinary Medicines (GALVmed), Doherty Building, Pentlands Park, Bush Loan, Edinburgh EH26 0PZ, UK

T. Marcotty
Veterinary Epidemiology and Risk Analysis - Research and Development (VERDI-R&D), Rue du Gravier 7, 4141 Sprimont, Belgium

G. Cecchi
Food and Agriculture Organization of the United Nations, Sub-Regional Office for Eastern Africa, Addis Ababa, Ethiopia

V. Delespaux
Faculty of Sciences and Bio-engineering Sciences, Vrije Universiteit Brussel, Pleinlaan 2, B-1050 Brussels, Belgium

Samia Djeffal
GSPA research Laboratory (Management of Animal Health and Productions), Institute of Veterinary Sciences, University Frères Mentouri Constantine 1, Constantine, Algeria
Institute of Veterinary and Agronomic Sciences, University Chadli Bendjedid, Eltarf, Algeria
Unité de recherche sur les maladies infectieuses et tropicales émergentes (URMITE), UM 63, CNRS 7278, IRD 198, INSERM 1095, IHU Méditerranée Infection, Faculté de Médecine et de Pharmacie, Aix-Marseille-Université, Marseille, France

Sofiane Bakour, Selma Chabou and Jean-Marc Rolain
Unité de recherche sur les maladies infectieuses et tropicales émergentes (URMITE), UM 63, CNRS 7278, IRD 198, INSERM 1095, IHU Méditerranée Infection, Faculté de Médecine et de Pharmacie, Aix-Marseille-Université, Marseille, France

Bakir Mamache
Institute of Veterinary and Agronomic Sciences, University Hadj Lakhdar, Batna, Algeria

Rachid Elgroud, Sana Hireche and Omar Bouaziz
GSPA research Laboratory (Management of Animal Health and Productions), Institute of Veterinary Sciences, University Frères Mentouri Constantine 1, Constantine, Algeria

Amir Agabou
PADESCA Research Laboratory, Institute of Veterinary Sciences, University Frères Mentouri, Constantine, Algeria

Kheira Rahal
Pasteur Institute, Medical Bacteriology Service, Algiers, Algeria

Umit Karademir
Department of Pharmacology and Toxicology, Faculty of Veterinary Medicine, University of Adnan Menderes, Isikli, Aydin, Turkey

Ibrahim Akin
Department of Surgery, Faculty of Veterinary Medicine, University of Adnan Menderes, Isikli Koyu, Aydin, Turkey

Hasan Erdogan and Kerem Ural
Department of Internal Medicine, Faculty of Veterinary Medicine, University of Adnan Menderes, Isikli Koyu, Aydin, Turkey

Gamze Sevri Ekren Asici
Department of Biochemistry, Faculty of Veterinary Medicine, University of Adnan Menderes, Isikli Koyu, Aydin, Turkey

Eduardo Casas, Guohong Cai, Karen B. Register and John D. Neill
USDA, ARS, National Animal Disease Center, Ames, IA 50010, USA

Larry A. Kuehn and Tara G. McDaneld
USDA, ARS, U.S. Meat Animal Research Center, Clay Center, NE 68933, USA

Ling-Cong Kong, Xia Guo, Zi Wang, Yun-Hang Gao, Bo-Yan Jia, Shu-Ming Liu and Hong-Xia Ma
College of Animal Science and Technology, Jilin Agricultural University, Changchun, China

Dechassa Tegegne, Amin kelifa, Mukarim Abdurahaman and Moti Yohannes
Department of Veterinary Microbiology and Public Health, College of Agriculture and Veterinary Medicine, School of Veterinary medicine, Jimma University, Jimma, Ethiopia

Dennis Muhanguzi, Godfrey Bigirwa, Ann Kitibwa, Grace Gloria Akurut, Sylvester Ochwo, Wilson Amanyire, Samuel George Okech and Robert Tweyongyere
College of Veterinary Medicine Animal Resources and Biosecurity, Makerere University, Kampala, Uganda

Albert Mugenyi
Coordinating Office for Control of Trypanosomiasis in Uganda, Ministry of Agriculture, Animal Industry and Fisheries, Plot 78, Buganda Road, Wandegeya, Kampala, Uganda

Maureen Kamusiime
Mercy Corps Uganda, Clock Tower, Kampala, Uganda

Jan Hattendorf
Swiss Tropical Institute, Socinstrasse 57, -4002 Basel, CH, Switzerland
University of Basel, Petersplatz 1, 4003 Basel, Switzerland

Claudia Mroz, Ute Ziegler, Martin Eiden and Martin H. Groschup
Institute of Novel and Emerging Infectious Diseases, Friedrich-Loeffler-Institut, Südufer 10, 17493 Greifswald - Isle of Riems, Germany

Mayada Gwida
Department of Hygiene and Zoonoses, Faculty of Veterinary Medicine, Mansoura University, Mansoura 35516, Egypt

Maged El-Ashker
Department of Internal Medicine and Infectious Diseases, Faculty of Veterinary Medicine, Mansoura University, Mansoura 35516, Egypt

Mohamed El-Diasty
Animal Health Research Institute-Mansoura Provincial Laboratory, Mansoura, Egypt

Mohamed El-Beskawy
Faculty of Veterinary Medicine, Mansoura University, Mansoura, Egypt

Yara M. Al-Kappany and Ibrahim E. Abbas
Parasitology Department, Faculty of Veterinary Medicine, Mansoura University, Mansoura, Egypt

Brecht Devleesschauwer
Department of Public Health and Surveillance, Scientific Institute of Public Health (WIV-ISP), Rue Juliette Wytsmanstraat 14, 1050 Brussels, Belgium

Pierre Dorny
Laboratory of Parasitology, Faculty of Veterinary Medicine, Ghent University, Merelbeke, Belgium Department of Biomedical Sciences, Institute of Tropical Medicine, Antwerp, Belgium

Malgorzata Jennes and Eric Cox
Laboratory of Immunology, Faculty of Veterinary Medicine, Ghent University, Merelbeke, Belgium

T. Martinello, C. Gomiero, S. Ferro, L. Maccatrozzo and M. Patruno
Department of Comparative Biomedicine and Food Science, University of Padua, Viale dell'Università 16, 35020, Legnaro – Agripolis, Padua, Italy

A. Perazzi, I. Iacopetti, F. Gemignani and G. M. DeBenedictis
Department of Animal Medicine, Production and Health, University of Padua, Padua, Italy

M. Zuin and E. Martines
Consorzio RFX, Padua, Italy

P. Brun
Department of Molecular Medicine, University of Padua, Padua, Italy

K. Chiers
Department of Pathology, Bacteriology and Poultry Diseases, University of Gent, Ghent, Belgium

J. H. Spaas
Global Stem cell Technology-ANACURA group, Noorwegenstraat 4, 9940 Evergem, Belgium

Slawomir Gonkowski, Krystyna Makowska and Jaroslaw Calka
Department of Clinical Physiology, Faculty of Veterinary Medicine, University of Warmia and Mazury, Oczapowski Str, 13 Olsztyn, Poland

M. Di Domenico, V. Curini, V. Di Lollo, M. Massimini, L. Di Gialleonardo and C. Cammà
Istituto Zooprofilattico Sperimentale dell'Abruzzo e del Molise "G. Caporale", Campo Boario, 64100 Teramo, Italy

A. Franco, A. Caprioli and A. Battisti
Istituto Zooprofilattico Sperimentale del Lazio e della Toscana "M. Aleandri", Via Appia Nuova 1411, 00178 Roma, Italy

Roberta Cimmino
Italian Buffalo Breeders Association, V. Petrarca 42/44, 81100 Caserta, Italy

Carmela M. A. Barone
Department of Agricultural Sciences, Federico II University, Via Università 133, 80055 Portici, Naples, Italy

Salvatore Claps and Domenico Rufrano
Research Centre for Animal Production and Aquaculture (CREA, S.S. 7 Appia, 85051, Bella Muro, PZ, Italy

Ettore Varricchio
Department of Sciences and Technologies, University of Sannio, V. Port'Arsa 11, 82100 Benevento, Italy

Mariangela Caroprese and Marzia Albenzio
Department of Agricultural Food and Environmental Sciences, University of Foggia, Via Napoli 25, 71122 Foggia, Italy

Pasquale De Palo
Department of Veterinary Medicine, University "Aldo Moro" of Bari, S.P. per Casamassima, km 3, Valenzano, 70010 Bari, Italy

Giuseppe Campanile and Gianluca Neglia
Department of Veterinary Medicine and Animal Production, Federico II University, V. F. Delpino 1, 80137 Naples, Italy

Tatjana Sattler
Institute for Veterinary Disease Control, AGES, Robert-Koch-Gasse 17, 2340 Mödling, Austria Clinic for Ruminants and Swine, University of Leipzig, An den Tierkliniken 11, 04103 Leipzig, Germany

Jutta Pikalo, Eveline Wodak, Sandra Revilla-Fernández, Adi Steinrigl, Zoltán Bagó and Friedrich Schmoll
Institute for Veterinary Disease Control, AGES, Robert-Koch-Gasse 17, 2340 Mödling, Austria

Ferdinand Entenfellner
Veterinary Practice Entenfellner, Bonnleiten 8, 3073 Stössing, Austria

Jean-Baptiste Claude, Floriane Pez and Maela Francillette
BioSellal, Bâtiment Accinov, 317 avenue Jean Jaurès, 69007 Lyon, France

Tadesse Eguale
Aklilu Lemma Institute of Pathobiology, Addis
Ababa University, Addis Ababa, Ethiopia

Annelies Michiels, Ioannis Arsenakis, Anneleen Matthijs and Dominiek Maes
Department of Reproduction, Obstetrics and Herd
Health, Faculty of Veterinary Medicine, Ghent
University, Salisburylaan 133, 9820 Merelbeke,
Belgium

Filip Boyen and Freddy Haesebrouck
Department of Pathology, Bacteriology and Avian
Diseases, Faculty of Veterinary Medicine, Ghent
University, Salisburylaan 133, 9820 Merelbeke,
Belgium

Geert Haesaert, Kris Audenaert and Mia Eeckhout
Department of Applied Biosciences, Faculty of
Bioscience Engineering, Ghent University, Campus
Schoonmeersen, Valentin Vaerwyckweg 1, 9000
Ghent, Belgium

Siska Croubels
Department of Pharmacology, Toxicology and
Biochemistry, Faculty of Veterinary Medicine, Ghent
University, Salisburylaan 133, 9820 Merelbeke,
Belgium

Index